PRINCIPLES OF RADIOLOGICAL HEALTH AND SAFETY

PRINCIPLES OF RADIOLOGICAL HEALTH AND SAFETY

JAMES E. MARTIN

CHUL LEE

WILEY-INTERSCIENCE

A JOHN WILEY & SONS, INC., PUBLICATION

For general information on our other products and services please contact our Customer Care Department within the U.S. at 877-762- 2974, outside the U.S. at 317-572-3993 or fax 317-572-4002. Wiley also publishes its books in a variety of electronic formats. Some content that appears in print, however, may not be available in electronic format.

Library of Congress Cataloging-in-Publication Data is available.

ISBN: 0-471-25429-0

Printed in the United States of America

10 9 8 7 6 5 4 3 2 1

To those who teach and inspire others, in particular:
Mrs. Johnnie Russell, Mrs. J. B. Smith, E. C. Webb,
J. W. Wheeler, G. Hoyt Whipple

CONTENTS

Preface   **xiii**

1   Introductory Concepts   **1**

Structure of Atoms   2
Nuclide Chart   8
Atom Measures   10
Energy Concepts for Atoms   12
Relativistic Energy   16
The Binding Energy of Nuclei   22
Summary   25
Additional Resources   26
Problems   27

2   Atom Structure and Energetics   **29**

Major Discoveries and Simple Tools   30
Theory of Electromagnetic Radiation—The Quantum   33
Discovery of the Atom's Structure   39
The Bohr Model of the Atom   43
Wave Mechanics—A Necessary Theory   47
Atom Systems   50
Summary   52
Problems   53

3   Radioactive Transformation   **55**

Transformation of Neutron-Rich Radionuclides   57

Transformation of Proton-Rich Radionuclides   60
Radioactive Transformation of Heavy Nuclei by
    Alpha Particle Emission   65
Gamma Emission   69
Decay Schemes   75
Rate of Radioactive Transformation   79
Radioactivity Calculations   83
Activity–Mass Relationships   87
Series Transformation—Radioactive Equilibrium   89
Additional Resources   97
Problems   98

4   Major Sources of Radiation   101

Production of X Rays   101
Natural Radiation   106
Radioactive Ores and Byproducts   116
Activation Products   121
Neutron Activation Calculations   130
Fission Product Radioactivity   135
Nuclear Reactor Designs   136
Fission Product Inventories   142
Summary   146
Problems   147

5   Radiation Interactions and Dose   149

Radiation Dose and Units   149
Radiation Dose Calculations   152
Interactions of Alpha Particles and Heavy Nuclei   153
Beta Particle Interactions and Dose   155
Photon Interactions and Dose   161
The Gamma Constant, Γ 171
Dosimetry for Bremsstrahlung   171
Summary   173
Additional Resources   174
Problems   174

6   Radiation Bioeffects and Risk   175

Radiobiology   175
Somatic Effects of Radiation   182
Genetic Effects of Radiation   184
Carcinogenic Effects of Radiation   185
Estimating Health Effects of Radiation   189
Radiation Risk Calculations   194

	Summary	195
	Additional Resources	196
	Problems	196

7	**Radiation Shielding**	**199**
	Shielding of Alpha-Emitting Sources	199
	Shielding of Beta-Emitting Sources	200
	Shielding of Photon Sources—"Good" and "Poor" Geometries	206
	Buildup Factors	216
	Gamma Flux for Distributed Sources	223
	Summary	229
	Additional Resources	230
	Problems	230

8	**Measurement of Radiation**	**233**
	Gas-Filled Detectors	233
	Crystalline Detectors and Spectrometers	237
	Portable Field Instruments	239
	Personnel Dosimeters	242
	Gamma Spectroscopy	244
	Laboratory Instruments	250
	Statistics of Radiation Measurements	257
	Propagation of Error	260
	Minimum Detection Levels	264
	The Chi-Square Test of a Detector System	270
	Summary	271
	Additional Resources	272
	Problems	272

9	**Internal Radiation Dosimetry**	**275**
	Absorbed Dose in Tissue	275
	Accumulated Dose	277
	Internal Dose—Medical Uses	279
	Factors in the Internal Dose Equation	280
	Deposition and Clearance Data	286
	Internal Radiation Dose Standards	292
	Biokinetics and Dose for Radionuclide Intakes	295
	Operational Determinations of Internal Dose	313
	Bioassay Determinations of Intake	314
	Tritium—A Special Case	320
	Summary	324
	Additional Resources	325
	Problems	325

10 Radiation Protection Standards   329

Evolution of Radiation Protection Standards   329
Current Radiation Protection Standards   337
ALARA-Based Standards for Specific Circumstances   339
Standards for Radionuclides in Air—40 CFR 60   341
Radionuclides in Drinking Water   341
Radon Guides   342
Uranium Mill Tailings   343
Naturally Occurring or Accelerator-Produced Radioactive
  Materials (NARM)   343
Site Cleanup Standards/Criteria   344
Summary   348
Problems   349

11 Radiation Protection Programs   351

Regulatory Authority for Radiation and Radioactive Materials   352
Radiation Protection Programs/Dose Limits   358
Compliance with Regulations   360
Shipment of Radioactive Materials   377
Summary   386
Additional Resources   387
Problems   388

12 Environmental Radiological Assessment   391

Elements of Environmental Models   393
Water Transport/Dispersion   395
Atmospheric Dispersion   396
Stack Effects   406
Nonuniform Turbulence—Fumigation, Building Effects   408
Puff Releases   417
Sector-Averaged χ/Q Values   418
Deposition/Depletion—Gaussian Plumes   422
Summary   430
Additional Resources   431
Problems   431

13 Radon—A Public Health Issue   435

Features of Radon and its Products   435
Health Effects of Radon   443
Radon Reduction Measures   447
Radon Minimization in New Construction   452
Radon Measurement   456
Summary   459

Additional Resources 461
Problems 461

14   Radioactive Wastes **463**

Legacy/Process Wastes 465
High-Level Radioactive Wastes 469
Low-Level Radioactive Wastes (LLRW) 471
Past Practices/Lessons Learned 473
Regulatory Control of LLRW (10 CFR 61) 479
Policy Implications/Status of LLRW 483
Summary 486
Additional Resources 487
Problems 488

Appendix **491**

A. Constants of Nature and Selected Particle Masses 491
B. Atomic Masses and Binding Energies—Selected Nuclides 493
C. Radioactive Transformation Data 505

Answers to Selected Problems **515**

Index **519**

PREFACE

This book is the outcome of teaching basic principles of radiological health (or radiation protection) to occupational and environmental health persons who need concepts of radiation protection should their career responsibilities call for it but do not expect to be specialists in the field. It is intended as a resource text for safety personnel who are increasingly expected to serve as radiation safety officers and manage protection programs that include radioactive materials. Real-world problems and the resource data to solve them are thus presented in a manner that can be understood and applied without specializing in radiological sciences, although the material could be used to communicate the elements of such specialization.

The book begins with a review (Chapter 1) of the basic structure of the atom as an energy system, which may be most useful for generalists or those with minimal science background. The major discoveries in nuclear physics are revisited in Chapter 2 in an attempt to recapture the insights grasped by those who discovered the laws of nature that govern radiant energy and atomic structure. Radioactive transformation and the major sources of radiation are addressed in Chapters 3 and 4 with an emphasis on the special condition of radioactive transformation (or disintegration) of atoms with excess energy since radioactivity is a major consideration in radiation protection.

Controls for radiation are based on its behavior, its potential to induce bioeffects, and means of assessing it. Interactions of radiation with matter are covered in Chapter 5, along with the corollary subjects of radiation exposure and dose and the various parameters that are needed to calculate them. Radiation dose is important because radiation has the potential to produce bioeffects by certain mechanisms; these are developed in Chapter 6, along with radiation risk coefficients. A primary means of controlling radiation exposure

is radiation shielding which is described in Chapter 7, and methods are provided for calculating exposure and dose for several common sources and
geometries. Chapter 8 builds upon and further amplifies the material discussed
in Chapters 5 and 7 to develop principles of radiation detection and the
methods and equipment used in radiation measurement.

Many public health issues involving radioactive materials are associated
with their ingestion in water and food or inhalation of radioactive aerosols.
Chapter 9 describes these mechanisms and the basic models for determining
internal radiation dose to various organs and tissues. Tables of data that relate
intakes and resultant doses are of immense practical use, and these are provided
for selected radionuclides in the same form as data in more extensive compendia should they need to be consulted for a specific circumstance. The evolution
of radiation policy and standards is summarized in Chapter 10 to provide
perspective on the ethical foundations of radiation control, and the design
and conduct of protection programs to meet these fundamental policies are
described in Chapter 11, along with implementing regulations. Potential and
actual exposures of workers and the public are dealt with in specialty chapters
on environmental radiological assessment (Chapter 12), radon in indoor environments (Chapter 13), and the broad public health aspects of radioactive waste
and its management (Chapter 14). These specialty topics were selected because
of current interest and because they tie together most of the elements of public
health concern for radiation protection.

A course in the fundamentals of radiological health that is based on this
book would be expected to include substantial treatment of the material in
Chapters 3 to 7 and Chapters 10 and 11, with selections from the other
chapters, all or in part, to develop needed background and to address specialty
areas of interest to instructor and student. Numerous data sets are provided in
the text on energy absorption coefficients, fission yields, dose equivalents, etc.,
and each is cross-referenced to compendia at the National Nuclear Data Center
at Brookhaven National Laboratory (www.nndc.bnl.gov). Appendices include
fundamental constants, nuclear masses, and data on modes of radioactive
transformation and emission energies (Appendix C) for some of the common
radionuclides encountered.

The units used in radiation protection that encompass the basic discoveries and applications of nuclear sciences to radiation safety have evolved over
the past hundred years or so. They continue to do so with a fairly recent
emphasis on Systeme Internationale (SI) units, a trend that is not entirely
accepted because U.S. standards and regulations for control of radiation and
radioactivity have continued to use conventional units. To the degree possible,
this book uses fundamental quantities such as electronic charge and voltage
(eV), transformations, time, distance, and the numbers of atoms or emitted
particles and radiations to describe nuclear processes, primarily because they
are basic to concepts being described, but partially to avoid the need to resolve
any conflict between SI units and conventional ones. Both sets of units are

defined as they apply to radiation protection, but in general the more funda-mental parameters are used.

Finally, in an undertaking of this type it is inevitable that undetected mistakes creep in and remain despite the best efforts of preparers and editors. Readers are encouraged to report such mishaps (email: jemartin@umich.edu) which will be noted along with other important updates on the web page: www-personal.umich.edu/~jemartin

We hope that this book helps each reader to understand and apply the basic principles of radiation protection to their individual circumstances.

James E. Martin, Ph.D., CHP
The University of Michigan

Chul Lee, MS
The University of Michigan

1

INTRODUCTORY CONCEPTS

"Nothing in life is to be feared. It is only to be understoood."
— *M. Curie*

Radiological Health (also known as Health Physics) brings together science, technology, human values, and public policy to provide safe levels of protection for workers and the public from radiation. It is also standard practice, which is perhaps unique for radiation protection, to optimize protection as far below established safe levels as reasonably achievable because it is presumed that some risk may well remain at any level of exposure and that this too should be avoided whenever practicable. Such an approach requires an understanding of

- the sources, behavior, and energy of emitted radiation(s);
- mechanisms by which radiant energy is deposited in various media to produce dose;
- biological effects and health risks that occur due to a given radiation dose; and
- application of controls to change the relations.

The emission and control of radiant energy, or radiation, generated by various devices and radioactive materials, is thus an interconnected system that is directly related to the unique properties of a radiation source, the types and properties of the radiation(s) emitted by it, the mechanisms by which radiation energy is deposited in various media, and their effects on biological systems.

1

STRUCTURE OF ATOMS

Four basic forces of nature control the dynamics (i.e., position, energy, work, etc.) of all matter, including the constituents of atoms–protons, neutrons, and electrons. These forces, along with their magnitude relative to gravity, are:

- gravity, which is an attractive force between masses $= G$;
- the weak force, which influences radioactive transformation $\cong 10^{24}$ G;
- the electromagnetic force, which exists between electric charges $\cong 10^{37}$ G;
- the nuclear force, which is strongly attractive between nucleons only $\cong 10^{39}$ G.

They range over some 40 orders of magnitude; however, two of these forces largely determine the energy states of particles in the atom (gravitational forces are insignificant for the masses of atom constituents, and the weak force is a special force associated with the process of radioactive transformation of unstable atoms):

1. the nuclear force between neutrons and protons, which is so strong that it overcomes the electrical repulsion of the protons (which is quite strong at the small dimensions of the nucleus) and holds the nucleus, along with its constituent protons and neutrons, together;
2. the force of electrical attraction between the positively charged nucleus and the orbital electrons, which not only holds the electrons within the atom but influences where they orbit.

The **nuclear force**, or strong force, is unique and a bit strange. It exists only between protons and neutrons or any combination of them; consequently, it exists only in the nucleus of atoms. The nuclear force is not affected by the charge on neutrons and protons, nor the distance between them. It is strongly attractive, so much so that it overcomes the natural repulsion between protons at the very short distances in the nucleus since it is about 100 times stronger than the electromagnetic force.

Nuclear Attraction

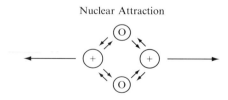

The **electromagnetic force**, on the other hand, exists between charged particles no matter where they are (a nucleus can also be thought of as a large charged particle although it contains several protons, each of which has a unit-positive charge). The electromagnetic force is inversely proportional to the square of the distance, r, between two particles with a charge of q_1 and q_2:

$$F_{em} = \frac{1}{4\pi\varepsilon_0} \cdot \frac{q_1 q_2}{r^2} = k_0 \cdot \frac{q_1 q_2}{r^2}$$

where the charges on each particle are expressed in coulombs and the separation distance r is in meters. The constant k_0 is for two charges in a vacuum and has the value $k_0 = 8.9876 \times 10^9\,Nm^2/C^2$. This fundamental relationship is called Coulomb's law after its developer, and is referred to as the coulomb force. If q_1 and q_2 are of the same sign (i.e., positive or negative), F will be a repulsive force; if they are of opposite signs, F will be attractive.

Atoms contain neutrons, protons, and electrons, and their number and array establish

- what the element is and whether its atoms are stable or unstable; and
- if unstable, how the atoms will emit energy (we will deal with energy later).

Modern theory has shown that protons and neutrons are made up of more fundamental particles, or quarks, but it is not necessary to go into such depth to understand the fundamental makeup of atoms and how they behave to produce radiant energy.

The **proton** has a reference mass of about 1.0. It also has a positive electrical charge of plus 1 (+1).

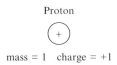

Proton

mass = 1   charge = +1

The **electron** is much lighter than the proton. Its mass is about 1/1800 (actually 1/1836) of the proton and it has an electrical charge of minus one (−1).

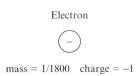

Electron

mass = 1/1800   charge = −1

The **neutron** is almost the same size as the proton, but slightly heavier. It has no electrical charge.

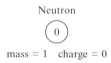

Neutron

mass = 1   charge = 0

When these basic building blocks are put together, as happened at the beginning of time, very important things become evident. First, a proton will attract a free electron to form an atom:

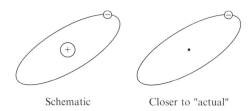

Schematic   Closer to "actual"

The resulting atom is electrically neutral. That is, each -1 charge on an orbital electron is matched by a $+1$ positively charged proton in the nucleus; i.e., nature requires each atom to have an equal number of each. The total atom (proton plus an electron) has a diameter of about 10^{-10} m (or 10^{-8} cm) and is much bigger than the central nucleus which has a radius of about 10^{-15} m or (10^{-13} cm); thus the atom is mostly empty space. The radius of the nucleus alone is proportional to $A^{1/3}$ where A is the atomic mass number of the atom in question or

$$r = r_0 A^{1/3}$$

The constant r_0 varies according to the element but has an average value of about 1.3×10^{-15} m, or femtometers. The femtometer (10^{15} m) is commonly referred to as a fermi in honor of the great Italian physicist and nuclear navigator, Enrico Fermi.

A free neutron is electrically neutral and, in contrast to a proton, an atom does not form; i.e., it is just a free neutron subject to thermal forces of motion.

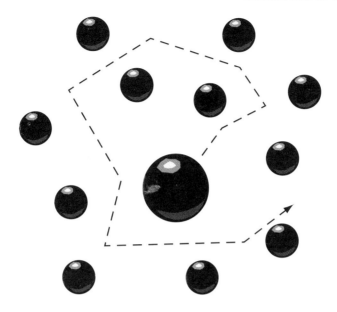

Likewise, two neutrons are also unaffected by any electrons present. How-ever, if left alone for a while, a free neutron will undergo transformation (commonly referred to as decay) into a proton and an electron; therefore, in a free state, the neutron, though not an atom (there being no orbiting elec-trons), behaves like a radioactive atom.

A neutron can, however, be bound with a proton to form a nucleus, and in this state it does not undergo radioactive transformation but will maintain its identity as a fundamental particle.

An electron will join with the proton–neutron nucleus to form an electrically neutral atom

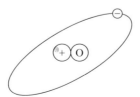

but it now weighs about twice as much as the other one because of the added neutron mass. And, if another (a second) neutron is added, a heavier one-proton atom is formed:

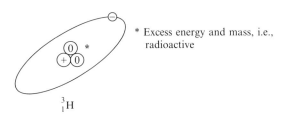

* Excess energy and mass, i.e., radioactive

This atom is the same electrically neutral atom (one proton balanced by one electron) we started with, but it weighs about three times as much due to the two extra neutrons, and because of the array of the particles in the nucleus it is an atom with excess energy; i.e., it is radioactive.

Each of these one-proton atoms is an atom of hydrogen because hydrogen is defined as any atom containing one proton balanced by one electron. Each atom has a different weight because of the number of neutrons it contains, and these are called isotopes (Greek: "iso" = same; "tope" = place) of hydrogen to recognize their particular features. The three isotopes of hydrogen are denoted by the following symbols:

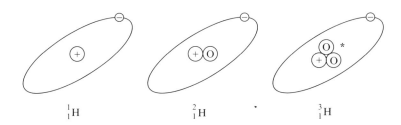

These three isotopes of hydrogen are identified as protium (or hydrogen), deuterium, and tritium; the first two are stable and exist in nature, but tritium is radioactive and will disappear through radioactive transformation unless replenished.

The symbols establish the nomenclature used to identify atoms: the subscript on the lower left denotes the number of protons in the atom; the superscript on the upper left refers to the mass number, an integer that is the sum of the number of protons and neutrons in the nucleus. It is common practice to leave off the subscript for the number of protons because the elemental symbol, H, defines the substance as hydrogen with only one proton. Almost all elements exist, or can be produced, with several different mass numbers yielding several isotopes. A particular substance is often identified by its element and the mass number of the isotope present, e.g., carbon-14 (^{14}C), hydrogen-3 (^{3}H, or tritium).

Two-Proton Atoms

If we try to put two protons together, the repulsive coulomb force between them at the very short distance required to form a nucleus is so great that it even overcomes the strongly attractive nuclear force between the protons; thus, an atom (actually a nucleus) cannot be assembled from just these two particles.

If, however, a neutron is added, it tends to redistribute the forces, and a stable nucleus can be formed. Two electrons will then join up to balance the two plus (+) charges of the protons to create a stable, electrically neutral atom of helium.

$^3_2$He

This atom is defined as helium because it has two protons. It has a mass of 3 (2 protons plus 1 neutron) and is written as helium-3 or ^{3}He. Because neutrons provide a cozy effect, yet another neutron can be added to obtain ^{4}He.

$^4_2$He

Although extra mass was added in forming ^{4}He, only two electrons are needed to balance the two positive charges. This atom is the predominant form of helium (isotope if you will) on earth, and it is very stable (we will see later that this same atom, minus the two orbital electrons, is ejected from some radioactive atoms as an alpha particle, i.e., a charged helium nucleus).

If yet another neutron is stuffed into helium to form helium-5 (^{5}He), the atom contains more mass than it can handle and it quickly breaks apart (in 10^{-21} s or so); it literally spits the neutron back out. There is just not enough room for the third neutron, *and* putting it in creates a highly unstable atom. In many cases (eg. for hydrogen), adding an extra neutron (or proton) to a nucleus only destabilizes it; i.e., it will exist for quite a while as an unstable, or radioactive, atom due to the "extra" particle mass. This is why it is important to know the "isotope" of a given element.

Three-Proton Atoms

Atoms with three protons can be assembled with three neutrons to form lithium-6 (^{6}Li) or with four neutrons, lithium-7 (^{7}Li),

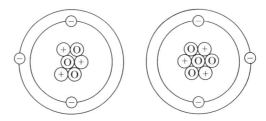

Since lithium contains three protons, it must also have three orbital electrons, but another orbit further away is required for the third electron because the first orbit can only hold two electrons (there is an important reason for this which is explained by quantum theory).

If we keep combining protons and neutrons we get heavier and heavier atoms, but they obey the same general rules. The ratio of neutrons to protons is fairly high in heavy atoms because the extra neutrons are necessary to distribute the nuclear force and moderate the repulsive electrostatic force between protons in such a way that the atoms stay together. The heaviest element in nature is ^{238}U with 92 protons and 146 neutrons; it is radioactive. The heaviest stable element in nature is ^{209}Bi with 83 protons and 126 neutrons. Lead with 82 protons is much more common in nature than bismuth and for a long time was thought to be the heaviest of the stable elements; it is also the stable endpoint of the radioactive transformation of uranium and thorium, two primordial, naturally occurring radioactive elements (see Chapter 4).

NUCLIDE CHART

This logical pattern of atom building can be plotted in terms of the number of protons and neutrons in each to create a **Chart of the Nuclides**, a portion of which is shown in Figure 1-1.

The Chart of the Nuclides contains basic information on each element, how many isotopes it has (atoms on the horizontal lines), and which ones are stable (shaded) or unstable (unshaded). A good example of such information is shown in Figure 1-2 for four isotopes of carbon, which has 6 protons, boron (5 protons), and nitrogen (7 protons). Actually there are 8 measured isotopes of carbon but these 4 are the most important. They are all carbon because each contains 6 protons, but each of the 4 has a different number of neutrons; hence they are distinct isotopes with different weights. Note that the two blocks in the middle for ^{12}C and ^{13}C are shaded, which indicates that these isotopes of carbon are stable (as are two shaded blocks for boron and nitrogen). The nuclides in the unshaded blocks (e.g., ^{11}C and ^{14}C) on each side of the shaded (i.e., stable) blocks are unstable simply because they do not have the right array of protons and neutrons to be stable (we will use these properties later to discuss radio-active transformation). The dark band at the top of the block for ^{14}C denotes that it is a naturally occurring radioactive isotope, a convention used for several other such radionuclides. The block to the far left contains information on naturally abundant carbon: it contains the chemical symbol C, the name of the element, and the atomic weight of natural carbon, or 12.0107 grams/mole,

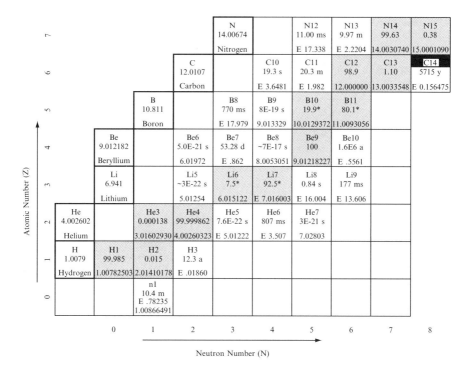

Figure 1-1. Part of The Chart of the Nuclides. (*Source*: *Chart of the Nuclides, Nuclide and Isotopes*, 15th Ed., 1996, General Electric Co.–Nuclear Energy.)

	N12 11.00 ms β+ 16.3 γ 4439, ... (3α) .192, ... E 17.338	N13 9.97 m β+ 1,190 E 2.2204	N14 99.63 σᵧ.080, .036 σₚ1.83, .85 14.0030740	N15 σᵧ 24.26 μb, 0.02mb 15.0001090	N16 7.13 s β−4.27, 10.44, ⋯ γ 6129, 7115, ... (α) 1.85, ⋯ E 10.419
C 12.0107 **Carbon** σₐ 3.5 mb, 1.6 mb	C11 20.3 m β+ .960 ε ω E 1.982	C12 98.90 σₐ3.5 mb, 1.6 mb 34.9688527	C13 1.10 σᵧ 1.4 mb, 1.6 mb 13.0033548	C14 5715 y β−.157 no γ σᵧ <1 μb E .156475	C15 2.450 s β−4.51, 9.82, ⋯ γ 5297.8, ⋯ E 9.772
	B10 19.9* σₐ384E1, 173E1 σᵧ.5, .2249 σₚ7 mb σₐ8 mb 10.0129371	B11 80.1* σᵧ5 mb, 2 mb 11.0093055	B12 20.20 ms β−13.37, ⋯ γ 4439, ⋯ (α) .2, ⋯ E 13.369	B13 17.4 ms β−13.4, ⋯ γ 3680 (n) 3.61, 2.40, ⋯ E 13.437	

Figure 1-2. Excerpt from the Chart of the Nuclides for the two stable isotopes of carbon (Z = 6) and its two primary radioactive isotopes, in relation to primary isotopes of nitrogen (Z = 7) and boron (Z = 5). (Adapted from GE, 1996.)

weighted according to the percent abundance of the two naturally occurring stable isotopes. The shaded blocks contain the atom percent abundance of ^{12}C and ^{13}C in natural carbon at 98.90 and 1.10 atom percent, respectively; these are listed just below the chemical symbol. Similar information is provided at the far left of the Chart of the Nuclides for all the elements.

The atomic weight of an atom is numerically equal to the mass, in unified mass units, or u, of the atom in question. The atomic mass is listed for all the stable isotopes (e.g., ^{12}C and ^{13}C) at the bottom of their respective blocks in the Chart of the Nuclides; other values for stable and unstable isotopes can be obtained from the listing in Appendix B.

ATOM MEASURES

Atoms are bound systems—they only exist when protons and neutrons are bound together to form a nucleus and when electrons are bound in orbits around the nucleus. The particles in atoms are bound into an array because nature forces atoms toward the lowest potential energy possible; when they attain it they are stable, and until they do they have excess energy and are thus unstable, or radioactive. Many health physics problems require knowledge of how many atoms there are in common types of matter, the total energy represented by each atom, and the energy of its individual components, or particles, which can be derived from the mass of each atom and its component particles. Avogadro's number and the atomic mass unit are basic to these concepts, especially the energy associated with mass changes that occur in and between atoms.

Avogadro's Number

In 1811, an Italian physicist, Amedeo Avogadro, reasoned that the number of atoms or molecules in a mole of any substance is a constant, independent of the nature of the substance; however, he had no knowledge of its magnitude, only that the number was very large. Because of this insight, the number of atoms or molecules in a mole is called Avogadro's number, N_A. It has the following value:

$$N_A = 6.0221367 \times 10^{23} \text{ atoms/mol}$$

Example 1-1: Calculate the number of atoms of ^{13}C in 0.1 gram of natural carbon.
Solution: From Figure 1-2, the atomic weight of carbon is 12.0107 g and the atom percent abundance of ^{13}C is 1.10%. Thus

$$\text{Number of atoms of } ^{13}C = \frac{0.1g \times N_A \text{ atoms/mol}}{12.0107g/mol} \times 0.011 = 5.5154 \times 10^{19} \text{ atoms}$$

Atomic Mass Unit (u)

Actual weights of atoms and constituent particles are extremely small and difficult to relate to, except to say that they are small. Therefore, a natural unit has been designated that approximates the weight of a proton or neutron for expressing the masses of particles in individual atoms (because the sum of these is the mass number, one of these must be a "mass unit"). The atomic mass unit, or amu, has been defined for this purpose, and the masses of stable and unstable isotopes of the elements contained in Appendix B and the Chart of the Nuclides are listed in unified mass units. To be precise, the unit is the unified mass unit, denoted by the symbol u, but by precedent most refer to it as the atomic mass unit, or *amu*.

One *amu*, or u, is defined as one-twelfth the mass of the neutral ^{12}C atom. One mole of ^{12}C by convention is defined as weighing exactly 12.000000 g. All other elements and their isotopes are assigned weights relative to ^{12}C. The *amu* was originally defined relative to oxygen-16 at 16.000000 g/mol but carbon-12 has proved to be a better reference nuclide, and since 1962 atomic masses have been based on the unified mass scale referenced to the carbon-12 mass at exactly 12.000000 g. The mass of a single atom of ^{12}C can be obtained from the mass of one mole of ^{12}C which contains Avogadro's number of atoms as follows:

$$m(^{12}C) = \frac{12.000000 \text{ g/mol}}{6.0221367 \times 10^{23} \text{ atoms/mol}} = 1.9926482 \times 10^{-23} \text{ g/atom}$$

This mass is shared by 6 protons and 6 neutrons; thus, the average mass of each of the 12 building blocks of the carbon-12 atom, including the paired electrons, can be calculated by distributing the mass of one atom of carbon-12 over the 12 nucleons. This quantity is defined as one *amu*, or u, and has the value

$$1u = \frac{1.9926482 \times 10^{-23}g}{12} = 1.66054 \times 10^{-24} \text{ g}$$

which is close to the actual mass of the proton (actually $1.6726231 \times 10^{-24}$ g) or the neutron (actually $1.6749286 \times 10^{-24}$ g). In unified mass units, the mass of the proton is 1.00727647 u and that of the neutron is 1.008664923 u. Each of these values is so close to unity that the mass number of an isotope is thus a close approximation of its atomic weight. These useful parameters and other constants of nature and physics are included in Appendix A.

Measured masses of elements are given in unified mass units, u, and are some of the most precise measurements in physics with accuracies to six or more decimal places. As we will see in Chapter 2, mass changes in nuclear processes represent energy changes; therefore, accurate masses, as listed in Appendix B, are very useful in calculations of energies of events in health physics.

ENERGY CONCEPTS FOR ATOMS

When we consider the energy of the atom, we usually focus on the energy states of the orbital electrons (referred to by many as atomic physics) or the arrangement of neutrons and protons in the nucleus (or nuclear physics). In any given atom, the electromagnetic force between the positively charged nucleus and the negatively charged electrons largely establishes where the electrons will orbit. An electron orbiting a nucleus experiences two forces: a centripetal (or pulling) force induced by the electromagnetic force between the electron and the positively charged nucleus, and a centrifugal force (one that falls away) due to its angular rotation. The array of neutrons and protons in the nucleus is similarly determined by a balancing of the nuclear force and the electromagnetic force acting on them.

Since atoms are arrays of particles bound together under the influence of the nuclear force and the electromagnetic force, the particles in atoms (or anywhere else for that matter) have energy states that are directly related to how the force fields act upon them. These concepts lead naturally to the question: what is energy? **Energy** is the ability to do work. O.K., what is work? Work is a force acting through some distance. This appears to be a circular argument which can perhaps be better illustrated with a macro-world example of lifting a rock that weighs 1 kilogram (2.2 pounds) from ground level onto a perch 1 meter (about 3.28 ft) high

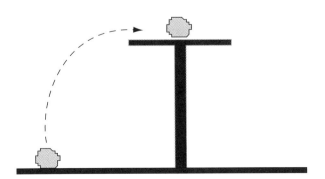

The work done by the lifter is obviously related to the amount of effort to lift the weight of the rock against the force due to gravity and how high it is to be lifted. Physics characterizes this effort as:

$$Work = Force \times height\ of\ perch$$
$$or\ W = mgh$$

where m is the mass of the rock in kilograms (2.2 pounds), g is the acceleration due to gravity (9.81 meters per second), and h is the height in meters. In this example, the amount of work done would be;

$$Work = 1\,kg \times 9.81 m/s^2 \times 1m$$
$$= 9.81\ newton \cdot meter = 9.81\ Joules$$

In this example, work is expressed in the energy unit of joules which represents a force of one newton ($kg\ m/s^2$) acting through a distance of 1 meter; i.e., energy and work are interrelated. The key concept is that 9.81 Joules represents an effort expended against gravity to raise a mass of 2.2 pounds (1 kg) over a distance of 3.28 feet (1 m); (i.e., work was done or energy was expended to get it there); if it were raised to another perch, one meter higher, the same amount of work would need to be done again.

The basic concepts of energy, work, and position associated with a rock can be extrapolated to atoms which are what they are because their particles exist at distinct energy levels. A rock on the perch can be thought of as "bound" with the ground; it has positive energy relative to the ground by virtue of the work done on it to get it up there. The rock on the perch represents stored work; if it is pushed off the perch and allowed to fall under the force of gravity, the stored work would be recovered when it hit the ground.

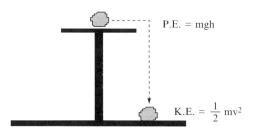

The stored energy (work $= mgh$) in the rock is called potential energy – it is sitting there ready to do work. As the rock falls back to the ground its potential energy is converted to kinetic energy, or energy due to its motion, as it falls under the influence of the gravitational force. At any point above the ground the falling rock has both potential energy (yet to be recovered) due to its height (or PE $=$ mgh) and kinetic energy due to its velocity or KE $= 1/2mv^2$. Its total energy at any point is the sum of the two, or

$$E_{\text{tot}} = mgh + \frac{1}{2}mv^2$$

where h and v are both variables. When the rock is at rest on the perch, all the energy is potential energy; when it strikes the ground, h is zero and all the energy is kinetic energy which is determined by its velocity. This set of conditions represents a total conversion of potential energy, mgh, into kinetic energy, $\frac{1}{2}mv^2$, such that

$$\frac{1}{2}mv^2 = mgh$$

which can be used to determine the velocity the rock would have after falling through the distance h and striking the ground, a quantity called the terminal velocity, expressed as

$$v = \sqrt{2gh}$$

This expression can also be used to determine the velocity at any intermediate height, h, above the ground, and because of the conservation of energy the relative proportions of potential energy and kinetic energy (always equal to the total energy) along the path of the falling rock.

Another good example of potential (stored) energy and kinetic energy is a pebble in a slingshot. Work is done to stretch the elastic in the slingshot so that the pebble has potential energy (stored work). When let go, the pebble is accelerated by the elastic returning to its relaxed position and it gains kinetic energy due to the imparted velocity, v. The main point of both of these examples is that a body (particle) with potential energy has a stored ability to do work; when the potential energy is released, it shows up in the motion of the body (particle) as kinetic, or "active", energy.

Potential energy can be positive or negative. A rock on the perch above the ground (or in a slingshot) has positive potential energy that can be recovered as kinetic energy by letting it go. But if the rock were in a hole, it would have negative potential energy of $-mgh$ relative to the ground surface.

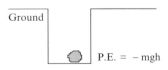

Energy would need to be supplied (i.e., do work on it) to get it out of the hole and back to the ground surface. Of course, the deeper the hole the more negative would be the potential energy and the more tightly bound it would be relative to the ground surface. Kinetic energy is, however, recovered as the rock falls into the hole and the amount is proportional to the distance it fell.

These same concepts can also be applied to the particles in an atom and are very important because the way atoms behave is determined by the potential energy states of its various particles. Particles that change potential energy states gain (or lose) kinetic energy which can be released from the atom either as a particle or as a wave. Perhaps the most important concept for particles in atoms is that they will always have a total energy which is the sum of the potential energy (PE), which is determined by their position, and the kinetic energy (KE) associated with their motion, or

$$E_{tot} = \text{Potential E} + \text{Kinetic E}$$

For example, an electron "free" of the nucleus will experience a decrease in potential energy depending on the position(s) it occupies in being bound near the nucleus. This phenomenon can be represented as perches in a hole that the electron might fall into with the release of energy

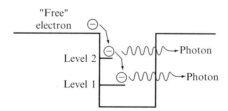

The perches in the hole represent negative energy states because it would take work to get them out and return them to the region where they would be considered as "free" electrons. The amount of energy required to get an electron out of an atom is determined by the energy level it occupies; this amount is also the amount of energy released as the electron goes from the surface (the "free" state) to level 1, or down to level 2.

These energy changes have been observed by measuring the photons (more about these later) emitted when electrons are disturbed in atoms of an element such as hydrogen, helium, neon, nitrogen, etc. The atom with bound electrons has been determined to be slightly lighter than without; the loss of mass exactly matches the energy of the photon emitted.

Checkpoints

Consideration of energy states yields two very important concepts directly applicable to atoms:

- an electron bound to an atom has less potential energy than if it were floating around "free"; and

- the process by which an atom constituent (electron, proton, or neutron) becomes bound causes the emission of energy as it goes to a lower potential energy state; the same amount of energy must be supplied to free the particle from its bound state.

These concepts can be extended in a similar way to protons and neutrons which are bound in the nucleus at different energy levels represented schematically as:

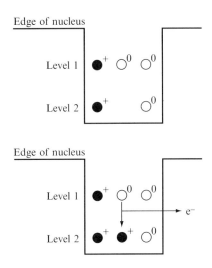

Rearranging the neutrons and protons to lower potential energy states yields energy in the form of an ejected particle (mass = energy), or pure electromagnetic radiation, or both. Processes which change the array of particles in the nucleus are fundamental to radioactive transformation which will be discussed in Chapter 3.

RELATIVISTIC ENERGY

The energy associated with atoms is governed by Einstein's special theory of relativity which states that mass and energy are one and the same. The rock on the perch gained potential energy from muscles doing the work necessary to lift it there; in an atom the energy of a particle is determined by the mass that is available to be transformed into energy as an electron or a nucleon (proton or neutron) undergoes a change in its bound state; i.e., as it becomes more tightly bound it gives up (or consumes) energy in relation to the change in its potential energy state.

The special theory of relativity developed by Einstein in 1905 describes the role of mass and energy and how mass changes at high speeds. In developing the theory of Special Relativity, Einstein showed that the mass, m, of a body varies with its speed, v, according to

$$m = \frac{m_0}{\sqrt{1 - (v^2/c^2)}}$$

where m_0, the rest mass, is the mass of the body measured when it is at rest and c is the velocity of light in a vacuum, which is a constant. This means that the mass of a body will increase with velocity; therefore, momentum (or mv) must be treated in a way that recognizes that mass changes as the velocity changes. In doing so, it is necessary to state Newton's second law precisely as he stated it; i.e., the net force on a body is equal to the time rate of change of the momentum of the body, or

$$F = \frac{d(mv)}{dt}$$

If the mass can be assumed to remain constant, this reduces to

$$F = m\frac{dv}{dt} = ma$$

which is the classical relationship used to calculate the dynamics of most objects in the macro world. $F = ma$ provides reasonably accurate calculations of the force on objects at low speeds (less than 10% of the speed of light) and is thus very practical to use. However, many particles in and associated with atoms move at high speeds, and Newton's second law must be stated in terms of the relativistic mass which changes with velocity, or

$$F = \frac{d(mv)}{dt} = \frac{d}{dt}\left(\frac{m_0 v}{\sqrt{1 - (v^2/c^2)}}\right)$$

In relativistic mechanics, as in classical mechanics, the kinetic energy, KE, of a body is equal to the work done by an external force in increasing the speed of the body by the value v in a distance ds:

$$KE = \int F\, ds$$

Using $ds = v\,dt$ and the relativistic generalization of Newton's second law, $F = d(mv)/dt$, this expression becomes:

$$KE = \int_0^s \frac{d(mv)}{dt} \cdot v\, dt = \int_0^{mv} v\, d(mv)$$

or,

$$\text{KE} = \int_0^v v\, d\left(\frac{m_0 v}{\sqrt{1 - v^2/c^2}}\right)$$

since m in relativistic terms is a function of the rest mass, m_0, which is a constant, and the velocity of the particle. If the term in the parenthesis is differentiated, and the integration performed,

$$\text{KE} = \frac{m_0 c^2}{\sqrt{1 - v^2/c^2}} - m_0 c^2$$

which reduces to

$$\text{KE} = mc^2 - m_0 c^2 = (m - m_0)c^2$$

The kinetic energy (KE) gained by a moving particle is due to the increase in its mass; it is also the difference between the total energy, mc^2, of the moving particle and the rest energy, $m_0 c^2$, of the particle when at rest. This is the same logical relationship between potential energy and kinetic energy we developed for a macro-world rock, but in this case the key variable is the change in mass of the moving body. Rearrangement of this equation yields Einstein's famous equivalence of mass and energy

$$E = mc^2 = \text{KE} + m_0 c^2$$

but more importantly it states that the total energy of a body is, similar to classical principles, the sum of the kinetic and potential energy at any given point in time and space with the important distinction that its potential energy is a property of its rest mass. Thus, even when a body is at rest it still has an energy content given by $E_0 = m_0 c^2$, so that in principle the potential energy inherent in the mass of an object can be completely converted into kinetic energy. Atomic and nuclear processes routinely convert mass to energy and vice versa, and because of these phenomena nuclear processes that yield or consume energy can be conveniently described by measuring the mass changes that occur. Such measurements are among the most accurate in science.

Although Einstein's concepts are fundamental to atomic phenomena, they are even more remarkable because when he stated them in 1905 he had no idea of atom systems, and no model of the atom existed. He had deduced the theory in search of the basic laws of nature that govern the dynamics and motion of objects. Einstein's discoveries encompass Newton's laws for the dynamics of macro-world objects but more importantly also apply to micro-world objects where velocities approach the speed of light; Newton's laws break down at these speeds, but Einstein's relationships do not. Einstein believed there was a more fundamental connection between the four forces of nature and he sought, without success, a unified field theory to elicit an even more fundamental law of nature. Even though his genius was unable to find the key to interconnect the four forces of nature, or to perhaps describe a unified force that encompassed them all, his brilliant and straightforward mass/energy concepts provided the foundation for later descriptions and understanding of the origins of atomic and nuclear phenomena, including the emission of radiation and its energy.

Checkpoints

Application of the concepts of special relativity leads to several important properties of mass and energy and their relationship to each other.

1. An increase in the kinetic energy of a body moving at relativistic speeds is due to an increase in its mass, which in turn is due to an increase in velocity, or

$$KE = mc^2 - m_0c^2 = (m - m_0)c^2$$

   where

$$m = \frac{m_0}{\sqrt{1 - v^2/c^2}}$$

2. The expression $E = mc^2$ represents the total energy of a body due to its motion (i.e., kinetic energy) plus that due to its rest mass (i.e., potential energy), or

$$E = mc^2 = KE + m_0c^2$$

Therefore, energy changes in an atom can be described by the change in mass that occurs during certain processes like radioactive transformation, interactions of bombarding particles, and fission and/or fusion of nuclei.

3. No body with rest mass can reach or exceed the speed of light because to do so it would be required to have infinite mass; i.e.,

$$m = \frac{m_0}{\sqrt{1 - v^2/c^2}} = \frac{m_0}{0} = \infty \text{ when } v = c,$$

   a feat that is impossible to accomplish.

4. Photons, which always travel at the speed of light must have zero rest mass; however, they have a mass equivalence associated with their energy. This energy can in turn be added to an atom in such a way that it changes

its mass, or it can be converted to a particle mass (e.g., an electron) in nuclear processes.

Electron Volt (eV)

The electron volt (eV) is a very useful practical unit for characterizing the energy in atoms, groups of atoms, or their constituent particles. The eV is defined as the increase in kinetic energy of a particle with one unit of electric charge (e.g., an electron) when it is accelerated through a potential difference of one volt. This can be represented schematically as

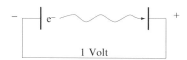

The stationary electron at the negative electrode in the figure has a potential energy of 1 eV; when released it will be repelled by the negatively charged electrode and attracted to the positively charged one, gaining in the process an energy of motion such that when it slams into the positive electrode all of the potential energy will be converted to kinetic energy. This energy, which is defined as one electron volt, is calculated by multiplying the unit charge of the electron by the voltage across the electrodes, i.e.,

$$1 eV = (q)(\Delta V) = (1.602177 \times 10^{-19} C)(1 Volt)$$
$$= 1.602177 \times 10^{-19} \text{ Volt} \cdot \text{Coulomb}$$
$$= 1.602177 \times 10^{-19} \text{ joule}$$
$$= 1.602177 \times 10^{-12} \text{ erg}$$

This relationship, $1 eV = 1.602177 \times 10^{-19}$ joule $= 1.602177 \times 10^{-12}$ erg, is used frequently in calculations of the amount of energy deposited in a medium. In absolute terms the eV is not very much energy. In fact, the energy of atomic changes is commonly expressed in keV (10^3 eV) and MeV (10^6 eV). The concept of representing the energy of small particles by the energy they possess in motion is, however, very useful in describing how they interact. For example, one can think of the energy of a 1 MeV beta particle or proton as each being a unit-charged particle that gained an acceleration equal to being subjected to a jolt of one million volts of electrical energy.

Example 1-2: An x-ray tube accelerates electrons from a cathode into a tungsten target anode to produce x rays. If the electric potential across the tube is 90 keV, what will be the energy of the electrons when they hit the target in eV, joules, and ergs?

Solution: $E = eV = 90,000$ eV, which is expressed in joules as

$$E = 90,000 \text{ eV} \times 1.6022 \times 10^{-19} \text{ joule/eV} = 1.442 \times 10^{-14} \text{ joules}$$

and in ergs

$$E = 1.442 \times 10^{-14} \text{ joules} \times 10^7 \text{ ergs/joule} = 1.442 \times 10^{-7} \text{ ergs}$$

Mass–Energy

Since the masses of isotopes of atoms and all the constituent particles are known to better than six decimal places, the energy changes in nuclear processes are readily determined by the mass changes (expressed in u) that occur. For this reason, a most useful quantity is the energy equivalent of the atomic mass unit, u, or

$$E = m_0 c^2 = \frac{1.66054 \times 10^{-27} \text{kg}/u(2.99792458 \times 10^8 \text{ m/s})^2}{1.6021892 \times 10^{-13} \text{ J/MeV}} = 931.502 \text{ MeV}/u$$

A similar calculation can be performed to determine the energy equivalent of the electron mass, or

$$
\begin{aligned}
E = m_0 c^2 &= 9.10953 \times 10^{-31} \text{kg} (2.99792458 \times 10^8 \text{ m/s})^2 \\
&= 8.187 \times 10^{-14} \text{ J}/1.6022 \times 10^{-13} \text{ J/MeV} \\
&= 0.511 \text{ MeV}
\end{aligned}
$$

Various nuclear processes occur in which electron masses are converted to 0.511 MeV photons or in which photons of sufficient energy are transformed to electron masses. For example, a photon with $E = hv > 1.022$ MeV can, in the vicinity of a charged body (a nucleus or an electron), vanish yielding two electron masses (a process called pair production) and kinetic energy, $KE = hv - 1.022$ MeV, that is shared by the particles thus produced. And in similar fashion, the amount of energy potentially available from the complete annihilation of one gram of matter would yield an energy of

$$E = mc^2 = 1\text{g} (2.99792458 \times 10^{10} \text{cm/s})^2 = 8.988 \times 10^{20} \text{g} \cdot \text{cm}^2/\text{s}^2$$

which is equal to 8.988×10^{20} ergs, or 8.988×10^{13} J, or about 25 million . This is an enormous amount of energy, and probably explains why scientists could not resist releasing the energy locked up in atoms after fission and fusion were discovered. Both fission and fusion cause atoms to become more tightly bound (see below) with significant changes in mass that is converted to energy and then released.

THE BINDING ENERGY OF NUCLEI

The equivalence of mass and energy takes on special significance for atoms which are bound systems in which the potential energy of the bound particles is negative due to the forces which hold the masses in the system together. The potential energy states of the particles that make up the atom are less than they would be if they were separate; i.e., if two masses, m_1 and m_2, are brought together to form a larger mass M, it will hold together only if

$$M < m_1 + m_2$$

This circumstance can be illustrated by introducing a quantity of energy, E_b, that is released as follows:

where E_b is the amount of energy released when the two masses are bound in such a way to decrease the rest-mass of the system. In other words, the amount of energy released in binding the masses together is simply the net decrease in the mass or

$$E_b = (m_1 + m_2 - M)c^2 = \Delta m c^2$$

E_b is called the binding energy (a negative quantity since $m_1 + m_2$ is greater than the mass M of the combined system) because it is responsible for holding the parts of the system together. It is also the energy that must be supplied to break M into separate masses, m_1 and m_2:

These relationships are perhaps the most important consequences of Einstein's mass–energy relation. They are most useful in understanding the energetics of particles and electromagnetic radiations, and are in fact fundamental properties of matter that are used constantly in atomic and nuclear physics.

Calculations of Binding Energy

The binding energy of an atom is determined from the mass differences of the total binding energy of bound atoms the assembled atom and the individual particles that make up the atom. For example, an atom of deuterium, ^{2}H, consists of a proton and a neutron with an orbital electron. It is one of the simplest bound stable atoms with a rest mass of 2.01410178 u. The mass of the proton is 1.00727647 u, the electron is 0.00054858 u, and the neutron is 1.008664904 u (note the eight-figure accuracy of the masses).

Example 1-3: Find the binding energy of the deuterium atom.
Solution: The mass of each individual particle is

$$1.00727647 \ u \ \text{(proton)}$$

$$1.00866491 \ u \ \text{(neutron)}$$

$$0.00054858 \ u \ \text{(electron)}$$

$$2.01648996 \ u \ \text{(total)}$$

which is less than the measured mass of deuterium at 2.01410178 u by 0.00238818 u; therefore the binding energy holding the deuterium atom together is

$$E_b = 0.0023881 \ u \times 931.502 \ \text{MeV}/u = 2.2246 \ \text{MeV}$$

which is released as photon energy as the constituents become arrayed in their bound (i.e., lower potential energy) states in the newly formed atom.

The calculation in Example 1-4 could be written as a nuclear reaction

$$_1^1\text{H} + {}_0^1 n \rightarrow {}_1^2\text{H} + \gamma + Q$$

where the mass of $_1^1$H is used instead of summing the masses of the proton nucleus and the orbiting electron. Q represents the net energy change, and in this case has a positive value of 2.225 MeV that is released as a gamma ray as the deuterium atom is formed. The binding energy of the orbital electron is implicitly ignored in this example; however, it is only about 13.6 eV which is negligible compared to 2.225 MeV.

Q-Value Calculations

Calculations of binding energy are a subset of Q-value calculations, so called because they involve a net energy change due to changes in the nuclear masses of the constituents. The Q-value represents the amount of energy that is gained when atoms change their bound states or that must be supplied to break them apart in a particular way. For example, in order to break ^{2}H into a proton and a neutron, an energy equal to 2.225 MeV (the Q-value) would need to be added to the ^{2}H atom, which in this case would be represented by the reaction

$$_1^1\text{H} + {}_0^1 n \rightarrow {}_1^2\text{H} + \gamma + Q$$

$$_1^2\text{H} + Q \rightarrow {}_1^1 \text{H} + {}_0^1 n$$

In this case, an energy equal to $Q = 2.225\,\text{MeV}$, usually as a photon, must be supplied to break apart the bound system (because it has negative potential energy). If the photon energy is greater than 2.225 MeV the excess energy will exist as kinetic energy shared by the proton and the neutron.

Example 1-4: Find the binding energy for tritium, which contains one proton, one electron, and two neutrons.

Solution: The constituent masses of the tritium atom are

$$
\begin{aligned}
\text{mass of proton} &= 1.00727647\ u \\
\text{mass of neutrons} &= 2\times\ 1.00866491\ u \\
\text{mass of electron} &= 0.0005485799\ u \\
\text{Total} &= 3.025155\ u
\end{aligned}
$$

which is larger than the measured mass of $3.016049\ u$ (see Appendix B) for ^3H by $0.0091061\ u$, and the total binding energy is

$$E_b = 0.009106u \times 931.502\,\text{MeV}/u = 8.48\,\text{MeV}$$

and since tritium (^3H) contains three nucleons, the average binding energy for each nucleon (E_b/A) is:

$$\frac{E_b}{A} = \frac{8.48\,\text{MeV}}{3} = 2.83\,\frac{\text{MeV}}{\text{nucleon}}$$

The Binding Energy per Nucleon (E_b/A) can be calculated from the isotopic masses of each of the elements (as listed in Appendix B), and when plotted for each of the stable elements plus uranium and thorium as shown in Figure 1-3, several significant features become apparent.

- The elements in the middle part of the curve are the most tightly bound, the highest being ^{62}Ni at 8.7945 MeV/nucleon and ^{58}Fe at 8.7921 MeV/nucleon.
- Certain nuclei, ^{4}He, ^{12}C, ^{16}O, ^{28}Si, and ^{32}S are extra stable because they are multiples of helium atoms (mass number of 4), with one important exception, ^{8}Be, which breaks up very rapidly (10^{-16} s or so) into two atoms of ^{4}He.
- Fission of a heavy nucleus like ^{235}U and ^{239}Pu produces two lighter atoms (or products) with nucleons that are, according to Figure 1-3, more tightly bound resulting in a net energy release.
- Fusion of light elements such as ^{2}H and ^{3}H produces helium with nucleons that are more tightly bound causing a release of energy that is even larger per unit mass than that released in fission.

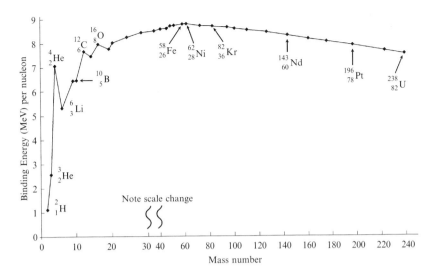

Figure 1-3. Curve of binding energy per nucleon vs atomic mass number (plotted from data in Appendix B).

SUMMARY

The dynamics of all objects in the universe from subatomic particles to stars and planets are governed by four fundamental forces of nature that have field strengths that vary by some 40 orders of magnitude: these are the gravitational, electromagnetic, weak, and nuclear (or strong) forces. The electromagnetic force fields of charged objects extend over all space as does the gravitational force field associated with the mass of an object, and as far as is known, the weak force and nuclear force exist only between nuclear particles. Although physicists have begun to delve deeper into the structure of matter and to develop and support a grand unification theory based on smaller and more energetic constituents (quarks, leptons, bosons, hadrons, etc), the purposes of radiation physics can be represented with simplified models based on electrons, protons, neutrons, and their respective energy states.

The dynamics of the forces of nature yield several important concepts related to atoms:

- Each atom has a nucleus which contains one or more protons (to establish the identity of the atom) and, except for hydrogen-1, one or more neutrons.
- Hydrogen, by definition, has one proton. Helium has two protons, lithium has three, and so on.
- Isotopes are atoms of the same element, but with different weights. Hydrogen (^1H) with a weight of one, deuterium (^2H) with a weight of two, and

tritium (^{3}H) with a weight of three are isotopes of hydrogen. Similarly, ^{3}He and ^{4}He are isotopes of helium, so named because it has two protons.

- Atoms left to themselves are electrically neutral. Electrons will form in orbits around a nucleus to balance the positive charge of each proton in the nucleus.

- The nucleus is very small when compared to the electron orbits, which are about 10^{-8} cm in size; the nucleus, which is at the center of the atom, is on the order of 10^{-13} cm in diameter; so the atom is mostly empty space.

- The nucleus is very small and therefore very dense; it contains essentially all the mass of the atom because the protons and neutrons in the nucleus each weigh about 1800 times more than the electrons orbiting about it.

- According to Avogadro, the number of atoms (or molecules) in a mole of a substance is the same, given by Avogadro's number, $N_A = 6.0221376 \times 10^{23}$ atoms/mol.

- The unified mass unit, u, has a mass of 1.66054×10^{-24} g and an energy equivalence of 931.502 MeV. The u, often referred to as the atomic mass unit, has a mass on the order of a neutron or proton (called nucleons because they are constituent parts of the nucleus of atoms); thus, the mass number, A, of an isotope is close to, but not exactly, the atomic mass of the individual atoms.

- The electron volt is a reference amount of energy that is used to describe nuclear and atomic events, and is defined as the energy that would be gained by a particle with a unit charge when it is accelerated through a potential difference of one volt. It is equivalent to 1.602177×10^{-19} Joules or 1.602177×10^{-12} ergs.

Einstein's theory of Special Relativity is applicable to atom systems where particles undergo interchangeable mass/energy processes. These processes yield "radiation" which can be characterized as the emission of energy in the form of particles or electromagnetic energy as atoms undergo change; the amount of energy emitted (or consumed) is, in accordance with the special theory of relativity, directly determined by the change in mass that occurs as atom constituents change potential energy states. These very exact values of mass allow the calculation of Q-values, or energy changes as atoms undergo change. Binding energy is one important result of such calculations; it denotes the amount of energy that is given off as constituents come together to form atoms, or alternatively, is a measure of energy that must be supplied to disengage consitituents from an atom.

ADDITIONAL RESOURCES

1 *Chart of the Nuclides, Nuclides and Isotopes*, 15th ed., 1996. Available from General Electric Company, 175 Curtner Avenue, M/C 397, San Jose, CA 95125, USA.

2   National Nuclear Data Center, Brookhaven National Laboratory, Upton,
    Long Island, NY 11973. Data resources are accessible through the internet
    at www.nndc.bnl.gov

PROBLEMS

1   How many neutrons and how many protons are there in: (a) ^{14}C, (b) ^{27}Al,
    (c) ^{133}Xe, and (d) ^{209}Bi?

2   If one were to base the atomic mass scale on ^{16}O at 16.000000 amu
    calculate the mass of the ^{16}O atom and the mass of one amu. Why is its
    mass on the ^{12}C scale different from 16.000000?

3   Calculate the number of atoms of $^1_1$H in one gram of natural hydrogen.

4   Hydrogen is a diatomic molecule, or H_2. Calculate the mass of one
    molecule of H_2.

5   Calculate the radius of the nucleus of ^{27}Al in meters and fermis.

6   A one gram target of natural lithium is to be put into an accelerator for
    bombardment of ^{6}Li to produce ^{3}H. Use the information in Figure 1-1 to
    calculate the number of atoms of ^{6}Li in the target.

7   A linear accelerator is operated at 700 kv to accelerate protons. What
    energy will the protons have when they exit the accelerator in (a) eV, and
    (b) Joules. If deuterium ions are accelerated what will be the corresponding
    energy?

8   A rowdy student in a food fight threw a fig (weight of 200 g) so hard that it
    had an acceleration of $10 \, m/s^2$ when it was released. What "fig Newton"
    force was applied?

9   Calculate the velocity required to double the mass of a particle.

10 Calculate from the masses in the mass table of Appendix B the total
    binding energy and the binding energy per nucleon of (a) beryllium-7, (b)
    iron-56, (c) nickel-62, and (d) uranium-238.

2

ATOM STRUCTURE AND ENERGETICS

"Subtle is the Lord, but Malicious He is Not."

– A. Einstein

Modern concepts of radiation and radioactivity, and the structure of the atom and its nucleus, are based on a remarkable series of discoveries that occurred from around 1895 to the demonstration of the fission chain reaction in uranium and plutonium in the 1940s. These discoveries, which encompass only about 50 years, quite literally changed the world forever; they begin with the following major events:

- Roentgen's discovery of x rays (1895),
- Becquerel's discovery of radioactivity shortly thereafter (1896),
- Thomson's discovery of the electron (1897),
- Planck's basic radiation law (1900),
- Einstein's theory of Special Relativity (1905),
- Rutherford's alpha-scattering experiments (1911), and
- Bohr's model of the atom (1913).

These phenomena and the information they provided made it clear that atoms were not solid little balls but had structure and properties that required new and elaborate means to describe them. We pick up three major threads in this tapestry of what is called modern physics: (1) discoveries that determined the substructure of atoms and their energy states; (2) the fundamentals of quantum emissions of radiation; and (3) combinations of these to describe the structure and dynamics of the atom system. The insight required to formulate fundamental theories to resolve conflicts between observed phenomena and prevailing ideas in physics not only explained the phenomena but provided new discoveries of the laws of nature that led to even greater discoveries. These insights are presented as a means to understanding concepts that are basic to radiation science as well as physics.

MAJOR DISCOVERIES AND SIMPLE TOOLS

The discovery of x rays, radioactivity, and the electron were prompted by the study of gases. As it was necessary to enclose a gas to study it, two simple tools were used: glassblowing techniques and the vacuum pump which had been recently developed. Such experiments eventually took the form of creating an electric potential across electrodes imbedded in closed glass tubes and observing how the discharge changed with pressure. When a low vacuum was produced in such a tube and an electric potential was applied, rays were found to emanate from the cathode and travel down the tube (Figure 2-1). Studies of these rays led to Roentgen's discovery of x rays in 1895 and Thomson's discovery of the electron in 1897.

When a cathode-ray tube is filled with a gas, a discharge will occur, emitting light containing wavelengths characteristic of the gas in the tube. But when such a tube is evacuated to very low pressure (made possible with a good vacuum pump), the glow changes and diminishes, and new phenomena are observed due to emanations from the cathode. At a pressure of about 10^{-3} atm, the tube produces a luminous glow filling the tube, as in a neon sign. Below 10^{-6} atm, the negative electrode, or cathode, emits invisible rays that propagate through the nearly empty space in the tube. These emanations were quite logically called cathode rays. In 1897, J.J. Thomson established that cathode rays are in fact small "corpuscles" of matter (or electrons) with a negative electric charge. Thomson used a specially designed cathode-ray tube to investigate and quantitate the deflection ("deflexion" per Thomson) of the rays by electric and magnetic fields. He is properly credited with the discovery of the electron, one of the basic constituents of matter, a discovery for which he received the Nobel Prize.

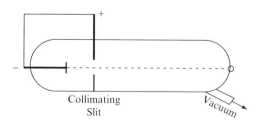

Figure 2-1. Cathode-ray tube.

Discovery of X Rays—1895

Wilhelm Conrad Roentgen, among others, studied cathode rays using a highly evacuated cathode-ray tube, called a Crooke's tube after its inventor. One evening in December 1895, in an attempt to understand the glow produced in such a tube, he covered it with opaque paper. Then one of those events that triggers great minds to discovery happened: he had darkened his laboratory to observe better the glow produced in the tube, and in the dim light he noticed flashes of light on a barium-platino-cyanide screen that happened to be near the apparatus. Because the tube (now covered with black paper) was obviously opaque to light emitted from the tube, he realized that the flashes on the screen must be due to emissions from the tube because they disappeared when the electric potential was disconnected.

Roentgen called the emissions x rays to denote their "unknown" nature, and in a matter of days went on to describe their major features. The most startling property was that the rays could penetrate dense objects and produce an image of the object on a photographic plate as shown in Figure 2-2, the classic picture of the bones in his wife's hand.

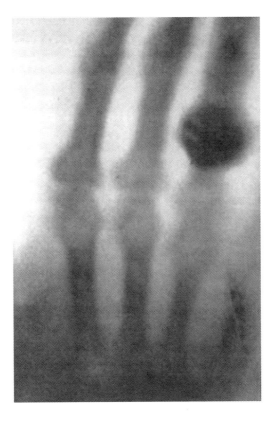

Figure 2-2. X-ray photograph of Frau Roentgen's hand.

He also discovered that x-radiation could produce ionization in any gas through which it passes, a property that is used to measure x-ray intensity. The x rays could be reflected, refracted, and diffracted; they are a form of electromagnetic radiation like light, only of much shorter wavelength.

Very few discoveries have been as important to human existence as Roentgen's x rays. The rays were used almost immediately in diagnosis and treatment of disease, uses that are even more common a century later, and new applications continue. Roentgen never considered patenting this remarkable invention; his was truly a gift to science and humankind.

Discovery of Radioactivity

X rays assumed great importance in the work of Becquerel who believed that fluorescence or phosphorescence might be the source of Roentgen's x rays since they appeared to emanate from the portion of the Crooke's tube that glowed the brightest. To test the theory, Becquerel placed a phosphorescent material on a tightly wrapped photographic plate and exposed it to sunlight to determine if the induced fluorescence perhaps contained Roentgen's x rays. He could have chosen many other fluorescent materials but happened to use a salt of potassium uranyl sulfate. When he developed the photographic film, the outline of the uranium crystal was clearly visible, which he assumed was due to the fluorescence induced in the crystal. Apparently he prepared several other photographic plates to repeat the experiment, but since the sky was cloudy for several days he placed the plates in the back of a drawer with the uranium crystals still attached. Even though the crystals had not been exposed to light, he developed the plates anyway (perhaps as a reference comparison for his earlier observation) and discovered that it had the same clear outline of the crystal on the plate as before but the image was even brighter than before (due, of course, to the longer exposure period). Aha! The images were much like those produced by Roentgen's rays but, since the crystals had not been exposed to sunlight, they could not be due to fluorescence or phosphorescence. Becquerel decided that the radiations affecting his plates originated from within the uranium salt itself and was spontaneous.

Marie S. Curie, along with her husband Pierre, studied this new property of materials and named it radioactivity. Her discovery of radium and polonium and her work on describing radioactivity far surpassed that of Becquerel. Nevertheless, Becquerel is credited with the discovery of radioactivity, an intrinsic property of certain substances. The catalyst for recognizing the radiation emitted was almost certainly Roentgen's discovery a few months earlier; otherwise, if Becquerel had thought to expose his plates to uranium, he might have just attributed the images to poor photographic emulsion as Goodspeed had done (Goodspeed's photographs of the clear outline of coins taken several years earlier exist; how he missed the discovery remains a mystery). Discovery requires a prepared mind, and Becquerel was fortunate enough through circumstances to stumble onto the phenomenon and to recognize it, although it took him a while.

Madame Curie, in a series of comprehensive experiments on the nature of radioactivity, established that the activity of a given material is not affected by any physical or chemical process such as heat or chemical combination, and that the activity of any uranium salt is directly proportional to the quantity of uranium in the salt. Further work by Rutherford and Soddy clearly established that radioactivity is a sub-atomic phenomenon (see Chapter 3). The emission of helium nuclei and electrons from radioactive atoms suggested that all atoms must be made up of smaller units; they could no longer be considered little round balls, and a theory of atomic structure that incorporated these discoveries was needed. Formulation of an adequate theory required two fundamental concepts not yet discovered: Planck's quantum theory of electromagnetic radiation and Einstein's equivalence of mass and energy (as discussed in Chapter 1).

THEORY OF ELECTROMAGNETIC RADIATION—THE QUANTUM

A warm body emits heat, which is a form of radiation. Most heat radiation is only felt but not seen because the human eye is not sensitive to these wavelengths. Objects can be heated enough to cause them to glow, first with a dull red color and, if heated still further, bright red, orange, and eventually "white hot." Intuitively the amount of energy emitted increases with temperature and the frequency of the emitted radiation, but what is the basic law that describes the relationship exactly? Physicists responded to this challenge in the late 1800s with startling and far-reaching insights into the nature of electromagnetic radiation. Providing a fundamental theory for this relatively benign phenomenon gave birth to the quantum theory, the greatest revolution in physical thought during the 20th century.

The problem of the spectral distribution of thermal radiation was solved by Planck in 1901 by means of a revolutionary hypothesis. Planck assumed, as did others, that the emitted radiation was produced by linear harmonic oscillators; however, he made the *ad hoc* assumption that each oscillator does not have an energy that can take on any value from zero to infinity, but instead has only discrete values which are intergral multiples of a base unit of energy, or

$$\epsilon = n\epsilon_0$$

where n is an integer and ϵ_0 is a discrete, finite amount, or *quantum*, of energy that must be proportional to the frequency ν of the radiation, or

$$\epsilon_0 = h\nu$$

where h is a new universal constant, called Planck's constant. When he used these postulates to derive the energy density for the emitted radiation, Planck obtained the distribution law for the energy density μ_r of thermal radiation emitted by a black body as

$$\mu_v = \frac{8\pi h v^3}{c^3} \cdot \frac{1}{(e^{hv/kT} - 1)}$$

or, in terms of wavelength,

$$\mu_\lambda = \frac{8\pi h}{\lambda^5} \cdot \frac{1}{(e^{hc/\lambda kT} - 1)}$$

This predictive relationship is plotted as the smooth curve in Figure 2-3 along with the curves based on theories developed by Wein and Rayleigh-Jeans. Whereas their distribution functions break down at either end of the measured spectrum, Planck's distribution law fits the experimental points exactly, and is obviously the more general and correct law for emission of radiant energy. It can be shown that it encompasses both the Rayleigh-Jeans and Wein formulas by expanding the exponential function in the denominator for $hv/kT \ll 1$ and for $hv/kT \gg 1$, respectively. Planck's law thus incorporates the principles observed by Wein and Rayleigh-Jeans, but its greatest significance is the key discovery that radiation is emitted in discrete units. Without it, theory and experiment did not agree. But much more importantly, the discovery states a basic law of nature that led to one of the greatest periods of thought in human history and science: the use of quantum mechanics to explain the structure of matter.

Planck determined the first value for h from the experimental data and was able to also calculate the best value to that date for k. Modern values of each are:

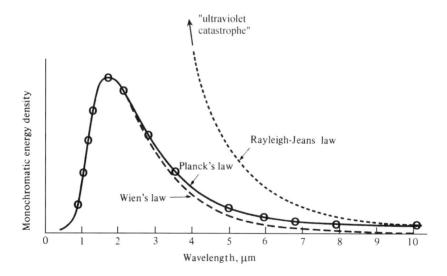

Figure 2-3. Planck's radiation law showing an excellent fit to measured data. Wein's law fits the data for short wavelengths; the Rayleigh-Jeans law holds for large wavelengths but exhibits "ultraviolet catastrophe" at short wavelengths.

$$h = 6.6260755 \times 10^{-34} \, \text{J} \cdot \text{s} = 4.1356692 \times 10^{-15} \, \text{eV} \cdot \text{s}$$

$$k = 1.380658 \times 10^{-23} \, \text{J/K}$$

These values are fundamental constants of nature, especially Planck's constant which appears in all descriptions of electromagnetic radiation and treatments of the atom by quantum mechanics.

The revolutionary nature of Planck's theory is contained in the postulate that the emission and absorption of radiation must be discontinuous processes. Any physical system capable of emitting or absorbing electromagnetic radiation is limited to a discrete set of possible energy values or levels; energies intermediate between these allowed values (which are integral multiples of hv) simply do not occur. This revolutionary theory applies to all energy systems including the energy states of particles in atoms, and because it does so it greatly influences the structure that an atom can have.

Checkpoints

- The quantum theory is a bold hypothesis which was made to bring theory and experiment into agreement;
- Discrete energy levels, which are a direct consequence of Planck's quantum hypothesis, are fundamental to understanding and using the physics of radiation protection for both atomic and nuclear phenomena; and
- The energy of any electromagnetic wave, an essential parameter in radiation protection, is provided by Planck's law as $E = hv$.

Confirmation of the Quantum Hypothesis

Planck himself did not immediately recognize the breadth and importance of his discovery, believing he had merely found an *ad hoc* formula for thermal radiation. Since the wave theory so clearly explained interference, diffraction, and polarization of light waves, it seems preposterous that electromagnetic radiation was a pulse of energy (or quanta). The truth of the quantum hypothesis, or the particle-like nature of radiation, occurred with Einstein's explanation of the photoelectric effect and Compton's theory in the scattering of x rays (see Chapter 5).

By adapting and extending Planck's quantum hypothesis, Einstein was able to explain Hertz's observation (in 1887) that shining light of a certain frequency on a metallic surface caused electrons to be emitted (for this, and obviously other contributions, he received a Nobel prize in 1921). He simply assumed that the electromagnetic radiation incident on the metallic surface consisted of Planck-like discrete bundles of energy, called *quanta*, which could interact as a "particle." If an electron absorbs a photon of energy, hv, it could be ejected from the metallic surface with a kinetic energy equal to hv minus an amount of energy, ϕ, which represented the degree to which the electrons were bound in

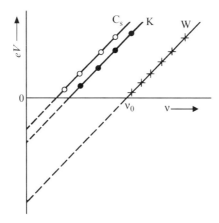

Figure 2-4. Kinetic energies of ejected photoelectrons vs frequency of the incident light where $KE = h\nu - \phi$. The work function energy ϕ is the y-axis intercept and the threshold frequency, ν_0, is obtained from the x-axis intercept.

the metallic surface. Einstein called this quantity the *work function*, which is different for each metal. The kinetic energy of the ejected electrons is

$$KE = h\nu - \phi$$

which is an equation for a straight line with a slope of h and a y-axis intercept of ϕ. The work function ϕ is determined by extrapolating the straight line to intercept the y-axis, and the threshold frequency, ν_0, is the x-axis intercept as shown in Figure 2-4.

The Electromagnetic Spectrum

The photon hypothesis, which was developed to explain blackbody radiation, represents one of the remarkable aspects of science, the broad applicability of a basic principle once it is discovered. Hertz demonstrated the existence of low frequency radio waves which greatly extended the low end of the spectrum of electromagnetic energy, and Roentgen's x rays extended it on the higher end. The simple expression for the energy of electromagnetic radiation, $E = h\nu$, applies across the full electromagnetic spectrum which, as now known and shown in Figure 2-5, covers more than 22 orders of magnitude.

Checkpoints—Quantized Energy

- A body that emits radiant energy (a furnace as heat, a light bulb as light, a radioactive element as gamma rays) undergoes changes that produce an electromagnetic wave that transmits energy from where it is produced (the energetic body) to some other place – all radiation is, therefore, a form of energy transmission.

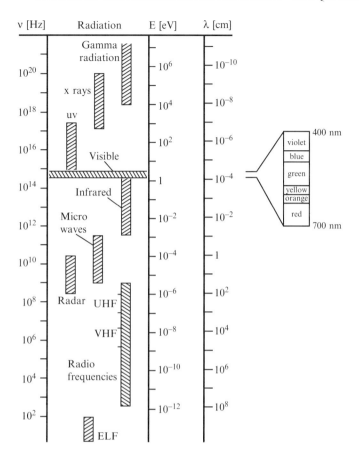

Figure 2-5. The electromagnetic spectrum with the wavelengths of the visible region amplified (Planck's radiation law $E = h\nu$ applies to wave phenomena over more than 22 orders of magnitude).

- Radiant energy emission occurs only as a pulse, not as a continuum of energy. An electron or particle in the atom can only exist at energy level 1 or level 2 (see Figure 2-6) but not in between, so that when an atom emits energy it does so in an amount that is the discrete difference between the two permitted states. For an electron, say, to become more tightly bound (i.e., achieve a lower potential energy state) the atom has to spit out a pulse of energy—the photon.
- The energy of a photon is directly proportional to its frequency. This constant of proportionality is one of the fundamental constants of nature and is, appropriately, called Planck's constant. All photons travel at the speed of light, which is another fundamental constant of nature; thus, the

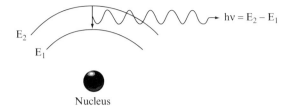

Nucleus

Figure 2-6. Discrete photon energy emission for permitted energy levels of an atomic electron; similar emissions occur for particles in the nucleus.

energy of a photon is expressed in terms of these two fundamental constants and its wavelength:

$$E = h\nu = \frac{hc}{\lambda}$$

That is, photons with long wavelengths have less energy than ones with very short wavelengths.

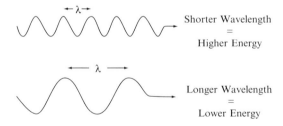

- Radiant energy (light, heat, x rays) is **pure** energy propagated along a wave in packets of energy or quanta that behave like a "particle."

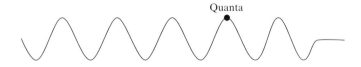

- Photons clearly have no rest mass, but their "particle" properties can be used to explain the photoelectric effect (Einstein, 1905) in which light of a certain frequency (wavelength) causes the emission of electrons from the surface of various metals.

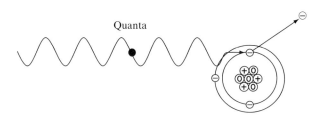

The photon, behaving like a particle, hits a bound electron and "knocks" it out of the atom (photocells depend on this principle).

- Compton scattering of x rays can be explained as interactions between "particle-like" photons and "free" electrons in which energy and momentum are conserved, and the calculated wavelength changes agree with experimental observations.
- Atoms that emit electromagnetic radiation must have quantized, or discrete, energy states for all particle arrangements: those involving electrons around the nucleus, and also the arrangement of neutrons and protons within the nucleus.

DISCOVERY OF THE ATOM'S STRUCTURE

The discoveries associated with x rays, radioactivity, quantized energy, and special relativity led to vital insights about the atom, its ultimate structure, and the dynamics of its components. The work of Thomson and Faraday established that electrons exist as a corpuscular unit that can be broken off of atoms and that these are all the same regardless of the element. Obviously the atom is made up of components, one of which is the electron and another which carries a positive charge since atoms are electrically neutral; this unit would likely be the remainder of the hydrogen atom, the lightest and simplest element, when the electron is stripped away or ionized. The expulsion of alpha particles (positively charged helium nuclei) during radioactive transformation also strongly suggested that atoms have components. An alpha particle must be a substructure of the atom, and its positive charge confirms that part of the atom carries a positive charge.

Atoms were first proposed by philosophy, not physics, as indivisible units. Early Greek philosophers had reasoned that if a small piece of matter were continually subdivided, one would eventually obtain a particle so small that it could not be divided any further and still have the same properties. They called these "pieces" atoms. Some 2,000 years later, John Dalton proposed a similar atomic theory of matter in an attempt to consolidate observations that chemicals combined in multiples of whole numbers, or definite proportions. These observations suggested to Dalton that all matter is made up of elementary particles (atoms) which retain their identity, mass, and physical properties in chemical reactions. Dalton's formulation, which is an important landmark in the development of modern atomic physics, described chemical reactions as a hook-and-eye connection of atoms that could of course be "hooked" and "unhooked" when materials combined or disassociated (Figure 2-7).

J.L. Gay-Lussac added additional evidence of the discreteness or atomic ature of matter with his law of combining volumes: when two gases combine to form a third, the ratios of the volumes are ratios of integers. For example, when hydrogen combines with oxygen to form water vapor, the ratio of the volume of hydrogen to that of oxygen is 2 to 1. Avogadro's remarkable hypothesis that a mole of any substance contains an equal number of atoms or molecules established that atoms are basic units of matter and that combining weights of substances contained a fixed number of these basic units.

The Atom Has Parts!

Faraday, an English physicist, established that a fundamental relationship exists between Avogadro's number, the unit of charge on the electron, and the amount of total charge necessary to deposit one gram equivalent weight of a metal on the cathode of an electrolysis experiment. Faraday showed in measurement after measurement that no matter what metal was used for the anode, it took exactly 96,485 coulombs to deposit one gram equivalent weight (or mole) of it on the cathode. This quantity is called the Faraday and is related to the unit of electrical charge by Avogadro's number as follows:

$$F = N_A e$$

where F is one faraday, N_A is Avogadro's number of atoms in one gram equivalent weight (or mole), and e is the charge on each electron liberated to flow through the circuit. Because the faraday was well established by Faraday's work, it would be possible to determine Avogadro's number or the charge on the electron if either were known.

Figure 2-7. Dalton visualized that atoms linked up in ways that resembled the hooks and eyes used on clothing.

Millikan's oil drop experiment established the electron's charge from which Avogadro's number could be determined from Faraday's constant. Later work on the scattering of x rays from crystals would provide a more precise determination of Avogadro's number, but from Faraday's work it was now clear that every atom contains electrons which are exactly alike, each carrying a definite amount of negative charge, and that each neutral atom contains just enough electrons to balance whatever positive charge it contains.

Checkpoint

- The Faraday, Avogadro's number, and the charge on the electron are three of those constants known as the fundamental constants of nature. Modern values of each are:

$$F = 96,485 \text{ C/mol}$$

$$N_A = 6.0221367 \times 10^{23} \text{ atoms/mol}$$

$$e = 1.60217733 \times 10^{-19} \text{ C}$$

Thomson's "Plum Pudding" Model

Because atoms are electrically neutral, any structure for the atom must describe a configuration whereby the negative charges of electrons and positive charges cancel each other. Thomson proposed a model, shown schematically in Figure 2-8, in which the charge and mass are distributed uniformly over a sphere of radius $\sim 10^{-8}$ cm, with electrons embedded in it (like plums in pudding) to make a neutral atom.

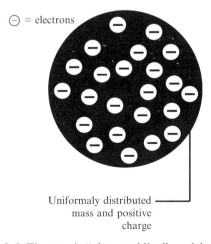

Figure 2-8. J. J. Thomson's "plum-pudding" model of the atom.

Unfortunately, Thomson's configuration of the atom did not produce a stable equilibrium, nor frequencies observed in optical spectra, nor the fact that the positive charges account for only about half the atom's weight.

Rutherford's Alpha-Scattering Experiments

Geiger (inventor of the Geiger counter) and Marsden, working with Rutherford, used alpha particles from radioactive substances to literally shoot holes in Thomson's distributed atom because it could not account for the scattering observed. They irradiated thin gold foils (about 10^{-4} cm thick) with alpha particles using an apparatus shown schematically in Figure 2-9. The whole apparatus was placed in a vacuum chamber to preclude interactions of the alpha particles with air molecules. Most of the particles passed through the foil with minimal deflection to produce a flash of light when they struck a zinc-sulfide screen (observed after sitting in the dark for a while to adapt their eyes). But, remarkably, some of the particles were observed to scatter at large angles, and one out of every 8,000 or so was scattered back towards the source.

Rutherford used the coulomb force equation to describe the repulsive force on a positively charged alpha particle (2e) by a positively charged atom (Ze), or

$$F = k \cdot \frac{q^1 q^2}{r^2} = k \frac{(2e)(Ze)}{r^2}$$

and observed that the large-angle scattering could only be explained if the separation distance r was on the order of 10^{-13} to 10^{-12} cm, suggesting that all of the positive charge—and hence almost all the mass of the atom—is confined to a very small volume at the center of the atom. Rutherford called this central volume the nucleus. Since the radius of the atom is about 10^{-8} cm, electrons could be assumed to be distributed in orbits that are more than 4 to 5 orders of magnitude larger than the nucleus itself.

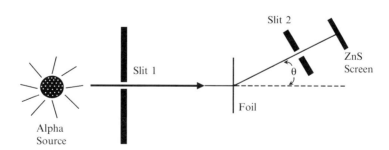

Figure 2-9. Experimental arrangement for studying scattering of alpha particles by thin metallic foils.

As brilliant as Rutherford's proposed model was, it had a fundamental flaw he could not explain. According to classical electromagnetic theory, an accelerated charge must radiate electromagnetic waves. An electron circling a hydrogen nucleus experiences a continuous acceleration v^2/r, and according to classical theory should radiate energy continuously; thus, r would steadily decrease and the electron would quickly (in about 10^{-8} second or so) spiral into the nucleus:

THE BOHR MODEL OF THE ATOM

The dynamic shown in Figure 2-10 corresponds to classical theory, but is wrong. The hydrogen atom is very stable, and it does not radiate electromagnetic waves unless stimulated to do so. Obviously, a new theory was needed, and Neils Bohr, a Danish physicist, provided it (in 1913) based on three postulations founded on a mix of classical physics and the energy-quantization ideas introduced by Planck:

Postulate I: An electron in an atom can move about the nucleus in discrete stationary states without radiating;

Postulate II: The allowed stationary states of the orbiting electrons are those for which the orbital angular momentum is an integral multiple of $h/2\pi$, or

$$L = n\frac{h}{2\pi}$$

where $n = 1, 2, 3, 4, \ldots$ represents the principal quantum number for discrete, quantized energy states.

Postulate III: An electron can jump from a higher potential energy state E1 to a lower energy orbit E2 (both of which are negative potential energy states for bound electrons) by emitting electromagnetic radiation of energy $h\nu$ where

$$h\nu = E_2 - E_1$$

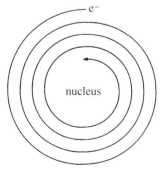

Figure 2-10. Rutherford's conundrum in which classical theory calls for the electron to spiral into the nucleus.

In postulate I, Bohr simply declared that an electron in its orbit did not radiate energy; it only radiates energy when it undergoes one of Planck's discontinuous quantum changes. This statement is completely arbitrary and represents Bohr's genius and his boldness—physics would just have to catch up. Postulate III is based on Planck's quantum hypothesis, but postulate II was unfounded and a bold step into the unknown; it would only be explained by deBroglie's hypothesis some 13 years later. Although it may appear that Bohr had no firm basis for his model of the hydrogen atom, he was able to take advantage of some significant building blocks:

1. Planck had established the discreteness of electromagnetic radiation, i.e., that it appears only as integral multiples, $n = 1, 2, 3, \ldots$ of a basic unit of energy;

2. Rutherford's experiments on the scattering of alpha particles clearly supported a nuclear model in which the positive charge and mass were contained in a nucleus of $\sim 10^{-12}$ cm radius at the center of the atom with the negative charges (electrons) distributed out around it to a radius of about 10^{-8} cm.

3. Balmer had observed a regularity in the wavelengths emitted by hydrogen and had derived an empirical relationship containing an integral term, n, much like the one in Planck's quantum theory:

$$\lambda = 3645.6 \frac{n^2}{(n^2 - 2^2)}$$

where n = 3, 4, 5.

Apparently, Bohr found some significance between the integral values 2 and n in these experimental relationships and Planck's quantum theory and incorporated it to explain the regularities observed for hydrogen spectra. He first used Rutherford's nuclear atom and assumed that orbiting electrons moved in circular orbits under the influence of two force fields: the coulomb attraction (a centripetal force here) provided by the positively charged nucleus and the centrifugal force induced by each electron due to its orbital motion.

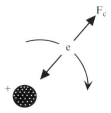

Figure 2-11. An electron in a Bohr orbit experiences a coulombic force of attraction towards the nucleus that is counterbalanced by the centrifugal force induced by its orbital motion.

Bohr assumed (Postulate I) that the electron is in constant circular motion at a radius, r_n, with velocity, v_n, and that the centrifugal force and the attractive coulomb force are equal and opposite each other:

$$\frac{m{v_n}^2}{r_n} = k\frac{q_1 q_2}{(r_n)^2}$$

The angular momentum of the electron from postulate II is

$$mvr_n = n\frac{h}{2\pi}$$

and since q_1 and q_2 are the same for hydrogen, the radius r_n can be calculated from these two equations as

$$r_n = \frac{(nh)^2}{(2\pi)^2 kmq^2} = n^2 r_1$$

where $n = 1, 2, 3, 4, \ldots$ and r_1 is the radius of the first orbit of the electron in the hydrogen atom, the so-called Bohr orbit. If the values of h, k, m and q^2 are inserted, $r_1 = 0.529 \times 10^{-8}$ cm, which agrees with the experimentally measured value, and since the quantum hypothesis limits values of n to integral values, the electron can only be in those orbits which are given by

$$r_n = r_1,\ (2)^2 r_1,\ (3)^2 r_1,\ (4)^2 r_1 \ldots$$

or

$$r_n = r_1,\ 4r_1,\ 9r_1,\ 16r_1 \ldots$$

These relationships can also be used to calculate the velocity of the electron in each orbit, but the most useful relation is for the energy levels of the electron. The total energy E_n of the electron in the nth orbit is the sum of its kinetic energy and its potential energy. That is:

$$E_n = \frac{m(v_n)^2}{2} + \left(-\frac{ke^2}{r_n}\right)$$

Substituting the equations for v_n and r_n and simplifying, the energy, E_n, for the electron in each orbit of quantum number n is:

$$E_n = -\frac{1}{n^2}\left(\frac{(2p)^2 k^2 q^4}{2h^2}\right) = -\frac{1}{n^2} 13.58\ \text{eV}$$

which for $n = 1$ is in perfect agreement with the measured value of the energy required to ionize hydrogen; i.e., to overcome the binding energy of the electron in its lowest potential state in a hydrogen atom. Thus, for other values of $n = 2$, 3, 4, ..., the allowed energy levels of the hydrogen atom can be calculated as shown in the energy level diagram in Figure 2-12. Bohr used these relationships to calculate the possible emissions (or absorption) of electromagnetic radiation for hydrogen. For $n = 3$, he obtained Balmer's series of wavelengths, and since the theory holds for $n = 3$, Bohr postulated that it should also hold for other values of n. The wavelengths calculated for $n = 1$, 2, 4, etc., were soon found providing dramatic proof of the theory. Bohr had put together several diverse concepts to produce a simple model of atomic structure, at least for hydrogen.

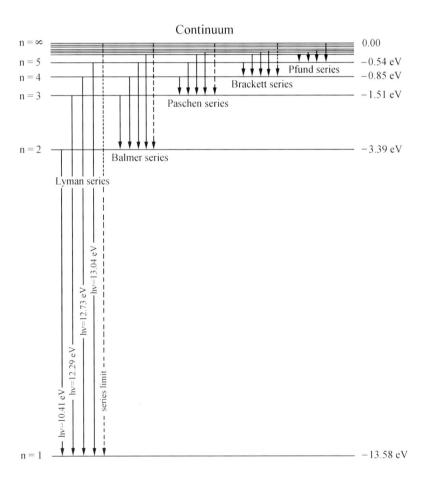

Figure 2-12. Relative energy states of electrons in the hydrogen atom and the corresponding radiation series named after those who discovered them.

WAVE MECHANICS—A NECESSARY THEORY

Although Bohr's second postulate [$mvr = n(h/2\pi)$] happened to pick out the stable orbits for hydrogen, it remained largely arbitrary even though it worked. But neither Bohr nor anyone else in 1913 was able to say why this particular condition should agree with the actual orbits observed (Why should mvr be directly proportional to n, for instance?). Some 13 years later, Louis deBroglie provided the key for explaining postulate II, which not only provided the foundation for Bohr's postulate but the essential property embodied in quantum mechanics, or wave mechanics; i.e., the wave motion of particles. deBroglie's proposal yielded the essential ingredient for describing the motion of particles under the force fields in the atom.

deBroglie Waves

In 1926, Louis deBroglie postulated that if Einstein's and Compton's assignment of particle properties to waves was correct, why shouldn't the converse be true, i.e., that particles have wave properties? This simple but far-reaching concept was proved by the diffraction (a wave phenomenon) of electrons (clearly particles) from a crystal containing nickel. From this he proposed that a particle such as an electron (or a car for that matter) has a wavelength, λ, associated with its motion of

$$\lambda = \frac{h}{p} = \frac{h}{mv}$$

and that as a wave it has momentum, p, with the value

$$p = \frac{h}{\lambda}$$

Because p is in the denominator in deBroglie's equation, larger p implies smaller λ. For the enormous momentum of macroscopic objects such as cars and people, the associated wavelength is so small that it cannot be measured or observed in any way. On the other hand, a single electron moving at 3×10^6 m/s (approximately its speed in a hydrogen atom) has a wavelength of 2×10^{-10} m, which is just about the diameter of an atom. Such a wavelength is easily measured with modern equipment.

Davisson and Germer inadvertently provided the experimental proof of deBroglie's hypothesis in studies of the intensity of scattered electrons from a nickel target, much as Rutherford had done earlier with alpha particles and Compton with x rays. They were looking for secondary electrons, not wave phenomena, from the nickel target which they had heated to clean it. When they inserted the heated target back into the experiment, the results were dramatically different; a peak (as shown in Figure 2-13b) now existed at an angle of 50° for electrons accelerated at 54 volts. Their prepared minds and deBroglie's hypothesis caused them to rethink their new observations; otherwise they might have simply thrown the target away and started over. They had observed diffraction of electrons and confirmed deBroglie's hypothesis!—electrons do not travel in a straight-lined circular orbit around a nucleus, but as a wave. A wave equation (the Schrödinger wave equation) could therefore provide a straightforward (though complicated) description of the position and dynamics of electrons (and other particles as well) in atoms. Thus was born the modern principles of wave mechanics.

Simple though it appears, deBroglie's hypothesis has consequences as significant and far-reaching as Einstein's famous equation $E = mc^2$. In Einstein's equation, energy and mass are in direct proportion to each other, with c^2 the constant of proportionality (a large constant by normal standards). In deBroglie's equations, the wavelength and the momentum of a particle are related to each other through Planck's constant (a small constant by normal standards). As Einstein's equation drew together two previously distinct concepts, energy and mass, so deBroglie's equation drew together apparently unrelated ideas, a wave property, λ, and momentum, p, which is a particle property. deBroglie's wave/particle behavior of electrons provided two important contributions to modern physics: (1) it provides a direct physical basis for Bohr's second postulate, $L = nh/2\pi$, and (2) it opened the door to description of the dynamics of particles by wave mechanics, perhaps the most revolutionary development in physical thought since Einstein's special theory of relativity. Diffraction patterns and other wave characteristics have also been observed for electrons, protons, neutrons, etc.

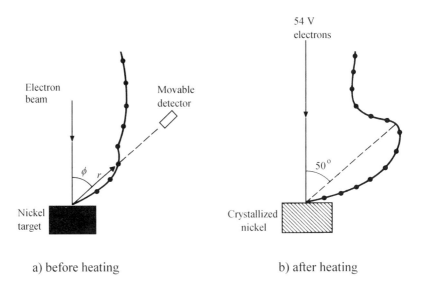

a) before heating   b) after heating

Figure 2-13. Distribution of electrons scattered from a nickel target before (a) and after (b) the nickel had been heated clearly demonstrates the presence of a diffraction peak.

deBroglie Waves and the Bohr Model

Bohr's second postulate can be derived from the fact that an electron travels at high speed some distance, r, from a nucleus with an associated deBroglie wavelength as shown in Figure 2-14.

If the electron is precisely in phase with itself, it will remain in phase as a "standing wave" as shown in Figure 2-14(a). The only values of r that allow it to remain in phase are when the circumference contains an integral number of wavelengths, or

$$2\pi r \equiv n\lambda$$

but by deBroglie's hypothesis electrons around a nucleus have a wavelength of h/mv and

$$mvr = n\frac{h}{2\pi}$$

And since the angular momentum L is equal to mvr,

$$L = n\frac{h}{2\pi}$$

which is the second postulate Bohr had stated on faith and insight with no idea about the wave nature of electrons. Confirmation of deBroglie's hypothesis allows the use of a wave equation to describe the way an electron, or any other particle, moves under a given force. Since the mechanics of the particle is described in terms of a wave, this type of mechanics is called wave mechanics, or since it also addresses quantum effects, quantum mechanics. Its use, however, requires complex mathematics.

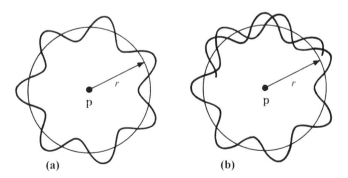

(a)　　　　　　　　**(b)**

Figure 2-14. An orbital electron positioned (a) such that its deBroglie wavelength, λ, will continue to exist; i.e., at a radius such that $n\lambda = 2\pi r$. If however, $n\lambda \neq 2\pi r$ destructive interference (b) occurs and it will cease to exist at that location.

ATOM SYSTEMS

Schrödinger's wave mechanical refinements confirm the exceptional insight of Bohr's original model of the atom for hydrogen; however, the Schrödinger equation and wave mechanics provide a more consistent and broadly applicable theory (Figure 2-15) of the orbits and energy states of electrons relative to the nucleus and the emissions of photons as electrons change energy states. The nuclear charge and mass remained uncertain until the discovery of the neutron, a very important piece which resolved much of the puzzle of the nucleus and the atom.

Discovery of the Neutron

Rutherford had postulated a "neutral" particle around 1920 s, but no evidence of its existence had been found until 1932. His penchant for shooting alpha particles at substances, which was repeated by others, eventually led to the discovery of this "neutral" particle. The Joliet–Curies, in particular, had irradiated all the light elements with alpha particles and discovered that they could produce radioactive substances artificially; however, alpha-particle bombardment of beryllium yielded a highly penetrating radiation which they postulated to be high-energy gamma radiation since the radiation was undeflected in magnetic fields, or

$$\,^{9}_{4}\,Be + \,^{4}_{2}\,He \rightarrow \,? + \gamma$$

Various materials, one of which was paraffin, were used to intercept it and determine its energy, yielding energetic protons which would have required the "radiation," if it were gamma rays, to have an energy on the order of 50 MeV. This was unlikely because the highest energy gamma rays observed thus far were at most a few MeV.

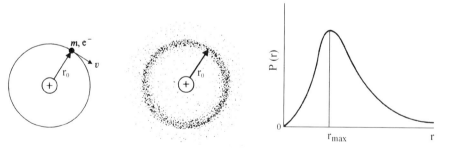

a) Bohr model of H   b) Wave-mechanical model of H   c) Probability of electron position

Figure 2-15. (a) The Bohr model of the hydrogen atom places the electron at r_1; (b) the wave mechanical model shows the relative probability of the electron's position; and (c) it is most likely to be found at $r = r_1$.

In 1932, Chadwick, who had worked with Rutherford and was aware of Rutherford's postulated neutral particle, repeated the Joliet–Curie experiments, but he postulated that the "radiation" was instead a neutral particle of about the same mass as the proton. In doing so, he was able to explain all the energies and momenta observed and to establish that the correct reaction for alpha irradiation of beryllium is

$$\_4^9 \text{Be} + \_2^4 \text{He} \rightarrow [\_6^{13}\text{C}^*] \rightarrow \_6^{12}\text{C} + \_0^1 \text{n}$$

He also calculated an approximate mass of the neutral particle, which he named the neutron. This discovery tied together so many aspects of nuclear physics that Chadwick was quickly (1936) awarded the Nobel Prize. The Joilet–Curies had been one of the first to produce and "observe" neutrons; they just did not know what they were.

The discovery of the neutron solved the puzzle of the mass of individual atoms and the presence of several isotopes of elements. A nucleus made up of Z protons provided part of the mass (a little less than half) and all the charge required to counterbalance the charge of the orbital electrons to produce a neutrally charged atom; the neutrons accounted for the balance of the mass, and since they had no charge they had no effect on electrical neutrality. Electrons could not exist in the nucleus, but with the discovery of the neutron's existence they were not necessary.

And, all of a sudden, physicists had a new source of projectiles to probe the nucleus of atoms. The absence of charge made them excellent projectiles for bombarding all elements, not just light ones, to see what happens; and marvelous things did. Fermi, in particular, set about to bombard all the known elements with neutrons with particular emphasis on producing elements beyond uranium, or transuranic elements. He also most certainly fissioned the uranium nucleus, one of the greatest discoveries of all time, but was so focused on producing transuranic elements that he missed it.

The Nuclear Shell Model

There is no one complete model of the nucleus. Instead, a shell model is used to describe the energy levels in the nucleus, and a liquid drop model best describes fission and other phenomena. The shell model of the nucleus is based on two primary assumptions:

- Each nucleon moves freely in a force field and occupies energy states that can be approximated as shells.
- The energy levels or shells in the nucleus are filled according to nuclear quantum numbers.

The shell model of the nucleus, in which nucleons exist in discrete energy states, corresponds well with the emission of gamma rays from excited nuclei. Gamma rays are photons of nuclear origin that are emitted when a proton or neutron moves to a more tightly bound (or lower potential energy) state. Gamma emission can thus be thought of as similar to the photon emission that occurs when an orbital electron changes to an allowed quantum state of lower potential energy.

SUMMARY

Roentgen's discovery of x rays in 1895 began a series of discoveries that led to explanation of the structure of matter. Radioactivity, which was discovered shortly thereafter, established that atoms contain components and provided a source of alpha particles for irradiation of foils and gases: in doing so Rutherford discovered that an atom must have a very small positively charged nucleus because the observed scattering of alpha particles requires a very small positively charged mass at the center of the atom. Bohr combined Rutherford's nuclear atom and Planck's quantum concept to describe the motion of electrons and the processes of emission and absorption of radiation; however, it was not quite precise, especially when applied to more complex atoms. Other discoveries led to a nebular model of the atom with electrons spread as waves of probability over the entire atom, a direct consequence of deBroglie's discovery of the wave characteristics of electrons and other particles, as shown in Figure 2-16.

Atoms are electrically neutral; i.e., the number of negatively charged electrons equals the number of positively charged protons in the nucleus. The nucleus has a radius of about 10^{-13} cm and contains Z protons, which contribute both mass and positive charge, and $A-Z$ neutrons (slightly heavier than

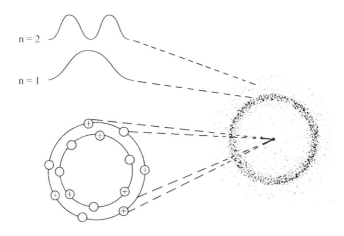

Figure 2-16. Very simplified model of a nebular atom consisting of an array of protons and neutrons with shell-like states within a nucleus surrounded by a cloud of electrons with three-dimensional wave patterns and also with shell-like energy states.

protons) which make up the balance of the mass and help to distribute the nuclear force. The electrons form a nebular cloud extending to a radius of about 10^{-8} cm which is some 4 to 5 orders of magnitude greater than the nuclear radius; thus the atom is mostly empty space. Electrons do not and, according to Heisenberg's uncertainty principle, cannot exist in the nucleus although they can be manufactured and ejected during radioactive transformation.

Three force fields affect the nucleus (the fourth, gravitational, exists but is negligible): the electrostatic force of repulsion between the protons; the strongly attractive nuclear force which acts uniformly on both neutrons and protons across the nucleus but nowhere else; and the weak force (relatively speaking) that influences radioactive transformation. The nuclear force is distributed uniformly across the nucleus; thus there is no center point towards which nucleons are attracted. A nuclear shell model is used to describe the dynamics and changes in energy states of nuclear constituents, and a liquid drop is used for describing nuclear fission in which the opposing nuclear and electrostatic forces compete to hold the nucleus together or to deform it such that fission can occur.

When a particle, such as an electron in an orbit, changes from one bound state to another of lower potential energy (i.e., a lower orbit), a discrete amount of energy is given up. When such a change occurs for electrons in their orbits the energy is emitted as a photon; when such changes occur in the nucleus, a discrete energy is emitted either as a particle or a photon because particles in atoms exist in quantized, or discrete, energy states, and energy changes can only occur in these same quantized (or discrete) amounts as particles take on different positions (energy states) in the atom.

The discovery of the neutron in 1932 led not only to a comprehensive model of the atom but provided a source of neutrally charged projectiles for production of a host of nuclear transmutations that yield radioactive activation and fission products that radiation physics needs to understand and control.

PROBLEMS

1   What is the energy, in electron volts, of a quantum of wavelength $\lambda = 5500$ Å?

2   When light of wavelength 3132 Å falls on a cesium surface, a photoelectron is emitted for which the stopping potential is 1.98 volts. Calculate the maximum energy of the photoelectron, the work function, and the threshold frequency.

3   Calculate the deBroglie wavelength associated with the following:
   (a) an electron with a kinetic energy of 1 eV
   (b) an electron with a kinetic energy of 510 keV
   (c) a thermal neutron (2200 m/s)
   (d) a 1,500 kg automobile at a speed of 100 km/h.

4   Calculate the deBroglie wavelength associated with (a) a proton with 15 MeV of kinetic energy, and (b) a neutron of the same energy.

3

RADIOACTIVE TRANSFORMATION

"...at the last Radiological Congress...held in Brussels in 1910...we decided to adopt the curie 'as the quantity of radium emanation in equilibrium with 10^{-8} gram of radium'. Madame Curie (who provided the International Radium Standard)...agreed (at first)...BUT—at an unearthly early hour the next morning, she informed (Rutherford and me) that...the use of the name 'curie' for so small a quantity of anything was altogether inappropriate...we compromised by letting her have her own way and adopted an experimental standard which was 10^8 times larger than anybody wanted; result—foreseen and duly anticipated—the 'curie' of emanation is now hardly ever mentioned and instead we have constant references to micromillicuries."

– B. Boltwood, January 20, 1921

Various substances have the remarkable property that the nuclei of their atoms spontaneously transform from one value of Z and N to another. This process is called radioactive transformation, which is a more accurate descriptor than the term radioactive decay, or disintegration, because the structure of the radioactive nucleus changes in such a way that it becomes an atom of a new element with the emission of radiation. These emissions include alpha particles, negatrons, positrons, electromagnetic radiation in the form of x rays or gamma rays and, to a lesser extent, neutrons, protons, and fission fragments.

Atoms undergo radioactive transformation because the constituents of the nucleus are not in the lowest potential energy states possible as required by the laws of nature. When such a state exists the atom is stable; if not, it has excess mass and energy, and one or more nuclear transformations must occur to produce a stable nucleus. The excess mass and energy is converted to a particle and/or electromagnetic radiation which is emitted from the atom, a process that can be thought of as a spontaneous nuclear reaction involving a single atom in which the Q-value is positive. Since the endpoint of radioactive transformation is to achieve stability, the dynamics of a radioactive transformation is determined by the relationship between an unstable, or radioactive, nuclide and those that are stable in nature.

Figure 3-1 is a plot, using the same axes as the Chart of the Nuclides, of the number of protons (Z) and the number of neutrons (N, or A–Z) in all stable nuclei. The stable nuclei are shown as black squares extending from $Z = 1$ for hydrogen up to $Z = 83$ for ^{209}Bi. Also shown in Figure 3-1 are the very long-lived nuclides ^{238}U and ^{232}Th which exist well above the end of the "line" of

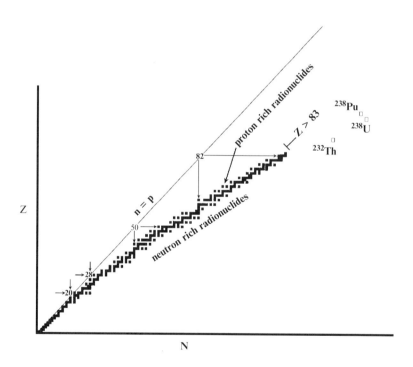

Figure 3-1. The "line" of stable nuclides on the Chart of the Nuclides relative to neutron-rich (below the "line"), proton-rich (above the "line"), and long-lived heavy ($Z > 83$) nuclei.

stable nuclei because they are quite prevalent in nature even though they and their transformation products are radioactive. Also shown is ^{239}Pu, a very important artificial radionuclide produced in nuclear reactors; it is above uranium (or transuranic). These long-lived radionuclides produce long chains of radioactive products before they become stable.

Although the plot of stable nuclei is called the "line" of stability, it is not a line but more a zigzag array with some gaps. The neutron and proton numbers are roughly equal for light nuclei; however, all of the heavy stable nuclei have more neutrons than protons to moderate the coulomb repulsion between the protons and to supply additional binding energy to hold the nucleus together. Interestingly, there are no stable nuclei with $A = 5$ or $A = 8$. A nucleus with $A = 5$, such as ^{5}He or ^{5}Li, will quickly (in 10^{-21} seconds or so) disintegrate into an alpha-particle and a neutron or proton because the helium nucleus, $^4_2$He^{++}, is particularly stable. A nucleus with $A = 8$, such as ^{8}Be, will quickly break apart into two helium nuclei; the energetics of existing separately as two helium nuclei (or as two alpha particles) is more favorable than being joined as ^{8}Be.

Regardless of their origin, radioactive nuclides can be grouped into three major categories around the "zigzag" line of stable nuclei that determine how they must undergo transformation to become stable:

- neutron-rich nuclei, which lie below the zigzag line of stable elements;
- proton-rich nuclei, which are above the line; and
- heavy nuclei with $Z > 83$.

TRANSFORMATION OF NEUTRON-RICH RADIONUCLIDES

Neutron-rich nuclei fall below the line of stable nuclei as shown in Figure 3-2 for ^{14}C and Figure 3-3 for ^{137}Cs and ^{60}Co, an activation product. Radioactive fission

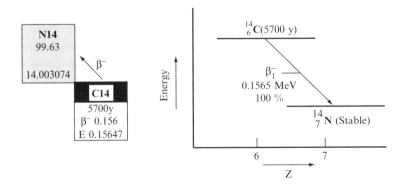

Figure 3-2. Position of ^{14}C relative to its transformation product, ^{14}N, on the Chart of the Nuclides and its decay scheme.

products are neutron-rich and exist below the "line" of stability because they are produced by fission of neutron-rich nuclei, but many others are the products of nuclear interactions that increase the ratio of neutrons to protons in the nucleus.

For neutron-rich nuclei to become stable, they need to reduce the number of neutrons in order to become one of the stable nuclides which are diagonally up and to the left on the Chart of the Nuclides. This process requires a reduction of one negative charge (or addition of a positive one) in the nucleus which occurs by the emission of a negatively charged electron (or beta particle). The transformation results in a slight diminution of mass (equal to the electron mass and the mass equivalence of the emitted energy); the atomic number, Z, increases by 1, and the neutron number, N, decreases by one. The mass change is small; so the mass number, A, remains the same. The only requirement for β^- emission is to reduce the ratio of neutrons to protons. For example, the transformation of ^{14}C (see Figure 3-2) by beta-particle emission can be shown by a reaction equation as follows:

$$^{14}_{6}C \rightarrow ^{14}_{7}N + ^{0}_{-1}e + \bar{\nu} + Q$$

The Q-value for this transformation is 0.156 MeV and is positive, which it must be for it to be spontaneous. This energy is distributed between the recoiling product nucleus (negligible) and the ejected electron, which has most of it. The transformation of ^{14}C results in a total nuclear change which is shown in Figure 3-2 as a shift upward and to the left on the Chart of the Nuclides to stable ^{14}N. Radioactive transformation of ^{14}C is quite simple compared to ^{60}Co and ^{137}Cs as shown in Figure 3-3 along with their positions on the Chart of the Nuclides. Two beta particles are emitted by each with dierent probabilities, and excitation energy is relieved by gamma emission, which occurs because of rearrangement of neutrons and protons in the nucleus to the lowest potential energy states possible following ¯ emission.

Radioactive transformations are typically displayed in decay schemes, which are a diagram of energy (vertical axis) versus Z (horizontal axis). Because the net result is an increase in atomic number and a decrease in total energy, transformation by β^- emission is shown by an arrow down and to the right on a plot of energy versus Z, also shown in Figure 3-2.

Decay schemes for simplicity do not typically show the vertical and horizontal axes, but it is understood that the transformation product is depicted below the radioactive nuclide (i.e., with less energy) and that the direction of the arrow indicates the change in atomic number Z. For similar reasons, gamma emission (Figure 3-3) is indicated by a vertical line (no change in Z) often with a wave motion.

Neutrino Emission accompanies the emission of beta particles which are emitted with energies ranging from just above zero MeV up to the maximum energy, E_{max}, available from the mass change that occurs. The observed spectrum of β^- energies (as shown in Figure 3-4) was most perplexing to physicists because the decrease in mass that occurs during the transformation would,

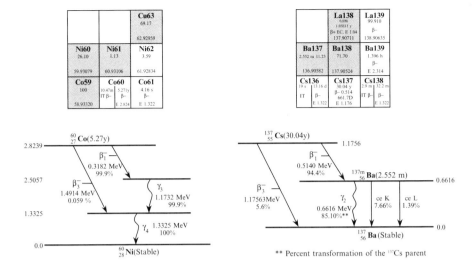

Figure 3-3. Artificial radionuclides of neutron-rich ^{60}Co and ^{137}Cs relative to stable nuclides and their transition by β^- emission.

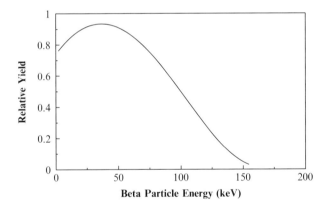

Figure 3-4. A continuous spectrum of beta-particle emission from ^{14}C.

because of the law of conservation of energy, require that all beta transformations yield the same energy change. For example, in the transformation of ^{14}C to ^{14}N the energy of the nucleus decreases by 0.156 MeV (the Q-value). Some of the emitted beta particles have this energy but most do not, being emitted at some smaller value with no observable reason since the "missing" energy was not detectable. The "missing" energy posed an enigma to physicists: is the well-established law of conservation of energy wrong, is Einstein's mass-energy relation wrong, or does something else account for the missing energy? The suggestion was made by Pauli and further developed by Fermi in 1934 that the emission of a beta particle is accompanied by the simultaneous emission of another particle, a neutrino (the "little one") with no charge and essentially no mass, that would carry off the rest of the energy. The neutrino (actually an antineutrino, $\bar{\nu}$, from ^{14}C transformation to conserve spin) is so evasive that its existence was not confirmed until 1956 when Reines and Cowan obtained direct evidence for it using an extremely high beta field from a nuclear reactor.

TRANSFORMATION OF PROTON-RICH RADIONUCLIDES

Proton-rich radionuclides exist above the line of stable nuclei (i.e., they are unstable due to an excess of protons). These nuclei are typically produced by bombarding stable elements with protons or deuterons, thereby increasing the number of protons in the nucleus of their atoms. Such interactions commonly occur with cyclotrons, linear accelerators, or other particle accelerators, and there are numerous (n, p) reactions in nuclear reactors that can produce them.

Proton-rich radionuclides achieve stability by a total nuclear change in which a positively charged electron mass, or positron, is emitted or an orbital electron is captured by the unstable nucleus. Both processes transform the atom to one with fewer protons and a decrease in the Z value; these are shown in decay schemes by an arrow down and to the left. Common proton-rich radionuclides are ^{11}C, ^{13}N, ^{18}F, and ^{22}Na, as shown in Figure 3-5 as they appear on the Chart of the Nuclides relative to stable nuclei. These nuclei will transform themselves to the stable elements ^{11}B, ^{13}C, ^{18}O, and ^{22}Ne respectively, which are immediately down and to the right (i.e., on the diagonal) on the chart. The proton number, Z, decreases by one; the neutron number, N, increases by one; and the mass number A does not change.

Positron Emission is often described as a proton being converted to a neutron and a positively charged electron; however, this description is not quite correct because the proton cannot supply the mass necessary to produce a neutron (which is heavier than the proton) plus the electron mass and the associated kinetic energy carried by the positron. Rather, the process occurs as a total change in the nucleus where the protons and neutrons rearrange themselves to become more tightly bound, thus supplying the energy needed to manufacture and eject a positron with its associated kinetic energy and a neutrino. Reaction equations for the transformation of ^{18}F and ^{22}Na (Figure 3-6) by positron emission are

$$^{18}_{9}\text{F} \xrightarrow{\beta^+} {}^{18}_{8}\text{O} + {}^{0}_{1}e + \nu + Q + \text{orbital electron}$$

$$^{22}_{11}\text{Na} \xrightarrow{\beta^+} {}^{22}_{10}\text{Ne} + {}^{0}_{1}e + \nu + Q + \text{orbital electron}$$

where two electron masses are required for the reactions to balance; i.e., the radioactive proton-rich nucleus must be two electron masses (i.e.,

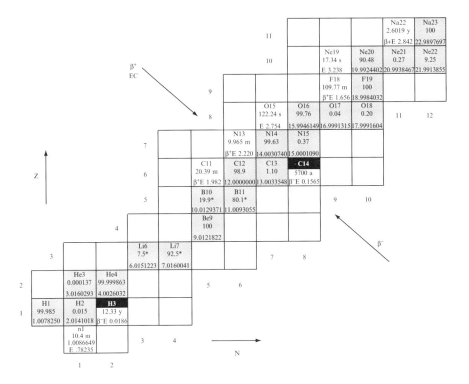

Figure 3-5. Proton-rich radionuclides ^{11}C, ^{13}N, ^{18}F, and ^{22}Na relative to stable nuclei. Also shown are neutron-rich naturally occurring ^{3}H and ^{14}C (denoted by the black line), percent abundance of stable (shaded) nuclei, half-lives for nuclides with time units, and the neutron which is radioactive when "free."

$Q \geq 1.022$ MeV) heavier than the product nucleus for transformation by positron emission to occur. One electron mass is the emitted positron; the other occurs because of the reduced charge on the ^{18}F or ^{22}Ne product nucleus, thus freeing one of the orbital electrons.

Positrons are also emitted as a spectrum of energies because a neutrino also shares some of the transformation energy; however, the spectrum contains more higher energy particles than a comparable spectrum of β^- particles (as shown in Figure 3-7) causing it to be skewed to the right. This occurs because the coulomb force field of the positively charged nucleus gives the positively charged electron an added "kick" as it leaves the nucleus. Figure 3-7 shows the different-shaped spectra for β^- emission and β^+ emission from radioactive transformation of ^{64}Cu, and even though the average β^+ and β^- energy is about $1/3E_{\max}$, the true average energy requires a weighted sum of the energies across the spectrum.

Annihilation Radiation always accompanies positron emission because the emitted positively charged electrons are antimatter in a matter world and they will be annihilated by negatively charged electrons. The interaction, as shown in

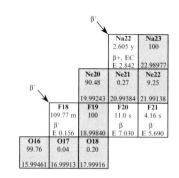

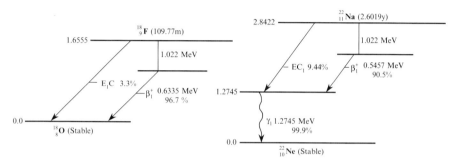

Figure 3-6. Proton-rich ^{18}F and ^{22}Na relative to stable nuclei and their transformation by β^+ emission, electron capture, and gamma-ray emission.

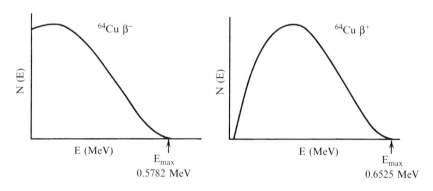

Figure 3-7. Comparison of β^- and β^+ energy spectrum from ^{64}Cu transformation showing the effect of the positively charged nucleus on spectral shape.

Figure 3-8, results in the formation, for a fleeting moment, of positronium, which is made up of the two electron masses with unique spin, charge neutralization, and energetics followed by the complete conversion of the two electron masses into two 0.511 MeV photons, or

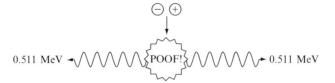

Figure 3-8. Annihilation of a positron and a negatron which converts the electron masses to electromagnetic radiation.

$$e^+ + e^- \rightarrow h\nu_1 + h\nu_2.$$

The two annihilation photons, $h\nu_1$ and $h\nu_2$, are emitted back-to-back or $180°$ from each other; they are commonly called gamma rays, but they do not originate from the nucleus and it is more accurate to refer to them as annihilation photons, or just 0.511 MeV photons.

Electron Capture (EC)   The decay schemes for ^{18}F and ^{22}Na show (see Figure 3-6) that positron emission occurs in 90% of the transformations of ^{22}Na, and 96.7% of those for ^{18}F. The balance occurs by electron capture which is another mechanism, and a competing one for these radionuclides, for reducing the number of protons; they compete with each other, but only if both can occur. The transformation of ^{59}Ni and ^{55}Fe, neither of which has sufficient excess mass to supply two electron masses to the product, occurs by capturing an orbital electron according to the decay schemes shown in Figure 3-9 and the reactions

$$^{59}_{28}\text{Ni} + ^{\ 0}_{-1}e \rightarrow ^{59}_{27}\text{Co} + \nu + Q$$

$$^{55}_{26}\text{Fe} + ^{\ 0}_{-1}e \rightarrow ^{55}_{25}\text{Co} + \nu + Q$$

As shown schematically in Figure 3-10, electron capture is possible due to the wave motion of the orbital electrons. The mass and charge of the captured electron are assimilated into the structure of the nucleus and its net positive

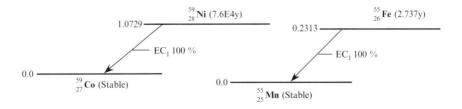

Figure 3-9. Decay scheme of ^{55}Fe and ^{59}Ni (each is followed by emission of characteristic x rays).

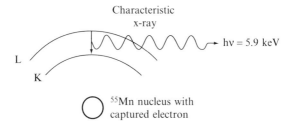

Figure 3-10. Schematic of orbital electron capture by a proton-rich nucleus.

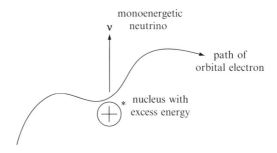

Figure 3-11. Emission of a 5.9 keV characteristic x ray by an atom of ^{55}Mn, the product of K-shell electron capture transformation of a nucleus of ^{55}Fe. The ^{54}Mn x-ray energies and frequencies from transformation of ^{55}Fe are 5.9 keV (24.54% for $k_{\alpha 1}$ and $k_{\alpha 2}$ x rays combined) and 6.5 keV (3.3.% for k_{β} x rays).

charge is reduced by one unit; i.e., Z decreases by 1, N increases by 1, and A, the mass number, remains the same. Since K-shell electrons are closest to the nucleus to begin with, most electron capture events (on the order of 90%) involve K-capture; however, it is probable for L-shell and M-shell electrons (and others as well) to come close enough for capture.

The capture of an electron from the K, L, or M shell leaves a vacancy that is immediately filled by electrons from higher energy levels with the emission of x rays characteristic of the product nuclide. For example, as shown in Figure 3-11, radioactive transformation of ^{55}Fe by electron capture results in the emission of x rays characteristic of ^{55}Mn, the transformed atom. These x rays have discrete energies (5.9 and 6.5 keV for k_{α} and k_{β} x rays from Mn), and the only way electron-capture transformations produce energy emission or deposition (radiation dose) is by these very weak x rays. They also represent the only practical means for detecting these radioactive nuclides.

Electron capture only occurs for proton-rich nuclei; conditions for its occurrence in neutron-rich or stable nuclei simply do not exist. Electron capture transformation is accompanied by emission of a monoenergetic neutrino in contrast to the spectrum of neutrino energies in β^- and β^+ transformations.

The energy of the monoenergetic neutrino is just the Q-value for the transformation minus any excitation energy that may be emitted in the form of gamma rays.

Example 3-1: What is the energy of the neutrino emitted in electron capture transformation of ^{22}Na to ^{22}Ne?

Solution: From Figure 3-6 electron capture of ^{22}Na is followed by emission of a 1.28 MeV gamma photon. The Q-value of the reaction is 2.84 MeV, thus

$$E_{\text{neutrino}} = 2.84 - 1.28 = 1.56 \,\text{MeV}$$

RADIOACTIVE TRANSFORMATION OF HEAVY NUCLEI BY ALPHA PARTICLE EMISSION

Unstable nuclei above $Z = 83$ (bismuth) include the long-lived naturally occurring elements of ^{238}U, ^{235}U, and ^{232}Th and their transformation products, and the artificially produced transuranic radionuclides such as ^{239}Pu, ^{241}Am, ^{242}Cm, etc. Although ^{238}U, ^{235}U, and ^{232}Th are radioactive, they are so long lived that they still exist after formation of the solar system some 4 to 14 billion years ago. For them to achieve stability, they must convert themselves to one of the stable isotopes of lead (^{206}Pb for ^{238}U, ^{207}Pb for ^{235}U, or ^{208}Pb for ^{222}Th), each of which is considerably lighter than the parent and has 8 to 10 fewer protons. The most efficient way for these nuclides to reduce both mass and charge is through the emission of He^{++} nuclei, or alpha particles, because each such transformation reduces both the proton number and the neutron number by 2 units. For example, ^{238}U atoms transform to stable ^{206}Pb atoms by decreasing the proton number by 10 ($Z = 92$ to $Z = 82$) and the neutron number by 22; ^{232}Th transforms to ^{208}Pb by reducing the proton number by 8 and the neutron number by 16.

Several alpha-particle emissions are required for ^{238}U, ^{232}U, TRUs, and other heavy nuclei to reach a stable end product; therefore, reduction of Z, N, and A for these heavy nuclides yields a long chain of transformation products (called series decay) before stability is reached, as shown in Figure 3-12 for transformation of ^{238}U to stable ^{206}Pb and ^{232}Th to ^{208}Pb. Figure 3-12 shows a hypothetical dashed "line of stability" extrapolated from the stable nuclei below to indicate how the path of transformation zig-zags back and forth before a stable configuration of N and Z is found. The intermediate alpha emissions yield unstable nuclei, and some of these emit beta particles if the neutron to proton ratio is "too high," creating a zig-zag path of transformation back and forth across the projected "line of stability" (no actual "line" exists above bismuth).

The alpha particle is a helium atom stripped of its electrons; it consists of two protons and two neutrons, and is, on the nuclear scale, a relatively large particle. The release of such a large particle causes the product nucleus to recoil in order to conserve momentum, as shown in Figure 3-13. A typical example is

$$^{226}_{88}\text{Ra} \xrightarrow{\alpha} \,^{222}_{86}\text{Rn} + {}^{4}_{2}\text{He} + 4.87 \,\text{MeV}$$

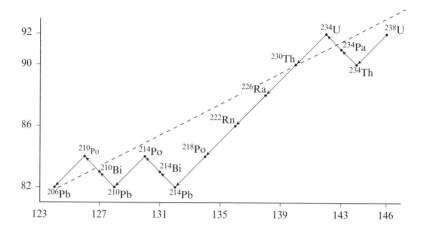

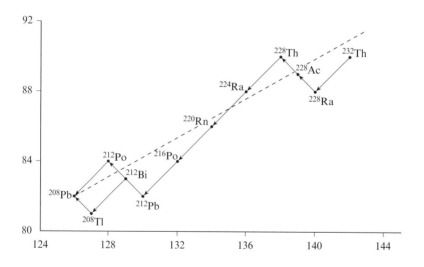

Figure 3-12. Change in proton numbers (Z) versus neutron numbers (N) for ^{238}U and ^{232}Th relative to an extrapolated "line of stability."

in which both Z and N decrease by 2 to yield the ^{222}Rn product while the mass number A decreases by 4. Although the total energy change for the reaction is 4.87 MeV, the alpha particle only carries 4.78 MeV; the balance of 0.09 MeV propels the radon nucleus as it recoils away from the center of the interaction. The kinetic energy of the ejected alpha particle is equal to the Q-value of the reaction minus the recoil energy imparted to the product nucleus, as shown in Example 3-2.

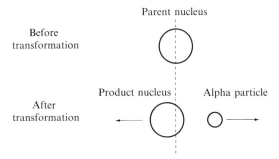

Parent nucleus

Before
transformation

Product nucleus | Alpha particle

After
transformation

Figure 3-13. Parent nucleus at rest before and after transformation; the emitted alpha particle and the product nucleus recoil in opposite directions to conserve linear momentum.

Example 3-2: Calculate (a) the kinetic energy of the alpha particle emitted when ^{226}Ra is transformed directly to the ground state of ^{222}Rn, and (b) the recoil energy of the ^{222}Rn product nucleus.

Solution: The Q-value of the transformation reaction is

$$Q = \text{mass of}\,^{226}\text{Ra} - (\text{mass of}\,^{222}\text{Rn} + \text{mass of}\,^{4}\text{He})$$
$$Q = 226.025406\,u - (222.017574 + 4.002603)\,u$$
$$Q = 0.005227\,u \times 931.502\,\text{MeV}/u$$
$$Q = 4.87\,\text{MeV}$$

The Q-value energy is shared between the product nucleus and the alpha particle, such that

$$Q = 1/2Mv_{\text{p}}^2 + 1/2mv_{\alpha}^2$$

Momentum is conserved and relativistic effects can be ignored; thus, $Mv_{\text{p}} = mv_{\alpha}$, and the kinetic energy, E_{α}, imparted to the alpha particle, is

$$E_{\alpha} = \frac{Q}{1 + \frac{m_{\alpha}}{M}} \tag{3-1}$$

where m_{α} and M are the masses (approximated by the mass numbers) of the alpha particle and the recoiling product nucleus, respectively. Therefore, the kinetic energy of the alpha particle is

$$E_{\alpha} = \frac{4.87\,\text{MeV}}{1 + 4/222} = 4.78\,\text{MeV}$$

and since energy is conserved, the recoil energy is

$$E_R = 4.87\,\text{MeV} - 4.78\,\text{MeV} = 0.09\,\text{MeV}.$$

Alpha particles resulting from a specific nuclear transformation are monoenergetic, and they have large energies, generally above about 4 MeV. The process of alpha-particle transformation is shown schematically as a decrease in energy due to the ejected mass and a decrease in the atomic number by two units on the decay scheme, or E–Z diagram, shown in Figure 3-14 for ^{226}Ra; its change in position on the Chart of the Nuclides when it is transformed to ^{222}Rn is also shown.

The decay scheme for ^{226}Ra shows two alpha particles of different energies. The first (4.78 MeV) occurs in 94.4% of the transformations and goes directly to the ground-state of ^{222}Rn; the other (4.6 MeV) occurs 5.6% of the time and leaves the product nucleus in an excited state. The excited nucleus reaches the ground state by emitting a gamma ray of 0.186 MeV with a yield of 3.59%, or 5.6% minus that relieved by internal conversion processes (see below). The two different alpha-particle energies and the discrete gamma ray suggest discrete energy states in the nucleus.

Transuranic (TRU) Radionuclides

Another special group of alpha-emitting heavy elements above $Z = 83$ are the transuranic elements (for $Z > 92$, hence the term transuranic) which are produced from ^{238}U by successive neutron activation. The process begins with production of ^{239}U followed by beta transformation to ^{239}Np which in turn yields ^{239}Pu. Increasingly heavier TRUs are produced by further neutron activation of ^{239}Pu and many of these undergo radioactive transformation by alpha-particle emission. Transformation of TRUs typically joins one of the naturally radioactive series in the process of achieving a path to stability.

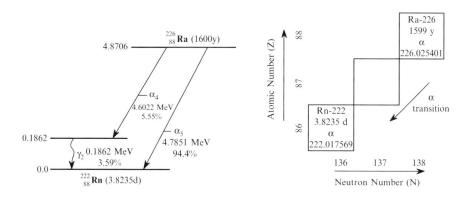

Figure 3-14. Decay scheme of ^{226}Ra and its transition to ^{222}Rn on the Chart of the Nuclides.

Typical TRUs are ^{239}Pu (rather notorious because of its use in nuclear weapons), ^{241}Am (used in anti-static devices and smoke detectors), and ^{252}Cf (a neutron emitter). All of these are well above the end of the curve of stable nuclei which requires alpha-particle emission to reach a stable endpoint at lead or bismuth. For example, ^{240}Pu (half-life of 6564 years) emits alpha particles with energies of 5.1237 MeV and 5.1682 MeV as shown in Figure 3-15.

GAMMA EMISSION

Radioactive transformation is a chaotic process that often leaves the transformed nucleus in an excited state in which the protons and neutrons in the shells of the nucleus are not in the most tightly bound state possible. This excitation energy will be emitted as electromagnetic radiation as the protons and neutrons in the nucleus are rearranged to the lower energy state(s) associated with the ground state of the nuclide. The shell model of the nucleus suggests discrete energy states for neutrons and protons, and it is this difference between energy states that is emitted as a gamma photon when rearrangement takes place. Thus, the emitted gamma ray is characteristic of that particular nucleus and can be used to identify a radioactive element; it can also be used to quantitate the amount present if its gamma yield is known.

Gamma emission, as important as it is in radiation protection, occurs only after the nucleus has transformed by alpha particle emission, negatron emission (see Figure 3-1), positron emission, or electron capture, all of which change the unstable atom to another atom. It is not precise to refer to this process as gamma decay (which connotes transformation) because gamma rays are produced only to relieve excitation energy; i.e., the atom is still the same atom after the gamma ray is emitted.

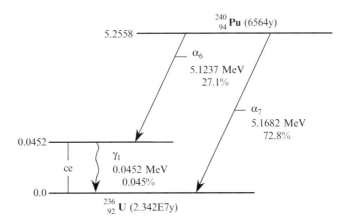

Figure 3-15. Radioactive transformation of ^{240}Pu to ^{236}U by alpha-particle emission.

Internal Transition (Metastable or Isomeric States)

Most excited states in atoms are relieved in less than 10^{-9}s or so by gamma emission; however, some excited states may exist long enough to be readily measurable, and in these cases the entire process of radioactive transformation may be thought of as two separate events. Such a delayed release of excitation energy represents a metastable or isomeric state; when radioactive transformation proceeds through a metastable state of some duration, it is known as internal transition. A good example of nuclear isomerism is ^{137m}Ba (Figure 3-3) which delays transition of ^{137}Cs to the ground state of ^{137}Ba with a half-life 2.552 minutes.

A widely used metastable radionuclide is ^{99m}Tc, which is an isomeric product of ^{99}Mo, as shown in Figure 3-16 with about 15% of the beta transitions omitted for clarity. Of the beta transitions of ^{99}Mo, 87.9% lead immediately to an energy state 0.142 MeV above the ground state of ^{99}Tc. This metastable (or isomeric) state, or ^{99m}Tc, is delayed considerably and is relieved to the ground state of ^{99}Tc by the emission of gamma radiation with a half-life of

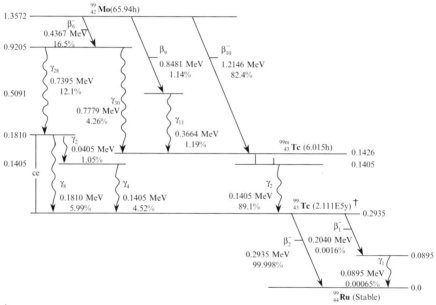

† NOTE   87.907% of transitions of ^{99}Mo produce the 0.1427 and in turn the 0.1405 level of ^{99m}Tc. Metastable ^{99m}Tc is relieved in 89.9% of its transitions by a 0.1405 MeV gamma ray; therefore, the 0.1405 Mev γ is produced in 78.32% of the transformation of ^{99}Mo. However if the ^{99}Mo parent and the ^{99m}Tc product are together and in equilibrium, the 0.1405 MeV gamma is produced in 82.84% of transformations of ^{99}Mo (this accounts for the 4.52% contribution through β6 and γ4)

Figure 3-16.   Decay scheme of ^{99}Mo through ^{99m}Tc to ^{99}Tc. The γ yields are for isomeric transition of ^{99m}Tc and thus total 100%; only 82.84 gamma rays (with energy of 0.14 MeV) are emitted per 100 transformations of ^{99}Mo. The ^{99}Tc end product is effectively stable.

6 hours. The activity of ^{99m}Tc is often stated in curies or becquerels to denote the gamma emission rate; this is not strictly correct because these units define the quantity of a radionuclide undergoing radioactive transformation to another element, which ^{99m}Tc does not do because it only emits excitation energy.

A ^{99}Mo source with its 66-h half-life can be fabricated into a generator of ^{99m}Tc, which is used extensively in diagnostic medicine. Eluting the ion-exchange column with saline solution selectively strips the ^{99m}Tc; however, care must be exercised to minimize the elution of ^{99}Mo in the process. The gamma emission rate of the eluted ^{99m}Tc is directly dependent on the activity of the parent ^{99}Mo, the fraction (87.9%) of beta-particle transformations that produce the metastable state of ^{99m}Tc, and the period of ingrowth that occurs between elutions. Calculation of these relative activities is done by series decay (see below).

Internal Conversion

Residual excitation energy following radioactive transformation is usually relieved by the emission of a gamma ray, but the excitation energy can also be transferred directly to one of the orbital electrons by internal conversion, a process that causes the electron to be ejected from the atom (Figure 3-17). Internal conversion competes with gamma emission and occurs because the wave motion of the orbital electrons brings them close to or through the nucleus where they pick up the excess energy; i.e., they are internally converted from an orbiting electron to an ejected electron to relieve the excitation energy remaining in the nucleus. When internal conversion occurs, a gamma photon is not emitted since the excitation energy in the nucleus is transferred directly to the orbital electron as it passes by; the process thus reduces the number of gamma rays that would normally be emitted, for example in the transition of ^{137m}Ba (Figure 3-18) which shows a significant percentage of internal conversion transitions in competition with the 0.662 MeV gamma.

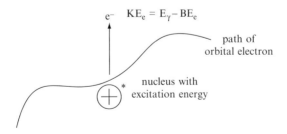

Figure 3-17. Relief of excitation energy by internal conversion. The excitation energy of the nucleus is transferred directly to an orbital electron which, due to its wave motion, can exist close to or in the nucleus where it can "pick up" the energy and be ejected. The kinetic energy of the internally converted electron is equal to that available for gamma emission minus the orbital binding energy of the ejected electron.

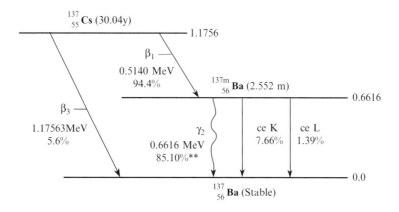

Figure 3-18. Gamma emission and internal conversion of ^{137m}Ba in the radioactive transformation of ^{137}Cs through ^{137m}Ba to ^{137}Ba.

When ^{137}Cs atoms undergo beta transformation, 94.4% go to an excited state of ^{137}Ba, indicated as ^{137m}Ba, with a metastable half-life of 2.552 minutes. The excitation energy remaining in the excited nucleus of ^{137m}Ba is then relieved by γ-rays which occur in 89.8% of emissions, by internal conversion of K-shell electrons which are ejected in 8.32% of ^{137m}Ba transitions, and by electron conversion from other shells in 1.91% of ^{137m}Ba transitions. That is, for each 100 transformations of ^{137}Cs, 85.1 gamma photons, 7.8 internally converted electrons from the K-shell, and 1.8 electrons from other shells will occur as ^{137m}Ba relieves its excitation energy. Internal conversion is most prominent for metastable radionuclides because more time is available for transfer of the excitation energy to an orbital electron, but it can (and does) occur for any atom that transitions through an excited state, just with a different (usually lower) probability. Conversion of K-shell electrons is most probable because the K-shell electrons are closer to the nucleus, but internal conversions can occur for L, M, and, on rare occasions, the more distant orbital electrons. As in electron capture, the process is completed by the emission of charateristic x rays.

The kinetic energy of the conversion electron is discrete; it is the difference between that available for emission of a gamma photon minus the binding energy of the electron in the particular shell that undergoes conversion. Internally converted electrons are monoenergetic because all of the energy states have definite values, and electron spectra will show the internally converted electrons as lines rather than a continuum. If, however, a radionuclide emits both beta particles and internally converted electrons, the internal conversion electrons appear as discrete lines superimposed upon the continuous beta spectrum. These discrete energy lines are of course associated with the electron shells from which they are ejected as shown in the beta spectra of ^{137}Cs and ^{203}Hg in Figure 3-19.

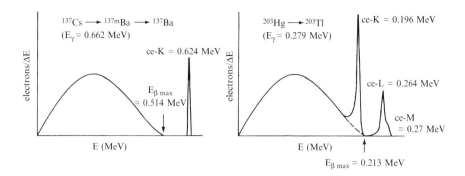

Figure 3-19. Internal conversion electron emission superimposed as discrete energy peaks upon continuous beta spectrum of ^{137}Cs and ^{203}Hg.

The internal conversion electrons from ^{137}Cs appear as a sharp peak at 0.624 MeV and above the 0.514 MeV endpoint energy of ^{137}Cs because the internally converted electrons have a kinetic energy of 0.662 MeV (the gamma excitation energy) minus the orbital binding energy of 0.038 MeV or 0.624 MeV which happens to be larger than the $E_{\beta, \max}$ of ^{137}Cs beta particles. This sharp peak of internally converted electrons makes ^{137}Cs a useful source for energy calibration of a beta spectrometer. The ^{203}Hg beta spectrum in Figure 3-19 also shows peaks of internally converted electrons; however, these are superimposed on the spectrum of beta particles and serve to extend it somewhat. Also notable is the combined peak for converted L- and M-shell electrons.

Spontaneous Fission

Another means by which some heavy unstable nuclei undergo transformation is spontaneous fission. In this case, a heavy nucleus has so much excess energy that it splits on its own to form two, and sometimes more, new products and several neutrons; i.e., it undergoes transformation without the addition of energy or a bombarding particle. The process yields the usual two fission fragments and 2 to 4 neutrons depending on the distribution probability of each fission. Radioactive transformation by alpha- or beta-particle emission may compete with spontaneous fission; for example, ^{252}Cf emits an alpha particle in about 97% of its transformations and neutrons (from spantaneous fission) in the other 3%, as shown in Figure 3-20. The neutron yield is 2.3×10^{12} n/s per g of ^{252}Cf, or 4.3×10^{9} n/s per Ci of ^{252}Cf. Although the spontaneous fission half-life is 87 years, the overall half-life of ^{252}Cf is 2.638 years because alpha transformation to ^{248}Cm is more probable than spontaneous fission.

The distribution of fission fragments from ^{252}Cf fission is shown in Figure 3-21 compared to that for ^{235}U. The fission product yield from spontaneous fission of ^{252}Cf is quite different from thermal neutron fission of ^{235}U, being

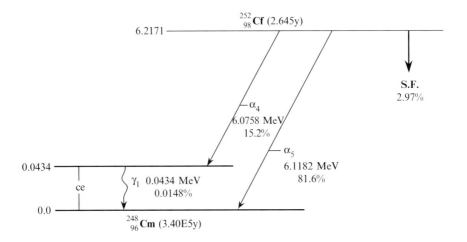

Figure 3-20. Decay scheme of ^{252}Cf showing spontaneous fission in competition with alpha transition.

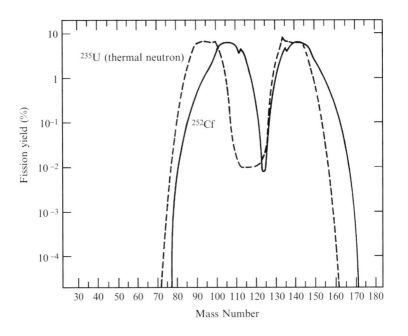

Figure 3-21. The mass number distribution of ^{252}Cf spontaneous fission fragments compared to thermal neutron fission of ^{235}U.

somewhat skewed for lighter fragments and containing a few products of slightly heavier mass. Isotopes of plutonium, uranium, thorium, and protactinium also experience spontaneous fission as do many other heavy nuclei. Most of these are not useful as neutron sources, but ^{252}Cf with a half-life of 2.645 years is commonly produced to provide a source of neutrons with a spectrum of energies representative of those associated with uranium or plutonium fission.

DECAY SCHEMES

The probability that a radionuclide will undergo transformation by one or more processes is constant, and it is possible to predict, on the average, the percentage of transformations that will occur by a given mode. These data are displayed in decay schemes, which range from very simple to enormously complex, depending on the radionuclide. The Chart of the Nuclides provides very basic information on whether an isotope is radioactive, the primary modes of radioactive transformation, the total transition energy, and the predominant energies of the emitted "radiations." Examples of these are shown in Figure 3-22 for ^{11}C, ^{13}N, and ^{14}C. Each is denoted as radioactive by having no shading; the half-life is provided just below the isotopic symbol; the principal (but not necessarily all) mode(s) of transformations and their energies are displayed just below the half-life, and the disintegration energy in MeV is shown at the bottom of each block. The emitted energies are given in MeV for particles; however, gamma energies are listed in keV presumably because gamma spectroscopy uses this convention.

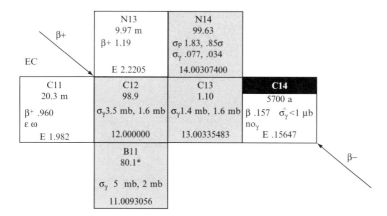

Figure 3-22. Excerpts from the Chart of the Nuclides showing basic transformation modes with principle energies, half-lives, and disintegration energies of ^{11}C, ^{13}N, and ^{14}C. Also shown are stable isotopes and their percent abundances, activation cross sections, and rest masses (u).

The most comprehensive information on decay schemes and the frequency of occurrence and energy of each mode of transition is provided by the National Nuclear Data Center which is operated for the US Department of Energy by Brookhaven National Laboratory. Diagrams can be constructed from such information as shown in Figure 3-23 for ^{51}Cr (and also for ^{137}Cs in Figure 3-24) along with a detailed listing of the frequency and energy of emitted particles and related electromagnetic energy. Similar diagrams and listings are contained in Appendix C for several other common radionuclides of interest to radiation protection. Also shown in Figures 3-23, 3-24 and 3-25, and in Appendix C are the mode of production of the radioisotope (upper left of the table of radiations) and the date compiled (at the upper right). Up-to-date information for other radionuclides can be obtained from the National Nuclear Data Center at www.nndc.bnl.gov.

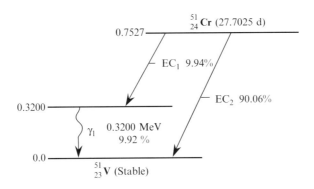

Radiation	Y_i (%)	E_i (MeV)
γ_1	9.92	0.3200
ce-K, γ_1	0.0167	0.3146
ce-L, γ_1	0.0016	0.3195a
ce-M, γ_1	2.58E-04	0.3200a
$K\alpha_1$ X-ray	13.4	0.0050
$K\alpha_2$ X-ray	6.8	0.0049
$K\beta$ X-ray	2.69	0.0054*
L X-ray	0.337	0.0005*
Auger-K	66.3	0.0044*

*Average Energy
aMaximum Energy For Subshell

Figure 3-23. Decay Scheme of ^{51}Cr and a listing of principal radiations emitted.

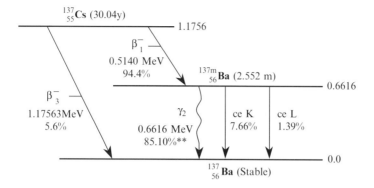

FP   August, 1997

Radiation	Y_i (%)	E_i (MeV)
β^-_1	94.40	0.1734*
β^-_3	5.60	0.4163*
γ_2	85.10**	0.6617
ce-K, γ_1	7.66	0.6242
ce-K, γ_1	1.39	0.6557
$K\alpha_1$ X-ray	3.61	0.0322
$K\alpha_2$ X-ray	1.96	0.0318
$K\beta$ X-ray	1.33	0.0364*
L X-ray	0.91	0.0045*
Auger-K	0.76	0.0264*
Auger-L	7.20	0.0367*

* Average Energy
** Photon yield per transformation of ^{137}Cs; photon yield from each transformation of $^{137}_{m}$Ba is 90.11%

Figure 3-24. Decay Scheme of ^{137}Cs through ^{137m}Ba to ^{137}Ba and a listing of principal radiations emitted.

Some decay schemes are quite complex, as shown in Figure 3-25 for ^{131}I which emits five distinct beta particles, each of which has its own particular yield and transformation energy. The beta-particle emission of $E_{\beta, max} = 0.61$ MeV is the most important; it occurs with a yield of 89.9% followed by immediate gamma emission through two routes including the prominent gamma-ray energy of 0.364 MeV. This 0.364 MeV gamma ray, which occurs for 81.7% of the transformations of ^{131}I, is often used to quantitate sources of ^{131}I.

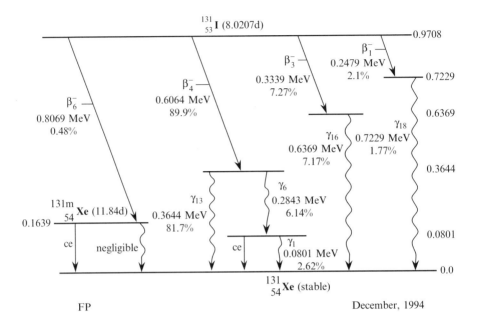

Figure 3-25. Decay scheme of ^{131}I and a listing of major radiations emitted and their yield.

Radiation	Y_i (%)	E_i (MeV)
β^-_1	2.10	0.0694*
β^-_3	7.27	0.0966*
β^-_4	89.9	0.1916*
β^-_6	0.48	0.2832*
γ_1	2.62	0.0801
ce-K, γ_1	3.54	0.0456
γ_6	6.14	0.2842
γ_{13}	81.7	0.3644
ce-K, γ_{13}	1.55	0.3299
ce-L, γ_{13}	0.246	0.3590[a]
γ_{16}	7.17	0.6370
γ_{18}	1.77	0.7229
$K\alpha_1$ X-ray	2.56	0.0298
$K\alpha_2$ X-ray	1.38	0.0295

*Average Energy
[a]Maximum Energy For Subshell
^{131m}Xe product, yield 1.18%; ^{131}Xe yield 98.8%

RATE OF RADIOACTIVE TRANSFORMATION

The transformation of radioactive atoms is not affected by natural processes such as burning, freezing, solidifying, or diluting them, and because of these features they can be used as energy sources in hostile environments such as outer space. Three concepts are important: (1) the activity, or transformation rate of a source, (2) a unit to describe the rate of transformation, and (3) a mathematical law that allows calculation of the amount and rate of radioactive transformation over time.

Activity

The activity of an isotope of a radioactive element, or its rate of radioactive transformation, is directly proportional to the number of atoms, N, in the source; it is defined mathematically as

$$\text{Activity} \equiv \lambda N \tag{3-2}$$

where the constant of proportionality, λ, is a probability factor, or disintegration constant, that describes what fraction of the number of atoms of a radionuclide will undergo transformation in a given amount of time, usually per second, minute, hour, etc. The disintegration constant can be assigned because, regardless of which mode of transformation occurs (α, β^-, β^+, EC, etc.), radioactive atoms undergo a consistent, continuous rate of transformation reflecting a very orderly process. The number of radioactive atoms, N, can be determined from the mass of the radioactive isotope and Avogadro's number (see Example 1-2).

Units of Radioactive Transformation

For many years the standard unit of radioactivity has been the curie (Ci), first defined as the emission rate of one gram of radium at the request of Madame Curie, the discoverer of radium, to honor her husband, Pierre, who was killed by a carriage in the streets of Paris. As measurements of the activity of a gram of radium became more precise, this definition led to a standard that varied which is unacceptable for a standard unit. The problem was subsequently eliminated by defining the curie as that quantity of *any* radioactive material that produces 3.7×10^{10} transformations per second.

Unfortunately, the curie is a very large unit for most samples and sources commonly encountered in radiation protection. Consequently, activity is many times stated in terms of decimal fractions of a curie, especially for laboratory or environmental samples. The millicurie (mCi), which is one-thousandth of a curie, and the microcurie (μCi), equal to one-millionth of a curie, correspond to amounts of radioactive materials that produce 3.7×10^7 (37 million) and 3.7×10^4 (37 thousand) transformations per second (t/s), respectively.

A nanocurie (nCi) is one billionth of a curie and corresponds to 37 t/s and is often used to report environmental levels of radioactivity as is the picocurie (pCi) which is one trillionth of a curie or 0.037 t/s (or 2.22 t/min). Despite its familiarity to radiation physicists, the curie is a somewhat inconvenient unit because of the necessity to state most measured activity rates in mCi, μCi, nCi, or even pCi. The Rutherford (Rd), defined as 10^6 t/s, was used for a time to provide a smaller decimalized unit, but it is, for all practical purposes, a defunct unit although it lives on somewhat in the new SI unit of megabecquerel (also 10^6 t/s) which honors a different founder.

The recent system of international units (SI) has defined activity in terms of the becquerel (Bq), in honor of the discoverer of radioactivity, as that quantity of any radioactive material that produces one transformation per second (t/s or tps). This unit suffers from being small compared to most radioactive sources, which in turn requires the use of a different set of prefixes, i.e., kBq, MBq, GBq, etc., for kilobecquerel (10^3 t/s), megabecquerel (10^6 t/s), and gigabecquerel (10^9 t/s), etc., respectively. Neither the curie nor the becquerel are easy to use despite the tradition of paying respect to great and able scientists or goals for pure units. Nonetheless, both are used with appropriate prefixes to adjust to reality.

Mathematics of Radioactive Transformation

Radioactive transformation is a statistically random process; therefore, if the number of such atoms is large, as it usually is, then the rate of transformation, or activity, is proportional to the number of radioactive atoms present, or

$$-\frac{dN}{dt} = \lambda N$$

where the constant of proportionality λ is the disintegration constant, and $-dN/dt$ is the rate of decrease (denoted by the minus sign) of the number, N, of radioactive atoms at any time t. Rearranging the equation gives an expression that can be integrated directly between the limits N_0 at $t = 0$ and $N(t)$ for any other time, t:

$$\int_{N_0}^{N(t)} \frac{dN}{N} = -\lambda \int_0^t dt$$

Integration and evaluation of the limits yields

$$\ln N(t) - \ln N_0 = -\lambda t$$

or

$$\ln N(t) = -\lambda t + \ln N_0 \tag{3-3}$$

which is an equation of a straight line with slope of $-\lambda$ and a y-axis intercept of $\ln N_0$. By applying the law of logarithms, this can also be written as

$$\ln\left(\frac{N(t)}{N_0}\right) = -\lambda t$$

And, since the logarithm of a number is the exponent to which the base (in this case e) is raised to obtain the number, the above expression is literally

$$\frac{N(t)}{N_0} = e^{-\lambda t}$$

or

$$N(t) = N_0 e^{-\lambda t} \tag{3-4}$$

If both sides are multiplied by λ, and recalling that activity $= \lambda N$, then

$$A(t) = A_0 e^{-\lambda t} \tag{3-5}$$

In other words, the activity, $A(t)$, at some time, t, of a source of radioactive atoms, all of the same species with a disintegration constant λ, is equal to the initial activity, A_0, multiplied by the exponential $e^{-\lambda t}$, where e is the base of the natural logarithm. If the activity of a particular radionuclide at a given time t (e.g., $t = 0$) and its probability of transformation (or its disintegration constant, λ) are known, then the activity that is present at any other time can be calculated exactly using Equation 3-5. If the activity, $A(t)$, is plotted against time, the exponential curve in Figure 3-26(a) is obtained, or if $\ln A(t)$ is plotted vs time, a straight line is obtained as shown in Figure 3-26(b). The activity does not go to zero in either curve except when t is infinite.

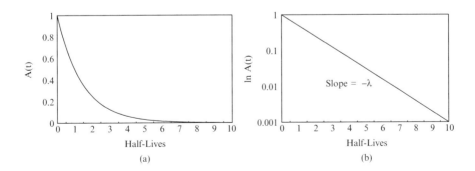

Figure 3-26. (a) Radioactive transformation rate (or activity) vs time, and (b) the natural logarithm of activity vs time, expressed as the number of elapsed half-lives.

Since it is difficult to think in exponential terms, the dynamics of radioactive transformation are described in terms of a half-life or half-period, $T_{1/2}$, which is the amount of time it takes for half of the atoms in a radioactive source to undergo transformation. The half-life of a radionuclide is thus

$$\frac{A(t)}{A_0} = \frac{1}{2} = e^{-\lambda T_{1/2}}$$

which can be solved by taking the natural logarithm of both sides, or

$$\ln 1 - \ln 2 = -\lambda T_{1/2}$$

or

$$T_{1/2} = \frac{\ln 2}{\lambda}$$

The disintegration constant, λ, which is required for calculating activity, follows directly from this relationship as

$$\lambda = \frac{\ln 2}{T_{1/2}}$$

which has units of reciprocal time (s^{-1}, min^{-1}, h^{-1}, y^{-1}) and is a value much less than 1.0.

Although the exponential law for radioactivity is derived from the premise of the disintegration constant, λ, and it is necessary for radioactivity calculations, it is standard practice to derive it from the known half-life of the substance. This expression for λ can be used in a general relationship for activity based on half-life where

$$A(t) = A_0 \, e^{-\left(\ln 2 / T_{1/2}\right)t}$$

The concept of half-life makes communication about radioactive transformation much easier because half-life, which is given in amounts of time, is much easier to relate to, but one needs to be mindful that the process is still exponential. It doesn't matter how much activity exists when the clock is started, the initial activity, A_0, will diminish by one-half after one half-life. If another half-life elapses, the activity will again decrease by one-half and yet again after another half-life, mathematically never reaching zero because the decrease is exponential. As shown in Figure 3-26(b), the activity of a source will diminish to 1% ($A/A_0 = 0.01$) after about 6.7 half-lives and to 0.1% ($A/A_0 = 0.001$) after 10 half-lives. It is common practice to assume that the activity of a source is zero after 10 half-lives, but this presumption should be used with caution because in fact it has just diminished to $0.001A_0$, which may or may not be significant.

RADIOACTIVITY CALCULATIONS

Many useful calculations can be made for radioactive substances using the relationships just derived; for example:

1. How radioactive will a substance be at some time, t, in the future?
2. How long will it take for a source to diminish to a specified level, perhaps one of no concern?
3. If the current activity is known, what was the activity at some previous time, such as when the source was shipped or stored?
4. What activity does a given mass of a radioactive substance have, or vice versa?
5. From a set of activity measurements, what is the half-life and disintegration constant of a substance?

These calculations can best be made with some basic starting information, and the equation for activity $A(t)$ and its decrease with time

$$A(t) = A_0\, e^{-\left(\ln 2/T_{1/2}\right)t}$$

which can be solved directly with scientific calculators that have an exponential function y^x and/or a natural logarithm function (i.e., a key labeled $\ln(x)$, $\ln x$, or $\ln$), as demonstrated in Example 3-3.

Example 3-3: A LLRW shipment contains 800 curies of Iodine-125 ($T_{1/2} = 60$ days $= 2$ months). How much will be left in 11 months?
Solution:

$$A(t) = A_0\, e^{-\left(\ln 2/T_{1/2}\right)t}$$

or

$$A(11\text{mo}) = 800 e^{(\ln 2/2)11} = 17.68\text{Ci}$$

Use of Curves of A(t)/A₀

A plot of the ratio $A(t)/A_0$ versus the number of half-lives that have elapsed, as shown in Figure 3-27, provides reasonable and straightforward determinations

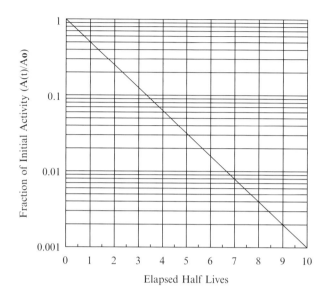

Figure 3-27. Semi-log plot of the fraction of activity, remaining, $A(t)/A_0$, vs the number of elapsed half-lives.

of activity change with time. The use of such a curve, which is subject to some uncertainty in reading it, is demonstrated in Examples 3-4, 3-5, and 3-6.

Example 3-4: If a reactor part contains 1000 curies of Cobalt-60 (half-life = 5.27 years), how much will exist in 40 years?
Solution: Figure 3-27 can be used. The number of elapsed half-lives in 40 years is 7.59. The fractional activity is approximately 0.0052; thus the activity at 40 years is

$$A(40\,\text{y}) = 1000 \text{ Ci} \times 0.0052 = 5.2 \text{ Ci}$$

Example 3-5: If a source of ^{137}Cs($T_{1/2} = 30.04$y) can be safely managed if it contains less than 0.01 Ci, how long would it take for a source containing 5 Ci of ^{137}Cs to reach this value?
Solution: First we determine what fraction 0.01 Ci is of the original 5 Ci, i.e., $A(t)/A_0 = 0.002$. From Figure 3-27, the number of elapsed half-lives that corresponds to 0.002 is 8.97. The time required for 5.0 Ci to diminish to 0.01 Ci is

$$t = 8.97 \times 30.04 \text{ y} = 269 \text{ y}$$

Example 3-6: Solve Example 3-3 using Figure 3-27.
Solution: At 11 months, 5.5 half-lives would have elapsed and the activity is between 25 Ci (5 half-lives) and 12.5 Ci for 6 half-lives (note that it is not

halfway between because radioactive transformation is exponential). The solution for 5.5 half-lives can be determined using the graph in Figure 3-27 and, assuming precise eyesight, the ratio $A(t)/A_0$ at 5.5 half-lives is determined to be

$$\frac{A(t)}{A_0} = 0.022$$

from which $A(t)$ is calculated to be 17.68 Ci.

Activity/Half-life Relationship

The activity of a radioactive source can be calculated by just dividing the initial activity by 2 a sufficient number of times to account for the number of half-lives elapsed, or

$$A(t) = \frac{A_0}{2^n}$$

where n is the number of half-lives that have elapsed during time t. If n is an integer, the activity after an elapsed time is calculated by simply dividing A_0 by 2 n times as shown in Example 3-7. If n is not an integer, then it is necessary to solve the relationship by other means (see below).

Example 3-7: A shipment of $^{125}I(T_{1/2} = 60$ days $= 2$ months) contains 800 Ci. What will be its activity (a) one year later, and (b) 11 months later?
Solution: (a) Since a year is 12 months and the half-life is 2 months, the elapsed time will be 6 half-lives. Dividing 800 by 2 six times yields

$$A(1y) = \frac{800 \text{ Ci}}{2^6} = 12.5 \text{Ci}$$

(b) For $t = 11$ months, 5.5 half-lives will have elapsed and

$$A(11\text{mo}) = \frac{800 \text{ Ci}}{2^{5.5}}$$

can be solved by use of the y^x function on an electronic calculator, which yields

$$A = \frac{800 \text{ Ci}}{45.2548} = 17.68 \text{ Ci}$$

Half-life Determination

The half-life of an unknown radioactive substance can be determined by taking a series of activity measurements over several time intervals and plotting the data. If the data are plotted as $\ln A(t)$ versus t and a straight line is obtained, it can be reasonably certain that the source contains only one radioisotope. The slope of the straight line is the disintegration constant, λ, and once the disintegration constant is known, the half-life can be determined directly.

Example 3-8: Determine the half-life of a sample of ^{55}Cr based on measured activities at 5-min intervals of 19.2, 7.13, 2.65, 0.99, and 0.27 t/s.
Solution: These data are plotted in Figure 3-28 by taking the natural logarithm of the activity rate at each time interval. A least-squares fit of the data yields the slope $-\lambda$ or

$$\lambda = 0.197 \, \text{min}^{-1}$$

from which the half-life is determined directly.

$$T_{1/2} = \frac{\ln 2}{0.197 \, \text{min}^{-1}} = 3.52 \, \text{min}$$

The technique used in Example 3-8 is also applicable to a source that contains more than one radionuclide with different half-lives. This can be readily determined by plotting the $\ln A(t)$ of a series of activity measurements with time as shown in Figure 3-29 for two radionuclides. If only one radionuclide were present, the semi-log plot would be a straight line; however, when two or more are present, the line will be curved as shown in Figure 3-29. The straight-line portion of the far end represents the longer lived component, and thus can be extrapolated back to time zero and subtracted from the total curve to yield a second straight line. The slopes of the two lines establishes the

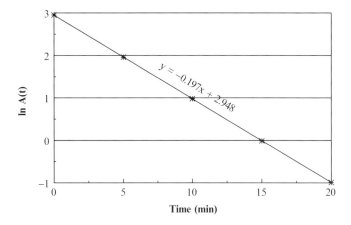

Figure 3-28. Linear least squares fit of $\ln A(t)$ vs time for a sample of ^{55}Cr.

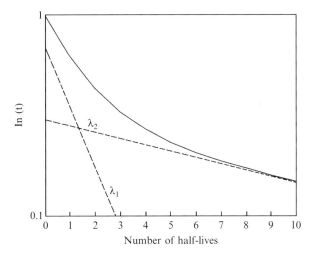

Figure 3-29. The activity of a radioactive source containing two radionuclides with two different half-lives plotted as ln of $A(t)$ versus time.

disintegration constants, λ_1 and λ_2, from which the half-lives of the two radionuclides can be calculated, possibly identifying them.

ACTIVITY–MASS RELATIONSHIPS

The basic equation for the radioactivity of a substance is:

$$A(t) = \lambda N = \frac{\ln 2}{T_{1/2}} N$$

where $T_{1/2}$ is the half-life, which is uniquely defined for each radioactive element, and N is the number of radioactive atoms which is directly proportional to the mass of the radioactive element; therefore, the activity of a radioisotope is directly proportional to the mass, m, of the radioactive material present, or

$$m(t) = m_0 e^{-\lambda t}$$

Example 3-9: If a sample of ^{24}Na($T_{1/2} = 15\,\mathrm{h}$) originally contains 48 g, how many hours will be required to reduce the amount of ^{24}Na in the sample to 9 g? **Solution:** Since activity $= \lambda N$, the activity of ^{24}Na is directly proportional to its mass. Therefore,

$$9\,g = 48\,ge^{-(\ln2/T_{1/2})t}$$

or

$$\ln\left(\frac{9\,g}{48\,g}\right) = -\frac{\ln2}{15\,h}\,t$$

and

$$t = 36.2\,h$$

Specific Activity is defined for a unit mass of pure radionuclide in units of activity per gram, and is perhaps most useful in determining the mass of radioactive material required to produce a given level of radioactivity. The specific activity of one gram of a radioactive isotope of atomic mass A when half-life is given in seconds is

$$SpA\left(\frac{Ci}{g}\right) = \frac{1.12824 \times 10^{13}s \cdot Ci/mol}{T_{1/2}(s)A\ g/mol}$$

and since specific activity is inversely proportional to the half-life of the radionuclide, accurate values are dependent on accurate half-life values, as shown in Example 3-10.

Example 3-10: What is the activity of one gram of ^{226}Ra if the radioactive half-life is 1600 years?
Solution: The number of atoms per gram of ^{226}Ra is

$$N = \frac{6.02214 \times 10^{23}\,atoms/mol}{226\ g/mol} = 2.665 \times 10^{21}\,atoms/g$$

The disintegration constant, λ, is

$$\lambda = \frac{\ln(2)}{T_{1/2}} = \frac{\ln(2)}{1600\ y \times 3.1557 \times 10^7 s/y} = 1.373 \times 10^{-11}s^{-1}$$

and the activity, A, is

$$A = \lambda N = (1.373 \times 10^{-11})s^{-1}(2.665 \times 10^{21}\,atoms/g) = 3.66 \times 10^{10}\,t/s \cdot g$$

which is slightly less than the current official definition of the curie ($3.700 \times 10^{10}t/s$) primarily because the half-life measurement has become more precise.

Example 3-11: Calculate the specific activity of ^{131}I which has a half-life of 8.0207 d.

Solution:

$$\text{Sp}A\left(\frac{\text{Ci}}{\text{g}}\right) = \frac{1.12824 \times 10^{13}\,\text{s}\cdot\text{Ci/mol}}{T_{1/2}(\text{s})A(\text{g/mol})}$$

$$= \frac{1.12824 \times 10^{13}\,\text{s}\cdot\text{Ci/mol}}{8.0207d \times 86,400\text{s/d} \times 131(\text{g/mol})} = 1.243 \times 10^{5}\,\text{Ci/g}$$

Values of specific activity for several radionuclides are listed in Table 3-1. These were calculated for current values of half-lives obtained from the National Nuclear Data Center (NNDC) using mass numbers rather than actual masses, except for ^{3}H. The specific activities for the radionuclides listed in Table 3-1 are for pure samples of the radioisotope, although it is recognized that no radioactive substance is completely pure since transformations produce new atoms which generally remain in the source. The term "intrinsic specific activity" is often used to designate that the activity is for a pure radioelement. Many radioactive samples are measured and reported in units of activity/gram; i.e., in the same units as specific activity. Such measurements are often referred to as the "specific activity" of the sample; however, it is more appropriate to refer to them (because the values are for non-pure samples) as concentrations in units of Bq or Ci per gram of sample. To do otherwise suggests that the sample is a pure element when in fact it contains other constituents.

SERIES TRANSFORMATION—RADIOACTIVE EQUILIBRIUM

Radioactive transformation of many radionuclides can yield a product which is also radioactive and which may in turn produce yet another radioactive product, and so on until stability is achieved. One of the products of a series of such transformations is often of more importance than the parent radionuclide that produces it. For example, ^{222}Rn, the product of the radioactive transformation of ^{226}Ra, can migrate while its ^{226}Ra parent remains fixed in soil. Similarly, the transformation product of ^{90}Sr is ^{90}Y which, because of a much higher beta energy, produces a higher beta dose and is easier to measure.

Radioactive Series Calculations can be used to determine the number of atoms of each member of a radioactive series at any time t by solving a system of differential equations for the rate of change of each product, $N_1, N_2, N_3, \ldots, N_i$ in accordance with each of the corresponding disintegration constants $\lambda_1, \lambda_2, \lambda_3, \ldots, \lambda_i$. Each series begins with a parent nuclide, N_1, which has a rate of transformation:

$$\frac{dN_1}{dt} = -\lambda_1 N_1$$

Table 3-1. Specific Activity of Selected Pure Radionuclides.

Radionuclide	$T_{1/2}$	SpA (Ci/g)	Radionuclide	$T_{1/2}$	SpA (Ci/g)
Hydrogen-3	12.3 y	9.64×10^3	Ruthenium-106	373.59 d	3.30×10^3
Carbon-14	5730 y	4.46×10^0			
Sodium-22	2.6088 y	6.23×10^3	Cadmium-109	462.0 d	2.59×10^3
Sodium-24	14.959 h	8.73×10^6	Iodine-123	13.27 h	1.92×10^6
Phosphorous-32	14.26 d	2.86×10^5	Iodine-125	59.402 d	1.76×10^4
			Iodine-131	8.0207 d	1.24×10^5
Sulfur-35	87.51 d	4.27×10^4	Barium-133	10.52 y	2.62×10^2
Chlorine-36	3.01×10^5 y	3.30×10^{-2}	Cesium-134	2.0648 y	1.29×10^3
Potassium-42	12.36 h	6.04×10^6	Cesium-137	30.07 y	8.69×10^1
Calcium-45	162.61 d	1.79×10^4	Barium-140	12.752 d	7.32×10^4
Chromium-51	27.704 d	9.25×10^4	Lanthanum-140	40.22 h	5.57×10^5
Manganese-54	312.12 d	7.76×10^3			
Iron-55	2.73 y	2.39×10^3	Cerium-141	32.5 d	2.85×10^4
Cobalt-57	271.79 d	8.44×10^3	Cerium-144	284.9 d	3.18×10^3
Iron-59	44.503 d	4.98×10^4	Promethium-147	2.6234 y	9.28×10^2
Nickel-59	7.6×10^4 y	7.99×10^{-2}			
Cobalt-60	5.2714 y	1.13×10^3	Europium-152	13.542 y	1.74×10^2
Nickel-63	100.1 y	5.68×10^1	Tantalum-182	114.43 d	6.27×10^3
Copper-64	12.7 h	3.86×10^6	Iridium-192	73.831 d	9.21×10^3
Zinc-65	244.26 d	8.23×10^3	Gold-198	2.69517 d	2.45×10^5
Gallium-67	3.26 d	5.98×10^5	Mercury-203	46.612 d	1.38×10^4
Arsenic-76	1.0778 d	1.60×10^6	Thallium-204	3.78 y	4.64×10^2
Bromine-82	35.3 h	1.08×10^6	Polonium-210	138.4 d	4.49×10^3
Rubidium-86	18.631 d	8.16×10^4	Radium-226	1600 y	0.99×10^0
Strontium-89	50.53 d	2.91×10^4	Thorium-232	1.41×10^{10} y	1.10×10^{-7}
Strontium-90	28.74 y	1.38×10^2	Uranim-233	1.592×10^5 y	9.64×10^{-3}
Yttrium-90	64.1 h	5.44×10^5	Uranium-235	$7.0E \times 10^8$ y	2.16×10^{-6}
Molybdenum-99	65.94 h	4.80×10^5	Uranium-238	4.468×10^9 y	3.36×10^{-7}
			Plutonium-239	24,110 y	6.21×10^{-2}
Technetium-99m	6.01 h	5.27×10^6	Plutonium-241	14.35 y	1.03×10^2
			Americium-241	432.2 y	3.43×10^0

Atoms of the second nuclide in the chain, N_2, will be produced by the transformation of N_1, but as soon as atoms of N_2 exist, they too can undergo transformation if they are radioactive; thus, the equation for the rate of change of atoms of N_2 must account for their rate of production by radioactive transformation of N_1 atoms (or $\lambda_1 N_1$) and their rate of removal (or $\lambda_2 N_2$) which is

$$\frac{dN_2}{dt} = \lambda_1 N_1 - \lambda_2 N_2$$

Similarly,

$$\frac{dN_3}{dt} = \lambda_2 N_2 - \lambda_3 N_3$$

and so on,

$$\frac{dN_i}{dt} = \lambda_{i-1} N_{i-1} - \lambda_i N_i$$

and if the series has a stable end product, as they all do, the stable atoms will appear at the rate of the last radioactive precursor.

It is useful to solve this system of equations for the first four members of the ^{222}Rn subseries of ^{238}U because of their public health importance (see Chapter 13). Radioactive transformation of naturally occurring ^{222}Rn requires transformation through several products before it reaches ^{210}Pb, a relatively long-lived end product, as shown in Figure 3-30. The number of atoms of N_1 is:

$$N_1(t) = N_1^0 e^{-\lambda_1 t}$$

where N_1^0 is the number of atoms of the parent at $t = 0$. This expression for N_1 can be inserted into the equation for dN_2/dt to give:

$$\frac{dN_2}{dt} = \lambda_1 N_1^0 e^{-\lambda_1 t} - \lambda_2 N_2$$

which, upon collecting terms, becomes

$$\frac{dN_2}{dt} + \lambda_2 N_2 = \lambda_1 N_1^0 e^{-\lambda_1 t}$$

This type of equation can be converted into one that can be integrated directly by multiplying through by an appropriate integrating factor, which for this form is always an exponential with an exponent that is the variable in the denominator of the derivative multiplied by the constant in the second term, or $e^{\lambda_2 t}$, thus

$$e^{\lambda_2 t} \cdot \frac{dN_2}{dt} + e^{\lambda_2 t} \lambda_2 N_2 = \lambda_1 N_1^0 e^{(\lambda_2 - \lambda_1)t}$$

The left side of this expression is exactly the time derivative of $N_2 e^{\lambda_2 t}$, which can be demonstrated by differentiation; thus

$$\frac{d}{dt}(N_2 e^{\lambda_2 t}) = \lambda_1 N_1^0 e^{(\lambda_2 - \lambda_1)t}$$

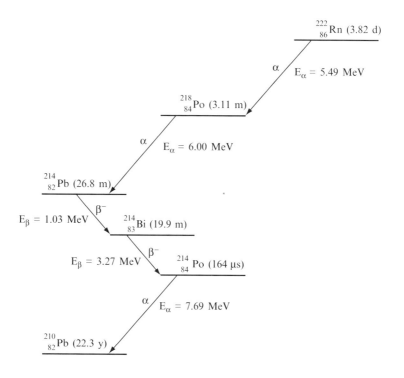

Figure 3-30. The ^{222}Rn subseries of the naturally occurring uranium-238 series.

which can be integrated directly to give

$$N_2 e^{\lambda_2 t} = \frac{\lambda_1}{\lambda_2 - \lambda_1} N_1^0 e^{(\lambda_2 - \lambda_1)t} + C$$

where C, the constant of integration, is determined by stating the condition that when $t = 0$, $N_2 = 0$; thus

$$C = \frac{\lambda_1}{\lambda_2 - \lambda_1} N_1^0$$

which upon simplification yields the solution for N_2 as a function of time as

$$N_2(t) = \frac{\lambda_1}{\lambda_2 - \lambda_1} N_1^0 (e^{-\lambda_1 t} - e^{-\lambda_2 t})$$

The number of atoms of the third kind is found by inserting this expression for N_2 into the equation for dN_3/dt and collecting terms

$$\frac{dN_3}{dt} + \lambda_3 N_3 = \frac{\lambda_2 \lambda_1}{\lambda_2 - \lambda_1} N_1^0 (e^{-\lambda_1 t} - e^{-\lambda_2 t})$$

for which the left side can be converted to the time derivative of $N_3 e^{\lambda_3 t}$ by multiplying through by the integrating factor $e^{\lambda_3 t}$. After the integration is performed and the constant of integration is evaluated, the equation for the number of atoms of N_3 with time is

$$N_3(t) = \lambda_1 \lambda_2 N_1^0 \left(\frac{e^{-\lambda_1 t}}{(\lambda_2 - \lambda_1)(\lambda_3 - \lambda_1)} + \frac{e^{-\lambda_2 t}}{(\lambda_1 - \lambda_2)(\lambda_3 - \lambda_2)} + \frac{e^{-\lambda_3 t}}{(\lambda_1 - \lambda_3)(\lambda_2 - \lambda_3)} \right)$$

Similarly, for the number of atoms of the fourth kind, the expression for $N_3(t)$ is inserted into the equation for dN_4/dt, and after the integration is performed the number of atoms of N_4 with time is:

$$N_4(t) = \lambda_1 \lambda_2 \lambda_3 N_1^0 \left[\frac{e^{-\lambda_1 t}}{(\lambda_2 - \lambda_1)(\lambda_3 - \lambda_1)(\lambda_4 - \lambda_1)} + \frac{e^{-\lambda_2 t}}{(\lambda_1 - \lambda_2)(\lambda_3 - \lambda_2)(\lambda_4 - \lambda_2)} \right.$$
$$\left. + \frac{e^{-\lambda_3 t}}{(\lambda_1 - \lambda_3)(\lambda_2 - \lambda_3)(\lambda_4 - \lambda_3)} + \frac{e^{-\lambda_4 t}}{(\lambda_1 - \lambda_4)(\lambda_2 - \lambda_4)(\lambda_3 - \lambda_4)} \right]$$

These equations yield the number of atoms of each of the first four members of a radioactive series that begins with a pure radioactive parent, i.e., there are no transformation products at $t = 0$.

Radioactive Equilibrium

The radioactivity of each member of a radioactive series can be obtained by multiplying the calculated number of atoms by the respective disintegration constant. The relative activities of the parent and the first product (commonly referred to as the daughter) in a series is

$$A_2(t) = \frac{\lambda_2}{\lambda_2 - \lambda_1} A_1^0 (e^{-\lambda_1 t} - e^{-\lambda_2 t})$$

which can be expressed in a simpler form by use of the identity

$$\frac{\lambda_2}{\lambda_2 - \lambda_1} = \frac{T_1}{T_1 - T_2}$$

or

$$A_2(t) = \frac{T_1}{T_1 - T_2} A_1^0 (e^{-\lambda_1 t} - e^{-\lambda_2 t})$$

which is somewhat easier to use. Each of these two equations provides an exact calculation of the activity $A_2(t)$ of the radioactive product of a radioactive parent. Although the calculation of $A_2(t)$ is not difficult with modern computing techniques, useful and practical simplifications can be obtained if certain conditions are met. These are known as secular equilibrium and transient equilibrium.

Secular Equilibrium occurs between a radioactive parent and its radioactive product if the period of observation (or calculation) of the activity of $A_2(t)$ is such that the activity of the parent nuclide remains essentially unchanged over the period of observation. When secular equilibrium exists, the activity of the radioactive product can be represented in simplified form as

$$A(t) = A_1^0(1 - e^{-\lambda_2 t})$$

i.e., the product activity will grow to a level that is essentially identical to that of the parent in about 6.7 half-lives of the product as shown in Figure 3-31. The buildup of ^{90}Y from the transformation of ^{90}Sr is a good example of secular equilibrium as is the ingrowth of ^{222}Rn from the transformation of ^{226}Ra. In both instances the activity of the product will increase until it is the same (i.e., it is in equilibrium) as that of the parent; 99% equilibrium occurs after about 6.7 half-lives of the product nuclide.

The simplified equation for secular equilibrium is very convenient and easy to use, but it is necessary to assure that the conditions for its appropriate use are met; if there is any uncertainty whether secular equilibrium exists, it is prudent to use the general expression which always gives the correct result for a radioactive parent undergoing transformation to a radioactive product.

Example 3-12: ^{90}Sr$(T_{1/2} = 28.78 \, \text{y})$ is separated from a radioactive waste sample and measured to contain $2 \, \mu\text{Ci}$ of ^{90}Sr. If measured again 10 days later, how much ^{90}Y$(T_{1/2} = 2.67\text{d})$ will be in the sample?
Solution: Since the activity of ^{90}Sr with a half-life of 28.78 years will barely change over the 10-day period, this is a clear case of secular equilibrium, therefore

$$A(t) = A_1^0(1 - e^{-\lambda_2 t})$$

or

$$A(t) = 2 \, \mu\text{Ci}(1 - e^{-(\ln(2)/2.67\text{d})\cdot 10\text{d}}) = 1.85 \, \mu Ci \, ^{90}Y$$

Transient Equilibrium can occur if the parent is longer-lived than the product $(\lambda_1 < \lambda_2)$, but the half-life of the parent is such that its activity diminishes appreciably during the period of consideration. If no product activity exists at $t = 0$, the activity of the product will change as shown graphically in Figure 3-32 for ^{132}Te$(T_{1/2} = 78.24 \, \text{h})$ undergoing transformation to ^{132}I$(T_{1/2} = 2.28 \, \text{h})$.

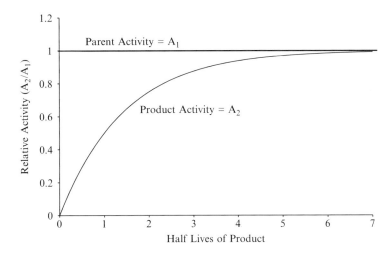

Figure 3-31. Example of secular equilibrium in which the parent activity remains essentially unchanged while the short-lived product (activity = 0 at t = 0) activity builds up to 99% of the parent activity in about 6.7 half-lives.

Transient equilibrium is achieved just after, but not before, the activity of the product nuclide ^{132}I crosses over the activity curve for ^{132}Te. The activity of the product is given by the general equation for the second member of a radioactive series; however, for values of t above 6.7 half-lives or so $e^{-\lambda_2 t} \approx 0$ such that

$$A_2(t) = \frac{\lambda_2}{\lambda_2 - \lambda_1} A_1^0 e^{-\lambda_1 t}$$

i.e. transient equilibrium exists and the activity of the product diminishes with the same half-life as the parent. After transient equilibrium is established, but not before, the product activity can be obtained by simply multiplying the parent activity at a given time t by the ratio $\frac{\lambda_2}{\lambda_2 - \lambda_1}$ or its identity $\frac{T_1}{T_1 - T_2}$.

   A Non-Equilibrium Condition exists when the parent has a shorter half-life than the product ($\lambda_1 > \lambda_2$); thus neither secular nor transient equilibrium is attained and it is necessary to calculate the product activity using the general-expression for the second member of a radioactive series. If a source initially contains just the parent nuclide, the number of product atoms will increase in proportion to transformations of the parent, and after passing through a maximum, the product activity will eventually be a function of its own unique half-life.

Time of Maximum Activity (Second Member of a Series)

Transient equilibrium (and also secular equilibrium) is not established until after the time at which the product radionuclide (or the daughter) reaches its maximum activity. Once the time of maximum activity is known, it can in turn be used in the general equation to determine the maximum amount of activity of the product. If the activity of the product is zero at $t = 0$, the activity, $A_2(t)$, as shown in Figure 3-32, will rise (grow in) and reach a maximum value at a time t_m. The time of maximum activity, t_m, is found by differentiating the general equation, setting it equal to zero, and solving for t_m as follows:

$$\frac{d(A_2)}{dt} = -\lambda_1 e^{-\lambda_1 t_m} + \lambda_2 e^{-\lambda_2 t_m} = 0$$

or

$$t_m = \frac{\ln(\lambda_2/\lambda_1)}{\lambda_2 - \lambda_1}$$

which is somewhat simpler when expressed in terms of half-lives, if they are in the same units

$$t_m = 1.4427 \frac{T_1 T_2}{T_1 - T_2} \ln \frac{T_1}{T_2}$$

This maximum occurs at the same time that the activities of the parent and daughter are equal if, and only if, the parent has only one radioactive product (see Figure 3-32).

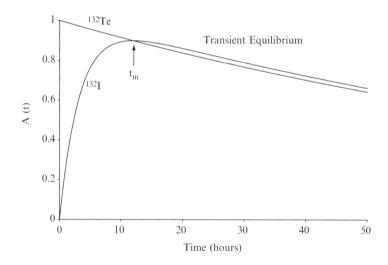

Figure 3-32. Transient equilibrium for initially pure ^{132}Te undergoing radioactive transformation to ^{132}I.

Example 3-13: Selenium-72 ($T_{1/2} = 8.4\,d$) produces radioactive arsenic-72 ($T_{1/2} = 1.083\,d$) by electron-capture transformation. If a pure sample containing $10\,\mu Ci$ of ^{72}Se is prepared and set aside, (a) when will the ^{72}As product reach a maximum in the sample; (b) what will be the activity at that point; and (c) what will be the activity at $2 \times t_m$?

Solution: (a) the time of maximum activity, t_m, is

$$t_m = 1.4427 \frac{8.4\ d \times 1.083d}{8.4d - 1.083d} \ln \frac{8.4d}{1.083d} = 3.675\ d$$

(b) The maximum activity of the product occurs at $t_m = 3.675\,d$ and is

$$A_2(3.675d) = \frac{8.4d}{8.4d - 1.083d} \times 10\mu Ci \left[e^{-\frac{\ln 2}{8.4d} \times 3.675d} - e^{-\frac{\ln 2}{1.083d} \times 3.675d} \right]$$

$$= 1.148 \times 10\,\mu Ci[0.7384 - 0.095232]$$

$$= 7.384\,\mu Ci$$

(c) The activity when $t = 2t_m$ or 7.35 d is obtained by realizing that for times $> t_m$ the activities of the parent and product are in transient equilibrium; therefore, the activity of the product at $t = 2 \times t_m$ is obtained by first determining the activity of the ^{72}Se parent at $t = 2 \times 3.675d = 7.35d$ and adjusting it by the ratio $\frac{T_1}{T_1-T_2}$, or

$$A_2(t) = \frac{T_1}{T_1 - T_2} A_1^0 e^{-\lambda t}$$

$$A_2(t = 7.35d) = \frac{8.4d}{8.4d - 1.083d} \times 10\,\mu Ci \left(e^{-\frac{\ln 2}{8.4d} \times 7.35d} \right)$$

$$= 10\,\mu Ci(1.148)(0.5453)$$

$$= 6.26\,\mu Ci$$

The calculation in Example 3-13(c) is made somewhat easier by establishing that transient equilibrium exists; however, if this is not known for certain, the general equation for the production of a radioactive product from a radioactive parent (i.e., for the second member of a radioactive series) will always provide the correct determination.

Acknowledgment

This chapter was compiled with the able and patient assistance of Major Arthur Ray Morton III, MS, US Army, and Chul Lee, MS, both graduates of the University of Michigan Radiological Health Program.

ADDITIONAL RESOURCES

1   Martin, J.E., *Physics for Radiation Protection* (New York: Wiley, 2000).

2 Krane, K.S., *Introductory Nuclear Physics* (New York: Wiley, 1988).
3 Evans, R.D., *The Atomic Nucleus* (New York: McGraw-Hill, 1955).
4 *The Chart of the Nuclides*, 15th ed. (San Jose, CA: General Electric Company, 1996).
5 The National Nuclear Data Center, Brookhaven National Laboratory, Upton, NY. Accessible on the World Wide Web at www.nndc.bnl.gov
6 MIRD: Radionuclide Data and Decay Schemes, Society of Nuclear Medicine, 136 Madison Ave., NY (1989).
7 ICRP—Report 38, Radionuclide Transformations—Energy and Intensity of Emissions (New York: Pergamon Press, 1978).

PROBLEMS

1 A sample of ^{134}La($T_{1/2}$ = 6.5 min) has an activity of 4 curies. What was its activity one hour ago?

2 Determine by Q-value analysis whether ^{65}Zn can undergo radioactive transformation by β^+, EC, or β^- transition.

3 A sample of material decays from 10 Ci to 1 Ci in 6 hours. What is the half-life?

4 An isotope of fission is ^{151}Sm, with a half-life of 90 years. How long will it take 0.05 g to decay to 0.01 g?

5 (a) How many grams of ^{210}Po($T_{1/2}$ = 138 days) does it take to produce an activity of 4 Ci; and (b) how many days will it take to reduce the 4 Ci to 1 Ci?

6 (a) What is the initial activity of 40 grams of ^{227}Th($T_{1/2}$ = 18.72 days), and (b) how many grams will be in the sample 2 years later?

7 In August, 1911, Mme. Marie Curie prepared an international standard of activity containing 21.99 mg of ^{226}RaCl$_2$: (a) Calculate the original activity of the solution, and (b) its activity as of August, 2002 if the half-life of ^{226}Ra is 1600 years.

8 (a) Calculate the specific activity of ^{210}Pb($T_{1/2}$ = 22.6 y), and (b) determine the activity per gram of a pure sample of ^{210}Pb that has aged 3 years.

9 A reference person weighs 70 kg and contains, among other elements, 18% carbon ($\sim 6.5 \times 10^{10}$ atoms of ^{14}C per gram of C) and 0.2% potassium. How many microcuries of ^{14}C($T_{1/2}$ = 5715 y) and of ^{40}K($T_{1/2}$ = 1.27 × 10^9 y) will be present in such a person?

10 What must be the activity of radiosodium (^{24}Na) compound when it is shipped from Oak Ridge National Laboratory so that upon arrival at a hospital 24 hours later its activity will be 100 mCi?

11 750 mg of ^{226}Ra is used as a source of radon which is pumped off and sealed into tiny seeds or needles. How many millicuries of radon will be available at each pumping if this is done at weekly intervals? How much time will be required to accumulate 700 mCi of the gas?

12 Regulations permit use of ^{3}H in luminous aircraft safety devices up to 4 Ci. How many grams of ^{3}H could be used in such a device? What change in luminosity would occur after 8 years of use?

13 What is the energy of neutrino(s) from electron-capture transformation of ^{55}Fe, and (b) what energy distribution do these neutrinos have?

14 Artificially produced ^{47}Ca($T_{1/2} = 4.54$d) is a useful tracer in studies of calcium metabolism, but its assay is complicated by the presence of its radioactive product ^{47}Sc($T_{1/2} = 3.35$d). Assume an assay based on beta-particle counting with equal counting efficiencies for the two nuclides, and calculate the data for and plot a correction curve that can be used to determine ^{47}Ca from a total beta count.

15 A source of radioactivity was measured to have 40, 14, 5.5, 2, and 0.6 dpm at 8-min intervals. Use a plot of the data to determine the half-life of the radionuclide in the source.

16 A mixed source was measured to produce the following activity levels at 3-min intervals: 60, 21, 8, 1, and 0.3 dpm. What are the half-lives of the radionuclides in the source?

17 10 mCi of ^{90}Sr is separated from a fission product mixture and allowed to set for ^{90}Y to grow in (a) What is the activity of ^{90}Y at 20 hours; (b) at 600 hours; and (c) at the time of maximum activity?

18 If ^{140}Ba($T_{1/2} = 12.75$d) is freshly separated from a fission product mixture and found to contain 200 mCi, (a) what is the activity of ^{140}La($T_{1/2} = 1.678$d) that will grow into the sample in 12 hours? (b) what is the maximum activity of ^{140}La that will ever be present in the sample and when will it occur?

19 Derive from first principles the equation for the third member of a radioactive series that begins with a radioactive parent containing N_i^0 atoms.

4

MAJOR SOURCES OF RADIATION

"The cause is hidden, but the effect is known."

– Ovid, ~20 B.C.

Radiation is a process of energy transmission from its place of origin to another point where it may undergo various interactions with matter. It is emitted as ejected particles or as "particle-like" pure-energy electromagnetic radiation by several major sources: x rays, naturally occurring radioactive materials, activation products, and fission products. Radiation protection for these major sources considers their modes of production, the key elements and materials, and how and why they emit radiant energy.

PRODUCTION OF X RAYS

An x-ray tube contains a cathode and a target, usually a high-Z material enclosed in a high vacuum as shown in Figure 4-1.

Electrons are boiled off of a cathode and accelerated through the evacuated space towards the positively charged anode (target) by a potential difference of several tens of thousands of volts. They strike the target with a kinetic energy of

$$KE = eV$$

A high vacuum is necessary to assure that the electrons reach the target without being absorbed by gas (air) molecules in between. As shown in Figure 4-2, the electrons are deflected by the intense force field of the positively charged nucleus of the target atom, causing them to accelerate as they are bent near the nucleus. Acceleration of a charged particle (the electron) causes it to emit electromagnetic

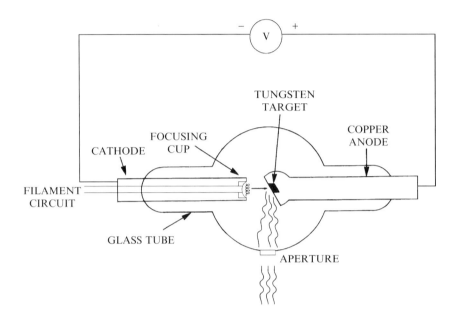

Figure 4-1. High-vacuum x-ray tube.

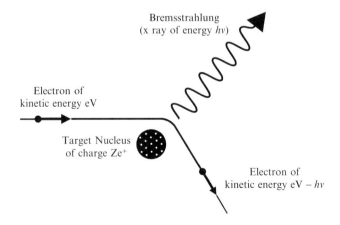

Figure 4-2. The production of x rays in which the deflection of accelerated electrons cause the emission of bremsstrahlung.

radiation, and since energy is lost in the process, the electron must slow down. Overall, the electron experiences a net deceleration, and its energy after being decelerated is eV – $h\nu$ where $h\nu$ appears as electromagnetic radiation. This process of radiation being produced by an overall net deceleration of the electrons is called *bremsstrahlung*, a German word meaning *braking radiation*. Roentgen called these radiations x rays to characterize their unknown status.

X-ray production occurs only when the electrons come close to a target nucleus and undergo significant deflection; most of them just ionize target atoms and about 98% of the kinetic energy of the accelerated electrons is lost as heat. The production of x rays is thus a probabilistic process, and since the electron can take any path including one in which all of its energy is lost, bremsstrahlung radiation is emitted in a continuous spectrum of energies, up to and including the accelerating energy eV, as shown in Figure 4-3 for targets of tungsten (W) and molybdenum (Mo), both of which are operated at 35 kV. X rays are also emitted in all directions, including absorption in the target, and because of this the tube is shielded except for a thin window through which a collimated beam of x rays can pass.

The maximum frequency, ν_{max}, of the emitted x rays is directly proportional to the maximum voltage, V, across the tube; it occurs only when the incoming electron loses all its energy in producing an x-ray photon of wavelength

$$\lambda_{min} = \frac{1.24 \times 10^{-4}}{V} \text{ cm}$$

where V is in volts. Measurement of λ_{min} can be done quite accurately and can be used to determine Planck's constant, h.

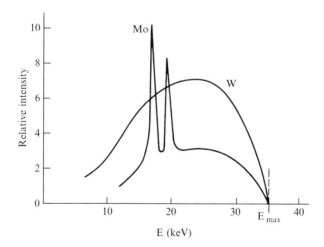

Figure 4-3. X-ray spectra of intensity I (dE) vs energy (keV) for tungsten (W) and molybdenum (Mo) targets both of which are operated at 35 keV.

Example 4-1: Calculate Planck's constant from the following data: x rays with a wavelength of 0.1377 nm are emitted from a copper target in an x-ray tube operated at 9,000 volts.

Solution: For maximum energy conversion:

$$h\nu_{max} = eV$$

$$h = \frac{eV}{\nu_{max}} = \frac{\lambda_{min}eV}{c}$$

$$= \frac{(9000 \text{ volts})(1.377 \times 10^{-10} \text{ m})(1.6022 \times 10^{-19} \text{ coul})}{3 \times 10^8 \, m/sec}$$

$$h = 6.618 \times 10^{-34} \text{ volt coulomb-sec}$$

$$= 6.618 \times 10^{-34} \text{joule-sec.}$$

The modern value for this universal constant is 6.625×10^{-34} joule-sec.

Characteristic X Rays

Figure 4-3 shows discrete lines superimposed on the continuous x-ray spectrum for a molybdenum target, but not for tungsten. Since the inner shell electrons in tungsten are tightly bound at 69.5 keV, the electrons accelerated through 35 kV do not have sufficient energy to dislodge them, but they can dislodge those in molybdenum which have a binding energy of 20 keV. Electromagnetic radiation is emitted when these shell vacancies are filled, and the emitted energy is just the difference between the binding energy of the shell being filled and that of the shell from whence it came. The electrons in each element have unique energy states; thus, these emissions of electromagnetic radiation are "characteristic" of the element, hence the term "characteristic x rays." They uniquely identify each element.

Characteristic x rays are designated by the shell that is filled; i.e., those that occur from filling K-shell vacancies are known as K-shell x rays, or simply K x rays. And if the K x ray is produced by a transition from the L_{III} subshell it is designated as a $K_{\alpha1}$ x ray, and if from the L_{II} subshell a $K_{\alpha2}$ x ray (transitions from the L_I are forbidden by the laws of quantum mechanics). In a similar fashion, transitions from the M, N, and O shells and subshells are known respectively as K_β, K_γ, K_δ, and so forth with appropriate numerical designations for the originating subshells.

L x rays are produced when the bombarding electrons knock loose an electron from the L shell and electrons from higher levels drop down to fill these L-shell vacancies; however, this occurs with lower probability than for the K shell. The lowest-energy x ray of the L series is known as L_α, and the other L x rays are labeled in the order of increasing energy, which corresponds to being filled by electrons from higher energy orbits. Characteristic x rays for M, N, . . . , shells are designated by this same pattern. The relative intensities of the characteristic x-ray peaks (see Figure 4-3) reflect the lower probabilities for interactions with electrons in the L, M, . . . , shells because the electrons are spread out over a larger volume, thus presenting a smaller target to incoming electrons.

A simplified (most subshell energies are omitted) schematic of electron shell energies in molybdenum is included in Figure 4-4 for determining the energies of characteristic x rays from the binding energies of each of the electron shells and subshells, as shown in Example 4-2. The energy differences due to the various subshells are small for most shells, but worth noting for the K and L shells especially for the higher Z elements.

Example 4-2: From the data in Figure 4-4, compute the energy of $K_{\alpha 1}$, $K_{\alpha 2}$, $K_{\beta 1}$, $K_{\beta 2}$, and L_α characteristic x rays.
Solution: The emitted x rays occur because electrons change energy states to fill shell vacancies. Each characteristic x ray energy is the difference in potential energy the electron has before and after the transition. For K x rays

$$K_{\alpha 1} = -2.42 - (-20.0) = 17.48 \text{ keV}$$
$$K_{\alpha 2} = -2.63 - (-20.0) = 17.37 \text{ keV}$$
$$K_{\beta 1} = -0.51 - (-20.0) = 19.49 \text{ keV}$$
$$K_{\beta 2} = -0.06 - (-20.0) \cong 20.00 \text{ keV}$$

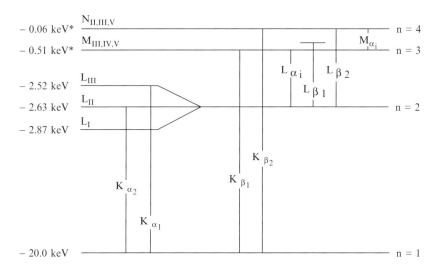

* Maximum energy for the shell; for subshell energy levels see Appendix C.

Figure 4-4. Simplified characteristic x-ray emissions from molybdenum.

For the L-shell vacancies it is presumed that the lowest energy subshell is filled from the maximum energy M shell, or

$$L_\alpha = -0.51 - (-2.52) = 2.01 \, \text{keV}$$

These and other potential permutations are included in listings of characteristic x rays associated with radioactive transformations, and are denoted as $K_{\alpha 1}$, $K_{\alpha 2}$, $L_{\alpha 1}$, $L_{\alpha 2}$, etc. (See Chapter 3 and Appendix C.)

NATURAL RADIATION

A natural radiation background exists everywhere and every natural substance contains some amount of radioactive material. Natural radiation sources represent a continuous exposure of beings on the earth and thus are a benchmark for consideration of levels of radiation protection and radioactivity, including those enhanced by human activities, and other sources as well.

Background Radiation

Creatures on the earth are continually exposed to external radiation from cosmic rays and a number of naturally occurring radionuclides that produce external gamma radiation or can be incorporated into the body to produce either a concentrated or uniformly distributed internal radiation dose. Human exposure from these sources, of course, varies widely depending on location and the surrounding environment; however, generally accepted average values are:

Source	(mrem/y)
Cosmic radiation	27
Cosmogenic nuclides	1
Terrestrial radiation	28
Nuclides in body	39

These listed doses do not include radon and its products, though it is common to do so, because of their wide variation and because the doses are almost exclusively to the lung. It is possible to convert lung dose due to radon and its products to an "effective" whole-body dose so it can be added to that due to the other sources; however, the distinct differences between doses should be recognized, especially the wide variations in radon exposures due to location and human behavior. Published estimates of total background exposure that include an "effective" dose due to radon are on the order of 300 mrem/y based on a national average radon level. Although it is proper to recognize radon exposure, it should be done warily.

Cosmic Radiation

The earth is bombarded continuously by radiation originating from the sun, and from sources within and beyond the galaxy. This cosmic radiation slams into the earth's upper atmosphere which provides an effective shield for beings below. Cosmic rays consist of high-energy atomic nuclei, some 87% of which are protons. About 11% are alpha particles, approximately 1% are heavier atoms which decrease in importance with increasing atomic number, and the remaining 1% are electrons. These "rays" have very high energies, some as high as 10^{14} MeV but most are in the range of 10 MeV to 100 GeV. As they strike the atmosphere they produce cascades of nuclear interactions that yield many secondary particles which are very important in the production of cosmogenic radionuclides.

Radiation dose rates at the earth's surface due to cosmic radiation are largely caused by muons and electrons, and both vary with elevation and with latitude, increasing with altitude as the thickness of the atmosphere decreases as shown in Figures 4-5 and 4-6, respectively. At geomagnetic latitude 55° N, for example, the absorbed dose rate in tissue approximately doubles with each 2.75 km (9000 ft) increase in altitude, up to about 10 km (33,000 ft). The global average cosmic-ray dose equivalent rate at sea level is about 24 mrem (240 mSv) per year for the directly ionizing component and 2 mrem (20 mSv) per year for the neutron component. Most of the US population is at the lower altitudes and cumulative exposures are correspondingly lower because of this. Except for short-term influences of solar activity, galactic cosmic radiation has been constant in intensity for at least several thousand years.

Cosmogenic Radionuclides

Several radionuclides of cosmogenic origin are produced when high-energy protons (87% of cosmic radiation) interact with constituents of the atmosphere. Showers of secondary particles, principally neutrons, from such interactions yield a number of such radionuclides, in particular ^{3}H, ^{7}Be, ^{14}C, and ^{22}Na (see Table 4-1), which are produced at relatively uniform rates. The high energy (> 1 MeV) secondary neutrons that produce these interactions originate in the atmosphere from cosmic-ray interactions, rather than coming from outer space (free neutrons are radioactive with a life of only about 12 min and they do not last long enough to travel from outer space). The concentration of neutrons increases with altitude, reaching a maximum concentration around 40,000 ft and then decreases at higher altitudes.

Oxygen in the atmosphere is essentially transparent to neutrons, but (n, p) interactions with ^{14}N to form ^{14}C are very probable. Neutron bombardment of ^{14}N in the atmosphere yields ^{14}C 99% of the time; the remaining 1% produces tritium through (n, t) reactions. Carbon-14 from ^{14}N$(n, p)^{14}$C reactions exist in the atmosphere as CO_2, but the main reservoir is the ocean. It has a half-life of 5715 years and undergoes radioactive transformation back to ^{14}N by

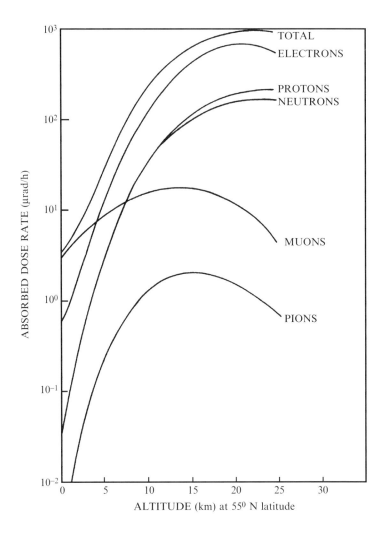

Figure 4-5. Absorbed dose rates from various components of cosmic radiation at solar minimum at 5 cm depth in a 30-cm slab of tissue (to convert to mGy/h, divide by 100).

beta-particle emission with a maximum energy of 156 keV (average energy of 49.5 keV). The natural atomic ratio of ^{14}C to stable carbon is 1.2×10^{-12} (corresponding to an activity of 0.226 t/s of ^{14}C per gram of carbon). For carbon weight fractions of 0.23, 0.089, 0.41, and 0.25 in the soft tissues, gonads, red marrow, and skeleton, annual average absorbed doses in those tissues are 1.3, 0.5, 2.3, and 1.4 mrem, respectively.

Tritium is also produced in the atmosphere mainly from $^{14}N(n, t)^{12}C$ and $^{16}O(n, t)^{14}N$ reactions. It exists in nature almost exclusively as HTO but may

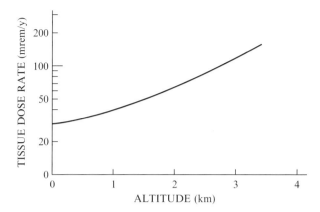

Figure 4-6. Variation of total (charged particles plus neutrons) annual cosmic-ray dose rate with altitude in the US.

Table 4-1. Global Distribution of Cosmogenic Radionuclides.

	^{3}H	^{7}Be	^{14}C	^{22}Na
Global inventory (PBq)	1300	37	8500	0.4
Distribution (%)				
Stratosphere	6.8	60	0.3	25
Troposphere	0.4	11	1.6	1.7
Land/biosphere	27	8	4	21
Mixed ocean layer	35	20	2.2	44
Deep ocean	30	0.2	92	8
Ocean sediments	–	–	0.4	–

be partially incorporated into organic compounds in some foods. Tritium has a half-life of 12.3 years and, upon transformation, releases beta particles with maximum energy of 18.6 keV (average energy 5.7 keV). The average concentration of cosmogenic tritium in environmental waters is 100 to 600 Bq/m^3 (3 to 16 pCi/l) based on a seven-compartment model (NCRP, 1979). If it is assumed that ^{3}H exists in the body, which contains 10% hydrogen, at the same concentration as surface water (i.e., 400 Bq/m^3 or 12 pCi/l); the average annual absorbed dose in the body is 1.2 mrem (0.012 μSv).

Except as augmented by man-made sources, ^{14}C and ^{3}H have existed for *eons* in the biosphere in equilibrium; i.e., the rate of production by neutron interactions in the atmosphere is equal to its subsequent removal by radioactive transformation. However, during the past hundred years or so the cosmogenic content of ^{14}C in the environment has been diluted by combustion of fossil fuels which releases CO_2 that does not contain ^{14}C (a phenomenon called the Suess effect after its discoverer). On the other hand, atmospheric nuclear-weapons tests and other human activities have added to the natural inventories of ^{3}H and ^{14}C. The net result of these human activities is that these nuclides no longer exist in natural equilibria in the environment; however, since atmospheric nuclear testing stopped for the most part in 1962, the concentration of ^{3}H is now returning to natural levels.

Beryllium-7, with a half-life of 53.4 days, and ^{10}Be($T_{1/2} = 1.6 \times 10^6$y) to a lesser extent, are also produced by cosmic-ray interactions with nitrogen and oxygen in the atmosphere. ^{7}Be undergoes radioactive transformation by electron capture to ^{7}Li with 10.4% of the captures resulting in the emission of a 478 keV gamma ray, which makes its quantitation relatively straightforward. Also, being a naturally occurring radionuclide, it is often observed in gamma-ray spectra of environmental samples at concentrations in temperate regions of about 3000 Bq/m^3 in surface air and 700 Bq/m^3 in rainwater. Average annual doses to adults are less than about 1.2 mrem to internal organs.

Sodium-22 with a half-life of 2.604 y is produced by spallation interactions between atmospheric argon and high-energy secondary neutrons from cosmic rays. A large fraction of ^{22}Na remains in the stratosphere where it is produced but nearly half the natural inventory is in the mixed layer of the ocean (Table 4-1). Annual average absorbed doses to the adult are about 1 mrad to the soft tissues, 2.2 mrad to red marrow, and 2.7 mrad to bone surfaces.

Neutrons associated with cosmic-ray interactions can also activate constituents of the sea or the earth; however, the flux at sea level is only about 30 n/cm$^2 \cdot$ h of which only 8 are thermal neutrons; therefore, production of radioactive materials by neutron capture in the earth's crust or the sea is minor compared to the atmospheric interactions. An important exception is the production of long-lived ^{36}Cl (NCRP 1975) by neutron activation of ^{35}Cl in the earth's crust.

The Natural Radioactive Series

Many of the naturally occurring radioactive elements are members of one of four long chains, or radioactive series, stretching through the last part of the Chart of the Nuclides. These series are named the uranium (^{238}U), actinium (^{235}U), thorium (^{232}Th), and neptunium (^{237}Np) series according to the radionuclides that serve as progenitor (or parent) of all the series products. With the exception of neptunium, the parent radionuclides are primordial in origin because they are so long-lived that they still exist some 4.5 billion years after the solar system was formed.

Primordial sources of ^{237}Np no longer exist because its half-life is only 2.1 million years. However, since the advent of nuclear reactors, ^{237}Np has once again become available primarily as the radioactive transformation product of alpha-emitting ^{241}Am, which is the product of ^{241}Pu (β^- transformation; $T_{1/2} = 14.4$ y) produced in nuclear reactors. Were it not for large-scale uses of nuclear fission, ^{237}Np and its series products would be insignificant. It has become, however, an important constituent of high-level radioactive waste.

The members of the two primary naturally radioactive series are listed in Tables 4-2 and 4-3 along with the principal emissions, the half-life, and the maximum energy and frequency of occurrence of emitted particles and electromagnetic radiations.

There are notable similarities in radioactive transformations in the radioactive series. For example, the uranium, thorium, and actinium (^{235}U) series each have an intermediate gaseous isotope of radon and each ends in a stable isotope of lead (the neptunium series has no gaseous product and its stable end product is ^{209}Bi). The sequences of transformation reduce both the proton number (and nuclear charge) by 10 and 8 units, the neutron number by 22 and 16 units, and the total mass by 32 and 24 units for ^{238}U and ^{232}Th, respectively. Each long-lived parent undergoes transformation by alpha-particle emission, which is followed by one or more beta transitions. Near the middle of each series there is a sequence of 3 to 5 alpha emissions, which are followed by one or two beta transformations; and there is at least one more alpha emission before stability is reached (Figure 4-7). These patterns produce a zig-zag path back and forth across a stability "line" extrapolated from the zig-zag line of stable nuclides below ^{206}Pb or ^{208}Pb (there are no stable nuclides above ^{209}Bi; thus, such a "line" is only theoretical). As the product nuclei move steadily away from the "line" the half-lives decrease as shown in the ^{238}U series for ^{230}Th, ^{226}Ra, ^{222}Rn, and ^{218}Po: 2.5×10^5 y, 8×10^4 y, 1.6×10^3 y, $3.82\,d$, and 3.11 min, respectively.

In practical terms, the uranium series contains several subseries in addition to ^{238}U itself. The more important ones are ^{230}Th, ^{226}Ra, ^{222}Rn, and ^{210}Pb and their subsequent products. ^{230}Th is strongly depleted from its precursors in seawater and enhanced in bottom sediments; thus, it can also be used to date such sediments. It has also been found in raffinates from uranium processing, and as such represents a separated source that produces ^{226}Ra and ^{222}Rn and their products. ^{226}Ra is often found separated from its precursors; for example, ^{226}Ra "hot spots" remain where uranium deposits have contacted sulfuric acid from natural oxidation of ferrous sulfide. Once released into natural waters, ^{226}Ra and other radium isotopes are mobile until scavenged or coprecipitated, e.g., the cement-like calcium carbonate "sinters" around some hot springs. When freshly prepared, and free of its decay products, ^{226}Ra produces minimal gamma radiation (a 0.19 MeV gamma ray is emitted in 3.59% of ^{226}Ra transformations). However, as ^{222}Rn and its succeeding gamma-emitting products grow into the source and reach equilibrium with it, the gamma intensity increases substantially.

Radon-222 is also a subseries when separated from ^{226}Ra, as it often is, because it is a noble gas. It is the only gaseous member of the ^{238}U series and is produced from ^{226}Ra; thus, it will grow into any radium source in just a few days. Because of its high intrinsic specific activity, ^{222}Rn is sometimes removed from its radium parent and concentrated into tubes and needles for implantation in malignant tissues for treatment of disease. For similar reasons, ^{222}Rn was

Table 4-2. The ^{238}U Series and Radiations of Yield, Y_i Greater than 1% (X rays, Conversion Electrons, and Auger Electrons are Not Listed).

Nuclide	Half-life	α E (MeV)	α Y_i (%)	β⁻ E (MeV)*	β⁻ Y_i (%)	γ E (keV)	γ Y_i (%)
^{238}U α	4.468E9 y	4.15	21				
		4.2	79				
^{234}Th β⁻	24.10 d			0.08	2.9	63.3	4.8
				0.1	7.6	92.4	2.8
				0.1	19.2	92.8	2.8
				0.2	70.3		
^{234m}Pa** IT	1.17 m			2.27	98.2	766	0.3
						1,001	0.84
^{234}Pa β⁻	6.75 h			22 β⁻s $E_{avg} = 0.224$ $E_{max} = 1.26$		1,313	18
						1,527	5.97
^{234}U α	2.457E5 y	4.72	28.4				
		4.78	71.4				
^{230}Th α	7.538E4 y	4.62	23.4				
		4.69	76.3				
^{226}Ra α	1600 y	4.6	5.55			186.2	3.6
		4.79	94.5				
^{222}Rn α	3.8235 d	5.49	99.9			510	0.08
^{218}Po** α	3.11 m	6	100.0				
^{214}Pb β⁻	26.8 m			0.19	2.35	53.2	1.11
				0.68	46.0	242	7.5
				0.74	40.5	295.2	18.5
				1.03	9.3	351.9	35.8
^{214}Bi** β⁻	19.9m			0.79	1.45	609.3	44.8
				0.83	2.74	768.4	4.8
				1.16	4.14	934.1	3.03
				1.26	2.9	1,120.3	14.8
				1.26	1.66	1,238.1	5.86
				1.28	1.38	1,377.7	3.92
				1.38	1.59	1,408.9	2.8
				1.43	8.26	1,729.6	2.88
				1.51	16.9	1,764.5	15.4
				1.55	17.5	2,204.2	4.86
				1.73	3.05	2,447.9	1.5
				1.9	7.18	nine other γs	
				3.27	19.9		
^{214}Po α	164.3 µs	7.69	99.99				
^{210}Pb β⁻	22.3 y			0.02	84	46.5	4.25
				0.06	16		
^{210}Bi** β⁻	5.013 d			1.16	100	1,764.5	15.4
^{210}Po α	138.376 d	5.3	100.0				
^{206}Pb	(Stable)						

* Maximum beta energy
** Branching occurs in 0.13% of ^{234m}Pa to ^{234}Pa by IT; 0.02% of ^{218}Po to ^{218}At by β⁻; 0.02% of ^{214}Bi to ^{210}Tl by α; 0.00013% of ^{210}Bi to ^{206}Tl by α
Source: NNDC 1995.

Table 4-3. The ^{232}Th Series and Radiations of Yield, Y_i, Greater than 1% (X Rays, Conversion Electrons, and Auger Electrons are Not Listed).

Nuclide	Half-life	α E (MeV)	α Y_i (%)	β⁻ E (MeV) *	β⁻ Y_i (%)	γ E (keV)	γ Y_i (%)
^{232}Th α	14.05E9 y	3.95	22.1				
		4.01	77.8				
^{228}Ra β⁻	5.75 y			0.02	40		
				0.04	60		
^{228}Ac** β⁻	6.15 h			0.45	2.6	209.3	3.88
				0.5	4.18	270.2	3.43
				0.61	8.1	328	2.95
				0.97	3.54	338.3	11.3
				1.02	5.6	463	4.44
				1.12	3	794.9	4.34
				1.17	31	835.7	1.68
				1.75	11.6	911.2	26.6
				2.08	10	964.8	5.11
						969.9	16.2
						1,588	3.3
^{228}Th α	1.9131 y	5.34	28.2			84.4	1.27
		5.42	71.1				
^{224}Ra α	3.62 d	5.45	5.06			241	3.97
		5.69	94.9				
^{220}Rn α	55.6 s	6.29	99.9				
^{216}Po α	0.145 s	6.78	100				
^{212}Pb β⁻	10.64 h			0.16	5.17	238.6	43.3
				0.34	82.5	300.1	3.28
				0.57	12.3		
^{212}Bi α 35.9% β⁻ 64.1%	60.55 m	6.05	25.13	0.63	1.87	727.3	6.58
		6.09	9.75	0.74	1.43	785.4	1.1
				1.52	4.36	1,621	1.49
				2.25	55.5		
^{212}Po	0.299 μs	8.79	100				
^{208}Tl β⁻ α	3.053 m			1.03	3.1	277.4	6.3
				1.29	24.5	510.8	22.6
				1.52	21.8	583.2	84.5
				1.8	48.7	763.1	1.8
						860.6	12.4
						2,615	99.2
^{208}Pb	(Stable)						

* Maximum beta energy
** Only gammas with yield greater than 2% listed
Source: NNDC 1995.

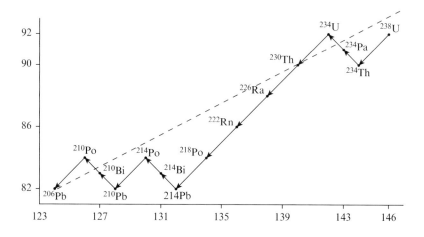

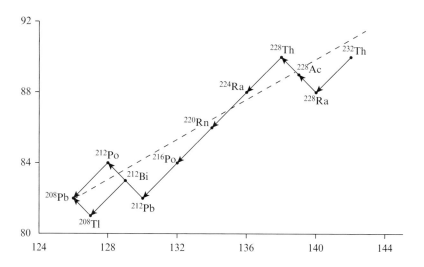

Figure 4-7. Comparative series transformation of ^{238}U and ^{232}Th relative to an extrapolated "line" of stability.

obtained from ^{226}Ra generators by early researchers to be mixed with beryllium to produce neutrons. Radon-222 has a half-life of 3.82 days and its principal transformation products are ^{218}Po, ^{214}Pb, ^{214}Bi, and ^{214}Po, which have half-lives ranging from 26.8 minutes to 164 microseconds. The gamma rays of ^{214}Bi have the highest yield and are the most energetic of the uranium series; thus the ^{222}Rn subseries is an important source of external radiation.

Lead-210, a long-lived radon transformation product, also heads a subseries of ^{238}U. Lead-210 itself and its ^{210}Bi and ^{210}Po products can be observed in significant concentrations in the atmosphere. The stable end product of this subseries, which is also the end of the ^{238}U series, is ^{206}Pb.

Thorium-232, which is the least mobile of the series radionuclides, has three subseries: ^{232}Th itself, ^{228}Ra, and ^{220}Rn. Thorium-232 exists naturally in the tetravalent state as a very stable oxide or in relatively inert silicate minerals, and it is strongly adsorbed on silicates. The subseries headed by ^{228}Ra yields ^{228}Ac, ^{228}Th, and ^{224}Ra which are generally in radioactive equilibrium. The third subseries is headed by ^{220}Rn (thorium emanation, or thoron) which has a 56 s half-life and which quickly forms transformation products down to stable ^{208}Pb, the longest-lived of which is 10.6-h ^{212}Pb. Although generally less important than ^{222}Rn in the uranium series, ^{220}Rn and its products can occasionally present radiation exposure situations if high concentrations of ^{228}Ra exist. An important product of ^{220}Rn (and ^{228}Ra itself because of the short half-life of ^{220}Rn) is ^{208}Tl which emits 2.62 MeV gamma rays, and the residues of thorium recovery can produce significant gamma exposures if not carefully managed.

Singly Occurring Primordial Radionuclides

Naturally occurring primordial radioactive nuclides also exist that are not members of any of the four series. There is no obvious pattern to the distribution of the long-lived, nonseries radioactive nuclei, major ones of which are listed in Table 4-4. Some of the half-lives are surprisingly long with very low specific activities, which makes detection very difficult; all are elements formed when the "big bang" occurred some 4.5 billion years ago.

The primordial radionuclides of ^{40}K and ^{87}Rb are of particular interest in radiation protection because of their presence in environmental media and their contribution to human exposure. Potassium-40 contributes about 40% of the exposure humans receive from natural radiation. It is present in natural potassium with an isotopic abundance of 0.0117 atom percent, has a half-life of 1.27×10^9 y, and undergoes radioactive transformation both by electron capture (10.67%) and emission of beta particles (89.27%) of maximum energy of 1.312 MeV (average energy 0.56 MeV). Electron capture is followed by emission of a 1461 keV gamma ray (in 10.67% of the transformations of ^{40}K), conversion electrons (0.3% of ^{40}K transformations), and the usual low-energy Auger electrons and characteristic x rays. The average elemental concentration of potassium in reference person (a man) is 2% which produces annual doses of 14 mrem to bone surfaces, 17 mrem on average to soft tissue, and 27 mrem to red marrow (UN 1982). It is also present in soil at an average concentration of 12 pCi per gram, which results in an annual whole-body dose equivalent of 12 mrem due to external gamma radiation (UN 1982).

Table 4-4. Naturally Occurring Primordial Radionuclides.

Nuclide	% Abundance	$T_{1/2}$ (y)	Emission(s)	Q (MeV)
^{40}K	0.0117	1.27×10^9	β^-, EC, γ	1.505, 1.311
^{50}V	0.25	1.4×10^{17}	β^-, EC, γ	2.208
^{87}Rb	27.83	4.88×10^{10}	β^-	0.283
^{113}Cd	12.22	9.3×10^{15}	β^-	0.316
^{115}In	95.7	4.4×10^{14}	β^-	0.495
^{123}Te	0.91	$> 1.3 \times 10^{13}$	EC	0.052
^{130}Te	33.87	1.25×10^{20}	$\beta^-\beta^-$	0.42
^{138}La	0.09	1.05×10^{11}	EC, β^-, γ	1.737, 1.044
^{142}Ce	11.13	5.0×10^{16}	α	?
^{144}Nd	23.8	2.1×10^{15}	α	1.905
^{147}Sm	15	1.06×10^{11}	α	2.31
^{152}Gd	0.2	1.08×10^{14}	α	2.205
^{74}Hf	0.162	2.0×10^{15}	α	2.496
^{176}Lu	2.59	3.8×10^{10}	β^-, γ	1.192
^{180}Ta	0.012	$> 1.2 \times 10^{15}$	EC, $\beta +$, γ	0.853, 0.708
^{186}Os	1.58	2.0×10^{15}	α	2.822
^{187}Re	62.6	4.4×10^{10}	β^-	0.00264
^{190}Pt	0.01	6.5×10^{11}	α	3.249

Source: NNDC 1996.

Rubidium-87 has a half-life of 4.88×10^{10} y and emits beta particles of maximum energy of 282 keV (average energy $\sim$ 79 keV). Its natural isotopic abundance is 27.84% and the mass concentrations in reference man range from 6 ppm in the thyroid to 20 ppm in the testes (ICRP 1975). It contributes an annual dose to body tissues of 1.4 mrem to bone-surface cells (UN 1982) and an annual effective dose equivalent of about 0.6 mrem.

RADIOACTIVE ORES AND BYPRODUCTS

Many ores are processed for their mineral content which enhances either the concentration of the radioactive elements in process residues or increases their environmental mobility; these processes result in materials that are no longer purely "natural." These and related issues enjoy a special category in radiation protection called NORM for Naturally Occurring Radioactive Material.

The presence of uranium and/or thorium, ^{40}K, and other naturally radioactive materials in natural ores or feedstocks is often the result of the same geochemical conditions that initially concentrated the main mineral-bearing ores (e.g., uranium in bauxite and phosphate ore). These technologically enhanced sources of naturally occurring radioactive material represent radiation sources that can cause generally low-level radiation exposures of the public, or in some localized areas, exposures significantly above the natural background. For example, processing of uranium-rich ores has created residues containing thorium, radium, and other transformation products of uranium.

Natural concentrations and levels of NORM vary widely as shown in Table 4-5. Uranium receives the most attention; it is quite prevalent in most environmental media (about 4 ppm by weight), and large amounts exist in the oceans at a concentration of about 0.3–2.3 mg per liter of seawater. Radium-226, a uranium transformation product, ^{40}K, and ^{232}Th and its products also exist in various rocks, soils, and minerals as shown in Table 4-5.

Removing a mineral from an ore can concentrate the residual radioactivity (called beneficiation). For example, bauxite is rich in aluminum, and its removal yields a residue in which uranium and radium originally present in the ore can be concentrated to 10 to 20 pCi of ^{226}Ra per gram in the residues, or "red mud", so called because its iron content (also concentrated) has a red hue. Similarly, processing phosphate rock in furnaces to extract elemental phosphorous produces a vitrified slag containing 10 to 60 pCi/g of ^{226}Ra which has been commonly used as an aggregate in making roads, streets, pavements, residential structures, and buildings. Such uses of high-bulk materials from ore processes raise risk/benefit tradeoffs of such uses; high volumes and relatively low but enhanced concentrations make such decisions challenging.

Uranium Ores

Well before Becquerel's discovery of radioactivity using a double salt containing uranium, it had been mined, principally in Czechoslovakia for use as a coloring agent in the glass industry. The search for uranium ore from which to recover radium stimulated the discovery of new deposits of pitchblende in the Belgian Congo, in Great Bear Lake, Canada, and eventually of carnotite ores in the Colorado Plateau of the US. Pitchblende is a brown-black ore which contains as much as 60 to 70% uranium as the oxide U_3O_8, commonly called "yellowcake" because of its golden color when purified. The US form is primarily yellow carnotite (hydrated potassium uranium vanadate) which was first processed for vanadium, then later to recover uranium when it became valuable.

Mining and processing of uranium ores has generated and continues to generate mining overburden wastes of about 38 million MT per year, which is added to a US inventory of about 3 billion MT yet to be reclaimed. The tailings from milling uranium ores contain ^{230}Th and ^{226}Ra in concentrations directly related to the richness of uranium in the original ores. Radon gas is continually produced by ^{226}Ra in the tailings, which requires that they be managed to preclude radon problems as well as ^{226}Ra exposures. Some milling processes

Table 4-5. Typical Concentrations of Naturally Occurring Radionuclides.

Mineral	Average Concentration (pCi/g)			
	^{40}K	^{238}U	^{226}Ra	^{232}Th
Igneous Rock				
Acidic	27	1.6	–	2.2
Intermediate	19	0.62	–	0.88
Mafic	6.5	0.31	–	0.3
Ultrabasic	4	0.01	–	0.66
Sedimentary Rock				
Limestone	2.4	0.75	–	0.19
Carbonate	–	0.72	–	0.21
Sandstone	10	0.5	–	0.21
Shale	19	1.2	–	1.2
Soil Type				
Serozem	18	0.85	–	1.3
Gray-brown	19	0.75	–	1.1
Chestnut	15	0.72	–	1
Chernozem	11	0.58	–	0.97
Gray forest	10	0.48	–	0.72
Sodpodzolic	8.1	0.41	–	0.6
Podzolic	4	0.24	–	0.33
Boggy	2.4	0.17	–	0.17
World average	10	0.7	–	0.7
US Coal	1.4	0.5	–	0.6
Building Materials				
Brick	16–20	–	1.4–2.6	1.0–3.4
Concrete	7–9	–	0.9–2.0	0.9–2.3
Plaster	< 2–10	–	0.1–0.6	<0.4–2.0
Granite	28–40	–	2.4–3	2.3–4.5
Limestone/Marble	~ 1	–	< 0.5	< 0.5

Adapted from Eisenbud 1987, UN 1977, UN 1982.

contain a large fraction of ^{230}Th; for example, raffinate streams, which contained much of the ^{230}Th, were sluiced to separate areas for storage and/ or further processing. These ^{230}Th residues represent minor radiation problems in themselves since ^{230}Th is a pure alpha emitter, but its radioactive transformation produces ^{226}Ra which will gradually produce a future source of external gamma radiation and radon gas. Modern management of uranium tailings is done under regulations that require stabilization for up to 1000 years in such a way that emissions of radon and its products are controlled. Older abandoned mill sites are being remediated to achieve the same requirements.

Fossil Fuels–Coal Ash

Coal contains uranium and thorium and their associated transformation products in concentrations that vary over two orders of magnitude depending upon the type of coal and where it is mined. Concentrations of ^{238}U and ^{232}Th range from 0.08 to 14 pCi/g and 0.08 to 9 pCi/g, respectively, and rarely exceed 10 pCi/g. Average (arithmetic) ^{238}U and ^{232}Th concentrations are about 0.6 and 0.5 pCi/g, respectively. Radium-226 concentrations in coal may be as low as a fraction of a picocurie per gram to as high as 20 pCi/g.

Utility and industrial boilers generate about 61 million MT of coal ash per year, nearly 20 million MT of which are used as an additive in concrete, as structural fill, and for road construction. Coal ash generally consists of fly ash, bottom ash, and boiler slags in amounts that are dependent upon the mineral content of the coal and the type of boiler. A reasonable average concentration of ^{226}Ra in coal ash is about 4 pCi/g.

Scales and Sludges–Oil and Gas Production

The petroleum industry generates 260,000 MT of pipe scale, sludge, and equipment or components contaminated with radium each year. Petroleum pipe scale may contain more than 400,000 pCi/g of ^{226}Ra. Scales and sludge with high concentrations are removed and stored in drums, but some equipment and associated contaminated components are also discarded. Typical concentrations of ^{226}Ra in scale are about 360 pCi/g, and about 75 pCi/g in sludge.

Water Treatment Sludge

Most water treatment wastes are believed to contain ^{226}Ra in concentrations comparable to those in typical soils. However, some water supply systems, primarily those relying on groundwater sources, may generate sludge with higher ^{226}Ra levels especially if their process equipment is effective in removing naturally occurring radionuclides from the water. A typical drinking water source contains about 8 pCi/L of ^{226}Ra which yields an average ^{226}Ra concentration of about 16 pCi/g in sludge after processing. About 700 water utilities in the US generate and dispose of 300,000 MT of sludge, spent resin, and charcoal beds in landfills and lagoons, or by application to agricultural fields.

Phosphate Industry Wastes

Mining of phosphate rock is the fifth largest mining industry by volume in the US, which has several major deposits principally in Florida (the world's largest phosphate rock producing area because of the richness and accesibility of the ore), North Carolina, Tennessee, eastern Idaho, northern Utah, western Wyoming, and southern Montana. These deposits occurred by geochemical processes in ancient seas that were also effective in depositing uranium; thus, some phosphate deposits, principally in Florida, contain significant amounts of uranium and its radioactive progeny. Uranium in phosphate ores found in the US ranges in concentration from 20 to 300 ppm (or about 7 to 100 pCi/g), while thorium occurs at essentially ambient background concentrations, between 1 to 5 ppm (or about 0.1 to 0.6 pCi/g).

Phosphate rock is processed to produce phosphoric acid and elemental phosphorus which are combined with other chemicals to produce phosphate fertilizers, detergents, animal feeds, other food products, and phosphorous chemicals. Over 90% of the phosphate rock is processed into fertilizers which contain about 8 pCi/g of ^{226}Ra; however, 86% of the uranium and 70% of the thorium originally present in the ore are entrained into the phosphoric acid product because the process used to extract phosphorous also entrains these elements.

Phosphogypsum is the principal waste byproduct; it is stored in waste piles called stacks and contains about 80% of the ^{226}Ra originally present in the phosphate rock. Sixty-three large stacks and three relatively smaller ones exist, mostly in Florida, and these contain some 8 billion MT. Phosphogypsum could be used in wall board or applied as a soil conditioner but such use is very small because it contains radium which most consumers would expect to be removed. Since large quantities of phosphate industry wastes are produced, these waste materials could present radiological exposures of individuals if the wastes were to become widely distributed in the environment.

Phosphoric acid scale, produced when phosphogypsum is physically separated from the phosphoric acid by large stainless steel filter pans, can be problematical because a scale is deposited on the pan and in ancillary piping and tanks. Radium-226 concentrations range from several hundred to as high as 100,000 pCi/g, but fortunately such scales represent only a few thousand m^3 per year.

Thorium Ores

Thorium ores have been processed since about 1900 to recover rare earth elements and thorium, a soft silvery metal used in specialized applications such as welding electrodes, gas lantern mantles, ultraviolet photoelectric cells, and in certain glasses and glazes. Thorium-232 is a naturally occuring radionuclide which heads the thorium series of radioactive transformation products (see Table 4-3); it is also a "fertile" material that can be transmuted by neutron irradiation into ^{233}U, a fissionable material. Concentrations of 1 pCi/g are typical of natural soils and minerals, but some areas of the world contain significant thorium; for example, Monazite sands in Brazil and India contain 4 to 8% thorium.

Residues from processing monazite and other thorium-rich ores to recover thorium or rare-earth mineral may still contain significant amounts of ^{232}Th, and huge inventories of these residual materials exist. Thorium-232 emits alpha particles to become ^{228}Ra($T_{1/2} = 5.75$ y) which will attain secular equilibrium with ^{232}Th some 40 years after ^{232}Th is purified. Radium-228 has a number of short-lived, gamma-emitting decay products; thus, even if radium is removed in processing, ^{232}Th will produce an external gamma flux after a few decades. If, however, ^{232}Th is removed from residues, the site can be returned to unrestricted uses after ^{228}Ra decays away, in about 40–50 years.

Concentrations of 5 pCi/g of ^{232}Th in equilibrium with its transformation products and spread over a large area with no soil cover will produce an estimated exposure of 15 mrem/y to an average person standing on the area. Whereas radon exposures can be significant for uranium wastes if elevated ^{226}Ra concentrations exist in soil and a structure is present, ^{220}Rn (thoron) in the ^{232}Th series is of minor significance because its very short (56 s) half-life precludes significant migration from thorium deposits.

ACTIVATION PRODUCTS

Activation products are a common source of radioactive materials; they occur from bombarding target materials with various particles such as neutrons, protons, deuterons, electrons, various atoms, and photons. The quantity and types of products produced depend on the energy and type of the bombarding particle, the target material, and the interaction probabilities of the various combinations.

Activation Cross Section denotes the "target size" or apparent area that an atom of a particular element presents to a bombarding particle such that a given reaction may take place. It is not specifically an area, but is best described as a measure of the probability that a reaction will take place; it is dependent on the target material and the features of the incident "particle," which may in fact be a photon. Schematically, the "cross sectional area" presented to the incoming projectile can be represented as where the energy, charge, mass, and deBroglie wavelength of the projectile, and the vibrational frequency, the spin, and the energy states of nucleons in the target atom influence whether a given scattering or absorption interaction will occur.

Cross sections for any given arrangement of projectile and target cannot be predicted directly by nuclear theory; thus, they are usually measured by determining the number of projectiles per unit area before and after they impinge on a target containing N atoms. When such measurements and calculations were first made for a series of elements, the researchers were surprised that the calculated values of σ were as large as observed. Apparently, he/she exclaimed that it was "as big as a barn," and the name stuck. One "barn", or b, is

$$b = 10^{-24}\,\text{cm}^2 \text{ or } 10^{-28}\text{m}^2$$

which is on the order of the nuclear radius squared. The cross section varies with incident particle energy, as shown in Figure 4-8 for neutrons on cadmium and boron. Such plots typically show a $1/v$ dependence of σ on neutron energy and resonance energies where the cross section is very high. The NNDC at

Brookhaven National Laboratory is the authoritative source for cross-section data for a host of nuclear interactions, including neutrons, protons, deuterons, alpha particles, and other projectiles.

Endoergic and Exoergic Nuclear Reactions

The energetics of nuclear reactions are determined by the Q-Value which represents the energy balance between the particles and targets in a nuclear interaction and the products of the reaction; it is calculated as

$$Q = [(M_X + m_x) - (M_Y + m_y)]931.502 MeV/u$$

where m_x, M_X, m_y, and M_Y represent the masses of the incident particle, target nucleus, product particle, and product nucleus, respectively, in atomic mass units u. Thus, the Q-value is readily calculated by subtracting the masses of the products from the masses of the reactants and converting the net mass change to energy (see Example 4-3).

If the Q-value of a reaction is negative, the reactions are "endoergic" and an amount of energy equal to the Q-value must be supplied by the bombarding particle for the reaction to occur. If, however, the Q-value is positive, the kinetic energy of the products is greater than that of the reactants and the reaction is "exoergic," i.e., energy is gained at the expense of the mass of the reactants.

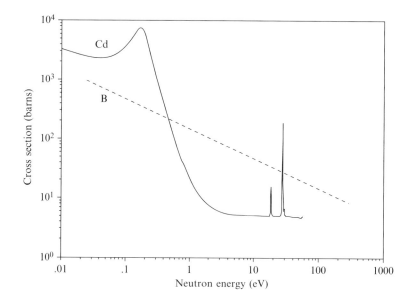

Figure 4-8. Neutron activation cross section vs energy for cadmium and boron.

An endoergic reaction actually requires that the bombarding particle have an energy somewhat greater than the Q-value to conserve momentum; i.e., its energy must exceed a threshold energy, E_{th}, for the reaction to occur

$$E_{th} = -Q\left[\frac{M_x + m_x}{M_x}\right]$$

where M_x is the the mass of the target atom and m_x is that of the incident particle. And, since the threshold energy depends on the Q-value, the Q-value is a practical determinant of whether a stated reaction can occur and whether energy would be produced or consumed.

Example 4-3: Calculate the Q-value for the ^{14}N(α, p)^{17}O reaction and the threshold energy.
Solution: The Q-value is obtained by subtracting the mass of the products from the mass of the reactants using the atomic masses from Appendix B:

Mass of Reactants	Mass of Products
14.003074 u for ^{14}N	16.999131 u for ^{17}O
4.002603 u for ^{4}He	1.007826 u for ^{1}H
18.005677	18.006957

which yields a mass difference of -0.001280 u; thus $Q = -1.19$ MeV, and the reaction is endoergic (i.e., energy must be supplied). The threshold energy, E_{th}, that the bombarding alpha particle must have for the reaction to occur is

$$Eth = -(-1.19\, MeV)(\frac{14.003074 + 4.002603}{14.003074}) = 1.53\, MeV$$

which can be supplied by almost any natural alpha source since such sources emit alpha particles in excess of 4 MeV. Very little error is introduced if calculations of E_{th} are done using the integer mass numbers rather than the exact masses.

Q-value calculations are usually done with the masses of the neutral atoms rather than the masses of the bombarding and ejected particles because the number of electrons is the same on both sides of the equation. By doing so, it is not necessary to account for the number of electrons, and the nuclear reaction equation balances readily. It is necessary, however, to examine the balance equation for a given reaction to assure that this is in fact the case. Two reactions involving radioactive transformation, which include the energy change Q, illustrate these factors:

$$^{14}_{6}C \rightarrow ^{14}_{7}N + ^{0}_{-1}e + Q$$

$$^{22}_{11}Na \rightarrow ^{22}_{10}Ne + ^{0}_{+1}e + \text{orbital } e + Q$$

The first reaction yields a positive Q-value (as all radioactive transformations must) of 0.156 MeV which is obtained by simply subtracting the ^{14}N mass from that of ^{14}C. It may appear that the beta-particle (electron) mass was left out, but it is in fact included because the mass of neutral ^{14}N includes 7 electrons, one more than the parent ^{14}C. By using the masses of the neutral isotopes, we have effectively "waited" until the product ^{14}N picks up an additional orbital electron to balance the net increase of one proton in the nucleus due to the beta-particle transformation of ^{14}C; i.e., the electron mass is included as though the beta particle stopped and attached itself to the ionized ^{14}N product. The binding energy of the electron in ^{14}N is in fact not accounted for but this is very small compared to the other energies involved. The Q-value energy thus represents kinetic energy given to the beta particle (recoil of the ^{14}N nucleus is negligible) due to the decrease in mass that occurred in the transformation reaction to yield 0.156 MeV.

This convenient circumstance does not exist, however, in the second reaction in which a positron is emitted. The proton number decreases by one unit, which requires that an electron drift off from the product atom. Using the neutral masses of the parent and product nuclei to calculate the Q-value leaves two electron masses unaccounted for (the orbital that drifts off and the emitted positron); thus, it is necessary to account for these in using the Q-value to determine if the transformation can occur. Positron emission, as in ^{22}Na transformation to ^{22}Ne, can occur only if the parent is at least two electron masses heavier than its product, which it is for ^{22}Na transformation to ^{22}Ne by positron emission.

Alpha-Particle Interactions

The reaction of alpha particles on nitrogen, illustrated by the Q-value calculation in Example 4-3, is one of the most important ever observed because it was the first demonstration of the transmutation of one element into another, or

$$^{14}_{7}N + ^{4}_{2}He \rightarrow [^{18}_{9}F] \rightarrow ^{17}_{8}O + ^{1}_{1}H$$

Rutherford believed that the alpha particle and the nitrogen nucleus had combined to form a compound nucleus which then broke apart into a new product nucleus and a proton. If transmutation had occurred, the alpha particle would disappear; but if it only knocked the proton out of the nitrogen nucleus, the alpha particle would still exist and the nuclear interaction could not be characterized as a transmutation. Blackett performed a classic experiment, documented in 20,000 cloud chamber photos, that clearly showed that the alpha particle had vanished and a new nucleus had formed; he received the

Nobel Prize in 1948 for providing the experimental proof of this fundamental precept.

Alpha-neutron (α, *n*) Reactions are not only an important source of neutrons but have historical significance because they led to Chadwick's discovery of the neutron in 1932 (and a Nobel Prize) and artificial radioactivity by the Joliet–Curies (and another Nobel Prize). Bombardment of beryllium by alpha particles produced a very penetrating radiation which the Joliet–Curies assumed to be high-energy gamma radiation, which was puzzling because the emissions from the reaction produced a stream of protons when directed at paraffin. Chadwick postulated that a neutral particle with mass close to that of the proton was emitted which produced elastic collisions with the hydrogen nuclei in the paraffin, and from the momenta observed he demonstrated that the emitted radiation was in fact a neutral particle (the neutron) produced by the reaction

$$\ce{^9_4 Be} + \ce{^4_2 He} \rightarrow [\ce{^{13}_6 C}] \rightarrow \ce{^{12}_6 C} + \ce{^1_0 n}$$

where $\ce{^1_0 n}$ is the symbol for the neutron.

Once it was known that neutrons existed and could be easily produced by irradiating beryllium with alpha particles, similar reactions were produced in other target materials such as the (α, *n*) reaction with aluminum

$$\ce{^{27}_{13} Al} + \ce{^4_2 He} \rightarrow \lfloor \ce{^{31}_{15} P} \rfloor \rightarrow \ce{^{30}_{15} P} * + \ce{^1_0 n}$$

which was the first observation of artificially produced radioactivity and gained the Joliet–Curies the Nobel Prize in 1935.

Transmutation by Protons and Deuterons

Although alpha particles can be used to produce a number of useful interactions, they are limited in application to light elements because the coulombic repulsion between their +2 charge and the high Z of heavy elements is just too great to overcome. Protons and deuterons with a single + charge are much more useful for nuclear interactions; however, since there are no natural sources of protons or deuterons, such reactions only became available with the invention of accelerators first by Cockcroft and Walton who accelerated protons using a voltage multiplier device. They could accelerate protons up to energies of about 0.8 MeV, which, though minuscule by today's standards, released two alpha particles from lithium by the reaction

$$\ce{^7_3 Li} + \ce{^1_1 H} \rightarrow \lfloor \ce{^8_4 Be} \rfloor \rightarrow \ce{^4_2 He} + \ce{^4_2 He}$$

This reaction has a certain historical interest because it was the first quantitative proof of the validity of the Einstein mass–energy relationship. The Q-value calculated from mass changes is +17.32 Mev, and the measured value of Q (from the energies of the incident protons and the emergent alpha particles) was found to be 17.33 MeV, thus demonstrating a genuine release of energy from the lithium atom at the expense of its mass. Many other nuclear transformations have since been studied in detail, and all validate the fundamental relationship between mass and energy.

Modern Linear Accelerators, as shown in Figure 4-9, introduce hydrogen ions (protons, deuterons, tritons) at the beginning of a series of drift tubes of varying lengths and separated by a series of fixed gaps, across which an alternating voltage is applied. Hydrogen ions are thus accelerated as they cross each gap, after which they enter successively longer drift tubes where their time of travel is just enough for the voltage to change as they enter the next gap to receive another "kick", thus increasing their energy. The length of the drift tubes increases along the particle path because the accelerated particles are travelling faster and faster as they proceed down the line.

Cyclotrons accelerate charged particles based on the curvilinear deflection they receive in moving through a magnetic field. As shown in Figure 4-10, a source of ionized protons, deuterons, or other atoms is introduced in the center of a gap which has an alternating electric field placed across it. The field gives the positively charged particle a "kick" towards the negatively charged pole which has a slit through which the particle passes into an evacuated zone called a "dee" because of its shape. A strong magnetic field perpendicular to the path of the particle deflects it into a circular path, bringing it back to the gap on the other side. During this passage, the electric field is reversed by the alternating voltage, and when the particle re-enters the gap, it receives another "kick" before it passes into the other "dee" where it is deflected once again back to the gap which by now has yet another reversed electric field. This series of field reversals at the gap provides a series of "kicks," increasing the velocity (and energy) of the particle at each pass through the gap. The ever-increasing circular path is convenient because it allows the particle to travel greater and greater distances in just the time required for the alternating field to switch. When the desired energy is reached, another magnetic field is used to draw off the accelerated particles and direct them to the target.

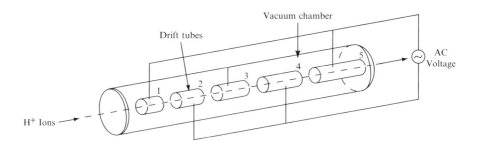

Figure 4-9. Schematic of a linear accelerator for protons and deuterons.

Magnetic Field

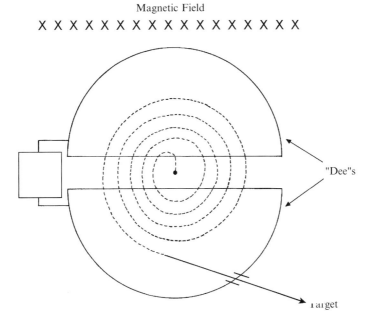

Figure 4-10. Schematic of cyclotron with a magnetic field directed into the paper. Ions introduced into the gap are accelerated numerous times by the alternating voltage and travel in ever-increasing circles in the "dee" space before being deflected toward a target.

Proton–Alpha Reactions can be used to produce radioactive substances for use in nuclear medicine by accelerating protons to high energies (on the order of 20 MeV) typically in a cyclotron. Two (p, α) reactions which produce positron emitters are

$$^{14}_{7}\text{N} +^{1}_{1}\text{H} \rightarrow \left[^{15}_{8}\text{O}\right] \rightarrow^{11}_{6}\text{C} +^{4}_{2}\text{He}$$
$$^{16}_{8}\text{O} +^{1}_{1}\text{H} \rightarrow \left[^{17}_{9}\text{F}\right] \rightarrow^{13}_{7}\text{N} +^{4}_{2}\text{He}$$

Proton–Neutron (p, n) Reactions can be used to transmute certain target elements to radioactive products with the production of neutrons. The effect of these transmutations is to increase the charge on the target nucleus by one unit, moving it above the line of stability on the Chart of the Nuclides with no change in the mass number. Examples of such reactions are

$$^{11}_{5}\text{B} +^{1}_{1}\text{H} \rightarrow \lfloor^{12}_{6}\text{C}\rfloor \rightarrow^{11}_{6}\text{C} +^{1}_{0}\text{n} + Q$$
$$^{18}_{8}\text{O} +^{1}_{1}\text{H} \rightarrow \lfloor^{19}_{9}\text{F}\rfloor \rightarrow^{18}_{9}\text{F} +^{1}_{0}\text{n} + Q$$

When these reactions are used to produce ^{11}C and ^{18}F, it is necessary to provide shielding for the emitted neutrons.

Proton–Gamma (p, γ) Reactions produce new products in which the target nucleus is left in an excited state which is relieved by emission of a gamma photon. Examples of (p,γ) reactions are:

$$\,^{7}_{3}\mathrm{Li} + \,^{1}_{1}\mathrm{H} \rightarrow \lfloor\,^{8}_{4}\mathrm{Be}\rfloor \rightarrow\,^{8}_{4}\mathrm{Be} + \gamma + Q$$

$$\,^{12}_{6}\mathrm{C} + \,^{1}_{1}\mathrm{H} \rightarrow \lfloor\,^{13}_{7}\mathrm{N}\rfloor \rightarrow\,^{13}_{7}\mathrm{N} + \gamma + Q$$

Proton–gamma interactions with lithium are particularly important because they yield photons with an energy of 17.32 MeV, which is far more energetic than those from naturally radioactive substances. These high-energy photons can in turn be used to produce nuclear interactions by a process called photo-disintegration.

Deuteron–Alpha (d, α) Reactions use deuterons, which like protons also have the advantage of a single + charge, allowing them to penetrate target nuclei more readily than alpha particles. They can also be accelerated up to energies of several MeV, typically in a cyclotron, to produce reactions such as

$$\,^{20}_{10}\mathrm{Ne} + \,^{2}_{1}\mathrm{H} \rightarrow \lfloor\,^{22}_{11}\mathrm{Na}\rfloor \rightarrow\,^{18}_{9}\mathrm{F} + \,^{4}_{2}\mathrm{He} + Q$$

The product, $^{18}\mathrm{F}$, is radioactive with a 109.77-min half-life and is widely used in nuclear medicine. These reactions are usually exoergic, i.e., they have positive Q-values.

The $^{11}\mathrm{C}$, $^{13}\mathrm{N}$, and especially $^{18}\mathrm{F}(T_{1/2} = 109.77\mathrm{min})$ produced in these reactions are positron emitters and are used for diagnosis of various disorders by tagging them onto pharmaceuticals that will take them to the region of interest in a patient. Since the positron emitters produce annihilation radiation, the differential absorption patterns of these photons can provide important medical information through a process known as positron emission tomography (PET). Since many of the light-element positron emitters are short lived, they need to be produced near where they will be used, and cyclotrons are effective for such purposes.

Neutron Interactions

Neutrons have proved to be especially effective in producing nuclear transformations because they have no electric charge and are not repelled (as are protons, deuterons, or alpha particles) by the positively charged nuclei of target atoms. Nuclear reactors provide copious quantities of neutrons; however, other sources are available from radioactive substances mixed with beryllium and/or deuterium or from (d,n) reactions produced in accelerators, for example

$$^{9}\mathrm{Be}(\alpha,\ n)^{12}\mathrm{C} \quad ^{9}\mathrm{Be}(d,\ n)^{10}\mathrm{B} \quad ^{2}\mathrm{H}(d,\ n)^{3}\mathrm{He} \quad ^{2}\mathrm{H}(\gamma,\ n)^{1}\mathrm{H}$$

And, some elements, notably $^{252}\mathrm{Cf}$, undergo spontaneous fission (Figure 3-20 and 3-21) which releases neutrons in sufficient numbers for use as a neutron source. Neutrons are generally classified according to their energy as thermal neutrons (0.025 eV) and those above thermal energy as intermediate or fast neutrons. Slow neutron interactions with target nuclei are quite effective in producing radioactive activation products.

Neutron activation products can be produced intentionally by placing the appropriate target material in a neutron flux, or such production can occur unintentionally as the byproduct of operating a source that emits neutrons. For example, the structural materials in a nuclear reactor contain various metals such as cobalt, which can be activated to ^{60}Co by the reaction:

$$^{59}\text{Co} + {}^1_0\text{n} \rightarrow [^{60}\text{Co} *] \rightarrow {}^{60}\text{Co} + \gamma + Q$$

which can be written as $^{59}\text{Co}(\text{n}, \gamma)^{60}\text{Co}$ and is known as a radiative capture reaction in which the excited nucleus emits its excess energy as gamma radiation.

Radiative Capture (n, γ) Reactions are the most common neutron interactions and they are quite useful in producing new products for research and medical use. The target nuclei are increased by one mass unit due to the addition of the neutron, and the Q-values are always positive. These radiative capture reactions shift the target atom to the right of the line of stability on the Chart of the Nuclides and the element remains the same since the Z does not change. One such reaction that occurs for air surrounding a neutron source is

$$^{40}_{18}\text{Ar} + {}^1_0\text{n} \rightarrow \left[{}^{41}_{18}\text{Ar*} \right] \rightarrow {}^{41}_{18}\text{Ar} + \gamma + Q$$

Other typical (n, γ) reactions are

$$^{115}_{49}\text{In} + {}^1_0\text{n} \rightarrow \left[{}^{116}_{49}\text{In*} \right] \rightarrow {}^{116}_{49}\text{In} + \gamma + Q$$

$$^{202}_{80}\text{Hg} + {}^1_0\text{n} \rightarrow \left[{}^{203}_{80}\text{Hg*} \right] \rightarrow {}^{203}_{80}\text{Hg} + \gamma + Q$$

Other (n, γ) reactions are used to produce radionuclides used in nuclear medicine, for example

$$^{98}_{42}\text{Mo} + {}^1_0\text{n} \rightarrow \left[{}^{99}_{42}\text{Mo*} \right] \rightarrow {}^{99}_{42}\text{Mo} + \gamma + Q$$

$$^{132}_{54}\text{Xe} + {}^1_0\text{n} \rightarrow \left[{}^{133}_{54}\text{Xe*} \right] \rightarrow {}^{133}_{54}\text{Xe} + \gamma + Q$$

^{99}Mo is the source of ^{99m}Tc, the most commonly used radionuclide in nuclear medicine; and ^{133}Xe is used for lung ventilation studies.

Transuranium (TRU) Elements are also produced by (n, γ) reactions in which uranium is bombarded with neutrons (first done by Fermi in 1934) to make and identify new elements with nuclear charge greater than 92. Neutron capture in ^{238}U yields

$$^{238}_{92}\text{U} + {}^1_0\text{n} \rightarrow \left[{}^{239}_{92}\text{U*} \right] \rightarrow {}^{239}_{92}\text{U} + \gamma + Q$$

The ^{239}U product undergoes radioactive transformation by beta-particle emission to produce ^{239}Np which does not exist in nature. Radioactive transformation of ^{239}Np by beta-particle emission forms an isotope of another new element with $Z = 94$, or ^{239}Pu($T_{1/2} = 24{,}100$ y) which can be fissioned in nuclear reactors and nuclear weapons.

Neutron–Alpha (n, α) Reactions produce a new product and an alpha particle, for example

$$_3^6\text{Li} + {}_0^1\text{n} \rightarrow \left[{}_3^7\text{Li}\right] \rightarrow {}_1^3\text{H} + {}_2^4\text{He}$$
$$_5^{10}\text{B} + {}_0^1\text{n} \rightarrow \left[{}_5^{11}\text{B}\right] \rightarrow {}_3^7\text{Li} + {}_2^4\text{He}$$

The cross sections for these two reactions are relatively high, and both are often used to detect neutrons. In one method, an ionization chamber is lined with boron, usually BF_3, which provides efficient capture of neutrons. The liberated alpha particles produce significant ionization which can be collected and amplified to produce an electronic signal allowing detection. The (n, α) reaction with ^{6}Li is also used to produce tritium, which is a key component of nuclear weapons and fusion energy.

NEUTRON ACTIVATION CALCULATIONS

The production of neutron activation products, both radioactive and stable, is shown schematically in Figure 4-11 in which a flux of ϕ neutrons/cm$^2 \cdot$ s is incident upon a thin target that initially contains N_1^0 atoms producing new atoms of N_2 (or N_3 if charged particle emission – CPE – occurs) at a rate

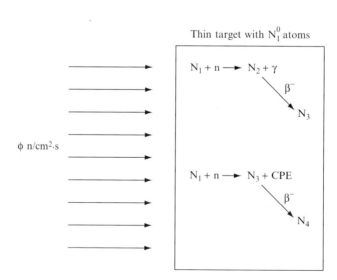

Figure 4-11. Schematic of neutron irradiation of a thin (greatly amplified) target material to produce new atoms of a radioactive product, N_2, by (n, γ) reactions or N_3 by CPE.

$$\text{Production of } N_2 = \phi\sigma N_1^0$$

where σ is the neutron-absorption cross section in barns for neutrons of a given energy.

As soon as atoms of N_2 are formed they can be removed by radioactive transformation and/or activation to a new product by the neutron flux; thus,

$$\text{Removal of } N_2 = \lambda_2 N_2 + \phi\sigma_2 N_2$$

If activation of N_2 can be ignored, which is usually (but not always) the case, the rate of removal of N_2 atoms is due only to radioactive transformation; therefore, for thin foils containing N_1^0 atoms, the rate of change of new atoms of N_2 with time is the rate of production minus the rate of removal, or

$$\frac{dN_2}{dt} = \phi\sigma_1 N_1^0 - \lambda_2 N_2$$

which can be solved by multiplying through by the integrating factor $e^{\lambda_2 t}$ and performing the integration to obtain

$$N_2(t) = \frac{\phi\sigma_1 N_1^0}{\lambda_2}(1 - e^{-\lambda_2 t})$$

and since Activity $= \lambda N$, the activity of the product is

$$A_2(t) = \lambda_2 N_2 = \phi\sigma_1 N_1^0(1 - e^{-\lambda_2 t})$$

Because of the factor $(1 - e^{-\lambda t})$ the activity of the product will reach an equilibrium (or saturation) value of $A_2(t) = \sigma\phi N$ in 7 to 10 half-lives as shown in Figure 4-12. These equations assume insignificant burn-up of target atoms and no significant alteration of the neutron flux over the target, both of which are conservative but realistic assumptions when thin foils are used for target materials.

Neutron cross sections are shown for each of these types of reactions on the Chart of the Nuclides as illustrated in the excerpts from it in Figures 4-13 and 4-14. Radiative capture (n, γ) cross sections are designated by σ_γ; CPE cross sections are designated as σ_p, σ_α, etc., where the subscript denotes the emitted particle. If the isotope fissions after absorption of a neutron, its cross section is denoted as σ_f.

Example 4-4: A thin foil of 0.1 g of tungsten is placed in a neutron flux of $10^{12} \text{n/cm}^2 \cdot \text{s}$ (a) What is the activity of ^{187}W after 24 h? (b) At saturation?
Solution: (a) The reaction equation to produce ^{187}W is

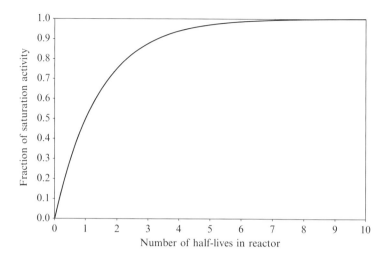

Figure 4-12. Ingrowth of activation product A_2 with time in a target containing N_1^0 atoms shown as a fraction of equilibrium activity.

$$^{186}W + {}_0^1 n \rightarrow [{}^{187}W] \rightarrow {}^{187}W + \gamma + Q$$

Thus only the ^{186}W portion of the foil yields ^{187}W, and from the Chart of the Nuclides the atom percent of ^{186}W in natural tungsten is 28.41%. The number of atoms of ^{186}W is

$$N_1^0 = \frac{0.1\,g \times 6.022 \times 10^{23}\ \text{atoms/mol}}{183.85\,g/mol} \times 0.2841 = 9.31 \times 10^{19}\ \text{atoms}$$

and the activation cross section $\sigma_\gamma = 38$ barns (see Figure 4-14, an excerpt of the Chart of the Nuclides). The activity at 24 h is

$$A(24h) = \phi \sigma N_1 (1 - e^{\frac{-\ln 2}{23.9\,h}\,24\,h})$$
$$= 10^{12}\,\text{n/cm}^2 \cdot s \times 38 \times 10^{-24}\,\text{cm}^2 \times 9.31 \times 10^{19}(1 - 0.5)$$
$$= 1.77 \times 10^9\,\text{d/s} = 47.8\text{mCi}$$

(b) At saturation the exponential term will approach zero (at about 10 half-lives or 240 h), and the saturation activity will be

$$A_{sat} = \phi \sigma N_1^0 = 3.54 \times 10^9\ \text{d/s} = 95.7\,\text{mCi}$$

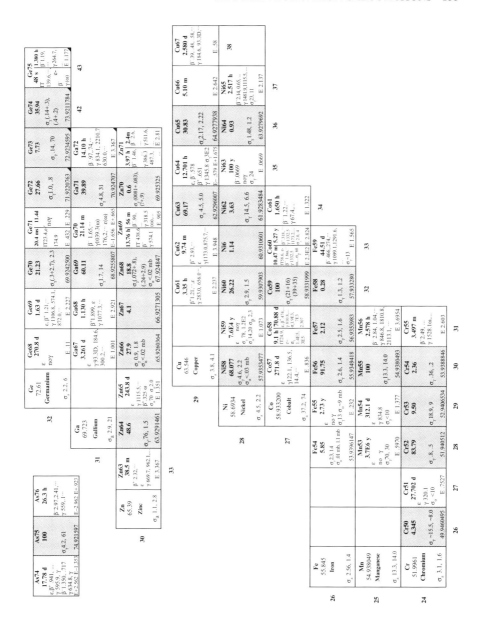

Figure 4-13. Excerpt from the Chart of Nuclides of light elements ($Z = 23$ to 33) that form activation products when irradiated with neutrons.

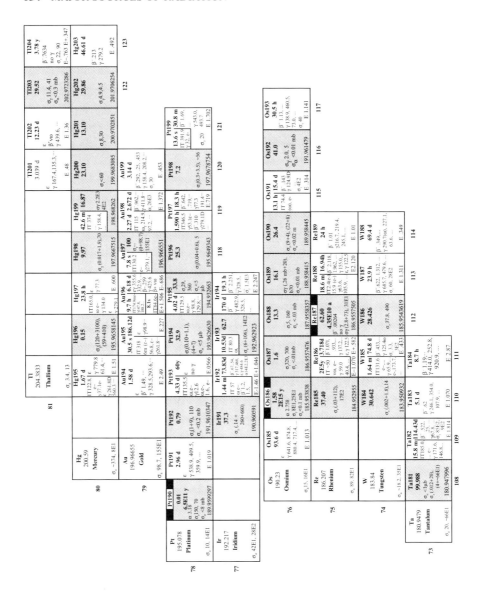

Figure 4-14. Excerpt from the Chart of Nuclides of heavy elements ($Z = 74$ to 81) that form activation products when irradiated with neutrons.

FISSION PRODUCT RADIOACTIVITY

The discovery of fission occurred from a concerted effort by several researchers, including Enrico Fermi, to produce transuranic elements by irradiating uranium with neutrons and other particles. Radioactive products with many different half-lives were found including barium-like radioactive products in precipitates that were thought to be artifacts by the physicists due to their rudimentary chemistry techniques. When Hahn, an expert chemist, and Strassmann showed that the products of neutron irradiation of uranium indeed contained radio-active barium, Hahn's former assistant Lise Meittner and her nephew Otto Frisch decided to apply Bohr's recent liquid drop model of the nucleus (shown schematically in Figure 4-15) to the results; and there it was! Uranium had split into lighter fragments, one of which was an isotope of barium which subse-quently transformed to lanthanum. They termed it nuclear fission because it resembled the division of cells (called fission) in biology. As soon as this insight was announced, physicists all over the world used ion chambers to observe the high rates of ionization produced by the fission fragments produced when uranium was irradiated with neutrons. Interestingly, Fermi had wrapped his uranium target in foil to just absorb the low-energy alpha particles emitted by uranium, with the expectation that any TRUs produced would emit higher energy alpha particles that would be detected. Unfortunately, the foil also absorbed the fission fragments; if he had not done so, this foremost physicist would also have been the discoveror of fission, a phenomenon to which he contributed so much. The discovery of nuclear fission not only changed physics, but because of the timing and location of its discovery, the world as well.

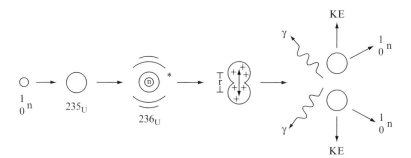

Figure 4-15. An absorbed neutron in ^{235}U produces an elongated nucleus of ^{236}U due to added excitation energy. If the centers of the "halves" reach a separation distance, r, fission is likely to occur, which it does 85% of the time in ^{235}U; if not, the nucleus becomes an atom of ^{236}U, returns to its original shape, and the excitation energy is emitted as a gamma photon, a process that occurs 15% of the time.

Nuclear fission can occur in various materials but is most important for ^{235}U, ^{238}U, and ^{239}Pu. As shown in Figure 4-15, absorption of a neutron by a ^{235}U nucleus produces an atom of ^{236}U with excitation energy which can be relieved by emission of a gamma photon or by fission; however, fission is much more likely and occurs 85% of the time. The excitation energy added by the neutron causes the ^{236}U nucleus to elongate and oscillate and if the asymmetrical "halves" separate by a distance, r, then repulsive forces exerted by the positive charges in each will cause the two parts to separate, or fission. A typical reaction, as shown in Figure 4-16, produces end products of ^{95}Mo and ^{139}La plus two high speed neutrons and prompt gamma rays; these are released almost instantaneously. The Q-value of the reaction is about 200 MeV, most of which is released as kinetic energy of the two products. Fissionable nuclei are very neutron rich and the fission products will be too; thus, they will emit beta particles and gamma rays until they reach stable end products (Figure 4-16).

NUCLEAR REACTOR DESIGNS

Nuclear reactors contain a number of basic systems as shown in Figure 4-17, enclosed in a vessel with various components and subsystems to enhance fissioning in the fuel, to control its rate, to extract heat energy, and to provide protection of materials and persons. Large amounts of radioactive fission products and activation products are produced which can accumulate in some reactor systems where they can cause radiation exposures of workers; some of the products may be released during routine operations, incidents, or in the form of radioactive wastes for potential radiation exposure of the public, depending on various pathways.

Reactor Fuel is UO_2 fabricated into ceramic pellets which have good stability and heat transfer characteristics. These pellets are loaded into thin zirconium alloy (zircalloy) tubes to make a fuel rod. A typical 1000 MWe plant contains about 50,000 such rods.

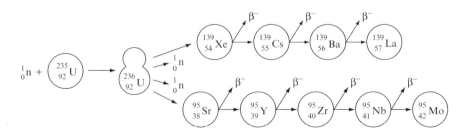

Figure 4-16. Uranium-235 fission and the series of radioactive transformations by beta emission of the fission products to stable end products when mass numbers 95 and 139 are produced.

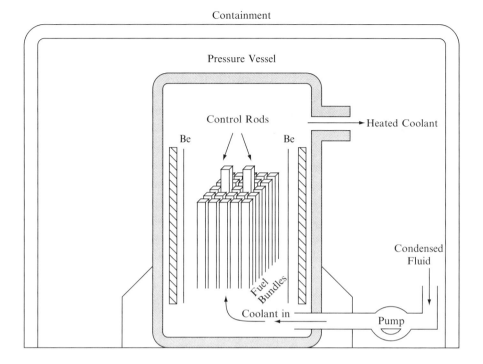

Figure 4-17. Schematic of major components of a nuclear reactor.

Coolant is required to remove the heat produced by fission and to maintain the fuel well below its melting point. The heat extracted into the coolant is used to produce steam either directly or through a heat exchanger. Coolant is typically water but can be a gas, a liquid metal, or another good heat transfer material.

Moderators are usually heavy water, light water, or graphite, and are used to moderate neutrons to thermal energies. A light-water moderator can also serve to cool the fuel and extract the heat produced; however, its use requires that the fuel be enriched to 3 to 4% to overcome neutron absorption by hydrogen in the water.

A reflector made of a light metal such as beryllium ($Z = 3$) is provided next to the core to reflect neutrons that otherwise might be lost back into the core where they can cause fission. Ordinary water is also a good reflector, and light-water-cooled reactors use this effect in managing the neutron population.

Control rods are movable pieces of cadmium or boron which are used to absorb and stabilize the neutron population. Withdrawing these rods increases the multiplication factor and hence the power level; insertion decreases it.

A pressure vessel encloses the entire fission reaction and must be strong enough to withstand the stresses of pressure and heat; a *thermal shield* is provided to absorb radiation and reduce embrittlement of the vessel; and a *biological shield* is placed outside the pressure vessel to reduce radiation exposure levels.

A containment structure encloses the entire reactor system to prevent potential releases of radioactivity to the environment, especially during incidents when fission products may escape. This may include a primary containment around the steam supply system and a secondary containment associated with the reactor building.

The Pressurized Water Reactor

The pressurized water reactor (PWR) was one of the first reactors designed to produce power, originally for use in submarines. As indicated in Figure 4-18, water enters the pressure vessel at a temperature of about 290 °C, flows down around the outside of the core where it helps to reflect neutrons back into the core, passes upward through the core where it is heated, and then exits from the vessel with a temperature of about 325 °C. This primary coolant water is maintained at a high pressure (2200 psi) so it will not boil.

Because the primary coolant is not allowed to boil, steam for the turbines must be produced in a secondary circuit containing steam generators (also shown in Figure 4-18). The heated coolant water from the reactor enters the steam generator at the bottom and passes upward and then downward through several thousand tubes each in the shape of an inverted U. The outer surfaces of

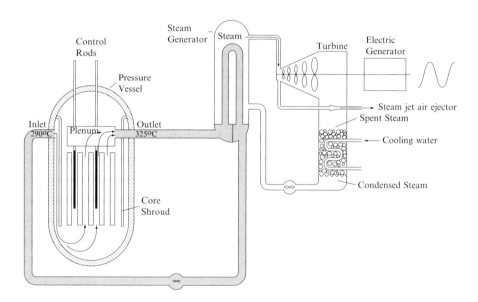

Figure 4-18. Schematic of a pressurized water reactor (PWR).

these tubes are in contact with lower-pressure and cooler feed water which extracts heat, boils, and produces steam that is routed to the turbine. The spent steam is condensed and pumped back to the steam generators to complete the secondary circuit.

Where the size of the reactor is an important consideration, as it is in a submarine, the fuel is enriched to over 90% in ^{235}U (often called "highly enriched uranium," or HEU) which makes it possible to have a smaller core and pressure vessel. Highly enriched uranium is expensive, however, and for stationary power plants where size is less of an issue, the fuel is only slightly enriched (from 2 to 4%). The fuel is uranium dioxide, UO_2, which is a black ceramic material with a high melting point of approximately 2800 °C; it is fabricated into small cylindrical pellets, about 1 cm in diameter and 2 cm long, which are loaded into sealed zircalloy tubes (called rods) about 4 m long. The rods are designed to contain the fission products, especially fission product gases, that are released from the pellets during reactor operation. The fuel rods are loaded into fuel assemblies and kept apart by various spacers to prevent contact between them. Otherwise, they may overheat and cause fission products to be released. In this context, the fuel tubes are also known as the fuel cladding. Fuel of this type will last about 3 years in the PWR before it is replaced and will deliver about 30,000 MW-d of thermal energy per ton of U.

The Boiling Water Reactor

The boiling water reactor (BWR) uses light water for both coolant and moderator, as shown in Figure 4-19. The water boils within the reactor which produces steam that is routed directly to the turbines, which can be a major advantage because it eliminates the need for a separate heat transfer loop. Since the BWR is a direct cycle plant, more heat is absorbed to produce steam than in a PWR; however, since the steam for the turbine is produced in the coolant which is in direct contact with the fuel, the water becomes radioactive in passing through the reactor core, requiring shielding of the steam piping, the turbines, the condenser, reheaters, pumps, pipings, etc. Cooling water in a BWR enters a chamber at the bottom and flows upward through the core. Voids are produced as the water boils, and by the time it reaches the top of the core the coolant is a mixture of steam, liquid water, and entrained radioactive isotopes of xenon, krypton, iodine, and other halogens in addition to disassociated oxygen and hydrogen. These gases will be exhausted by the steam jet air ejector after they pass through the turbine and must be held for decay before release to the environment.

The fuel in a BWR is slightly enriched UO_2 pellets in sealed tubes, and the core configuration is similar to that of a PWR. However, BWR control rods are always placed at the bottom (in a PWR they are at the top). Since steam voids exist in the coolant at the upper portion of the core, the effect of a given control-rod movement on the fission reactions and overall reactor power will be greater in the lower part of the core. A typical BWR produces saturated steam at about 290 °C and 7 MPa and has an overall efficiency of 33 to 34%.

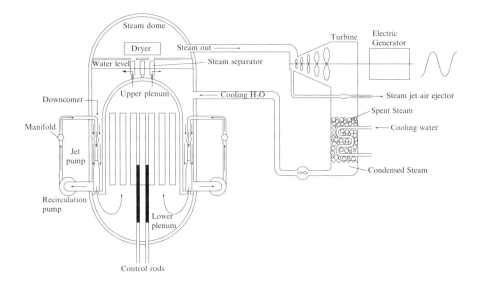

Figure 4-19. Schematic of boiling water reactor.

The pressure in a BWR is approximately half that in a PWR so that thinner pressure vessels can be used, but because the power density (watts/cm^3) is smaller, the pressure vessel must be made larger which tends to equalize the costs for each.

Radioactivity Releases from Nuclear Reactors

Release of radioactivity during routine reactor operations is a complex process governed by reactor type, power level and operating history, fuel performance and overall system cleanliness, and the particular waste processing and cleanup systems used. The fuel is the most important barrier to release of fission products; it consists of ceramic pellets that are loaded into fuel rods of zircalloy tubing and welded on each end, as shown in Figure 4-20. As fissioning proceeds, gaseous and volatile fission products migrate slowly through the ceramic matrix into the "gap" between the pellets and the zircalloy tube walls. As more and more atoms of these products are produced, they build up pressure which may drive them through any imperfections in the zircalloy cladding. The rods are thin (0.024 to 0.034 inches) to promote heat transfer and minimize effects on neutron flux, and because of this and the large number of rods some imperfections are to be expected. Since these imperfections and the stress of heating and cooling the rods could cause volatile fission products to leak into the coolant, fuel manufacture is subjected to intense quality control.

Reactor design and type also influence the release of radioactive materials to internal systems and the environment. The direct-cycle design of the BWR

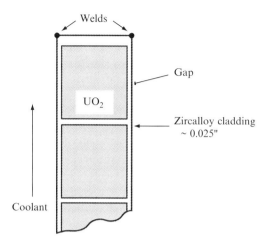

Figure 4-20. UO$_2$ fuel pellets in zircalloy cladding.

introduces any entrained radionuclides into the various cleanup systems whereas these same materials would remain in the primary coolant of the PWR and would be released only if steam generator leaks occur. These features are reflected in the capacities and designs of auxiliary systems for the two reactor types. These general patterns of environmental releases for BWRs and PWRs and the influence of these elements are shown schematically in Figure 4-21.

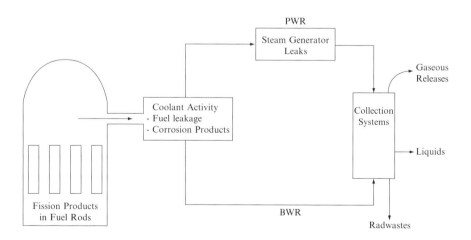

Figure 4-21. Major routes by which mobile fission and activation products are produced and released to the environment from boiling water (BWR) and pressurized water (PWR) reactors.

FISSION PRODUCT INVENTORIES

Thermal neutron fission of ^{235}U yields over 60 primary fission products that are very neutron rich and are formed far below the line of stable nuclides. They are thus very radioactive and emit several negatively charged beta particles before reaching a stable end product. The mass number remains the same even though the element changes, thus "fission product chains" are formed that contain well over 1200 different products, several of which are shown in Figure 4-22. Several important fission product radionuclides such as ^{131}I, ^{133}Xe, and ^{140}Ba-La are formed well down the chains for mass numbers 131, 133, and 140, respectively.

The fission yield of a given mass number is expressed as a percentage of the fissions that occur. These are listed in Figure 4-22 for ^{235}U (below each product) and for ^{239}Pu (shown in parentheses below each). Each of the listed fission yields for the element is given in cumulative percent yield, and it is important to note that each yield value may contain a portion due to independent yield production (the independent yield for each nuclide in a chain can be obtained by subtracting the cumulative yield of the most immediate precursor.) If precursors are short lived, as is often the case, the cumulative yield can be used directly in calculations of fission product inventories.

Figure 4-23 is a plot of the cumulative fission yield versus mass number for ^{235}U and ^{239}Pu. The area under each curve is 200% since the fissioning of 100 fissionable atoms yields 200 new atoms. Figure 4-23 also shows that most fission is asymmetrical and that the masses of the major fission products consist of a "light" group, dominated by mass numbers from 80 to 110, which includes the important fission products ^{85}Kr, ^{90}Sr, ^{95}Zr-Nb, and ^{99}Tc and a "heavy" group with mass numbers from 130 to 150 that includes ^{131}I, ^{137}Cs, and xenon nuclides. The most abundant mass numbers are 95 (with a percentage yield of about 6.5%) and 134 (at 7.87%).

Many fission products of interest to radiation protection have long half-lives, and because of this their effective yields are the same as the cumulative fission yield. Several of these and their cumulative thermal fission yields in ^{235}U are:

Nuclide	^{235}U thermal fission yield (%)
^{85}Kr	0.2834
^{90}Sr	5.7823
^{99}Tc	6.1092
^{129}I	0.5434
^{137}Cs	6.1288

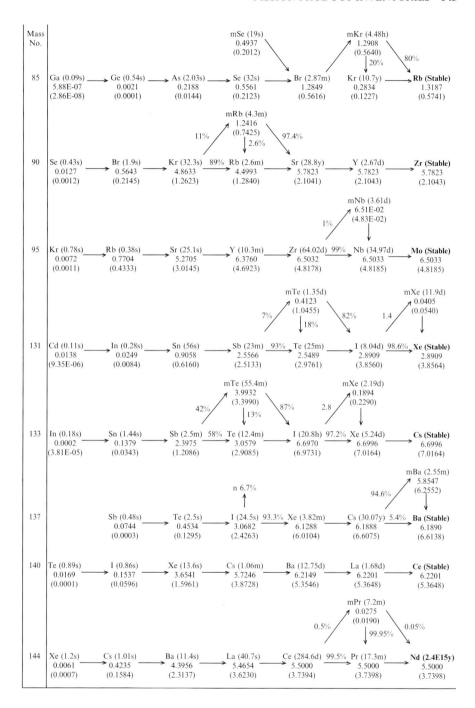

Figure 4-22. Percent fission yields of different mass numbers due to thermal fission of ^{235}U and ^{239}Pu.

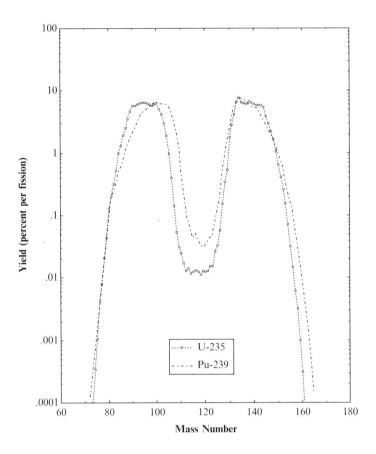

Figure 4-23. Thermal neutron fission product chains for selected mass numbers.

Calculation of Fission Product Inventories

The number of atoms of each fission product that is produced is directly related to the number of fissions that occur which is a function of the power level of a reactor or the explosive yield of a nuclear explosion. Since each fission of a ^{235}U atom yields about 200 MeV, 3.12×10^{16} fission/s are required to maintain a one-megawatt thermal power level, and it takes 2.7×10^{21} fissions to produce one megawatt-day of thermal energy. For nuclear weapons based on fission, one kiloton of explosive yield corresponds to 1.19×10^{23} fissions.

The total number of fissions (or fission rate) and the fission yield, Y_i, of the particular product provide the number of atoms of any given fission product, N_i

$$N_i = \# \text{ of fissions} \times Y_i$$

Tabulations of Y_i account for the fact that two new atoms are formed from each atom that fissions; however, Y_i is usually listed as a percentage and it is necessary to convert it to a fractional value when calculating the number of fission product atoms produced. It is also necessary to account for removal of atoms of N_i by radioactive transformation or activation in determining the inventory of atoms of N_i at any given time, or

$$\frac{dN_i}{dt} = \text{(amount produced)} - \text{(amount removed)}$$

For a nuclear reactor that has a power level of P megawatts of thermal energy (MWt), the fission rate is 3.12×10^{16} fissions/s per MW(t); therefore, the rate of change of N_i is

$$\frac{dN_i}{dt} = [(3.12 \times 10^{16} f/s \cdot MW) \times P(MW) \times Y_i] - [\lambda_i N_i - \sigma_i \phi N_i]$$

If neutron activation of the product is minimal (i.e., $\sigma_i \approx 0$) and if the number of atoms of fission product i is zero ($N_i = 0$) at $t = 0$, the number of atoms of N_i at any time t after operation begins at a sustained power level P [in MW(t)] is

$$N_i(t) = \frac{3.12 \times 10^{16} PY_i}{\lambda_i}(1 - e^{-\lambda_i t}) \text{ atoms}$$

and since activity $= \lambda_i N_i$

$$A_i(t) = 3.12 \times 10^{16} \, P Y_i (1 - e^{-\lambda_i t}) \, t/s$$

or

$$A_i(t) = 8.43 \times 10^5 \, P Y_i \, (1 - e^{-\lambda_i t}) \, \text{curies}$$

Example 4-5: Calculate the activity of ^{131}I in a 3,000 megawatt reactor (thermal) at (a) 8 days after startup, and (b) at saturation.
Solution: (a) From Figure 4-22, the cumulative fission yield of ^{131}I in ^{235}U is 0.029 (2.9%). By assuming that all precursors have reached equilibrium with ^{131}I, the activity of ^{131}I at 8 days is

$$A(8d) = 8.43 \times 10^5 (3000)0.029(1 - e^{-\frac{0.693*8}{8}})$$
$$= 3.67 \times 10^7 Ci^{131}I$$

(b) and at saturation

$$A(\text{saturation}) = 8.43 \times 10^5 (3000)0.029(1 - 0)$$
$$= 7.33 \times 10^5 Ci^{131}I$$

Use of the cumulative fission yield of the nuclide of interest is appropriate for determining the amount of a fission product because the half-lives of all the precursors are short compared to the reactor operating time; however, this is not always the case. For example, short-lived ^{140}La($T_{1/2} = 1.678$ d) in the chain for mass number 140 must await the transformation of longer-lived ^{140}Ba($T_{1/2} = 12.75$ d), and for operating periods on the order of days ^{140}La will not be in equilibrium. After several months of reactor operation, ^{140}La will have reached an effective equilibrium with ^{140}Ba and it is then reasonable to presume that the activity of ^{140}La is the same as that of ^{140}Ba.

SUMMARY

A vast array of radiation sources exists for potential exposure of workers and the public. The major ones can be summarized as

- X rays which are produced primarily for medical purposes but also for use in industry and research. These rays emanate from the electron field of target atoms and consist of photons of various energies typically ranging from a few tens of keV up to several MeV, although most uses occur between 70 and 140 keV.
- Naturally occurring radiation associated with cosmic rays and naturally occurring primordial radioactive materials, many of which are incorporated into environmental media. Uranium and thorium head long chains of radioactive materials that have masses well above the heaviest stable elements and contain several subseries, e.g., radon and its products. Many of these massive radionuclides exist in minerals and ores and are characterized as alpha emitters but they also emit beta particles and gamma rays or produce products that do. Lighter primordial materials also exist, principally ^{40}K which is incorporated into tissue and exposes all living beings. These lighter materials and uranium and thorium were formed when the earth was formed and persist because of their long half-lives.
- Tritium, ^{14}C, ^{7}Be, ^{22}Na, and ^{36}Cl which exist as part of a group of shorter-lived naturally occurring radionuclides due to the continual bombardment of the atmosphere by cosmic rays. Most of these are neutron-rich and transition by beta emission, but some are proton-rich nuclides that transition by positron emission and electron capture. These are in equilibrium in the environment because of a steady rate of cosmic ray interactions in the atmosphere. Collectively, the various naturally occurring radiations and radioactive materials produce a dose to persons at sea level of about 100 mrem (1 mSv) per year, excluding radon and its products.
- Activation products which are produced by bombarding various materials primarily with neutrons, but also alpha particles, protons, deuterons, and to a lesser degree other projectiles (e.g., tritons and other charged ions of light elements). These activation products can be produced both above the line

of stable nuclei (e.g., ^{11}C, ^{13}N, ^{18}F from proton and deuteron interactions) and below it (^{14}C, ^{36}Cl, ^{51}Cr, ^{60}Co, etc., from neutron interactions). Transuranic radionuclides are a special category of activation products that are artificially produced by successive bombardment of uranium with neutrons. Some of these radionuclides (^{239}Pu, ^{241}Pu) are fissionable with neutrons and/or undergo spontaneous fission (^{252}Cf), and since they have masses well above stable lead and bismuth, they are also radioactive.

- Fission products that are produced when ^{235}U and ^{239}Pu atoms split. These comprise over 1200 radioactive nuclides and they are neutron-rich because the fissioning nuclei are neutron-rich; most are short lived, although a number of long-lived radionuclides (e.g., ^{85}Kr, ^{90}Sr, ^{99}Tc, ^{129}I, ^{137}Cs, etc.) are produced that require attention in radioactive waste management. Many of the products of fission are used in medical and research activities and are thus controlled in these environments for their potential to expose workers and, if released, the public.

In sum, radiation and radionuclides are ubiquitous, and exposure to various ones, especially those that occur naturally, are certain. Exposure from other sources is also possible to varying degrees depending on circumstances. Awareness, living habits, and different policies and controls can modify some of these exposures somewhat, but not completely. The challenge to a modern technological society is to achieve an optimum balance between these possibilities and their circumstances.

PROBLEMS

1. Calculate the maximum activity (d/s), assuming no source depletion, that could be induced in a copper foil of 100 mg exposed to a thermal neutron flux of 10^{12}n/cm$^2 \cdot$ s. (*Note:* Natural copper consists of 69.1% ^{63}Cu and 30.9% ^{65}Cu).

2. A sample of dirt (or CRUD) irradiated for 4 h at 10^{12}n/cm$^2 \cdot$ s produces a disintegration rate of 10,000 dpm of ^{56}Mn. How much manganese is in the sample?

3. Measurement of ^{129}I can be done by neutron activation. Describe the nuclear reaction for the technique and recommend an irradiation time to optimize the sensitivity of the procedure using a constant thermal neutron flux.

4. Calculate the activity of ^{203}Hg and ^{197}Hg in 10 g of HgO irradiated uniformly in a research reactor at a flux of 10^{12}n/cm$^2 \cdot$ s for 10 days.

5. A nuclear reactor of 1000 MW(e) and a generation efficiency of 32% has run continuously for several months. What is the equilibrium level of ^{133}Xe in the core?

6. A nuclear reactor operates for 3 years at a power level of 3300 MW(t) (about 1000 MWe) after which all the fuel is removed (a problem core) and

processed soon after releasing all of the noble gases. How many curies of ^{85}Kr would be released to the earth's atmosphere?

7   What is the equilibrium inventory of ^{140}Ba in the core of a nuclear reactor with a power level of 1200 MW(e) and a generation efficiency of 33%? What is a reasonable estimate of the ^{140}La inventory? Explain.

8   High-level radioactive waste standards for spent nuclear fuel are based on inventories of various radionuclides. (a) What is the inventory of ^{137}Cs in 1000 metric tons of spent fuel based on a burnup of 33,000 megawatt-days per ton over 3 years? (b) What would be the ^{137}Cs inventory based on an average power level for the three-year period?

9   What is the inventory of ^{99}Tc for the fuel in problem 8? Show whether it is necessary to use average power level in this calculation, and explain why.

10 A 20 KT nuclear fission device was tested in the open atmosphere in country X. How many curies of ^{131}I would be produced and presumably available for release to the atmosphere?

5

RADIATION INTERACTIONS AND DOSE

"Bodies of things are safe 'till they receive a force which may their proper thread unweave."

– Lucretius, ~ 60 B.C.

When radiation is emitted, regardless of what type it is, it produces various interactions that deposit energy in the medium that surrounds it; if it occurs in the living tissue of individuals, the end-point effects will be biological changes, most of which, if not all, are undesirable. The interactions by which charged particles and photons are attenuated and absorbed are fundamental to the determination of radiation exposure and dose, the units used to define them, and calculations of radiation exposure and dose for various sources.

RADIATION DOSE AND UNITS

The term *radiation dose*, or simply *dose*, is defined carefully in terms of two key concepts: (1) the energy deposited per gram in an absorbing medium, principally tissue, which is the absorbed dose, and (2) the damaging effect of the radiation type, which is characterized by the term effective dose equivalent. A related term is radiation exposure, which applies to air only and is a measure of the amount of ionization produced by x and gamma radiation in air. Each of these is defined in the conventional system of units, which in various forms and refinements have been used for several decades, and in the newer SI system. SI units are gradually replacing the conventional units which continue to be firmly imbedded in governmental standards and regulations, at least in the US.

Radiation Absorbed Dose

Since different radiations penetrate to different depths in tissue, radiation dose has been specified in regulations for three primary locations:

- The shallow dose, which is just below the dead layer of skin which has a density thickness of $7 \, \text{mg/cm}^2$, or an average thickness of $70 \, \mu\text{m}$;
- The eye dose just below the lens of the eye with a density thickness of $300 \, \text{mg/cm}^2$; and
- The deep dose, which occurs at a depth of 1 cm in tissue (density thickness of $1000 \, \text{mg/cm}^2$) primarily to account for highly penetrating radiation such as x or gamma rays or neutrons.

The conventional unit for absorbed dose is the **rad**, which is an acronym for **r**adiation **a**bsorbed **d**ose, and is equal to the absorption of 100 ergs of energy in one gram of absorbing medium, typically tissue:

$$1 \text{rad} = 100 \, \text{ergs/g of medium}$$

The SI unit of absorbed dose is the gray (Gy) and is defined as the absorption of one joule of energy per kilogram of medium, or

$$1 \, \text{Gy} = 1 \, \text{J/kg}$$
$$= 100 \, \text{rads}$$

A milligray is abbreviated mGy and is 100 millirads, which is about the amount of radiation one receives from natural background, excluding radon. This is a convenient relationship for translating between the two systems of units.

Radiation Dose Equivalent

The dose equivalent, denoted by H, is defined as the product of the absorbed dose and a factor, Q, the quality factor, that characterizes the damage associated with each type of radiation, or

$$H(\text{dose equivalent}) = D(\text{absorbed dose}) \times Q(\text{quality factor})$$

In conventional units, the unit of dose equivalent is the rem which is calculated from the absorbed dose as

$$\text{rems} = \text{rads} \times Q.$$

and in SI units

$$\text{Sieverts} = \text{Gy} \times Q$$

where $Q = 1.0$ for x rays, gamma rays and electrons, $Q = 2$ to 10 for neutrons of different energies, and $Q = 20$ for alpha particles and fission fragments. The Sievert corresponds to 100 rems and is a very large unit for radiation dose equivalent; thus it is often necessary to abbreviate it to mSv or, in some cases, to μSv to describe the radiation dose equivalent (or rate) received by workers and the public.

The definition of dose equivalent is necessary because different radiations produce different amounts of biological damage even though the deposited energy may be the same. Biological effects depend not only on the total energy deposited but also on the way in which it is distributed along the path of the radiation. Radiation damage increases with the linear energy transfer (or LET) of the radiation; thus, for the same absorbed dose, the biological damage from high LET radiation (e.g., α-particles, neutrons, etc.) is much greater than from low LET radiation (β-particles, γ-rays, x rays, etc.).

Radiation Exposure—The Roentgen (R)

The unit of radiation exposure is the Roentgen (abbreviated as R), which is defined only for air and applies only to x rays and gamma rays up to energies of about 3 MeV. It can be conveniently measured directly by collecting the electric charge produced by x rays or gamma rays in air whereas that which occurs in a person cannot be measured. The R is not appropriate for describing energy deposition from particles or for energy deposition in the body.

The Roentgen was originally defined as "...that amount of x or gamma radiation such that the associated corpuscular emission produces in 0.001293 g of air (1 cm^3 of air at atmosphere pressure and 0°C) 1 electrostatic unit (1 esu $= 3.336 \times 10^{-10}$ coulomb) of charge of either sign." The ionization produced by the associated corpuscular emissions is due to photoelectric, compton, and if applicable, pair production interactions, and corresponds to

$$1\,R = \frac{1\,\text{esu} \cdot 3.336 \times 10^{-10}\,\text{C/esu}}{0.001293\,\text{g}} \times 10^3\,\frac{\text{g}}{\text{kg}}$$

$$= 2.58 \times 10^{-4}\,\text{C/kg of air}$$

The modern definition of the Roentgen (or R) is based on this value; i.e., 1 R is that amount of x or gamma radiation that produces 2.58×10^{-4} coulombs of charge in 1 kg of air. A milliroentgen, denoted as mR, is 0.001 R and the unit mR/h is conveniently used for exposure rate.

Since the Roentgen is determined in air, it is not a radiation dose, but with appropriate adjustment a measured value of exposure can be converted to dose. The production of an ion pair requires 33.97 eV on average; therefore, the R corresponds to an energy deposition in air, the absorbing medium, of 87.64 ergs/g of air (see Example 5-1).

Example 5-1: From the original definition of the Roentgen, determine the energy deposition in air associated with 1 R of exposure: (a) in ergs per gram of air, and (b) in Joules per kilogram of air.

Solution: (a) Since $1\,R = 1$ esu in air at STP (0.001293 g),

$$Exp(ergs/g) = \frac{1\,esu}{0.001293\,g} \times 33.97\,eV/ip \times \frac{1.6022 \times 10^{-12}\,erg/eV}{1.6022 \times 10^{-19}\,C/ip}$$
$$\times\, 3.336 \times 10^{-10}\,C/esu = 87.64\,erg/g \cdot R$$

(b) In SI units $1\,R = 2.58 \times 10^{-4}\,C/kg$

$$Exp(J/kg) = 2.58 \times 10^{-4}\,C/kg \times 33.97\,J/C = 8.764 \times 10^{-3}\,J/kg \cdot R$$

The interactions of photons in a medium are described by two specific terms: **Kerma** (**k**inetic **e**nergy **r**eleased in **ma**terial) which represents the kinetic energy transferred to charged particles as they traverse the medium and **Linear Energy Transfer, or LET**, which is defined as the energy imparted *to the medium*. The LET applies only to the energy imparted to the medium at or near the site of photon collisions and can be different from that lost in traversing the medium.

Exposure is defined in the SI system as the production of one coulomb of charge in one kg of air, or

$$X = 1 \text{ coulomb/kg of air}$$

The X unit is a huge unit equal to 3,876 R; consequently, most exposure measurements are made and reported in R which seems appropriate for the discoverer of x rays.

RADIATION DOSE CALCULATIONS

Radiation dose can be calculated if two things are known: (1) the number of "radiations" that impinge on the mass of medium being irradiated, and (2) the amount or rate of energy deposited per unit of mass. The general procedure for such calculations is:

1. First establish the number of radiation pulses (particles or photons) per unit area (i.e., the flux) that impinges upon a medium of known density.
2. For particles, which have a finite range in a medium, establish the mass of medium in which the energy is dissipated, and distribute the total energy in the mass to determine energy deposited per gram of medium.
3. For photons, determine the energy deposition in a unit depth (e.g., 1 cm) from the pattern(s) of interaction probabilities.

All emitted radiations must be considered in this process, and adjustments should be made for any attenuating medium between the source and the point of interest.

The Inverse Square Law states that the flux of radiation emitted from a "point" source is inversely proportional to r^2, where r is the distance from the source. A point source of activity $S(t/s)$ emits radiation uniformly in all directions, and these will pass uniformly through an area equal to that of a sphere of radius r that encloses the source; thus the flux $\phi(\#/cm^2 \cdot s)$

$$\phi(\#/cm^2 \cdot s) = \frac{S(t/s)\,Y_i}{4\pi r^2}$$

where S (t/s) is the radiation emission rate of the source and Y_i is the fractional yield per transformation of each emitted radiation.

Example 5-2: What is the photon flux produced by a 1 mCi point source of ^{137}Cs at a distance of 100 cm?
Solution: ^{137}Cs($T_{1/2} = 30.04$ y) emits 0.662 MeV gamma rays through ^{137m}Ba($T_{1/2} = 2.52$ m) in 85% of its transformations. The gamma flux is

$$\phi(\gamma/cm^2 \cdot s) = \frac{1\,\text{mCi} \times 3.7 \times 10^7 t/s.\text{mCi} \times 0.85\ \gamma/t}{4\pi(100\,\text{cm})^2} = 2.5 \times 10^2 \gamma/cm^2 \cdot s$$

INTERACTIONS OF ALPHA PARTICLES AND HEAVY NUCLEI

Alpha particles are helium nuclei with a charge of plus two since they are ejected from the nucleus of a radioactive atom without orbital electrons. This relatively high charge combined with the large kinetic energy they possess produces considerable ionization in passing through matter. When the alpha particle has dissipated its kinetic energy by ionizing the absorbing medium, it then picks up orbital electrons to become a neutral helium atom. The energy transfer per mm of path length is large and the depth of penetration in most absorbing media is on the order of μm; even in air the maximum range of alpha particles is only a few centimeters. Heavy recoil nuclei and fission fragments, such as alpha particles and their recoil nuclei, are also highly charged and they produce very dense patterns of ionized atoms in an absorbing medium.

The Range of Alpha Particles is the same for all particles emitted by a source since they are emitted monoenergetically, as shown in Figure 5-1. Alpha particle ranges have been measured in and reported in units of centimeters of air, which is roughly the same as its energy in MeV, as shown in Figure 5-2. An empirical fit of the data in Figure 5-2 shows a relationship between range and energy as

$$R_a = 0.325E^{3/2} \quad \text{or} \quad E = 2.12\,R_a^{2/3}$$

where R_a is in cm of standard air and E is in MeV.

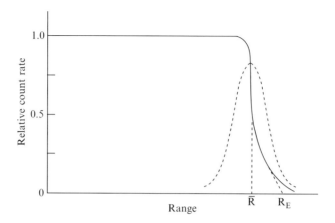

Figure 5-1. The range in air of a collimated monoenergetic source of alpha particles showing straggling that is normally distributed about a mean range $\bar{R}$. An extrapolated range, R_E, can be obtained by extending the straight-line portion of the curve to the x axis.

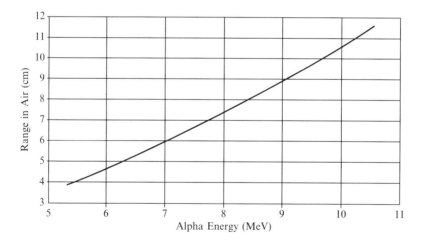

Figure 5-2. Range vs energy of alpha particles in air.

The range of an alpha particle in any other medium can be determined from the range in air, R_a, by adjusting it by the ratio of the densities of each. For tissue ($P_t = 1.0$), which is important for absorbed dose determinations, the range is

$$R_t = \frac{\rho_a}{\rho_t} R_a$$

Thus a 5.0 MeV alpha particle would have a range in tissue of 4.65×10^{-3} cm or about 46.5μm. Calculated ranges of alpha particles in tissue can be used directly in calculations of radiation dose.

BETA PARTICLE INTERACTIONS AND DOSE

Beta particles lose energy in traversing a medium in four ways: direct ionization, delta rays from electrons ejected by ionization, production of bremsstrahlung, and Cerenkov radiation. Although each mechanism can occur, the most important ones are direct ionization and bremsstrahlung production.

Energy Loss by Ionization

The kinetic energy of beta particles (or positrons) and their negative (or positive) charge is such that coulombic forces dislodge orbital electrons in the absorbing medium to create ion pairs. The particles thus lose energy by ionizing the absorbing medium, and the energy carried by these excited electrons is absorbed essentially at the interaction site. If the beta particles eject K-, L-, or M-shell electrons from atoms in the absorbing medium, characteristic x rays will also be emitted as these vacancies are filled. These too are likely to be absorbed nearby.

Since beta particles and orbital electrons are about the same size, beta particles are deflected through a rather tortuous path as they traverse a medium, as shown in Figure 5-3a for air. The path length is quite long and the energy transfer per unit path length is relatively low. A 3-MeV β^- particle has a range in air of over 1000 cm and produces only about 50 ion pairs/cm of path.

Delta rays are often formed along the ionization tracks of beta particles because some of the ionized electrons are ejected with so much energy (on the order of 1 keV or so) that they can ionize other atoms in the medium. These secondary ionizations form a short trail of ionization extending outward from the main path of the beta particle as shown in Figure 5-3b and tend to widen the track of energy deposition.

Energy Losses by Bremsstrahlung

Since beta particles are high-speed electrons, they produce bremsstrahlung, or "braking radiation," in passing through matter, especially if the absorbing medium is a high-Z material. In passing near a nucleus they experience acceleration due to the deflecting force and give up energy by photon emission, which in turn reduces the speed of the beta particle by an amount that corresponds to

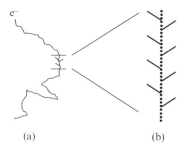

(a)   (b)

Figure 5-3. (a) Range and ionization path of a beta particle in an absorbing medium; (b) amplified segment of ionization track showing delta ray tracks produced by ejected electrons.

the energy lost to the bremsstrahlung photon. The process is the same one that occurs in the target material of an x-ray tube as illustrated in Figures 4-1 and 4-2. The amount of energy, if any, that a given electron loses by bremsstrahlung production depends on the path it takes toward a target nucleus and the amount of deflection, which is proportional to the nuclear charge (Z) of the target, that occurs. Since the electrons can approach the nucleus of a target atom from many different angles, the spectrum of bremsstrahlung produced consists of a continuous band of energies. Consequentially, when a beta source is configured in a way (e.g., in a high-Z shield) that the bremsstrahlung flux is appreciable, it is necessary to consider it as a special case of photon emission (see Chapter 7).

Range vs Energy of Beta Particles

Since the amount of beta absorption is related to its energy, the amount of absorber required to just stop all of the particles emitted by a source is also a measure of the beta-particle energy. The range of beta particles emitted by a given beta source is determined by placing different thicknesses of an absorber such as aluminum between the source and a detector and measuring the amount of thickness (in mg/cm²) required to just stop all the beta particles. This measured range (in mg/cm² of aluminum or other similar absorber) is then used to determine the energy by an empirical range–energy curve as shown in Figure 5-4.

Empirical relations between beta-particle energy in MeV and the range R in mg/cm² have been derived from data such as that shown in Figure 5-4 for $0.01 \leq E \leq 2.5\,\text{MeV}$ as

$$R = 412E^{1.265-0.0954\ln E}$$

or alternatively for $R \leq 1200\,\text{mg/cm}^2$

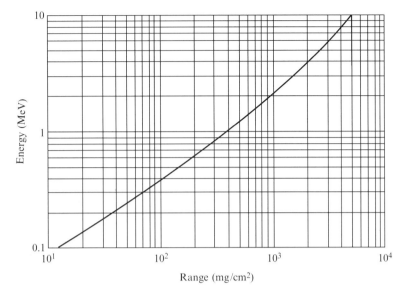

Figure 5-4. Equivalent range of electrons in mg/cm² of low-Z absorbers.

$$\ln E = 6.63 - 3.2376(10.2146 - \ln R)^{1/2}$$

If $E_{\beta\,max} \geq 0.6$ MeV, a somewhat simpler empirical relationship between E in MeV and R in mg/cm² is provided by Feather's rule, or

$$R = 542E - 133$$

Radiation Dose from Beta Particles

When it is possible to establish a flux of beta particles impinging upon a mass of medium, then beta dose calculations are straightforward if the "beta absorption" coefficient and the average beta-particle energy are known. The average beta-particle energy for each radionuclide can be obtained from the listed radiations in Appendix C, which are weighted average values of $\bar{E}$ for each beta emitter. The rule of thumb, $\bar{E} = 1/3E_{\beta\,max}$, can also be used as a first approximation, but accurate calculations require weighted values (as listed in Appendix C). Approximate relationships for the "beta absorption" coefficient have been determined for air and tissue specifically because of their utility in radiation protection; these are

$$\mu_{\beta,\ air} = 16(E_{\beta\,max} - 0.036)^{-1.4}$$
$$\mu_{\beta,\ tissue} = 18.6(E_{\beta\,max} - 0.036)^{-1.37}$$

For any other medium,

$$\mu_{\beta,\,i} = 17(E_{\beta\,max})^{-1.14}$$

where values of μ_β are in units of cm^2/g (based on measurements with low-Z absorbers, usually aluminum) and $E_{\beta\,max}$ is in units of MeV.

Example 5-3: What is the beta dose rate in tissue 1 m from a one curie point source of ^{32}P($E_{\beta\,max} = 1.710$ MeV and $\bar{E}_\beta \cong 0.695$ MeV) assuming no attenuation of the beta particles as they traverse the medium?
Solution: The beta flux is determined for the point source by the inverse square law as

$$\phi_\beta = \frac{3.7 \times 10^{10}\text{t/s} \cdot \text{Ci}}{4\pi(100\text{ cm})^2}$$
$$= 2.94 \times 10^5 \beta/\text{cm}^2 \cdot \text{s}$$

The beta absorption coefficient in tissue for ^{32}P beta particles is

$$\mu_{\beta,\,tissue} = 18.6(1.71 - 0.036)^{-1.37} = 9.183\text{ cm}^2/\text{g}$$

and the β dose rate is

$$D_\beta = \frac{2.94 \times 10^5 \beta/\text{cm}^2 \cdot \text{s} \times 0.695\text{ MeV}/\beta \times 9.183\text{ cm}^2/\text{g} \times 1.6022 \times 10^{-6}\text{ erg/MeV} \times 3600\text{ s/h}}{100\text{ ergs/g} \cdot \text{rad}}$$
$$= 108.43\text{ rads/h}$$

Most practical problems of beta radiation dose require adjustments of the energy flux. In Example 5-3 for instance, the dose rate from a point source would be more accurate if absorption of beta particles in the one-meter thickness of air were also considered because some absorption occurs in the intervening thickness of air before reaching the person. This is done by adjusting the energy flux for air attenuation using $\mu_{\beta,\,air}$ for the 1.71 MeV beta particles of ^{32}P, which is

$$\mu_{\beta,\,air} = 16(1.71 - 0.036)^{-1.4} = 7.78\text{ cm}^2/\text{g}$$

The beta flux after traversing a thickness x of air is then

$$\phi_\beta = \phi_\beta^o e^{-(\mu_{\beta,\,air})(\rho x)}$$

The density thickness ρx, which is a function of temperature and pressure, for 1 m of air at 20 °C and 1 atm is 1.205×10^{-3} g/cm$^3 \times 100$ cm and the dose rate is

$$D_\beta(x) = D_\beta^\circ e^{-\mu_{\beta, \, air}(\rho x)}$$

$$= 108.43e^{-7.78 \, \text{cm}^2/\text{g} \times 0.001205 \, \text{g/cm}^3 \times 100 \, \text{cm}}$$

$$= 42.46 \, \text{rads/h}$$

A further adjustment to Example 5-3 is appropriate to account for absorption of beta particles in the dead layer of skin with a density thickness of $0.007 \, \text{g/cm}^2$; i.e., to calculate the dose where it matters, to living tissue. The dose rate just below the $7 \, \text{mg/cm}^2$ skin layer is known as the "shallow dose." To make this adjustment, the beta energy flux is again calculated considering air attenuation and the density thickness of tissue ($0.007 \, \text{g/cm}^2$), and the beta absorption coefficient in tissue for ^{32}P beta particles ($9.183 \, \text{cm}^2/\text{g}$). Since all other factors remain the same,

$$D = 42.46e^{-\mu_{\beta, \, t}(\rho x)}$$

$$= 42.46e^{-9.183 \, \text{cm}^2/\text{g} \times 0.007 \, \text{g/cm}^2}$$

$$= 39.82 \, \text{rads/h}$$

To summarize, if the flux of beta particles is known, the general expression for the beta dose rate at the shallow depth in tissue after traversing a thickness x of air is

$$D_\beta(\text{rads/h}) = 5.768 \times 10^{-5}\phi_\beta \times \bar{E}_\beta \times \mu_{\beta, \, t}[e^{-\mu_{\beta, \, a}(\rho x)}][e^{-\mu_{\beta, \, t}(0.007 \, \text{g/cm}^2)}]$$

where ϕ is in units of $\beta/\text{cm}^2 \cdot \text{s}$, E_β is the average β energy in MeV, $\mu_{\beta, \, a}$ and $\mu_{\beta, \, t}$ are in units of cm^2/g for beta particles of $E_{\beta, \, max}$, ρx is the density thickness of air, and $0.007 \, \text{g/cm}^2$ is the density thickness of the dead skin layer.

Beta Dose from Contaminated Surfaces

If a floor, wall, or other solid surface is uniformly contaminated with a beta emitter, the beta dose rate is, as usual, determined by the energy flux, which in turn is influenced by beta particles that penetrate into the surface but are scattered back out to increase the beta flux reaching a receptor near the surface. For contaminated solid surfaces, therefore, the flux will be determined by the area contamination level (e.g., $(t/\text{cm}^2 \cdot \text{s})$), a geometry factor (usually 1/2), and a backscatter factor which can be obtained from Table 5-1 for various beta energies and materials. It is noticeable that the backscatter factor is highest for low-energy beta particles on dense (or high-Z) materials.

Example 5-4: What is the beta dose rate to skin and at the shallow depth: (a) at contact from a large copper plate uniformly contaminated with $10 \, \mu\text{Ci/cm}^2$ of ^{32}P, and (b) at a distance of $30 \, \text{cm}$?
Solution: (a) The beta flux is

Table 5-1. Backscatter Factors for Beta Particles on Thick Surfaces.

Beta Energy (MeV)	Carbon	Aluminum	Concrete*	Iron	Copper	Silver	Gold
0.1	1.040	1.124	1.19	1.25	1.280	1.38	1.5
0.3	1.035	1.120	1.17	1.24	1.265	1.37	1.5
0.5	1.025	1.110	1.15	1.22	1.260	1.36	1.5
1.0	1.020	1.080	1.12	1.18	1.220	1.34	1.48
2.0	1.018	1.060	1.10	1.15	1.160	1.25	1.40
3.0	1.015	1.040	1.07	1.12	1.125	1.20	1.32
5.0	1.010	1.025	1.05	1.08	1.080	1.15	1.25

* BSFs for concrete based on arithmetic average of BSFs for aluminum and iron.
Adapted from Tabata et al.

$$\phi_\beta = 10 \, \mu Ci/cm^2 \times 3.7 \times 10^4 t/s \cdot \mu Ci$$
$$= 3.7 \times 10^5 \, MeV/cm^2 \cdot s$$

and since half of the beta particles would be directed into the surface, the geometry factor is 0.5. The backscatter factor for ^{32}P beta particles ranges between 1.18 and about 1.30 based on the average energy, which by interpolation in Table 5-1 yields a backscatter factor of 1.24. Therefore, the surface-level beta dose rate in tissue at the air/tissue interface is

$$D_\beta(rad/h) = 5.768 \times 10^{-5} \phi_\beta \times \bar{E}_\beta \times \mu_{\beta,t} \times G \times BSF$$
$$= (5.768 \times 10^{-5})(3.7 \times 10^5 \beta/cm^2 \cdot s)(0.695 \, MeV/\beta)(9.183 \, cm^2/g) \times 0.5 \times 1.24$$
$$= 84.45 \, rads/h$$

and the beta dose at the shallow depth in tissue is

$$D_{\beta, \, sh} = 84.45 e^{-9.183 \, cm^2/s \times 0.007 \, g/cm^2} = 79.2 \, rads/h$$

(b) At 30 cm, the beta particles will be attenuated by 30 cm of air, $\mu_{\beta, \, a} = 7.78 \, cm^2/g$, $P_a = 0.001205$, and the dose rate at the surface of the skin is

$$D_{\beta, \, skin} = 84.45 \, rads/h \cdot e^{-7.78 \, cm^2/g \times 0.001293 \, g/cm^3 \times 30 \, cm}$$
$$= 63.75 \, rads/h$$

and at the shallow depth, a dose rate of

$$D_\beta, \, sh = 63.75 e^{-9.183 \, cm^2/g \times 0.007 \, g/cm^2}$$
$$= 59.78 \, rads/h$$

Contamination of Skin or Clothing can also produce a shallow dose from beta particles, e.g., contamination of the skin or from contaminated protective clothing in contact with the skin. If an area energy flux can be established, calculation of the shallow dose is straightforward by assuming a geometry factor of 0.5 (one-half the activity goes outward) and adjusting the energy flux to account for beta-energy absorption in the dead layer of the skin. There will be some backscatter of beta particles by the tissue layer below the skin but because tissue is a low-Z material this is only a few percent and can be ignored for most radiation protection situations.

Example 5-5: Estimate the shallow dose rate from five mL of solution containing $10\,\mu Ci/mL$ of ^{32}P spilled onto the sleeve of a worker's lab coat and distributed uniformly over an area of about $50\,cm^2$.
Solution: Uniform absorption of the solution onto the lab coat would produce an area contamination of $1\,\mu Ci/cm^2$. Assuming no attenuation of beta particles in the fabric, one-half of the beta particles emitted would impinge on the skin, and the beta flux just reaching the basal skin layer would be

$$\phi_{\beta,\,E} = 1\mu Ci/cm^2 \times 3.7 \times 10^4\,t/s \cdot \mu Ci \times 0.5[e^{-\mu_{\beta,\,t}\,0.007}]$$
$$= 1.73 \times 10^4 \beta/cm^2 \cdot s$$

and the beta dose rate at the shallow depth of tissue is

$$D_{\beta,\,sh} = (5.768 \times 10^{-5})(1.73 \times 10^4 \beta/cm^2 \cdot s)(0.695\,MeV/\beta)(9.183\,cm^2/g)$$
$$= 6.39\,rads/h$$

A similar calculation can be made for direct contamination of the skin, although in most practical situations it is more difficult to determine the area contamination. The skin contamination may be estimated by careful measurements with a thin-window probe or by washing the contaminated area with a swab or liquid and measuring the activity removed.

PHOTON INTERACTIONS AND DOSE

Photons originate in interesting ways; they simply appear when it is necessary to carry off excess energy (such as bremsstrahlung production, relief of excitation energy following radioactive transformation, or in nuclear interactions). The dichotomy of being both a wave and a particle is intriguing; however, their particle properties largely determine their absorption and energy deposition, which occurs by three principal modes: the photoelectric effect, the Compton effect, and pair production. Photons also undergo Rayleigh scattering, Bragg scattering, photodisintegration, and nuclear resonance scattering; however, these result in negligible attenuation or energy deposition and can generally be ignored for purposes of radiation protection.

Photoelectric Interactions

Since photons have "particle-like" properties, a low-energy photon can, by a process known as the photoelectric effect, collide with a bound orbital electron and eject it from the atom with an energy equal to that of the incoming photon, $h\nu$, minus the binding energy of the electron in its particular orbit. The interaction must occur with a bound electron to conserve momentum, and it usually (but not always) occurs with one of the inner-shell electrons due to their greater electron density. Since a vacancy is created in the electron shell, a characteristic x ray, typically from filling the K-shell, will also be emitted. For purposes of dosimetry, it is generally assumed that the kinetic energy of the ejected electron is absorbed in the medium where photoelectric absorption occurs. Characteristic x rays that are produced are also very likely to be absorbed in the medium, typically by another photoelectric interaction or by ejecting other orbital electrons within the same atom by the "Auger effect"; these too are absorbed immediately in the medium.

The Photoelectric Absorption Coefficient, τ, is a function of the atomic number Z of the absorbing material (generally related to the density, ρ, of the absorbing medium) and the energy of the radiation as follows:

$$\tau \cong \text{constant}\,\frac{Z^5}{E^3}$$

i.e., photoelectric absorption is most pronounced for low-energy photons (less than 0.5 MeV) in high-Z materials.

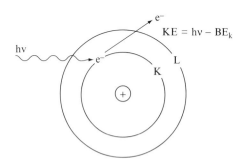

Figure 5-5. Schematic of the photoelectric effect.

Compton-Effect Interactions

Compton scattering interactions are especially important for gamma rays of medium energy (0.5 to 1.0 MeV), and, for low-Z materials such as tissue, can be the dominant interaction mechanism for photons down to about 0.1 MeV. Compton scattering involves a collision between a photon and a "free" or very loosely bound electron in which a part of the energy of the photon is imparted to the electron, as shown in Figure 5-6. Both energy and momentum are conserved in the collision, and the Compton-scattered photon emerges from the collision in a new direction and with reduced energy and increased wavelength, or

$$\lambda' - \lambda = 2.4264 \times 10^{-10}(1 - \cos\theta)\,\text{cm}$$

where the change in wavelength (and decrease in energy) of the photon is determined only by the scattering angle. Energy transfer to the recoiling electron is the most important consequence of Compton interactions since it will be absorbed locally to produce radiation dose. The fraction of the photon energy, $h\nu$, that is transferred to the Compton electron is shown in Figure 5-7; the value $\bar{E}_{tr}$ is the average value due to the randomness of Compton interactions in a medium and $E_{tr,\ max}$ is the maximum energy that can be transferred for a given energy due to a scattering angle of 180°.

The Compton Interaction Coefficient, σ, is determined by the electron density of the absorbing medium and is directly related to Z and inversely proportional to E as follows:

$$\sigma \cong \text{constant}\,\frac{Z}{E}$$

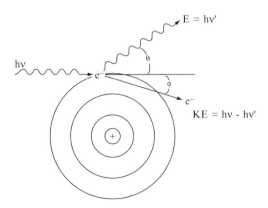

Figure 5-6. Compton scattering with a "free" electron.

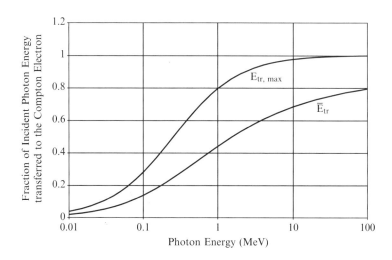

Figure 5-7. Fraction of incident photon energy ($h\nu$) transferred to the Compton electron.

It consists of two components:

$$\sigma = \sigma_a + \sigma_s$$

where σ is the total Compton interaction coefficient, σ_a is the Compton absorption coefficient for photon energy lost by collisions with electrons, and σ_s is the loss of energy due to the scattering of photons out of the beam.

Pair-Production Interactions

When a high-energy (> 1.022 MeV) photon interacts with the strong electromagnetic field surrounding a nucleus as shown in Figure 5-8, its energy can be converted into a pair of electron masses, an electron and a positron. Pair production is a dramatic confirmation of Einstein's special theory of relativity in which the pure energy of the photon is converted into two electron masses; their shared kinetic energy ($h\nu - 1.022$) is quickly dissipated in the medium.

Pair-production interactions are accompanied by the emission of two annihilation photons of 0.511 MeV each, which are also shown in Figure 5-8. The positron will exist as a separate particle as long as it has momentum and kinetic energy. However, when its energy has been fully dissipated, being anti-matter in a matter world, it will interact with a negatively charged electron, forming for a brief moment a "neutral particle" of "positronium," which then vanishes yielding two 0.511 MeV photons; i.e., mass becomes energy. These two photons

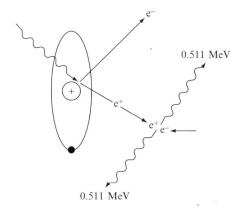

0.511 MeV

0.511 MeV

Figure 5-8. Schematic representation of a pair-production interaction for a photon of $h\nu > 1.022$ MeV and the formation of two 0.511 MeV photons when the positron is annihilated with an electron.

will either be absorbed by photoelectric and/or Compton interactions or will escape the medium.

The absorption of high-energy photons thus yields a complex pattern of energy emission and absorption in which the pure energy of the photon produces an electron and a positron which deposit $h\nu - 1.022$ MeV of kinetic energy along a path of ionization, followed in turn by positron annihilation with a free electron to produce two new photons of 0.511 MeV which may or may not interact in the medium.

The Pair Production Interaction Coefficient, κ, is proportional to the square of the atomic number, Z, for photons with energy greater than 2×0.511 MeV (the energy required to form an electron–positron pair) and has the following relationship:

$$\kappa \cong \text{constant } Z^2 \, (E - 1.022)$$

where Z is the atomic number, and E is the photon energy in MeV.

The Linear Attenuation Coefficient, μ, is the sum of the coefficients of each of the principal modes of photon interaction in a medium, or

$$\mu = \tau + \sigma + \kappa$$

and since it represents the cumulative effect of each of the interaction processes for photons, all of which are probabilistic, it can be used to determine the decrease in photon flux ϕ_o by the exponential relationship

$$\phi(x) = \phi_o e^{-\mu x}$$

where $\phi(x)$ is the flux of photons after traversing a distance x through an absorbing medium and ϕ_o is the initial flux.

In simplest terms, μ represents the probability of interaction per unit distance in an absorbing medium. It is synonymous with the radioactive disintegration constant, λ, which expresses the probability of transformation of radioactive atoms per unit time. The exponential relationship for photon absorption suggests that, unlike charged-particle interactions, complete absorption of a beam of photons never really occurs, but in a practical sense essentially all of the photons can be presumed to be absorbed if thick absorbers are used. Each of the photon absorption coefficients τ, σ, and κ increase with Z, which confirms that dense materials are effective absorbers of photons.

The Energy Absorption Coefficient, μ_{en}, accounts for the amount of energy lost due to the various processes that carry energy out of the medium. Such processes are Compton-scattered photons that do not interact, bremmstrahlung from high-energy recoil electrons, annihilation radiation if $h\nu > 1.022\,\mathrm{MeV}$, and characteristic x rays that do not interact. These depleting mechanisms and their effect on μ are shown in Figure 5-9. These losses are accounted for by subtracting coefficients for them from the total absorption coefficient μ as follows:

$$\mu_{en} = \mu - (\sigma_s + \text{other low probability interactions})$$

Consequently, μ_{en}, which is a coefficient of energy absorption only, is smaller than the attenuation coefficient μ. Its most useful form is the mass energy absorption coefficient, μ_{en}/ρ (in units of cm^2/g), which is used to determine radiation dose when a flux of x or gamma rays is known or can be determined.

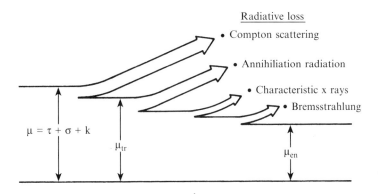

Figure 5-9. Relationship of the photon attenuation coefficient, μ, the energy transfer coefficient, $\mu_{tr} = \mu - \sigma_s$, and the radiative loss processes that propagate energy out of the absorbing medium to yield the energy absorption coefficient, μ_{en}.

Table 5-2 lists values of μ_{en}/ρ for those materials of importance for determining radiation dose for photons; also included are tabulations of the linear attenuation coefficient, μ, which can be used to determine the mass attenuation coefficient, μ/ρ, for the medium by dividing by the density.

Checkpoints

Several things are noteworthy about the exponential relationship for photon absorption:

1. Since the radiation-absorbed dose rate is proportional to the photon flux and the energy of each photon, the energy absorption coefficient, μ_{en}, determines how much of the incident flux is deposited in the medium;
2. Values of μ_{en} are based only on the energy absorbed in the medium; energy lost from the medium due to Compton-scattered photons, bremsstrahlung, and other radiative processes has been subtracted;
3. The most useful form of the energy absorption coefficient is $\mu_{en}/\rho (\text{cm}^2/\text{g})$, which is used for determining energy deposited in the absorbing medium; and
4. Values of μ and μ_{en}/ρ depend on the absorbing medium and the energy of the photons; thus, extensive tabulations of each are needed (as provided in Table 5-2) for calculations of flux changes and or exposure/dose calculations.

Radiation Dosimetry for Photons

The transfer and absorption of energy in a medium occurs in two stages as shown in Figure 5-10: first, a photon interaction transfers kinetic energy to an electron; and second, each ejected electron deposits energy in the medium through excitation and ionization all along its path to produce an absorbed dose. The kinetic energy released in the material, or kerma, occurs at the point of interaction; the absorbed dose occurs, however, along the path of the recoiling electrons extending over their entire range, R_e, in the medium.

As illustrated in Figure 5-10, energy is transferred to an electron by photoelectric and Compton interactions, but not all of it is retained in the medium since the ejected electrons will produce bremsstrahlung that can be radiated away (a small fraction of energy may also be lost in the form of characteristic x rays that are not absorbed after photoelectric interactions). Pair-production interactions yield a positron and an electron which deposit their kinetic energy $(h\nu - 1.022\,\text{MeV})$ by similar processes, and if both annihilation photons are absorbed in the medium, then most of the photon energy will be deposited. The mass energy absorption coefficient, μ_{en}/ρ, accounts for all the processes of

Table 5-2. Photon Attenuation (μ) and Mass Energy–Absorption (μ_{en}/ρ) Coefficients for Selected Materials (J. H. Hubbell and S. M. Seltzer).

Energy (keV)	Dry Air (Sea Level) ($\rho = 0.001205$ g/cm³) μ (cm⁻¹)	μ_{en}/ρ (cm²/g)	Tissue, (ICRU) ($\rho = 1.00$ g/cm³) μ (cm⁻¹)	μ_{en}/ρ (cm²/g)	Muscle (ICRP) ($\rho = 1.05$ g/cm³) μ (cm⁻¹)	μ_{en}/ρ (cm²/g)	Cortical Bone (ICRP) ($\rho = 1.92$ g/cm³) μ (cm⁻¹)	μ_{en}/ρ (cm²/g)
10	0.0062	4.742	4.937	4.564	5.6238	4.964	54.7392	26.80
15	0.0019	1.334	1.558	1.266	1.7777	1.396	17.3414	8.388
20	0.0009	0.5389	0.7616	0.5070	0.8615	0.564	7.6819	3.601
30	4.26E-4	0.1537	0.3604	0.1438	0.3972	0.1610	2.5555	1.070
40	2.99E-4	0.0683	0.2609	0.0647	0.2819	0.0719	1.2778	0.4507
50	2.51E-4	0.0410	0.2223	0.0399	0.2375	0.0435	0.8145	0.2336
60	2.26E-4	0.0304	0.2025	0.0305	0.2150	0.0326	0.6044	0.1400
70*	2.10E-4	0.0255	0.1899	0.0263	0.2006	0.0276	0.4866	0.0905
80	2.00E-4	0.0241	0.1813	0.0253	0.1914	0.0262	0.4280	0.0690
100	1.86E-4	0.0233	0.1688	0.0250	0.1778	0.0254	0.3562	0.0459
150	1.63E-4	0.0250	0.1490	0.0273	0.1567	0.0275	0.2842	0.0318
200	1.49E-4	0.0267	0.1356	0.0294	0.1426	0.0294	0.2513	0.0300
300	1.29E-4	0.0287	0.1175	0.0316	0.1235	0.0316	0.2137	0.0303
400	1.15E-4	0.0295	0.1051	0.0325	0.1105	0.0325	0.1902	0.0307
500	1.05E-4	0.0297	0.0959	0.0327	0.1008	0.0327	0.1732	0.0307
600	9.71E-5	0.0295	0.0887	0.0325	0.0932	0.0325	0.1600	0.0305
662*	9.34E-5	0.0293	0.0853	0.0323	0.0896	0.0323	0.1538	0.0303
800	8.52E-5	0.0288	0.0779	0.0318	0.0818	0.0318	0.1403	0.0297
1,000	7.66E-5	0.0279	0.0700	0.0307	0.0736	0.0307	0.1261	0.0288
1,173*	7.05E-5	0.0271	0.0644	0.0299	0.0677	0.0299	0.1160	0.0279
1,250	6.86E-5	0.0267	0.0626	0.0294	0.0658	0.0294	0.1127	0.0275
1,333*	6.62E-5	0.0263	0.0605	0.0290	0.0636	0.0290	0.1090	0.0271

1,500	6.24E-5	0.0255	0.0570	0.0281	0.0599	0.0281	0.1026	0.0262
2,000	5.36E-5	0.0235	0.0489	0.0258	0.0514	0.0258	0.0885	0.0242
3,000	4.32E-5	0.0206	0.0393	0.0226	0.0413	0.0226	0.0719	0.0215
4,000	3.71E-5	0.0187	0.0337	0.0204	0.0354	0.0205	0.0625	0.0198
5,000	3.31E-5	0.0174	0.0300	0.0189	0.0315	0.0190	0.0566	0.0186
6,000	3.04E-5	0.0165	0.0274	0.0179	0.0288	0.0179	0.0525	0.0179
6,129*	3.01E-5	0.0164	0.0271	0.0178	0.0285	0.0178	0.0520	0.0178
7,000*	2.83E-5	0.0159	0.0255	0.0171	0.0267	0.0170	0.0495	0.0174
7,115*	2.82E-5	0.0158	0.0253	0.0170	0.0265	0.0170	0.0492	0.0173
10,000	2.46E-5	0.0145	0.0219	0.0155	0.0230	0.0155	0.0444	0.0164

* Coefficients for these energies were interpolated using polynomial regression.

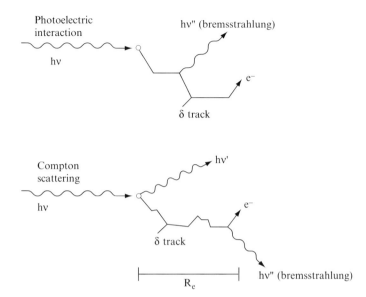

Figure 5-10. Energy transfer and absorption is a two-stage process: first, a photon ($h\nu$) interacts with an atom of the medium, transferring all (photoelectric effect) or some of its energy (Compton scattering) to an electron or, in the case of pair production, to the created electron and positron; and second, the kinetic energy of the electrons is then deposited in the medium mostly by ionization, although some may be lost as bremsstrahlung and/or other radiative processes.

photon energy transfer and absorption illustrated in Figure 5-10 and as such is directly applicable to calculations of radiation absorbed dose.

Calculations of radiation exposure or dose are straightforward if two values are known: first, the photon fluence, fluence rate, or flux (photons/cm² · s); and second, the mass energy absorption coefficient μ_{en}/ρ for photons of a specified energy in a medium. The product of the photon flux, μ_{en}/ρ, and the photon energy can in turn be used to determine exposure in air or energy deposition (i.e., radiation absorbed dose) in an absorbing medium as illustrated in Example 5-6.

Example 5-6: Determine the radiation dose rate to muscle tissue at 1 m from a point source of ^{137}Cs.
Solution: As shown in Appendix C the γ yield of 0.662 MeV gamma rays from ^{137}Cs is 0.85/t; thus a one-curie point source of ^{137}Cs produces a photon flux at 1 m

$$\phi = \frac{3.7 \times 10^{10}\text{t/s} \cdot \text{Ci} \times 0.85\gamma/\text{t}}{4\pi(100\,\text{cm})^2} = 2.503 \times 10^5 \gamma/\text{cm}^2 \cdot \text{s}$$

From Table 5-2, the mass energy absorption coefficient, μ_{en}/ρ, for 0.662 MeV photons in muscle is $0.0323\,\mathrm{cm}^2/\mathrm{g}$; therefore, the absorbed dose rate in muscle tissue is

$$D = \frac{2.5 \times 10^5\,\gamma/\mathrm{cm}^2 \cdot \mathrm{s} \times 0.0323\,\mathrm{cm}^2/\mathrm{g} \times 0.662\,\mathrm{MeV}/\gamma \times 1.6022 \times 10^{-6}\,\mathrm{erg}/\mathrm{MeV} \times 3600\,\mathrm{s}/\mathrm{h}}{100\,\mathrm{ergs}/\mathrm{g} \cdot \mathrm{rad}} \times 1.0 \frac{\mathrm{rem}}{\mathrm{rad}}$$

$$= 0.31\,\mathrm{rem}/\mathrm{h} \cdot \mathrm{Ci} \ @ \ 1\,\mathrm{m}$$

THE GAMMA CONSTANT, Γ

The dose rate from a point source of a gamma-emitting radionuclide (as illustrated in Example 5-6 for one curie of ^{137}Cs is a unique value because the energy and yield of each emitted photon are constant for each radionuclide. Therefore, a gamma ray constant (Γ) can be calculated; such values have been tabulated in Table 5-3 for several common gamma emitting radionuclides by summing the exposure rate for each photon emitted. For example, ^{60}Co emits two gamma rays, one at 1.332 MeV in 100% of transformations and one at 1.173 MeV in 99.9% of transformations; Γ was determined for each gamma ray and the two were added together (see Problem 5–5).

The gamma constant Γ is of considerable practical value because tabulated values can be readily adjusted for sources of different activity by simple multiplication, and for different distances by the inverse square law. The most useful form of the gamma constant in radiation protection is in units of rem/h @1 m per curie, but it has also been tabulated as R/h @1 m per Ci, perhaps due to precedent, and occasionally in coulombs/kg of air allowing conversion to R or the X unit of exposure.

DOSIMETRY FOR BREMSSTRAHLUNG

Assessments of radiation exposure due to breamsstrahlung from beta sources can be done conservatively by assuming that each bremsstrahlung photon has an energy equal to the maximum energy of the beta particles emitted by the source; however, it is probably more reasonable to assign the average beta particle energy to each bremsstrahlung produced. When beta particles are absorbed in tissue (which has a low atomic number), less than 1% of the interactions produce bremsstrahlung, and many of those that do are likely to escape the tissue medium because their probability of interaction is also low in this low-Z medium. Consequently, energy losses due to bremsstrahlung production in tissue can generally be ignored in radiation dose calculations except for very high energy (greater than several MeV) electrons. Bremsstrahlung production is, however, an important consideration in shielding beta sources (see Chapter 7) especially if high-Z materials are used in shields. In heavy elements such as lead ($Z = 82$), the energy loss by radiation can be as much as 10% or so for 2-MeV beta-particles; in aluminum ($Z = 13$) the radiation loss is much less for most beta-particle energies available from β^- emitters.

Table 5-3. The Gamma Constant, Γ (rem/h·Ci @1 m), for One Curie of Selected Radionuclides.

Nuclide	Γ (rem/h @ 1 m)	Nuclide	Γ (rem/h @ 1 m)
Antimony-124	1.067	Iodine-131	0.283
Arsenic-76	0.545	Iodine-132	1.427
Barium-133	0.455	Iridium-192	0.592
Barium-La-140	0.165*	Iron-59	0.662
Beryllium-7	0.028	Manganese-54	0.448
Bromine-82	1.619	Manganese-56	0.924
Cadmium-115m	0.013	Mercury-203	0.122
Calcium-45	0.585	Molybdenum-99	0.113
Cerium-141	0.042	Nickel-65	0.273
Cerium-Pr-144	0.023	Niobium-95*	0.480
Cesium-134	0.999	Potassium-40	0.082
Cesium-137	0.319	Radium-226	0.005
Chromium-51	0.023	Ruthenium-103	0.273
Cobalt-57	0.053	Ruthenium-Rh-106	0.138*
Cobalt-58	0.614	Scandium-46	1.047
Cobalt-60	1.370	Silver-110m	1.652
Copper-64	0.132	Sodium-22	1.147
Europium-152	0.744	Sodium-24	1.938
Europium-154	0.756	Strontium-85	0.759
Fluorine-18	0.054	Tin-113	0.179
Gallium-67	0.111	Tungsten-187	0.329
Gold-198	0.292	Vanadium-48	1.701
Iodine-123	0.277	Yttrium-88	1.783
Iodine-125	0.275	Zinc-65	0.330
Iodine-130	1.403	Zirconium-Nb-95*	0.465*

*Short-lived transformation product in equilibrium with longer-lived parent.

Example 5-7: A researcher transferred 100 mCi of ^{90}Sr − Y into 10 mL of water in a plastic bottle and set it at the back of the laboratory bench about a meter away. When reminded of potential bremsstrahlung after the experiment, an estimate was needed of the tissue dose at 1 meter.

Solution: Although the solution contained both ^{90}Sr and ^{90}Y, the latter is the most important contributor of bremsstrahlung due to its high energy, $E_{\beta Max} = 2.28$ MeV. The fraction of the beta energy converted to bremsstrahlung in the water solution is, by interpolation in Table 7-1, 0.81%; therefore, since the average ^{90}Y beta energy is 0.9337 MeV (Appendix C), the amount of beta energy converted to photons is

photon $E = 100\,\text{mCi} \times 3.7 \times 10^7 \text{t/s} \cdot \text{mCi} \times (0.81 \times 10^{-2}) \times 0.9337\,\text{MeV}$
$$= 3 \times 10^7\,\text{MeV/s}$$

And if it is conservatively assumed that each bremsstrahlung photon has an energy equal to $E_{\beta\text{Max}}$, and that the bottle approximates a point source, the photon flux at 100 cm is

$$\phi_{bs} = \frac{3 \times 10^7\,\text{MeV/s}}{2.28\,\text{MeV}/h\nu 4\pi (100\,\text{cm})^2}$$
$$= 105\,h\nu/\text{cm}^2 \cdot \text{s}$$

each with 2.28 MeV of energy, for which μ_{en}/ρ in tissue (Table 5-2) is $0.025\,\text{cm}^2/\text{g}$; therefore, the tissue dose rate is

$$\text{Dose} = \frac{105\,h\nu/\text{cm}^2 \cdot \text{s} \times 2.28\,\text{MeV}/h\nu \times 0.025\,\text{cm}^2/\text{g} \times 1.6022 \times 10^{-6}\,\text{erg/MeV} \times 3600\,\text{s/h}}{100\,\text{erg/g} \cdot \text{rad}}$$
$$= 3.45 \times 10^{-4}\,\text{rad/h}$$

SUMMARY

Radiation interacts in an absorbing medium to deposit energy which is defined as radiation absorbed dose (rad); the dose equivalent (in rem or Sv) is obtained from the absorbed dose by considering the damaging effect of the radiation type, or the quality factor. A related term is radiation exposure, which applies only to air and is a measure of the amount of ionization produced by x and gamma radiation in air.

Radiation dose can be calculated if two quanties are known: the number of "radiations" per unit area (the flux) that impinge on a mass; and the energy deposition per unit mass, which is just the energy absorption coefficient. For particles, the thickness of the mass in which the energy is dissipated is the maximum range of the particles; however, beta-particle absorption is often based on a beta absorption coefficient, $\mu_\beta(\text{cm}^2/\text{g})$. Since photons are highly penetrating due to their probabilistic pattern of interactions, a unit depth (e.g., 1 cm) is used to characterize energy deposition which is determined by the mass energy absorption coefficient, $\mu_{en}/\rho(\text{cm}^2/\text{g})$. These energy deposition determinations require consideration of all and adjustments should be made for any attenuating medium between the source and the point of interest.

Acknowledgments

Many of the data resources in this chapter came from the very helpful people at the National Institute of Standards and Technology, in particular Dr. John H. Hubbell and his colleagues M.J. Berger and S. M. Seltzer.

ADDITIONAL RESOURCES

1   Martin, J.E., *Physics for Radiation Protection* (New York: Wiley, 2000).
2   Hubbell, J.H. and Seltzer, S.M., Tables of X-Ray Attenuation Coefficients and Mass Absorption Coefficients 1 keV to 20 MeV for Elements Z = 1 to 92 and 48 Additional Substances of Dosimetric Interest, NISTIR 5632, National Institute of Standards and Technology, Gaithersburg, MD, 1995.
3   Tabata, T., R. Ito, and S. Okabe, Backscatter Factors for Beta-Particles, *Nacl. Instrum. Meth.* 94, 509, 1971.

PROBLEMS

1   What energy must: (a) an alpha particle have to just penetrate the dead layer of a person's skin if the density thickness is $7\,\mathrm{mg/cm^2}$; (b) a beta particle?

2   What would be the depth of penetration of $^{218}\mathrm{Po}$ alpha particles ($E = 6.0\,\mathrm{MeV}$) in tissue?

3   A solution containing $10\,\mu\mathrm{Ci}$ of $^{32}\mathrm{P}$ was spilled on a wooden table top but was not discovered until it had dried. If the area of contamination was a circle 30 cm in diameter and the solution dried uniformly over it, what would be the β flux and the tissue dose rate just above the table top? How would these change if the surface were stainless steel?

4   For $^{203}\mathrm{Hg}$, determine: (a) the gamma constant (rem/hr · Ci at 1 m); and (b) the absorbed dose rate in tissue at 50 cm from a 1 mCi source.

5   Determine Γ (rem/hr·Ci @1 m) for $^{60}\mathrm{Co}$ which emits 1.17 MeV and 1.33 MeV γ rays assuming each is emitted 100% of the time.

6   A source of 1 MeV gamma rays produces a uniform flux of $10^6\,\gamma/\mathrm{cm^2}\cdot\mathrm{s}$ at a distance of 150 cm. (a) What is the absorbed dose rate to a person at that distance, and (b) at 200 cm?

6

RADIATION BIOEFFECTS AND RISK

"...repeated small doses of soft x rays, when applied to human tissues, produce gradual changes therein which may cause such to develop malignant features."
— *H. Coldwell and S. Russ, 1915*

The deposition of radiation energy in living systems induces chemical reactions and molecular alterations in individual cells which may lead to three main effects: teratogenic or birth effects due to cells exposed *in utero*, genetic effects due to aberrations that may be transferred to future generations of individuals, and increased cancer risk to exposed individuals and/or populations. Radiation is teratogenic because it can induce or increase the incidence of congenital malformations during the growth and development of an embryo; it is mutagenic because genetic mutations can be produced in either somatic (body) or germ (reproductive) cells; and it is carcinogenic because damage to DNA may be perpetuated in cell division, eventually resulting in uncontrolled cell growth or malignancy.

RADIOBIOLOGY

The average human has about a trillion (10^{12}) cells of many different sizes and shapes that are of two general types: somatic cells and germ cells. Somatic cells make up the organs, tissues, and other body structures. Germ cells, which are also called gametes, determine how a new individual will turn out because they carry the hereditary material of the species. Damage to somatic cells is limited to the individual exposed whereas damage to the gametes can be transmitted to future generations which can potentially damage the offspring of an individual.

Cells are very complex, and even though each one contains several major entities, the key components for consideration of radiation bioeffects (as shown in very simplified form in Figure 6-2) are

- the **cell membrane** which encloses the entire cell;
- the **cytoplasm** in which energy for the cell to function is produced; and
- the **nucleus**, which is a large, generally spherical body that functions as the control center of the cell; it contains the chromatin.

The cytoplasm comprises the bulk of the cell volume; however, the nucleus is the most important part affected by radiant energy deposition. The cytoplasm is where cell function takes place; e.g., the manufacture of proteins and production of energy (which occurs in the mitochondria). Of particular importance

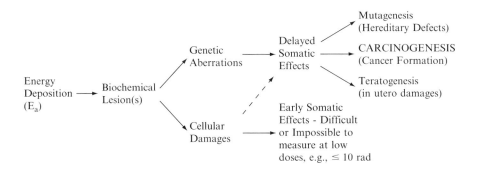

Figure 6-1. Major events in human low-dose radiobiology.

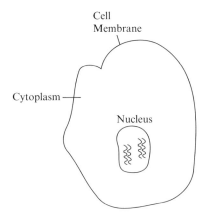

Figure 6-2. A typical somatic cell.

are the ribosomes which direct various chemical operations and manufacture the proteins which determine the special function (skin, liver, etc) of the cell and hence its type. Proteins are required for cell growth and for the repair of any damage that occurs to the cell.

The nucleus is the nerve center of the cell that controls cell function, and it is also the "computer memory" that contains the master program instructions that are transmitted to new generations of cells that are produced. The nucleus contains the **chromatin**, the genetic material of the cell. Every living cell was produced by one that already existed through a process of cell division; i.e., a cell makes an exact copy of itself so that the function it performs in an organ or other tissue can continue. All cells have a finite life and thus must be continually replaced. During normal cell function (i.e., when it is not dividing), the chromatin which controls protein synthesis appears as a confused mass of strands of DNA (deoxyribonucleic acid); however, when the cell starts to reproduce itself (a process called mitosis), these strands become untangled and coil into a fixed number of bundles, or **chromosomes**. The chromosomes are in turn duplicated during mitosis or cell division such that the new (or reproduced) cells have the necessary identifying information to carry out their function in a normal way not only in the next generation but in each succeeding one as well. After mitosis, the chromosomes uncoil and return to their former tangled state, but they always exist as specific entities in the cell, and if radiation deposits energy in such a way that chromosomes are broken or genes on them are damaged, then a defect can be perpetuated into future generations of cell growth.

Repair may occur in damaged cells, or badly injured cells may be replaced through mitosis of healthy cells. If too many cells in an organ are damaged, the organ may be damaged beyond repair and its function will be lost. On the other hand, damaged cells that are perpetuated through several generations of cell reproduction may eventually affect organ function or in the worst case may grow uncontrollably and produce malignant tumors. Such effects can also be caused by other agents, and effects that occur cannot be directly related to radiation although it is one potential cause. In terms of radiation effects, cell death, as long as tissue function is retained, is preferable to unrepaired damage because dead cells are easily removed from the body.

Physical and Chemical Effects

Radiation bioeffects are broadly classified as direct effects and indirect effects. **Direct Effects** of radiation usually result in the breaking of bonds of molecules by being struck by the impinging particle or photon or by one of the secondary electrons they may produce. These direct effects occur, therefore, along the narrow path of the impinging radiation or along the path of ejected electrons. Such effects can be substantial for high LET (linear energy transfer) radiations such as neutrons and alpha particles which deposit relatively large amounts of concentrated energy in the cells through which they pass. The direct effects of these radiations are thus more damaging than beta particles or x and gamma rays, which are sparsely ionizing radiations. Whatever effects do occur necessarily depend upon which cellular structures are damaged and on their importance to cell function.

Indirect Effects of radiation occur because the deposition of radiant energy in cells, which contain 70 to 80% water, produces a large number of secondary electrons with energies of 10 to 70 eV, well above the energy ($\sim$7.4 eV) required to ionize matter. Radiation interactions in cells thus yield ionized water molecules which in turn form free hydroxyl radicals (OH), excited electrons (e^-), and excited water molecules (H_2O^+ and H_2O^*), all of which are highly reactive. Since these are produced in abundance, are free to move, and are often attracted to molecular structures, the area of interaction is much larger and the overall indirect effects of radiation are greater than the direct effects. And since this predominant effect produces chemicals, the biological effects of radiation are essentially the same as those of various chemicals. The occurrence of the ionizing event, the formation of highly reactive chemical components, and their reactions with the surrounding medium all occur on the order of a millisecond or so and can quickly (within minutes) affect the process of cell division. If the damage occurs in the GI tract or the CNS, effects on these systems would be expressed on the order of days (see acute radiation syndrome discussion). Effects such as cancer, which require many generations of cell divisions to appear, generally require years to develop. Although the mechanisms may differ, the net effect of free radical formation is diminished cell function, or more importantly, disturbance of the master code in DNA that can produce genetic defects or late somatic effects.

Cellular DNA exists as a double helix of linked molecules, and is perhaps the most important part of the cell from the standpoint of radiobiology. It is a very large macromolecule that contains the information that governs the growth and development of a cell, information that is replicated and transferred to each new cell produced. The structure of DNA takes the form of a twisted ladder (the double helix) with the sides made up of alternating molecules of deoxyribose (a sugar) and phosphate. Each rung (or step) of the ladder is made up of two of four possible nitrogenous bases: adenine, guanine, thymine, or cytosine. Although either of these bases can attach to the deoxyribose molecules that make up the twisted sides of the ladder, the only possible paring for the rungs are adenine/guanine (referred to as purines) or thymine/cytosine (or pyrimidines). No other bondings are possible; thus each cell has a unique code.

Various combinations of phosphate/deoxyribose and one of the four bases are possible (each particular combination is called a nucleotide) and these are strung together by one of the possible base pairs as shown in Figure 6-3 and 6-4, forming the familiar double helix. When cell division occurs to replicate a new cell, this long, double-helix molecule separates along the center much like a zipper, and since only one base molecule (supplied as basic building blocks in food) can join each of the protruding base molecules, these line up as prescribed forming two identical DNA molecules, each of which is carried over to the new cell; thus, the master code for each is retained so that each new cell will carry out its required function. This fascinating process is the miracle of life.

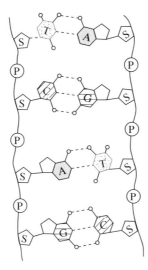

Figure 6-3. The DNA macromolecule.

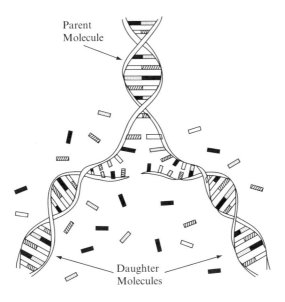

Parent
Molecule

Daughter
Molecules

Figure 6-4. The separation of DNA along the center of the double helix to reproduce an exact copy of itself.

Since DNA is the nerve center for cell function and also the master program for prescribing the behavior in replicated cells, damage to it can yield serious consequences. Such damage can occur, as shown in Figure 6-5, in one of four ways:

- a single strand break,
- a double strand break,
- double strand breaks followed by cross linking, and
- breaking of the base pairs on a rung or a change in or removal of one of them.

Because of the chemical bonding that occurs and the rigorous prescription of the ordering of connected molecules, repair of damage to DNA can be expected and in fact has been observed. But perfect repair cannot be assured, and it is for this reason that effects due to radiation damage to DNA are presumed to be probabilistic (stochastic), i.e., without threshold.

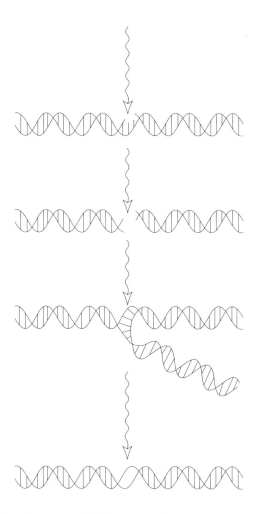

Figure 6-5. Radiation damage to DNA by strand break or disturbance of its base pairs.

Radiosensitivity

Cells have different lifetimes related to their functions: white blood cells live about 13 days, red blood cells about 120 days, liver cells about 500 days, and nerve cells, which live for about 100 years, do not reproduce and thus cannot be replaced. These lifetimes are consistent with the law of Bergonie and Tribondeau who found that radiosensitivity of tissue is directly proportional to the reproduction rate of the cells it contains and inversely with their degree of differentiation. For example, white blood cells are quite sensitive to radiation whereas nerve cells are much less so. Cancer treatment by radiation takes advantage of the sensitivity of rapidly dividing cells; it is possible to kill cancer cells while normal tissue cells can recover and continue their normal function.

The small intestine is quite radiosensitive, whereas the stomach and esophagus are much less so. Germ cells are highly radiosensitive, while other cells of the reproductive system are relatively radioresistant. In terms of the nervous system, the spinal cord and peripheral nerves are highly radioresistant, but the brain is more sensitive than often supposed. The eye is highly susceptible to irreversible damage by radiation, especially the lens. The heart and blood vessel system are damaged seriously only after very high doses of radiation. Skin is easily damaged but has a tremendous capacity for repair, and is less likely to develop fatal cancer. Although the marrow of the bone is very radiosensitive, the bone cells, fibers, and salts are relatively radioresistant. The liver is relatively radioresistant compared to other organs, while the breast in females is one of the more radiosensitive organs with respect to cancer induction.

Other physical factors that influence the occurrence of radiation effects are:

- the sensitivity and age of individuals (a select group may be more or less sensitive for a given effect than a heterogeneous group);
- the type of radiation, in particular whether it is high or low LET radiation since a smaller dose of high LET radiation is required to achieve a given degree of effect than for low LET radiation (little recovery from sublethal damage is observed for exposure to high LET radiation; however, there is some recovery of sublethal damage);
- the dose rate (the effects of high dose rates are 2 to 10 times those at low dose rates);
- the total energy absorbed per gram of tissue (i.e., the total dose), especially in a small tissue mass (large amounts of absorbed dose will override the considerations of whether it occurs with high or low LET);
- whether the radiation dose is delivered all at once or fractionated so that sublethal damage can be repaired or replaced (i.e., a dose that would be lethal if given in a short time may not be lethal if given over a long time); and
- partial body dose versus whole body irradiation (a dose delivered as partial body irradiation is generally less severe than the same dose given uniformly to the whole body).

Sensitizers can be used to enhance the effects of radiation since radiation causes indirect chemical reactions. Selected chemicals can be injected just prior to or during irradiation to increase its cell-killing effect. For example, halogenized pyrimidines can be incorporated into DNA to amplify the radiation effect by as much as a factor of two; i.e., one rad with the sensitizer present will produce the same effect that would require two rads without it.

Radioprotectors are chemical compounds that diminish the effects of a given radiation dose. Molecules containing a sulfhydryl group (S+H) such as cysteine and cysteamine can diminish the effect by as much as a factor of two, theoretically allowing a person to sustain a radiation dose that might otherwise be very damaging or lethal. Unfortunately, what works for animals or a cell mixture turns out to be of little practical value for protecting humans because the levels required are quite toxic; i.e., the cure may be worse than the disease.

Stochastic and Non-Stochastic Effects

Biological effects of radiation are classified broadly as either stochastic or nonstochastic in nature; i.e., probabilistic or deterministic. Stochastic effects are cancers and genetic mutations whose probability of occurrence is determined by the random probability of radiation interactions that deposit damaging energy in cells. Any dose, therefore, has a certain probability, however low, of causing stochastic effects due to injury to a single cell or a small number of cells that is propagated through cell division. Once the effect occurs its severity is independent of the dose that set in motion the events that led to the expression of the effect.

Nonstochastic (or deterministic) effects of radiation are predictable ones that require the dose to be above a dose threshold before they occur. Examples of nonstochastic effects are nonmalignant skin damage (erythema), the formation of cataracts (opaqueness of the lens of the eye), hematological effects (changes in the composition of the blood), and impairment of fertility. The severity of these effects is directly related to the dose delivered, and the effects produced are due to collective injury of a substantial number of cells in the tissue. Even though nonstochastic processes require some sort of threshold dose before they become manifest, as in the case of cataract formation, the severity of the injury increases with dose. For example, cataracts occur at an acute dose of about 200 rads to the lens of the eye or an accumulated chronic exposure of about 500 rads, with the severity of the effect increasing with dose once the threshold is exceeded. The latent period is several years, and although the effect can be corrected surgically, prudent radiation protection policy is to use controls that prevent exposures from reaching the threshold (the recommended occupational limit for the lens of the eye is 15 rads/y).

SOMATIC EFFECTS OF RADIATION

Somatic effects occur in the individual who is exposed and can be classified as either (1) acute effects which appear within days to a few weeks after exposure, or (2) late somatic effects which occur due to chronic or low doses and which manifest themselves well after exposure, usually several years later.

Acute Radiation Effects present clinically observable symptoms in the exposed person and their severity increases with dose as detailed in Table 6-1. Nausea, vomiting, loss of appetite, and fatigue occur to varying degrees for doses above about 100 rads (1 Gy) with the most important effect being damage to the blood-forming organs. The first signs may appear within a few days, depending upon the dose, and the total effect may not develop for a few weeks. Severe changes occur when the dose is greater than 200 rads (2 Gy), and above 300 rads (3 Gy) death becomes more and more probable due to bone-marrow damage (or bone-marrow death). At doses between 500 and 2000 rads (5–20 Gy), symptoms appear within hours and death often occurs within a week or so due to damage to the lining of the intestinal tract (GI death).

Table 6-1. Acute Effects of Whole Body Radiation Doses > 100 Rads (1 Gy).

DOSE	EFFECTS
100–200 rads (1–2 Gy)	• Mild symptoms; unnoticeable except in special clinical tests • Some nausea and vomiting possible; increased heartbeat • At ≈ 200 rads, miscarriages of early pregnancies; temporary sterility
200–400 rads (2–4 Gy)	• Nausea and vomiting in a few hours • Mild weakness, fatigue for days to 3 weeks • After 3–4 weeks, sore throat, fever, general weakness, loss of hair, and in some cases development of purplish blotches on skin as white blood cell count drops • Recovery begins in 5th week; complete by 3–6 months
400–600 rads (4–6 Gy)	• Nausea and vomiting soon after; decrease in white blood cell count more severe • Most persons die from infections and internal bleeding • Bone marrow transplants may save the patient
600–1000 rads (6–10 Gy)	• Damage to the gastrointestinal (GI) tract resulting in severe vomiting, stomach cramps, and diarrhea • Within 1 week nausea and vomiting return, progressing to bloody diarrhea • Shock and death occur within 10–14 days after exposure due to destruction of the lining of the GI tract
>1000 rads (10 Gy)	• Damage to CNS (Central Nervous System) producing burning sensation minutes after exposure, quickly followed by severe nausea and vomiting and occasionally loss of consciousness • Within the hour the victim experiences loss of muscular coordination, mental confusion, and prostration • Death occurs within approximately 20–40 hours

Radiation doses above about 2000 rads (20 Gy) are fatal within a day or two due to breakdown of the central nervous system (an effect that is termed CNS death). It is extremely rare for persons working with radiation to receive radiation exposure above 100 rems (1 Sv); thus, acute effects of radiation occur primarily with radiation accidents.

Late Somatic Effects are associated with exposures well below 100 rems, and in this range, few, if any, harmful effects are readily apparent. Nevertheless, low-level doses of ionizing radiation are the focus of great concern because of their potential to cause cell damage which can eventually lead to congenital abnormalities, genetic defects, and cancer. Other late somatic effects include damage to tissues such as cataracts, potential shortening of life span, and sterility which may be either permanent or temporary. Life-shortening may also occur but information on this effect in humans is sparse; with the exception of tumor induction, the effect is at most a decrease in life by 10 days per rad of exposure (Bushong). By comparison, a person's life span may be reduced 5 minutes for each cigarette smoked or by 2 days for each ounce one is over-weight.

Teratogenic Effects are another form of somatic effects of radiation due to malformations of cells, tissues, or organs of a fetus exposed *in utero*. The effect is well known; children born to highly irradiated mothers exhibit abnormally small heads (microencephally), and mental retardation and decreased IQ have been observed in children exposed *in utero* at Hiroshima and Nagasaki. The greatest risk to a fetus occurs for exposure between the eighth and fifteenth week of pregnancy when the nervous system is undergoing the most rapid differentiation and proliferation of cells. Data on the teratogenic effects of radiation are difficult to obtain, and especially so at dose rates near background levels where any effects produced may be masked by effects produced by other agents or background radiation.

GENETIC EFFECTS OF RADIATION

Very little quantitative data are available on radiogenic mutations (genetic effects) in humans, particularly from low-dose exposures because (1) these mutations are interspersed over many generations, (2) some are so mild they are not noticeable, and (3) some mutagenic defects are so similar to nonmutagenic effects that they are not recorded as mutations. Mutations in dominant genes give rise to damage in the offspring of the first generation. Damage to offspring caused by recessive genes occurs only if the same altered gene is received from each parent. Unless genetic changes occur frequently, recessive damage will not show up for many generations.

The natural spontaneous mutation rate in humans is about 10% which hampers the study of radiation-induced mutations; and since such changes are not unique to radiation, it only serves to increase the frequency of an effect that may be so small even for high doses that it has proved difficult to observe. There is general agreement that the probability of inducing genetic changes increases linearly with dose and without threshold; however, the exact rate is uncertain because accurate studies require the use of large numbers of subjects over many generations, which is very difficult since large numbers are seldom available and the time between generations is so long. Statistically significant genetic effects have not been observed in children born to survivors of the atomic bombings of Hiroshima and Nagasaki due to parental exposure; however, the data show genetic changes which correlate with radiation exposure. Based on observed mutagenic changes (principally chromosome damage), it has been estimated that an average acute dose of about 200 rems of ionizing radiation will double the existing spontaneous mutation rate (the doubling dose in mice is about 30 to 40 rems).

Consequently, most of the information supporting the mutagenic character of ionizing radiation has been based on extensive studies in Drosophila and in experimental animals, mostly mice (UNSCEAR 77,82; NAS 72,80). Mutation rates calculated from these studies are extrapolated to humans (because the basic mechanisms of mutations are believed to be the same in all cells) and form the basis for estimating the genetic impact of ionizing radiation on humans (NAS 80, UNSCEAR 82). The International Commission on Radiation Protection and Measurements (ICRP) has estimated the risk of serious genetic effects in the first two generations after exposure to be about $4 \times 10^{-3}/\text{Sv}(4 \times 10^{-5}/\text{rem})$.

CARCINOGENIC EFFECTS OF RADIATION

Cancer, like genetic and teratogenic effects, occurs spontaneously in the population, and those that may be induced by radiation are indistinguishable from those that occur naturally. It is a leading cause of death with a fatality rate of about 18.9% and an overall cancer incidence rate in the US of about 25%. Radiogenic cancer is by far the greatest risk of radiation exposure and is typically 10 to 100 times larger than genetic risk. Even among heavily irradiated populations, the number of cancers and genetic defects is not known with either accuracy or precision simply because of sampling variability. Exposed groups (principally the Japanese) have not been followed for their full lifetime, so that information on ultimate effects is yet to be determined.

Evidence that radiation causes cancer in humans is extensive and is based on increases above the natural incidence in several groups of exposed persons. X rays were found to produce biological effects within one month of Roentgen's discovery, and the first human cancer attributed to x rays (skin cancer) was reported in 1902, some 6 years later. By 1911, 94 cases of radiation-related skin cancer and 5 cases of leukemia in humans had been reported, and radiation-related cancers had also been observed in experimental animals.

Cancer induction is believed to occur by a two-stage process: initiation and promotion. In the first or *initiation* stage, a lesion (that is, some injury or other deleterious effect) is produced in the DNA of one or a number of cells, but the cell(s) do not undergo uncontrolled division right away because of inherited hormonal, immunologic, or other protective agents. If, however, a promotion mechanism occurs some time later (the second stage), these protective mechanisms may fail to function, allowing the initiated cells to multiply out of control. Radiation is recognized as both an initiator and a promoter. If the effects of a given radiation dose are limited to initiation, subsequent events could promote the initial change into cancer. Examples of such events are viral infections, chemical irritants, failure of the immunological system, aging, and, of course, other radiation doses (or the same dose).

The cell or cells triggered into malignancy during the promotion stage are generally not the same cells initiated during irradiation but are the progeny of the original cells, often many generations removed from the initiating event. The induced effect is thus conferred on the irradiated cell and then carried through many generations of cell reproduction, suggesting a **chromosomal origin** of cancer.

The two-stage theory of cancer results in a *latent period* of several years between the time of irradiation and the appearance of many types of cancers. It also accounts for the fact that a predisposition for cancer seems to be inherited. If this is true, there may be segments of the population that are especially susceptible to radiation-induced cancer, or alternatively some who are especially resistant to cancer induced by radiation, or other agents as well.

Studies of Radiogenic Cancer

Data on radiogenic cancer in humans have been reviewed by the United Nations Scientific Committee on the Effects of Atomic Radiation (UNSCEAR 82, 88), and by the National Academy of Sciences Advisory Committee on the Biological Effects of Ionizing Radiations (NAS-BEIR Committee). The latter has provided three such reviews, in 1972 (BEIR I), in 1980 (BEIR III), and most recently in 1990 (BEIR V). In its 1972 BEIR I Report, the BEIR Committee determined that cancer was the most important effect of low-dose ionizing radiation instead of genetic effects which had dominated concern for radiation exposure up to that point; it has reaffirmed this major conclusion in subsequent reviews. All of the major studies have indicated that cancer may be induced by radiation in nearly all human tissues with two significant characteristics:

- leukemia (due to an abnormal increase in white blood cells) has a relatively short latency period and appears within 2 to 5 years of exposure; and
- solid tumors that occur in various tissues have a long latency period, usually 10 to 35 years (NAS 80, UNSCEAR 82).

These findings are based on several large studies of populations exposed to high levels of radiation as follows:

1. *Atomic Bomb Survivors*—Persons who survived the atomic bomb explosions at Hiroshima and Nagasaki, Japan, were exposed to whole-body external radiation doses from background levels up to more than 200 rads. The

Table 6-2. Cancer Mortality in Japanese Bomb Survivors based on BEIR V (Cember 1996).

Group	Number	Cancer Deaths	%
Exposed	41,719	3,435	8.3
Control	34,273	2,499	7.3
Excess	(0.01)(41,719)	417	—

population has been studied since 1950, and contains a control group of 34,273 unexposed persons and 41,719 exposed persons in which 417 excess fatal cancers have been observed (see Table 6-2). The excess cancer mortality above that expected based on the control group includes leukemia, thyroid, breast, lung cancer, esophageal and stomach cancer, colon cancer, cancer of urinary organs, and multiple myeloma. These same cancers were observed in the control group; therefore, the excess cannot be identified in specific persons who were exposed but only statistically indicated.

2. *Ankylosing Spondylitis*—A large group of patients were given x-ray therapy doses of 100 rads or more as treatment for ankylosing spondylitis of the spine during the years 1934 to 1954. British investigators have studied this group since about 1957 and have shown excess cancers in irradiated organs, including leukemia, lymphoma, lung and bone cancer, and cancer of the pharynx, esophagus, stomach, pancreas, and large intestine (UNSCEAR 77, NAS 80). Since individual doses are generally not well known, risk factors per unit of dose are difficult to derive, but it is clear that high doses caused excess cancers in this group of patients.

3. *Mammary Exposure*—Several groups of women who were exposed to x rays during diagnostic radiation of the thorax or during radiotherapy for conditions involving the breast have been studied. Most of the groups have been followed only for a relatively short time (about 15 years); however, a significant increase in the incidence of breast cancer has been observed (UNSCEAR 77). The dose that produced these effects averaged about 100 rads; a significant difference in the incidence of breast cancer was observed for those exposed from the front versus those from the back.

4. *Medical Treatment of Benign Conditions*—Several groups of persons who were medically treated with x rays to alleviate some benign conditions such as ringworm of the scalp have been studied. Excess cancer has developed in many of the organs irradiated (e.g., breast, brain, thyroid, and probably salivary glands, skin, bone, and pelvic organs) following doses ranging from less than 10 to more than 100 rads (UNSCEAR 77). Excess leukemia has also occurred in some groups. The follow-up period for most groups has been short, often less than 20 years.

5. *Thorotrast Studies*—Thorotrast, which is colloidal thorium dioxide, was injected into patients as an x-ray contrast medium. Since thorium is naturally radioactive, it and its transformation products caused cumulative doses of tens to hundreds of rads in the patients that received the injections, although the exact dose an individual patient received is generally uncertain. Patients in Denmark, Portugal, Japan, and Germany have been studied for about 40 years and patients in the US for about 10 years (UNSCEAR 77, NAS 80). An increased incidence of liver, bone, and lung cancer has been reported in addition to increased anemia, leukemia, and multiple myeloma.

6. *Diagnostic X-ray Exposure During Pregnancy*—Effects of x-ray exposure of the fetus during pregnancy have been studied in Great Britain since 1954, and several retrospective studies have been made in the US since that time (NAS 80, UNSCEAR 77). Increased incidence of leukemia and other childhood cancers may be induced in populations exposed to absorbed doses of 0.2 to 20 rads *in utero* (NAS 80, UNSCEAR 77).

7. *Underground Miners*—Studies of excess cancer mortality in US underground miners exposed to elevated levels of radon started in the 1950s and 1960s. Groups that have worked in various types of mines, including uranium and fluorspar, are being studied in the US, Canada, Great Britain, Sweden, China, and Czechoslovakia. Most of the miners studied had received high exposures to radon and its progeny, and had an increased incidence of lung cancer generally proportional to exposure levels (NAS 98–BEIR VI).

8. *Radium-Dial Painters*—Workers who ingested ^{226}Ra while painting the dials of clocks and various instruments with luminous paints containing radium have been studied for 35 to 40 years. They have shown an excess of leukemia and osteosarcoma for cumulative average doses from 200 to 1700 rads. Patients who received injections of ^{226}Ra or ^{224}Ra for medical purposes have been studied for 20 to 30 years (NAS 72,80) and have shown a similar excess of these cancers.

Despite the unfortunate circumstances surrounding radiation exposures of humans, the incidence of disease associated with such exposures provide direct evidence of effects attributable to radiation dose, at least for doses above 20 rems or so. The data show that the total number of cancers that eventually develop varies consistently with dose, age at exposure, and latency period. The studies also show that cancer can occur in all tissues and that high-LET radiations (alpha radiations or neutrons) represent a higher cancer incidence rate than low-LET radiations (x rays or gamma rays). The overwhelming body of epidemiological data in humans makes it unnecessary to extrapolate from animal studies as is the case for many substances presumed to be carcinogens. Radiation-induced cancers in animals are relevant, however, to the interpretation of human data (NAS 80) and contribute additional evidence on the mechanisms by which cancer occurs.

The most important data set for determining cancer incidence per rem (or rad) is the atomic bomb survivors. The Japanese survivors have been followed since about 1950 and will be for the rest of their lives, a period well in excess of 50 years after the doses were received. These data comprise a cohort of 41,719 exposed persons grouped into categories according to doses received and a control group of 34,273 persons. As of 1990, 3,435 cancer deaths had occurred in the exposed group versus 3,018 expected indicating that 417 are statistically attributable to radiation, as shown in Table 6-2. The cancer incidence rates versus dose in this population are the most reliable because the doses are known more accurately than in other studies, the exposure period is well known, the populations in the exposed and control group are quite uniform due in part to Japanese cultural patterns, and medical followup has been thorough and continuous for both groups.

The derivation of cancer risk coefficients per unit of dose is thus based primarily on the Japanese data; data from other human exposures corroborate the more definitive conclusions that can be drawn from data from the Japanese bomb survivors. It is noteworthy that most current radiation risk factors and policies derived therefrom are based on these 417 excess cancers that occurred over a period in excess of 50 years in the most highly exposed group of persons in human history. It is perhaps understandable, when this fact is noted, that so much controversy exists about the risks of radiation exposure at low doses and low dose rates.

ESTIMATING HEALTH EFFECTS OF RADIATION

The effects of radiation on human health are known more quantitatively than the effects of most other environmental pollutants because of the rather large body of human data. Even with human data, estimating the probability of fatal cancer, serious genetic effects, and other detrimental health effects resulting from exposure to ionizing radiation is complex. The process is even more difficult because radiogenic cancer and radiation-induced genetic defects do not occur very frequently compared to the natural baseline incidences, which are about 25% and 10%, respectively, and radiation-induced cancers are not unique to radiation exposure (as contrasted to mesothelioma from asbestos exposure); thus they cannot be distinguished from cancers produced by other causes. The extent of the cancer-inducing effect of radiation must be deduced from the observed increase (often small) of cancer above the natural incidence rate in groups of persons exposed to radiation. This is especially difficult at low doses where the effect, if it exists at all, may be hidden in the statistical fluctuations of the normal incidence of cancer in the group under consideration.

Leukemia was observed in Japanese bomb survivors who received whole-body doses in excess of 40 rads. Excess leukemia occurred as early as 3 years after exposure, thus demonstrating that this is an important consequence of acute radiation exposure. The relative risk factor for leukemia ranges from 3.89

to 6.40 for 100 rads of exposure with a best estimate at 90% confidence of 4.92. This relative risk factor means that for those persons exposed to 100 rads of whole-body radiation the leukemia rate is 4.92 times the spontaneous rate in a similar but unexposed population, and since the normal spontaneous rate is quite low this represents a significant and observable increase.

Radiation Risk Factors

Estimates of radiation risks differ from one group of radiation scientists to the next because the data on the effects of human exposure are subject to a number of interpretations. Each group uses essentially the same original data on doses and incidence, principally the Japanese survivor data, but each group constructs a risk model based on its observations of trends in the data. The observed excess cancer rates are for persons exposed above 20 rads or so and are demonstrable epidemiologically. These excess rates constitute a series of points with uncertainty bounds as shown in Figure 6-6. Most exposures of interest are much lower where effects are not demonstrated; thus, it is necessary to construct an appropriate model to assess risks at low doses and dose rates. A **Relative Risk Model**, which applies a risk factor to the normal cancer rate for specified age groups, can be used and since the normal rate increases with age, a relative risk model yields a larger number of projected cancers than an **Absolute Risk Model** which is presumed to be constant (cancers/rem) over all ages. The BEIR V Committee has determined, based on available evidence, that a relative risk model best represents the distribution of radiogenic cancers in a uniform population.

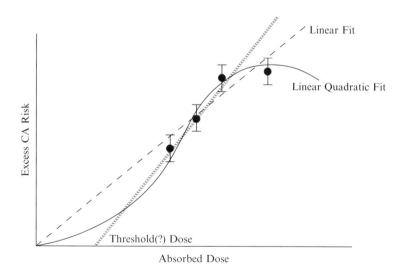

Figure 6-6. Excess cancer risk with increased total absorbed dose showing linear and linear-quadratic fits to observed excess incidence.

The data points in Figure 6-6 are based on reliable data on risk versus dose and demonstrate that risk generally increases with the absorbed dose. The high-dose data point turns downward because extensive radiation damage kills cells outright, and these cells do not survive to produce cancer. Measured risk of cancer have not been determined for doses of radiation below the plotted data points in Figure 6-6; the exact shape of the risk-dose curve in the low-dose region is not known, but may be determined by extrapolating or extending a line through the observed data points downward to low doses. Such an extra-polation is problematic and, perhaps because of this difficulty, controversial as well.

For example, a best-fit line can be drawn through the observed points in a way that it intersects the x axis suggesting that a dose is obtained below which it could be presumed that no effects occur (i.e., a threshold). Demonstration of such a point of intersection is very difficult, and perhaps because of it, most extrapolation models presume that effects that occur at high doses also occur at low doses only with lower probability. Extrapolation of the data points by a straight-line model yields a risk projection that is linear with dose. If, however, the extrapolation is done by drawing a best-fit curve through the points and zero risk and zero dose using a quadratic equation, a straight-line portion exists at low doses (the linear-quadratic relationship). The linear-quadratic model may thus account for repair of cellular damage and would project cancer risks that are lower by a factor of about 3 than the straight-line model. The linear-quadratic dose-response model assumes that events leading to "lesions" or permanent changes in cellular DNA require the formation of interacting pairs of "sublesions," one of which may be repaired before it can interact to form a lesion, with the probability of such repair increasing with time.

Despite continuing analysis and reanalysis of the Japanese experience, con-sensus on the dose response at environmental levels (about 100 mrad per year) is lacking. Most radiation risk models presume, based on available human data and the thesis that the same mechanisms that cause cancer at high doses also exist at low doses, that any dose of radiation involves a risk to human health that increases with dose, whether or not it is measurable. This presump-tion may be prudent but it is not demonstrable with available data. Regardless, there appears to be a growing consensus that radiation standards should be based on a linear extrapolation model because it is unlikely to understate the risk.

Risk factors from the original dose/risk data, primarily based on the Japan-ese survivors, are summarized in Table 6-3. The NAS BEIR estimates are for lifetime exposure and lifetime expression of induced cancers (NAS 72, NAS 80, NAS 90) whereas the UNSCEAR 77 report was based on a 40-year absolute risk model, which yielded smaller risk estimates. The BEIR V Committee determined that most of the leukemia observed at Hiroshima was caused by neutrons, and obtained mortality estimates (for lifetime exposure) of about 200 fatalities per 10^6 person-rems for an absolute risk projection and about 770 fatalities per 10^6 person-rem for a relative risk projection.

Table 6-3. Comparison of Estimates of the Risk of Fatal Cancer from a Lifetime Exposure at 1 Rem/Year (Low-LET Radiation).

Study	Cases per 10^6 person-rads	Risk Model
BEIR I (a)	667	Relative Risk
BEIR III (b)	403	Relative Risk
BEIR III (d)	169	Relative Risk
BEIR III (c)	158	Absolute Risk
BEIR I (c)	115	Absolute Risk
BEIR III (d)	67	Absolute Risk
UNSCEAR 77	165–300	Doses > 100 rad
UNSCEAR 82	75–175	Low doses
BEIR V (e)	770	Relative Risk
BEIR V (f)	509	Relative Risk
EPA 98 (g)	575	Relative Risk

(a) BEIR-I relative risk model.
(b) Absolute risk model for bone cancer and leukemia; relative risk model for all other cancer.
(c) Linear dose response, absolute risk model.
(d) Linear-quadratic dose response (Table V-4, BEIR III).
(e) Average of males and females without DDREF.
(f) Dose and Dose Rate Effectiveness Factor (DDREF) of 2.0 applied to all cancer incidence except for leukemia and breast cancer.
(g) EPA updating of risk data in Federal Guidance Report 13; endorsed by Science Advisory Board.

A dose and dose-rate effectiveness factor (DDREF) of 2.0 to 10 has been introduced in risk projections by the BEIR and UNSCEAR committees to account for the fact that the biological response per unit dose at low doses and dose rates from low-LET radiation may be overestimated if one extrapolates from observations made at high, acutely delivered doses. The EPA risk factor used a DDREF of at least 2.0 except for 1.0 for breast cancer risk which was derived from US data at low dose-rates. "Low dose" and "low dose-rate" are defined as < 0.2 Gy and < 0.1 mGy min^{-1}, respectively. By incorporating these considerations, the EPA and its Science Advisory Board has obtained a summed whole-body risk coefficient of 575 fatal cancers per 10^6 person-rem (5.75×10^{-4} per rem).

Most of the major radiation risk estimates have been based on whole-body doses received instantaneously; however, observed cancers in the Japanese survivors have been reported by tissue site, and it is now possible with these data and corroborating evidence from other studies to project organ risk factors, as shown in Table 6-4. The risk coefficients in Table 6-4 are age-averaged values for an assumed stationary US population of males and females and are based on statistically significant excess cancer mortality for leukemia, esophagus, stomach, colon, liver, lung, breast, ovary, urinary tract, and multiple myeloma. The data are based on observations that encompass 10 to 40 years after exposure for solid tumors and 5 to 40 years for leukemia. Risk factors for the thyroid and breast were based primarily on results of epidemiological studies of medical exposures of these organs because the data appear to be better.

The data in Table 6-4 incorporate a DDREF of 1.0 for breast cancer and leukemia and 2.0 for all other cancers. Also included in Table 6-4 are lethality fractions for each type of radiogenic cancer, which except for thyroid cancer, are generally assumed to be the same as for the totality of other cancers at that site (e.g., about 10 percent of thyroid cancers are fatal, but almost 100 percent of liver cancers are fatal). On the average, females have twice as many total cancers as fatal cancers following radiation exposure, and males have 1.5 times as many (NAS 80). Although many of the radiation-induced cancers are not fatal, it is difficult to weigh the intrinsic effects of impaired life relative to radiation exposure, and perhaps for these reasons most current assessments consider only the risk of fatal cancers.

Table 6-4. Age-averaged Site-specific Fatal Cancer Risk Factors per Rem from Low-dose, Low-LET Radiation.

Site	Risk/rem		Combined genders	
	Males	Females	Lethality (%)[a]	Risk/rem
Esophagus	7.30×10^{-6}	1.59×10^{-5}	95	1.17×10^{-5}
Stomach	3.25×10^{-5}	4.86×10^{-5}	90	4.07×10^{-5}
Colon	8.38×10^{-5}	1.24×10^{-4}	55	1.04×10^{-4}
Liver	1.84×10^{-5}	1.17×10^{-5}	95	1.50×10^{-5}
Lung	7.71×10^{-5}	1.19×10^{-4}	95	9.88×10^{-5}
Bone	9.40×10^{-7}	9.60×10^{-7}	70	9.50×10^{-7}
Skin[b]	9.51×10^{-7}	1.05×10^{-6}	0.2	1.00×10^{-6}
Breast	0.00	9.90×10^{-5}	50	5.06×10^{-5}
Ovary	0.00	2.92×10^{-5}	70	1.49×10^{-5}
Bladder	3.28×10^{-5}	1.52×10^{-5}	50	2.38×10^{-5}
Kidney	6.43×10^{-6}	3.92×10^{-6}	65	5.15×10^{-6}
Thyroid	2.05×10^{-6}	4.38×10^{-6}	10	3.24×10^{-6}
Leukemia	6.48×10^{-5}	4.71×10^{-5}	99	5.57×10^{-5}
Residual[c]	1.35×10^{-4}	1.63×10^{-4}	71	1.49×10^{-4}
Total	4.62×10^{-4}	6.83×10^{-4}		5.75×10^{-4}

[a]Lethality fractions (mortality-to-morbidity ratios), expressed in percent, are from Tables B-19 of and B-20 of ICRP Publication 60 (ICRP 1991).
[b]At least 83% of skin cancers are basal cell carcinomas ($\sim 0.01\%$ lethality) and the remainder are squamous cell carcinomas ($\sim 1\%$ lethality).
[c]Residual is a composite of all radiogenic cancers that are not explicitly identified by site.
Source: EPA Science Advisory Board.

Uncertainties in Risk Models for low doses and low dose-rates are difficult to quantify but are reasonably well understood qualitatively. Sampling variability could lead to sizable errors in estimates of excess relative risk, particularly for sites showing relatively small numbers of excess cancer deaths, and random errors in individual dose estimates may be as much as 25 to 45%, which could overestimate the average dose in the high-dose groups and, assuming a linear dose-response function, slightly underestimate the dose response. Sources of uncertainty in the Japanese bomb survivors include: (1) bias in gamma ray estimates; (2) uncertainty in the characterization of radiation shielding by buildings; and (3) uncertainty in the biological effectiveness of neutron doses.

RADIATION RISK CALCULATIONS

Whole-Body Irradiation Risks can be estimated using one of the risk factors listed in Table 6-3, each of which is expressed as the number of fatal radiogenic cancers per 10^6 person-rems of low-LET dose. The choice of a risk coefficient is highly dependent on one's persuasion towards a particular model of radiation bioeffects and on the circumstances of exposures to be assessed. All such models have been developed by highly reputable scientific committees that have studied the basic data. The US Environmental Protection Agency (EPA) uses a risk factor of 575 fatal radiogenic cancers per 10^6 person-rems $(5.75 \times 10^{-4}/\text{rem})$ to a uniform population of males and females based on updated data from the NAS–BEIR V Committee and other sources.

The Background Radiation Risk, unlike cigarette smoking, auto accidents, and other common risks, is neither voluntary nor readily avoidable; therefore, it has been used as a benchmark for estimated risks due to radiation exposure. The average dose to the US population from background radiation is about 90 to 100 mrem per year due to three major low-LET components: cosmic radiation, which averages about 28 mrad per year in the US; terrestrial sources, such as uranium, thorium, and radium in soil, which contribute an average of 26 mrad per year; and about 39 mrad per year of low-LET dose due to naturally occurring internal emitters. Extremes in background radiation dose do occur (see Chapter 4); however, a population-weighted analysis indicates that 80% of the US population receives annual radiation doses that are between 75 mrad and 115 mrad per year, minus any contribution from radon and its products.

Example 6-1: Estimate the 70-year lifetime radiation risk to a typical individual due to natural radiation exposure of 100 mrem/y.
Solution: An exposure for 70 years at 100 mrem per year yields an average lifetime dose of 7.0 rems, assuming a quality factor of 1.0. The risk of fatal cancer per person based on EPA's recommended risk factor is estimated to be

$$\frac{575 \text{ fatalities}}{10^6 \text{person-rem}} \times 7\text{rem} = 4 \times 10^{-3}$$

Vital statistics data indicate that the probability of dying of cancer in the US due to all causes is about 18.9%, suggesting that about 2% of all US cancer is due to background radiation.

Organ/Tissue Risks can be estimated using the risk factors in Table 6-4, as shown in Example 6-2 for breast-cancer risk due to mammography examinations. Mammography examinations are generally administered for asymptomatic women over 50 and the risk of diagnostic chest x rays is often used as a risk communication comparison for various sources of radiation exposure.

Example 6-2: What is the breast cancer risk due to mammography if the average dose to the female breast is about 300 mrem of low-LET x radiation? **Solution:** The risk factor from Table 6-4 is 9.9×10^{-5}/rem; thus, the breast cancer risk for each examination is about 3×10^{-5}, which is an additive risk depending on the number of examinations received in a person's lifetime.

Perspective on Radiation Risk is a difficult social concept that involves equity, fairness, and a willingness or unwillingness to accept chosen versus imposed risks. Although radiation risk models now allow relatively straightforward calculations of risks associated with radiation exposure, it is most difficult to communicate their relevance. One mechanism has been to compare the estimated increase in risk for a given type of exposure with other common risks, shown in Table 6-5, fully aware that they are different and the public perceives them differently.

Table 6-5. **Activities which Increase Risk of Death by 10^{-6}/y.**

Activity	Nature of Risk
Smoking 3 cigarettes (US)	Cancer, heart disease
Drinking 1/2 liter of wine	Cirrhosis of liver
Spending 1 hour in a coal mine	Black lung disease
Traveling 300 miles by car	Accident
Flying 6,000 miles by jet	Cancer due to cosmic rays
One chest ray (about 10 mrem)	Cancer caused by radiation

SUMMARY

The biological effects of radiation are known fairly well, and certainly well enough that it can be regulated and controlled with a degree of confidence that is perhaps greater than for other agents that affect humankind. The principal risk of radiation exposure is radiogenic cancer which is 10 to 100 times estimated genetic risks for the same dose. Genetic effects have been demonstrated only in animal studies, and although they are presumed to occur in humans, the major radiation studies in humans have been unable to statistically document their presence above natural levels. And since populations receive radiation doses on the order of just a few tens of millirems each year, any associated risks from cancer, life-shortening, or congenital defects have also not been demonstrated. Studies have tried, without success, to show a change in the health status of persons exposed to different levels of natural background radiation which varies considerably around the globe (people in Florida receive about 60 mrem/y, those in Colorado receive about 150 mrem/y, and persons in parts of Brazil and India receive more than 10 times these amounts with no observable difference in cancer incidence).

Radiation risks due to exposures of workers and the public can be estimated, but all such estimates are based on calculations using extrapolated risk models that produce results that are statistically small when compared with other societal health risks (e.g., improper diet, lack of exercise, uncontrolled consumption of alcohol, and use of tobacco products). Even though research on radiation effects, especially for the Japanese survivors, continues in order to develop the best information possible, public policies regarding radiation exposure presume that any dose of radiation, be it from x rays, research, or nuclear energy applications, carries with it some risk. Although radiation risks due to low doses of radiation on the order of tens of millirem per year can only be estimated as a statistical increase within the population being considered, a widely accepted policy is to avoid unnecessary and unproductive radiation exposure but to maintain a balanced perspective on the various sources of radiation exposure and their potential risks.

Acknowledgment

Suellen Cook contributed a first draft of this chapter which has incorporated material provided by Wendy Drake and J. Puskin and N. Nelson of the Radiation Bioeffects Group at the US Environmental Protection Agency.

ADDITIONAL RESOURCES

1 Publications of the Committee on Biological Effects of Ionizing Radiation (BEIR) of the National Academy of Sciences, in Particular BEIR I (1992) and BEIR V (1990).
2 Reports of the United Nations Scientific Committee on the Effects of Atomic Radiation (UNSCEAR), 1977, 1982, 1988, 2000.
3 Cember, H., *Introduction to Health Physics*, 3rd ed. (New York: McGraw-Hill, 1996).

PROBLEMS

1 A worker accidentally received an unknown but high dose of external gamma radiation. Vomiting and diarrhea occurred soon after and his

condition stabilized; however, after about a week he experienced bloody diarrhea. Based on these symptoms, what was the likely dose received?

2 A couple moved to Leadville, Co (elevation about 10,000 ft) to achieve a more relaxed lifestyle; however, the background radiation level due to cosmic rays and mountainous terrain is about 3 times their previous sea-level location of 80 mrem/y. If they reside in Leadville for 30 years, what increase in radiogenic risk does the change in lifestyle cost them?

3 The average dose to workers in nuclear power plants is 600 mrem/y. If the average career is 30 years, what is the average radiogenic increase in cancer risk per worker?

4 An old thorium processing site is reclaimed for a park but with a residual radiation level of 20μrem/h compared with an average exposure rate of 5μrem/h outside the park. If a local group of 100 persons spends an average of 4 hours per day at the park, what additional radiogenic cancer risk would over a 30-year period?

5 A public health agency proposes to do an x-ray screening program for a serious disease with a 1% frequency in the population; however, the program will result in doses to the thyroid of each person of 100 mrads every year for 20 years: (a) what average individual radiothyroid cancer risk would be projected; and (b) how does this compare to the total cancer risk currently experienced by each individual?

6 A nuclear facility that will provide 1200 jobs for the local population is proposed and its operation will be controlled to limit the exposure to an individual in the offsite environs to 5 mrem/y. What risk would be imposed to an individual for 25 years of operation; and (b) what would it be if the dose level were to be controlled at 25 mrem/y?

7 For each of the problems 2 to 5, construct an argument, based on risk and other considerations, of whether the activity involved should occur.

7

RADIATION SHIELDING

"Much of what we know of the Universe comes from information transmitted by photons."

– John Hubbel, 1995

Various materials, placed between a radiation source and a receptor, can affect the amount of radiation transmitted to the receptor. Such effects are due to attenuation and absorption of the emitted radiation within the source itself, in material used for encapsulation, or in a shield. The design, use, and effectiveness of shields, which are a form of radiation control, are directly related to how the various radiations interact in a medium (as described in Chapter 5). Radiation shielding is a very complex discipline but fortunately simplified configurations of point sources, line sources, and area and volume sources can be used conservatively to address most of the situations encountered in practical radiation protection.

SHIELDING OF ALPHA-EMITTING SOURCES

Alpha particles are easy to shield since their relatively large mass and charge limit their range in most media to a few tens of microns and to just a few centimeters in air. Consequently, most shielding problems for alpha emitters involve fixing the material in place so that it cannot become a source of contamination taken into the body by touch or as airborne particles. A layer of paint or other fixatives can be used to reduce even highly contaminated areas to nondetectable levels. Caution and continuous vigilance through inspections and surveys are required to assure that the fixative remains intact, especially on floors or high-use areas.

In a similar vein, the very short range of alpha particles makes their detection very difficult because they must be able to penetrate detector coverings or windows to reach the sensitive volume of the detector and be registered. It is easy to miss highly contaminated areas by failing to get close enough (most alphas are absorbed in a few cm of air), or by use of the wrong detector or one with a window that is too thick.

SHIELDING OF BETA-EMITTING SOURCES

Beta particles have a limited range in most media because they are charged particles and will dissipate all their energy as they traverse a medium by ionization and bremsstrahlung production. The path of ionization can be rather long and tortuous, but even so the mean range of most beta particles is no more than a few meters in air, and range is only a few mm in dense media. Because of this behavior, all of the beta particles emitted by a source can be stopped by determining the maximum range of the highest energy beta particles and choosing a thickness of medium that matches or exceeds the range. Most beta shields are based on this very practical approach, usually with a little more thickness just to be sure (see Example 7-2).

Attenuation of Beta Particles is readily achieved by most absorbers, including air, as shown by the curve plotted in Figure 7-1 as the number of beta particles measured versus absorber thickness. The straight-line part of the curve, which is typical of most beta source/absorber configurations, suggests that beta particle attenuation can be represented as an exponential function for absorber thicknesses less than the maximum range of the beta particles in a particular medium, or

$$I(x) = I_0 e^{-\mu_{\beta,i}(\rho x)}$$

where $\mu_{\beta,i}(g/cm^2)$ is a function of the maximum beta particle energy, I_0 is the initial intensity (or number) of beta particles, $I(x)$ is the intensity (or number) observed after penentrating an absorber of thickness x, and ρx is the density thickness (g/cm²) of the absorber. Values of $\mu_{\beta,i}$ are determined (and expressed) in units of cm²/g; therefore, it is necessary to specify the absorber as a density thickness ρx with units of g/cm². The absorption of beta particles is not truly exponential, or probabilistic, as the curve for absorption would suggest, but is an artifact produced by a combination of the continually varying spectrum of beta energies and the scattering of the particles by the absorber.

Approximate relationships for the "beta absorption" coefficient have been determined for three types of absorbers (air, water or tissue, and solid materials) because of their utility in radiation protection. For air

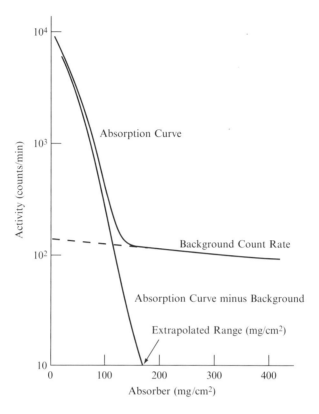

Figure 7-1. Decrease in measured activity of a beta-particle source vs mg/cm² of absorber thickness. Subtraction of the extrapolated background curve produces a new curve of activity versus absorption for just the source, and this curve is extrapolated to obtain the range (mg/cm²) of beta particles of maximum energy.

$$\mu_{\beta,\,air} = 16(E_{\beta\,max} - 0.036)^{-1.4}$$

For water (or tissue which is equivalent)

$$\mu_{\beta,\,tissue} = 18.6(E_{\beta\,max} - 0.036)^{-1.37}$$

And, for a solid medium, i,

$$\mu_{\beta,\,i} = 17(E_{\beta\,max})^{-1.14}$$

where values of μ_β are in units of cm²/g (based on measurements with low-Z absorbers, usually aluminum) and $E_{\beta\,max}$ is in units of MeV.

Example 7-1: Beta particles with $E_{\beta\,max}$ of 2.0 MeV produce a flux of 1000 beta particles/cm$^2 \cdot$ s incident on an aluminum ($\rho = 2.7$ g/cm^3) absorber 0.1 mm thick. What will be the flux of beta particles reaching a receptor just beyond the foil?
Solution: The beta absorption coefficient μ_β is

$$\mu_\beta = 17 \times (2)^{-1.14}$$
$$= 7.714\,\text{g/cm}^2$$

The density thickness of the aluminum absorber is

$$\rho x = 2.7\,\text{g/cm}^3 \times 0.01\,\text{cm} = 0.027\,\text{g/cm}^2$$

Therefore, the number of beta particles penetrating the foil is

$$I(x) = I_0 e^{-\mu_{\beta(\rho x)}} = 1000 e^{-7.714\,\text{cm}^2/\text{g} \times 0.027\,\text{g/cm}^2}$$
$$= 812.0\,\beta\text{s that reach the receptor.}$$

Range-Energy of Beta Particles

The range of beta particles in a medium is directly related to their maximum energy. A curve like the one in Figure 7-1 is constructed by observing the change in activity as absorber thickness is added between the source and a detector, which must have a thin window to prevent absorption of the beta particles before they can be counted. Aluminum is the absorber medium most often used for determining ranges of beta particles, but most absorber materials will have essentially the same electron density because, with the exception of hydrogen, the Z/A ratio varies only slowly with Z.

As shown in Figure 7-1, the counting rate, when plotted against absorber thickness (mg/cm^2 or g/cm^2) on a logarithmic scale, decreases as a straight line, or very nearly so, over a large fraction of the absorber thickness, eventually tailing off into another straight-line region represented by the background, which is always present. It is very important to extrapolate the background curve and subtract it from the measured curve to create another curve that represents absorption due to beta particles alone. The point where the corrected beta absorption curve meets the x axis is the range, $R_{\beta,\,max}$, traversed by the most energetic particles emitted. The measured range can be used to determine the maximum energy of the beta source (and vice versa) by use of the curve in Figure 5-5. Although mathematical relationships can be used to calculate ranges and energies for low-energy electrons, they can generally be obtained more accurately from the empirical absorption curve in Figure 5-5.

It is important to consider all the attenuating materials that may be present between the beta source and the point of interest. For example, as shown in Figure 7-2, a given configuration may contain, in addition to the shield material

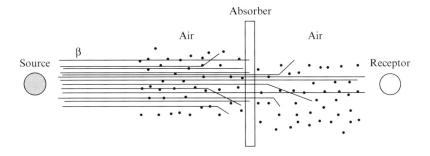

Figure 7-2. Schematic of various absorbing materials between a beta-emitting source and a layer or volume of interest.

enclosing the beta emitter, a thickness of air, and yet another layer between the air gap and the sensitive volume of interest (e.g., a dead layer of skin or a detector window). And, of course, geometry factors for a given configuration and a backscatter factor if appropriate also influence the dose rate from a source. These, with the exception of backscatter, tend to add additional conservatism in the effectiveness of beta shields, but not always if bremsstrahlung is considered.

Bremsstrahlung Effects for Beta Shielding

Since beta particles are high-speed electrons, they produce bremsstrahlung, or "braking radiation" in passing through matter, especially if the particles are energetic and the absorbing medium is a high-Z material. Bremsstrahlung production is minor in air or tissue, both low-Z materials, but can be significant for shielding designs where more dense materials are typically used. It is necessary, therefore, to know the fraction of the beta-particle energy that is converted to photons in order to account for them in dosimetry or radiation shielding. Although empirical relations have been developed for the fraction of the total electron energy that is converted to photon production, measured values of radiation yields for monoenergetic electrons are the most practical. These data are listed in Table 7-1 for monoenergetic electrons for several absorbers, and illustrate the effects of changes in electron energy and Z of absorber. Once the fraction of beta energy converted to photons is determined, it can be used in the usual way in radiation protection determinations (see Example 5-7).

Assessments of radiation exposure due to bremsstrahlung can be done conservatively by assuming that each bremsstrahlung photon has an energy equal to the maximum energy of the beta particles emitted by the source. The average beta particle energy could also be assigned to each bremsstrahlung produced which would increase the bremsstrahlung photon flux by about 3, but each

Table 7-1. Percent Radiation Yield for Electrons of Initial Energy E on Different Absorbers.

E (MeV)	Absorber (Z)					
	Water	Air	Al (13)	Cu (29)	Sn (50)	Pb (82)
0.100	0.058	0.066	0.135	0.355	0.658	1.162
0.200	0.098	0.111	0.223	0.595	1.147	2.118
0.300	0.133	0.150	0.298	0.795	1.548	2.917
0.400	0.166	0.187	0.368	0.974	1.900	3.614
0.500	0.198	0.223	0.435	1.143	2.224	4.241
0.600	0.229	0.258	0.501	1.307	2.530	4.820
0.700	0.261	0.293	0.566	1.467	2.825	5.363
0.800	0.293	0.328	0.632	1.625	3.111	5.877
0.900	0.325	0.364	0.698	1.782	3.391	6.369
1.000	0.358	0.400	0.764	1.938	3.666	6.842
1.250	0.442	0.491	0.931	2.328	4.340	7.960
1.500	0.528	0.584	1.101	2.720	4.998	9.009
1.750	0.617	0.678	1.274	3.113	5.646	10.010
2.000	0.709	0.775	1.449	3.509	6.284	10.960
2.500	0.897	0.972	1.808	4.302	7.534	12.770
3.000	1.092	1.173	2.173	5.095	8.750	14.470

photon would have about one-third the energy and the two tend to offset each other. Example 7-2 demonstrates the process based on photons equal to $E_{\beta\,max}$.

Example 7-2: One curie of $^{32}P(E_{\beta\,max} = 1.71\,\text{MeV})$ is dissolved in 50 mL of water for an experiment. If it is to be kept in a polyethylene ($\rho = 0.93$) bottle, (a) how thick should the wall be to stop all the beta particles emitted by ^{32}P, and (b) what thickness of lead would be required to assure that the dose equivalent rate due to bremsstrahlung photons will be less than 1 mrem/h at 1 m?
Solution: (a) From Figure 5-5, the maximum range of 1.71 MeV beta particles is $810\,\text{mg/cm}^2$ or $0.81\,\text{g/cm}^3$; therefore, the thickness of polyethylene required to absorb 1.71 MeV beta particles is

$$\frac{0.81\,\text{g/cm}^2}{0.93\,\text{g/cm}^3} = 0.87\,\text{cm}$$

(b) Essentially all of the bremsstrahlung can be assumed to be produced in the water solution since the thin polyethylene walls are of similar density. From Table 7-1, the fraction of 1.71 MeV beta particles converted to photons due to absorption in water is about 6×10^{-3}, and since the average beta energy for ^{32}P (Appendix C) is 0.695 MeV, the photon-energy emission-rate fraction is

$$E_{rad} = 3.7 \times 10^{10} \, t/s \times 0.695 \, \text{MeV}/t \times 6 \times 10^{-3}$$
$$= 1.55 \times 10^8 \, \text{MeV/s due to bremsstrahlung production}$$

If this fraction of the total beta energy is assumed to produce photons equal to the maximum beta energy of 1.71 MeV, a conservative assumption, then the photon emission rate is

$$\frac{1.55 \times 10^8 \, \text{MeV/s}}{1.71 \, \text{MeV/photon}} = 9.1 \times 10^7 \, \text{photons/s}$$

or 7.24×10^2 photons/cm$^2 \cdot$s at 1 m, each of which is presumed to have an energy of 1.71 MeV. From Table 5-1 (or Table 7-2 for H$_2$O, a reasonable surrogate), μ_{en}/ρ for 1.71 MeV photons in tissue is 0.027 cm^2/g by interpolation; therefore, this flux produces an energy absorption rate in tissue of

$$E_{ab} = 7.24 \times 10^2/\text{cm}^2 \cdot \text{s} \times 1.71 \, \text{MeV} \times 0.027 \, \text{cm}^2/\text{g} \times 1.6022 \times 10^{-6} \, \text{erg/MeV} \times 3600 \, \text{s/h}$$
$$= 0.193 \, \text{ergs/g} \cdot \text{h}$$

which corresponds to an absorbed dose rate of 1.93 mrad/h, or a dose equivalent rate of 1.93 mrem/h. Also, from Table 7-2, the attenuation coefficient, μ, for 1.71 MeV photons in lead is 0.565 cm^{-1}, and the thickness of lead required to attenuate the flux to 1 mrem/h is

$$1 \, \text{mrem/h} = (1.93 \, \text{mrem/h}) \, e^{-0.565x}$$

and x = 1.16 cm.
If the effect of the Compton scattered photons (an effect known as buildup) is considered, which would be appropriate, iterative calculations will result in a somewhat thicker shield.

Beta Shield Designs must account for the yield of bremsstrahlung photons. High-Z materials such as lead (Z = 82) should not be used to shield high activity sources such as ^{32}P($E_{\beta \, max} = 1.71 \, \text{MeV}$) and ^{90}SrY($E_{\beta \, max} = 2.28 \, \text{MeV}$) because 8 to 12% of the emitted energy could be in the form of bremsstrahlung photons. Since the bremsstrahlung yield is higher in high-Z materials, the most practical approach for shielding beta sources is to use a layer of plastic, aluminum, or other low-Z material to absorb all the beta particles while minimizing bremsstrahlung production, then to add a layer of lead or other dense material to absorb any bremsstrahlung and characteristic x rays that are produced in the beta shield, as shown in Example 7-2.

SHIELDING OF PHOTON SOURCES—"GOOD" AND "POOR" GEOMETRIES

As discussed in Chapter 5, photon interactions with matter are very different from those of charged particles. When x or gamma rays traverse matter, some are absorbed, some pass through without interaction, and some are scattered as lower energy photons in directions that are quite different from those in the primary beam. The attenuation of a photon beam by an absorber is characterized as occurring in "narrow beam geometry" or "broad beam geometry," as shown in Figure 7-3.

Narrow Beam Geometry (also called "good geometry") exists when every photon that interacts is either absorbed or scattered out of the primary beam such that it will not impact a small receptor some distance away. When it exists, only those photons that have passed through the absorber without any kind of interaction will reach the receptor, and each one of these photons will have all of its original energy. This situation exists when the primary photons are confined to a narrow beam and the receptor (e.g., a detector) is small and located far enough away that scattered photons have a sufficiently large angle that they leave the beam without hitting the receptor. Readings taken with and without the absorber in place will yield the fraction of photons removed from the narrow beam, by whatever process. Attenuation coefficients for photons of a particular energy are determined under narrow beam conditions.

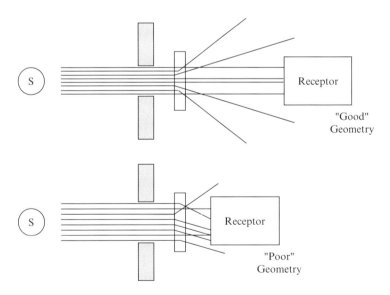

Figure 7-3. Attenuation of a photon beam in "narrow beam geometry" in which all interactions reduce the number of photons reaching a small receptor, and in "broad beam geometry" in which scattered photons also reach the receptor.

Broad Beam Geometry (also called "poor geometry") exists in most practical conditions when tissue is exposed or a shield is used to attenuate a photon source. It occurs when a significant fraction of scattered photons also reach the receptor of interest, as shown in Figure 7-3. Typical broad beam geometry configurations are a source enclosed by an absorber (e.g., a "point" source in a lead pig), a shielded detector, or any other condition where a "broad beam" of photons strikes an absorber. Scattered photons will be degraded in energy according to the angle through which they are scattered, and the flux of the photons reaching a receptor will consist of many different energies, in addition to those originally present in the primary beam. The pattern of energy deposition in the receptor (e.g., a person or a detector), will be governed by the energy distribution and the exact geometrical arrangement used. The amount of energy deposited in an absorber or a receptor under such conditions is very difficult to determine analytically.

Photons incident on an absorber under narrow beam geometry conditions either interact to be absorbed or scattered completely out of the beam; i.e., an either/or reaction which is a function of probability. The interaction probability per unit path length in an absorption medium is a constant; thus, the number of photons, $I(x)$, traversing a medium of thickness x and reaching a receptor, as shown in Figure 7-3, is proportional to the number incident on the absorber, which can be expressed mathematically as a decreasing function with thickness of absorber,

$$-\frac{dI}{dx} = \mu I$$

where the constant of proportionality μ is the total attenuation coefficient of the medium for the photons of interest. This relationship between beam intensity and attenuation coefficient is valid for photoelectric, Compton, and, for photons of sufficient energy, pair production interactions. It also holds for each mode of attenuation separately, or in combination with each other; i.e., τ, σ, κ, or the sum of all. For a narrow beam of photons of monoenergetic energy the number $I(x)$ penetrating an absorber of thickness x (i.e., without interaction in the medium) is found by rearranging and integrating

$$\int_{I_0}^{I(x)} dI/I = -\mu \int_0^x dx$$

to yield

$$\ln I(x) - \ln I_0 = -\mu x$$

or

$$\ln I(x) = -\mu x + \ln I_0$$

which is an equation of a straight line with a slope of $-\mu$ and a y-axis intercept (i.e., with no absorber) of $\ln I_0$. This can be simplified by the law of logarithms to

$$\ln \frac{I(x)}{I_0} = -\mu x$$

and since the natural logarithm of a number is the exponent to which the base e is raised to obtain the number, this expression translates to

$$\frac{I(x)}{I_0} = -e^{-\mu x}$$

or

$$I(x) = I_0 e^{-\mu x}$$

The Attenuation Coefficient, μ, is dependent on the particular absorber medium and the photon energy, values of which are listed in Table 7-2 for several common absorbers used for shielding photons. Also included are mass energy absorption coefficients, μ_{en}/ρ, for use when needed (μ/ρ or μ_{en} may be obtained from the data by dividing or multiplying by ρ, as appropriate). Values of μ generally increase as the Z of the absorber increases as shown in Figure 7-4 for lead (a high-Z material), iron, and aluminum (low-Z). High-Z materials can significantly increase photoelectric interactions for low-energy photons and pair-production interactions for high-energy photons, and materials can be selected to take advantage of this effect.

Absorption Edges can occur, as shown in Figure 7-4, due to strong resonance absorption when the energy of the incoming photon is the exact amount required to remove a K electron entirely from the atom. The attenuation (or absorption) cross section increases sharply at this energy, hence the term "absorption edge." As long as the photon energy is below this amount, there can be no resonance absorption by K-shell electrons, and the absorption cross sections due to photoelectric and Compton interactions will decrease with increasing photon energy. Even at the exact $K_{\alpha 1}$ energy, resonance absorption is still impossible, since there is only enough energy to move a K-shell electron up to the L shell, which by the Pauli principle is full; therefore, resonance absorption occurs only when the photon energy is enough to eject the electron from the atom; i.e., it equals the electron binding energy. As photon energies increase above the absorption edge, the attenuation coefficient once again decreases. Similar interactions produce L-absorption edges and M-absorption edge which are also observed in Figure 7-4 for lead (these edges also occur in other materials but they are difficult to observe).

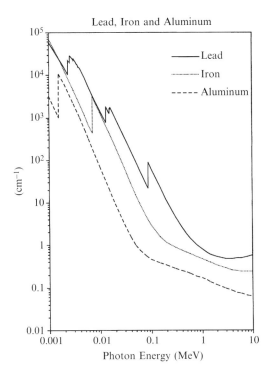

Figure 7-4. Variation of μ versus Z which also illustrates the effect of Z on resonance absorption to create absorption "edges." The K-absorption edge is prominent at 88 keV for lead, and the substructures of the L shell (shown in detail) and the M shell are also visible.

Shielding of "Good Geometry" Photon Sources

Calculations of photon attenuation by various absorbers are relatively straightforward as shown in Example 7-3 if the photon intensity and energy are known and ideal narrow beam conditions exist. When these conditions exist,

$$I(x) = I_0 e^{-\mu x}$$

where I_0 is the initial photon intensity (usually expressed as a fluence or flux), $I(x)$ is the photon intensity after passing through an absorber of thickness x, and $\mu(\text{cm}^{-1})$ is the total attenuation coefficient, which accounts for all interaction processes, including scattering reactions, that remove photons from the beam. Such calculations are always done on the basis of narrow-beam conditions and then adjusted by a buildup factor that accounts for scattered photons (see below).

Table 7-2. Photon Attenuation (μ) and Mass Energy-Absorption (μ_{en}/ρ) Coefficients for Selected Materials (J. H. Hubbell and S. M. Seltzer).

Energy (keV)	Dry Air (Sea Level) ($\rho = 0.001205 \, g/cm^3$)		Water, Liquid ($\rho = 1.00 \, g/cm^3$)		Aluminum ($\rho = 2.699 \, g/cm^3$)	
	μ (cm^{-1})	μ_{en}/ρ (cm^2/g)	μ (cm^{-1})	μ_{en}/ρ (cm^2/g)	μ (cm^{-1})	μ_{en}/ρ (cm^2/g)
10	0.0062	4.742	5.329	4.944	70.795	25.43
15	0.0019	1.334	1.673	1.374	21.4705	7.487
20	0.0009	0.5389	0.8096	0.5503	9.2873	3.094
30	4.26E-4	0.1537	0.3756	0.1557	3.0445	0.8778
40	2.99E-4	0.0683	0.2683	0.0695	1.5344	0.3601
50	2.51E-4	0.0410	0.2269	0.0422	0.9935	0.1840
60	2.26E-4	0.0304	0.2059	0.0319	0.7498	0.1099
70*	2.10E-4	0.0255	0.1948	0.0289	0.6130	0.0713
80	2.00E-4	0.0241	0.1924	0.0272	0.5447	0.0551
100	1.86E-4	0.0233	0.1707	0.0255	0.4599	0.0379
150	1.63E-4	0.0250	0.1505	0.0276	0.3719	0.0283
200	1.49E-4	0.0267	0.1370	0.0297	0.3301	0.0275
300	1.29E-4	0.0287	0.1186	0.0319	0.2812	0.0282
400	1.15E-4	0.0295	0.1061	0.0328	0.2504	0.0286
500	1.05E-4	0.0297	0.0969	0.0330	0.2279	0.0287
600	9.71E-5	0.0295	0.0896	0.0328	0.2106	0.0285
662*	9.34E-5	0.0293	0.0862	0.0326	0.2024	0.0283
800	8.52E-5	0.0288	0.0787	0.0321	0.1846	0.0278
1,000	7.66E-5	0.0279	0.0707	0.0310	0.1659	0.0269
1,173*	7.05E-5	0.0271	0.0650	0.0301	0.1526	0.0261
1,250	6.86E-5	0.0267	0.0632	0.0297	0.1483	0.0257
1,333*	6.62E-5	0.0263	0.0611	0.0292	0.1435	0.0253
1,500	6.24E-5	0.0255	0.0575	0.0283	0.1351	0.0245
2,000	5.36E-5	0.0235	0.0494	0.0261	0.1167	0.0227
3,000	4.32E-5	0.0206	0.0397	0.0228	0.0956	0.0202
4,000	3.71E-5	0.0187	0.0340	0.0207	0.0838	0.0188
5,000	3.31E-5	0.0174	0.0303	0.0192	0.0765	0.0180
6,000	3.04E-5	0.0165	0.0277	0.0181	0.0717	0.0174
6,129*	3.01E-5	0.0164	0.0274	0.0180	0.0713	0.0173
7,000*	2.83E-5	0.0159	0.0258	0.01723	0.0683	0.0170
7,115*	2.82E-5	0.0158	0.0256	0.0172	0.0680	0.0170
10,000	2.46E-5	0.0145	0.0222	0.0157	0.0626	0.0165

Table 7-2. (*continued*)

Iron ($\rho = 7.874\,\text{g/cm}^3$)		Copper ($\rho = 8.96\,\text{g/cm}^3$)		Lead ($\rho = 11.35\,\text{g/cm}^3$)	
μ (cm^{-1})	μ_{en}/ρ (cm^2/g)	μ (cm^{-1})	μ_{en}/ρ (cm^2/g)	μ (cm^{-1})	μ_{en}/ρ (cm^2/g)
1343.304	136.90	1934.464	148.4	1482.31	124.7
449.4479	48.96	663.4880	57.88	1266.66	91.0
202.2043	22.60	302.7584	27.88	980.1860	68.99
64.3778	7.251	97.8432	9.349	344.1320	25.36
28.5747	3.155	43.5635	4.163	162.9860	12.11
15.4173	1.638	23.4125	2.192	91.2654	6.740
9.4882	0.9555	14.2733	1.2900	56.9884	4.1490
6.2318	0.5836	9.2401	0.7933	35.0670	2.6186
4.6866	0.4104	6.8365	0.5581	27.4557	1.9160
2.9268	0.2177	4.1073	0.2949	62.9812	1.9760
1.5465	0.0796	1.9864	0.1027	22.8589	1.0560
1.1496	0.0483	1.3969	0.0578	11.3330	0.5870
0.8654	0.0336	1.0026	0.0362	4.5752	0.2455
0.7402	0.0304	0.8434	0.0312	2.6366	0.1370
0.6625	0.0291	0.7492	0.0293	1.8319	0.0913
0.6066	0.0284	0.6832	0.0283	1.4165	0.0682
0.5821	0.0280	0.6555	0.0279	1.2419	0.0587
0.5275	0.0271	0.5918	0.0268	1.0067	0.0464
0.4720	0.0260	0.5287	0.0256	0.8061	0.0365
0.4329	0.0251	0.4840	0.0246	0.7020	0.0315
0.4213	0.0247	0.4714	0.0243	0.6669	0.0299
0.4071	0.0243	0.4553	0.0239	0.6369	0.0285
0.3845	0.0236	0.4303	0.0232	0.5927	0.0264
0.3358	0.0220	0.3768	0.0216	0.5228	0.0236
0.2851	0.0204	0.3225	0.0202	0.4806	0.0232
0.2608	0.0199	0.2973	0.0199	0.4764	0.0245
0.2477	0.0198	0.2847	0.0200	0.4849	0.0260
0.2407	0.0200	0.2785	0.0203	0.4984	0.0274
0.2403	0.0200	0.2781	0.0203	0.5002	0.0276
0.2370	0.0202	0.2757	0.0206	0.5141	0.0287
0.2368	0.0203	0.2755	0.0207	0.5161	0.0289
0.2357	0.0211	0.2780	0.0217	0.5643	0.0318

(*continues*)

Table 7-2. (*continued*)

Energy (keV)	Polyethylene ($\rho = 0.93\,\text{g/cm}^3$)		Concrete, ($\rho = 2.3\,\text{g/cm}^3$)		Concrete, Barite ($\rho = 3.35\,\text{g/cm}^3$)	
	μ (cm^{-1})	μ_{en}/ρ (cm^2/g)	μ (cm^{-1})	μ_{en}/ρ (cm^2/g)	μ (cm^{-1})	μ_{en}/ρ (cm^2/g)
10	1.9418	1.781	47.04	19.37	357.45	99.60
15	0.6930	0.4834	14.61	5.855	120.63	33.63
20	0.4013	0.1936	6.454	2.462	55.44	15.27
30	0.2517	0.0593	2.208	0.7157	18.60	4.912
40	0.2116	0.0320	1.163	0.2995	39.70	4.439
50	0.1938	0.0244	0.7848	0.1563	22.35	3.206
60	0.1832	0.0224	0.6118	0.0955	13.88	2.266
70*	0.1754	0.0221	0.5131	0.0638	9.04	1.630
80	0.1695	0.0227	0.4632	0.0505	6.593	1.211
100	0.1599	0.0242	0.3997	0.0365	3.759	0.7138
150	0.1427	0.0279	0.3303	0.0290	1.482	0.2659
200	0.1304	0.0303	0.2949	0.0287	0.8603	0.1369
300	0.1132	0.0328	0.2523	0.0297	0.4891	0.0641
400	0.1013	0.0337	0.2250	0.0302	0.3698	0.0447
500	0.0925	0.0339	0.2050	0.0303	0.3119	0.0372
600	0.0855	0.0338	0.1894	0.0302	0.2762	0.0334
662*	0.0823	0.0335	0.1822	0.0299	0.2635	0.0325
800	0.0751	0.0330	0.1662	0.0294	0.2324	0.0295
1,000	0.0675	0.0319	0.1494	0.0284	0.2048	0.0274
1,173*	0.0621	0.0310	0.1374	0.0276	0.1851	0.0256
1,250	0.0604	0.0305	0.1336	0.0272	0.1810	0.0254
1,333*	0.0584	0.0300	0.1291	0.0268	0.1744	0.0248
1,500	0.0550	0.0291	0.1216	0.0260	0.1647	0.0240
2,000	0.0471	0.0268	0.1048	0.0240	0.1439	0.0223
3,000	0.0376	0.0233	0.0851	0.0212	0.1231	0.0208
4,000	0.0320	0.0209	0.0740	0.0195	0.1135	0.0204
5,000	0.0283	0.0192	0.0669	0.0184	0.1085	0.0205
6,000	0.0257	0.0179	0.0620	0.0176	0.1059	0.0207
6,129*	0.0254	0.0178	0.0616	0.0175	0.1056	0.0207
7,000*	0.0236	0.0169	0.0585	0.0171	0.1047	0.0210
7,115*	0.0235	0.0168	0.0582	0.0170	0.1046	0.0211
10,000	0.0199	0.0151	0.0524	0.0162	0.1051	0.0221

Table 7-2. (*continued*)

Leaded Glass ($\rho = 2.23$ g/cm^3)		Glass ($\rho = 2.23$ g/cm^3)		Gypsum ($\rho = 2.69$ g/cm^3)	
μ (cm^{-1})	μ_{en}/ρ (cm^2/g)	μ (cm^{-1})	μ_{en}/ρ (cm^2/g)	μ (cm^{-1})	μ_{en}/ρ (cm^2/g)
640.04	98.21	38.02	16.42	124.62	39.7700
532.25	69.91	11.63	4.828	39.3384	12.4700
408.53	52.52	5.122	1.995	17.2598	5.3520
143.37	19.27	1.781	0.5684	5.5530	1.5870
67.98	9.198	0.9680	0.2361	2.6427	0.6645
38.15	5.118	0.6739	0.1235	1.5928	0.3403
23.9035	3.152	0.5390	0.0765	1.1212	0.2001
15.0524	1.930	0.4623	0.0523	0.8594	0.1256
11.6252	1.458	0.4215	0.0423	0.7323	0.0929
26.2235	1.498	0.3695	0.0321	0.5808	0.0573
9.641	0.8042	0.3097	0.0273	0.4402	0.0344
4.8640	0.4508	0.2779	0.0276	0.3827	0.0306
2.0507	0.1934	0.2384	0.0289	0.3220	0.0298
1.2340	0.1114	0.2127	0.0295	0.2857	0.0299
0.8888	0.0768	0.1939	0.0296	0.2598	0.0299
0.7078	0.0592	0.1792	0.0294	0.2398	0.0296
0.6602	0.0551	0.1723	0.0292	0.2293	0.0294
0.5238	0.0425	0.1573	0.0287	0.2102	0.0288
0.4301	0.0347	0.1413	0.0277	0.1888	0.0279
0.3803	0.0307	0.1300	0.0269	0.1743	0.0270
0.3624	0.0293	0.1264	0.0265	0.1687	0.0266
0.3467	0.0281	0.1221	0.0261	0.1633	0.0262
0.3239	0.0264	0.1151	0.0253	0.1537	0.0254
0.2841	0.0237	0.0992	0.0234	0.1328	0.0235
0.2539	0.0228	0.0805	0.0207	0.1089	0.0210
0.2449	0.0234	0.0700	0.0190	0.0957	0.0196
0.2438	0.0244	0.0633	0.0180	0.0874	0.0187
0.2462	0.0254	0.0587	0.0172	0.0820	0.0181
0.2468	0.0255	0.0581	0.0171	0.0815	0.0181
0.2504	0.0263	0.0553	0.0167	0.0783	0.0178
0.2510	0.0264	0.0549	0.0166	0.0779	0.0178
0.2671	0.0288	0.0496	0.0158	0.0719	0.0173

* Coefficients for these energies were interpolated using polynomial regression.

Example 7-3: If a narrow beam of 2000 monoenergetic photons of 1.0 MeV is reduced to 1000 photons by a slab of copper 1.31 cm thick, determine the total linear attenuation coefficient of the copper slab for these photons.
Solution: (a) Since $I/I_0 = e^{-\mu x}$, μ is determined by taking the natural logarithm of each side of the equation, or

$$\ln(0.5) = -\mu(1.31 \text{ cm})$$

$$\mu = \frac{0.69315}{1.31 \ cm} = 0.5287 \ cm^{-1}$$

Half-Value and Tenth-Value Thickness

As was done in Chapter 3 for radioactivity, it is also useful to express the exponential attenuation of photons in terms of a half-thickness, $x_{1/2}$, or Half-Value Layer (HVL). The HVL (or the half-value thickness) is the thickness of absorber required to decrease the intensity of a beam of photons to one-half its initial value, or

$$\frac{I(x)}{I_0} = \frac{1}{2} = e^{-\mu x_{1/2}}$$

which can be solved for $x_{1/2}$ to yield

$$x_{1/2} = HVL = \frac{\ln 2}{\mu}$$

Half-value thicknesses (Table 7-3) can be used in calculations of photon attenuation in much the same way half-life is used for radioactive transformation, (see Example 7-4). And, in similar fashion, a tenth-value thickness or layer (TVL) is

$$x_{1/10} = \frac{\ln 10}{\mu} = \frac{2.3026}{\mu}$$

Example 7-4: If the HVL for iron is 1.47 cm for 1 MeV photons, and the dose rate from a source is 800 mrem/h, calculate: (a) the thickness of iron required to reduce the dose rate to 200 mrem/h; and (b) the thickness of iron required to reduce it to 150 mrem/h.
Solution: (a) It is observed that $\frac{800}{2^n} = 200$ when $n = 2$; therefore, 2 HVLs, or 2.94 cm of iron, will reduce 800 mrem/h to 200 mrem/h.

(b) Since, $I = I_0 e^{-\mu x}$, the thickness of iron required to reduce an 800 mR/h exposure rate to 150 mR/h is

$$150 = 800 e^{-\mu x}$$

where μ is determined from Table 7-2 as 0.47 cm^{-1}.

Table 7-3. Half-value Layers (in cm) vs Photon Energy for Various Materials.

Energy (MeV)	Lead (11.35 g/cm^3)	Iron (7.874 g/cm^3)	Aluminum (2.699 g/cm^3)	Water (1.00 g/cm^3)	Air (0.001205 g/cm^3)	Stone Concrete (2.30 g/cm^3)
0.1	0.011	0.237	1.507	4.060	3.726×10^3	1.734
0.3	0.151	0.801	2.464	5.843	5.372×10^3	2.747
0.5	0.378	1.046	3.041	7.152	6.600×10^3	3.380
0.662	0.558	1.191	3.424	8.039	7.420×10^3	3.806
1.0	0.860	1.468	4.177	9.802	9.047×10^3	4.639
1.173	0.987	1.601	4.541	10.662	9.830×10^3	5.044
1.332	1.088	1.702	4.829	11.342	1.047×10^4	5.368
1.5	1.169	1.802	5.130	12.052	1.111×10^4	5.698
2.0	1.326	2.064	5.938	14.028	1.293×10^4	6.612
2.5	1.381	2.271	6.644	15.822	1.459×10^4	7.380
3.0	1.442	2.431	7.249	17.456	1.604×10^4	8.141
3.5	1.447	2.567	7.813	19.038	1.747×10^4	8.828
4.0	1.455	2.657	8.270	20.382	1.868×10^4	9.366
5.0	1.429	2.798	9.059	22.871	2.094×10^4	10.361
7.0	1.348	2.924	10.146	26.860	2.449×10^4	11.846
10.0	1.228	2.940	11.070	31.216	2.817×10^4	13.227

Calculated from Attenuation Coefficients listed in Table 7-2 (Data provided by Hubbell and Seltzer, 1995.)

$$\ln\frac{150}{800} = -0.47x$$

$$x = 3.55\,\text{cm}$$

Shielding of "Poor Geometry" Photon Sources

Most practical situations in radiation protection do not satisfy the ideal conditions of narrow beam geometry, and a calculated value of $I(x)$ based on the attenuation coefficient, μ, which is determined under narrow-beam geometry conditions will thus underestimate the number of photons reaching the receptor. The design of shields based on an assumed narrow beam (i.e., for tabulated values of μ) will not be thick enough. This circumstance is best dealt with by including a buildup factor, B, in the attenuation equation

$$I(x) = I_0 B e^{-\mu x}$$

where B, which is greater than 1.0, accounts for photons scattered towards the receptor from regions outside the primary beam. Experimentally determined values of B for photons of different energies absorbed in various media are listed in Table 7-4 for point sources (earlier compilations also included buildup factors for broad beams, but these are out of date). The buildup factor B is dependent on the absorbing medium, the photon energy, the attenuation coefficient for specific-energy photons in the medium, and the absorber thickness x. The latter two are depicted in Table 7-4 as μx which is dimensionless and is commonly referred to as the number of mean free paths or the number of relaxation lengths, the value that reduces the initial flux (or exposure) by $1/e$ (the mean free path can be thought of as the mean distance a photon travels in an absorber before it undergoes an absorption or scattering interaction that removes it from the initial beam). It is also clear from Table 7-4 that B can be quite large, especially for low-energy photons, and that calculations of the radiation exposure associated with a beam of photons would be significantly in error if it were not included. Care must be exercised, however, to select the correct set of values for the source being evaluated.

BUILDUP FACTORS

Estimates of photon fields for broad beam conditions are made by first using "good geometry" conditions and then adjusting the results to account for the buildup of scattered photons, as follows:

- First, determine the total attenuation of the beam by calculating the change in intensity for the energy/absorber combination as

Table 7-4. Exposure Buildup Factors for Photons of Energy E vs μx (for Various Absorbers).

μx	Energy (MeV)									
	0.1	0.5	1	2	3	4	5	6	8	10
Al										
0.5	1.91	1.57	1.45	1.37	1.33	1.32	1.28	1.26	1.22	1.19
1.0	2.86	2.28	1.99	1.78	1.68	1.62	1.54	1.49	1.41	1.35
2.0	4.87	4.07	3.26	2.66	2.38	2.19	2.04	1.94	1.76	1.64
3.0	7.07	6.35	4.76	3.62	3.11	2.78	2.54	2.37	2.11	1.93
4.0	9.47	9.14	6.48	4.64	3.86	3.38	3.04	2.81	2.46	2.22
5.0	12.1	12.4	8.41	5.72	4.64	3.99	3.55	3.26	2.82	2.52
6.0	14.9	16.3	10.5	6.86	5.44	4.61	4.08	3.72	3.18	2.83
7.0	18.0	20.7	12.9	8.05	6.26	5.24	4.61	4.19	3.55	3.14
8.0	21.3	25.7	15.4	9.28	7.1	5.88	5.14	4.66	3.92	3.46
10.0	28.7	37.6	21.0	11.9	8.83	7.18	6.23	5.61	4.68	4.12
15.0	51.7	78.6	37.7	18.9	13.4	10.5	9.03	8.09	6.64	5.87
20.0	81.1	137	57.9	26.6	18.1	14.0	11.9	10.7	8.68	7.74
25.0	117	213	81.3	34.9	23.0	17.5	14.9	13.3	10.8	9.74
30.0	159	307	107	43.6	28.1	21.0	18.0	16.0	13.0	11.8
Fe										
0.5	1.26	1.48	1.41	1.35	1.32	1.3	1.27	1.25	1.22	1.19
1.0	1.4	1.99	1.85	1.71	1.64	1.57	1.51	1.47	1.39	1.33
2.0	1.61	3.12	2.85	2.49	2.28	2.12	1.97	1.87	1.71	1.59
3.0	1.78	4.44	4	3.34	2.96	2.68	2.46	2.3	2.04	1.86
4.0	1.94	5.96	5.3	4.25	3.68	3.29	2.98	2.76	2.41	2.16

(continues)

Table 7-4. (*continued*)

μx	Energy (MeV)									
	0.1	0.5	1	2	3	4	5	6	8	10
5.0	2.07	7.68	6.74	5.22	4.45	3.93	3.53	3.25	2.81	2.5
6.0	2.2	9.58	8.31	6.25	5.25	4.6	4.11	3.78	3.24	2.87
7.0	2.31	11.7	10.0	7.33	6.09	5.31	4.73	4.33	3.71	3.27
8.0	2.41	14.0	11.8	8.45	6.96	6.05	5.38	4.92	4.2	3.71
10.0	2.61	19.1	15.8	10.8	8.8	7.6	6.75	6.18	5.3	4.69
15.0	3.01	35.1	27.5	17.4	13.8	11.9	10.7	9.85	8.64	7.88
20.0	3.33	55.4	41.3	24.6	19.4	16.8	15.2	14.2	12.9	12.3
25.0	3.61	79.9	57.0	32.5	25.4	22.1	20.3	19.3	18.2	18.1
30.0	3.86	108	74.5	40.9	31.7	27.9	25.9	25.1	24.5	25.7
Sn										
0.5	1.35	1.32	1.33	1.27	1.29	1.28	1.31	1.31	1.33	1.31
1.0	1.38	1.61	1.69	1.57	1.56	1.51	1.55	1.54	1.6	1.57
2.0	1.41	2.15	2.4	2.17	2.07	1.96	1.97	1.94	2.04	2.05
3.0	1.43	2.68	3.14	2.82	2.64	2.45	2.43	2.38	2.51	2.61
4.0	1.45	3.16	3.86	3.51	3.25	3.0	2.54	2.87	3.05	3.27
5.0	1.47	3.63	4.6	4.23	3.92	3.6	3.52	3.43	3.69	4.09
6.0	1.49	4.14	5.43	5.03	4.68	4.29	4.19	4.09	4.45	5.07
7.0	1.5	4.64	6.27	5.87	5.48	5.04	4.93	4.83	5.34	6.26
8.0	1.52	5.13	7.11	6.74	6.32	5.84	5.74	5.65	6.36	7.69
10.0	1.54	6.13	8.88	8.61	8.19	7.65	7.63	7.63	8.94	11.5
15.0	1.58	8.74	13.8	14.0	13.8	13.5	14.1	14.9	19.7	29.6

mfp										
20.0	72.1	40.7	26.4	23.5	21.1	20.5	20.1	19.1	11.4	1.61
25.0	168	79.7	43.9	36.2	30.6	28.1	26.9	24.5	14.0	1.64
30.0	377	150	69.3	53.0	42.1	36.6	34.2	30.0	16.5	1.66

Pb

mfp										
0.5	1.28	1.3	1.26	1.25	1.21	1.23	1.21	1.2	1.14	1.51
1.0	1.51	1.51	1.42	1.41	1.36	1.4	1.4	1.38	1.24	2.04
2.0	2.01	1.9	1.73	1.71	1.67	1.73	1.76	1.68	1.39	3.39
3.0	2.63	2.36	2.08	2.05	2.02	2.1	2.14	1.95	1.52	5.6
4.0	3.42	2.91	2.49	2.44	2.4	2.5	2.52	2.19	1.62	9.59
5.0	4.45	3.59	2.96	2.88	2.82	2.93	2.91	2.43	1.71	17.0
6.0	5.73	4.41	3.51	3.38	3.28	3.4	3.32	2.66	1.8	30.6
7.0	7.37	5.39	4.13	3.93	3.79	3.89	3.74	2.89	1.88	54.9
8.0	9.44	6.58	4.84	4.56	4.35	4.41	4.17	3.1	1.95	94.7
10.0	15.4	9.73	6.61	6.03	5.61	5.56	5.07	3.51	2.1	294
15.0	50.8	25.1	13.7	11.4	9.73	8.91	7.44	4.45	2.39	5800
20.0	161	62.0	26.6	19.9	15.4	12.9	9.98	5.27	2.64	1.33×10^5
25.0	495	148	49.6	32.9	23.0	17.5	12.6	5.98	2.85	3.34×10^6
30.0	1470	344	88.9	52.2	32.6	22.5	15.4	6.64	3.02	8.87×10^7

U

mfp										
0.5	1.27	1.28	1.24	1.23	1.19	1.2	1.19	1.17	1.11	1.04
1.0	1.49	1.48	1.38	1.37	1.32	1.35	1.35	1.31	1.19	1.06
2.0	1.97	1.85	1.66	1.64	1.6	1.64	1.65	1.53	1.3	1.08
3.0	2.56	2.27	1.98	1.94	1.89	1.95	1.95	1.73	1.39	1.1
4.0	3.31	2.78	2.33	2.27	2.21	2.28	2.25	1.9	1.45	1.11
5.0	4.26	3.39	2.74	2.63	2.55	2.62	2.56	2.07	1.52	1.12
6.0	5.43	4.11	3.19	3.04	2.93	2.99	2.88	2.23	1.58	1.13
7.0	6.9	4.96	3.71	3.49	3.33	3.38	3.19	2.38	1.63	1.14
8.0	8.73	5.97	4.28	3.99	3.76	3.78	3.51	2.52	1.68	1.14

(*continues*)

Table 7-4. (*continued*)

μx		Energy (MeV)								
	0.1	0.5	1	2	3	4	5	6	8	10
10.0	1.16	1.77	2.78	4.17	4.64	4.72	5.14	5.68	8.61	13.9
15.0	1.18	1.96	3.35	5.84	7.06	7.72	9.1	11.0	20.8	43.4
20.0	1.2	2.11	3.82	7.54	9.8	11.6	15.1	20.1	48.6	131
25.0	1.22	2.23	4.23	9.27	12.8	16.5	23.7	35.4	110	385
30.0	1.23	2.33	4.59	11.0	16.0	22.5	36.0	60.4	244	1100
H₂O										
0.5	2.37	1.6	1.47	1.38	1.34	1.31	1.28	1.27	1.23	1.2
1.0	4.55	2.44	2.08	1.83	1.71	1.63	1.56	1.51	1.43	1.37
2.0	11.8	4.88	3.62	2.81	2.46	2.24	2.08	1.97	1.8	1.68
3.0	23.8	8.35	5.5	3.87	3.23	2.85	2.58	2.41	2.15	1.97
4.0	41.3	12.8	7.68	4.98	4	3.46	3.08	2.84	2.46	2.25
5.0	65.2	18.4	10.1	6.15	4.8	4.07	3.58	3.27	2.82	2.53
6.0	96.7	25.0	12.8	7.38	5.61	4.68	4.08	3.7	3.15	2.8
7.0	137	32.7	15.8	8.65	6.43	5.3	4.58	4.12	3.48	3.07
8.0	187	41.5	19.0	9.97	7.27	5.92	5.07	4.54	3.8	3.34
10.0	321	62.9	26.1	12.7	8.97	7.16	6.05	5.37	4.44	3.86
15.0	938	139	47.7	20.1	13.3	10.3	8.49	7.41	5.99	5.14
20.0	2170	252	74.0	28	17.8	13.4	10.9	9.42	7.49	6.38
25.0	4360	403	104	36.5	22.4	16.5	13.3	11.4	8.96	7.59
30.0	7970	594	139	45.2	27.1	19.7	15.7	13.3	10.4	8.78
Air										
0.5	2.35	1.6	1.47	1.38	1.34	1.31	1.29	1.27	1.23	1.2
1.0	4.46	2.44	2.08	1.83	1.71	1.63	1.57	1.52	1.43	1.37

2.0	11.4	4.84	3.6	2.81	2.46	2.25	2.09	1.97	1.8	1.68
3.0	22.5	8.21	5.46	3.86	3.22	2.85	2.6	2.41	2.15	1.97
4.0	38.4	12.6	7.6	4.96	4	3.46	3.11	2.85	2.5	2.26
5.0	59.9	17.9	10.0	6.13	4.79	4.07	3.61	3.28	2.84	2.54
6.0	87.8	24.2	12.7	7.35	5.6	4.69	4.12	3.71	3.17	2.82
7.0	123	31.6	15.6	8.61	6.43	5.31	4.62	4.14	3.51	3.1
8.0	166	40.1	18.8	9.92	7.26	5.94	5.12	4.57	3.84	3.37
10.0	282	60.6	25.8	12.6	8.97	7.19	6.13	5.42	4.49	3.92
15.0	800	134	47.0	20	13.4	10.3	8.63	7.51	6.08	5.25
20.0	1810	241	72.8	27.9	17.9	13.5	11.1	9.58	7.64	6.55
25.0	3570	385	103	36.2	22.5	16.7	13.6	11.6	9.17	7.84
30.0	6430	567	136	45	27.2	19.9	16.1	13.6	10.7	9.11
Concrete										
0.5	1.89	1.57	1.45	1.37	1.33	1.31	1.27	1.26	1.22	1.19
1.0	2.78	2.27	1.98	1.77	1.67	1.61	1.53	1.49	1.41	1.35
2.0	4.63	4.03	3.24	2.65	2.38	2.18	2.04	1.93	1.76	1.64
3.0	6.63	6.26	4.72	3.6	3.09	2.77	2.53	2.37	2.11	1.93
4.0	8.8	8.97	6.42	4.61	3.84	3.37	3.03	2.8	2.45	2.22
5.0	11.1	12.2	8.33	5.68	4.61	3.98	3.54	3.25	2.81	2.51
6.0	13.6	15.9	10.4	6.8	5.4	4.6	4.05	3.69	3.16	2.8
7.0	16.3	20.2	12.7	7.97	6.2	5.23	4.57	4.14	3.51	3.1
8.0	19.2	25.0	15.2	9.18	7.03	5.86	5.09	4.6	3.87	3.4
10.0	25.6	36.4	20.7	11.7	8.71	7.15	6.15	5.52	4.59	4.01
15.0	44.9	75.6	37.2	18.6	13.1	10.5	8.85	7.86	6.43	5.57
20.0	69.1	131	57.1	26.0	17.7	13.9	11.6	10.2	8.31	7.19
25.0	97.9	203	80.1	33.9	22.5	17.4	14.4	12.7	10.2	8.86
30.0	131	290	106	42.2	27.4	20.9	17.3	15.2	12.2	10.6

$$I(x) = I_0 e^{-\mu x}$$

where $I(x)$ is the unscattered intensity (flux, exposure, etc.) and μ is the linear attenuation coefficient (cm^{-1})

• Second, a buildup factor, chosen for the particular photon energy/ absorber combination, is selected to account for unscattered photons, or

$$I_b(x) = B I_0 e^{-\mu x}$$

where B is obtained for the absorber in question, the photon energy, and the particular value of μx (or number of mean free paths (mfp)). It is usually necessary to interpolate between the energies and the μx values (sometimes both) to obtain the proper value. This procedure is illustrated in Examples 7-6 and 7-7.

Example 7-5: A beam of 1.0 MeV gamma rays is emitted from a point source and produces a flux of $10,000 \, \gamma/cm^2 \cdot s$. If 2 cm of iron is placed in the beam, what is the best estimate of the flux after passing through the shield?
Solution: First, determine the flux based on narrow-beam conditions. The atteunation coefficient for 1 MeV photons in iron is $0.472 \, cm^{-1}$ and the value of μx (or mfp) for 2 cm of iron is 0.944; therefore, the attenuated unscattered flux would be

$$I(x) = I_0 e^{-\mu x} = (10,000 \, \gamma/cm^2 \cdot s) \, e^{-0.472 \times 2 \, cm} = 3,890 \, \gamma/cm^2 \cdot s$$

This needs to be adjusted by the buildup factor which is determined by interpolation from Table 7-4 as ~ 1.8, and the best estimate of $I_b(x)$ is

$$I_b(x) = 1.8 \times 3,890 \cong 7,000 \, \gamma/cm^2 \cdot s$$

Example 7-6: A fluence of $10^5 \gamma/cm^2$ of 1.5 MeV photons strikes a 2 cm–thick piece of lead. What is the best estimate of the total energy that reaches a receptor beyond the lead shield?
Solution: The linear attenuation coefficient μ for 1.5 MeV photons in lead is $0.59247 \, cm^{-1}$, and for narrow beam conditions

$$I(x) = 10^5 \, \gamma/cm^2 e^{-(0.59247)(2)}$$
$$= 3.06 \times 10^4 \, \gamma/cm^2$$

From Table 7-4, the buildup factor for 1.5 MeV photons in lead for $\mu x = 1.185$ is found by interpolation to be 1.45, and

$$I_b(x) = B I_0 e^{-\mu x} = 1.45(3.06 \times 10^4) = 4.43 \times 10^4 \, \gamma/cm^2$$

The buildup fluence will contain primary-beam photons and scattered photons of lower energy; however, despite the presence of lower-energy scattered photons, the best estimate of the energy fluence is, conservatively,

$$I_b(x)_E = 4.43 \times 10^4 \, \gamma/cm^2 \times 1.5 \, MeV = 6.65 \times 10^4 \, MeV/cm^2$$

GAMMA FLUX FOR DISTRIBUTED SOURCES

Although many radiation sources can, with its ease and utility, be represented as a point or approximate point source, many "real world" exposure conditions cannot. Typical examples are a long pipe or tube containing radioactive material which approximates a line source, a contaminated area that is representative of a disc or infinite planar source, and various volume sources. Practical approaches can be used to integrate the photon flux from a collection of small "points" or "areas" spread over these geometries, and once the flux is determined it can be used in the usual way to calculate radiation exposure.

Line Sources

Line sources can many times be practically considered as "infinitely long" with respect to a point P, located a distance x away from a lineal source as shown in Figure 7-5a. The differential flux $d\phi_\ell$ at a point P located a distance r from a point-sized element dl that emits S_L gamma rays per cm of length is

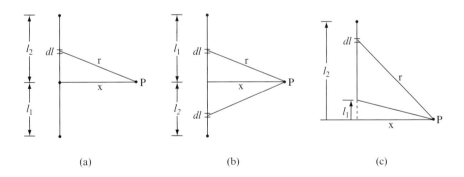

Figure 7-5. Schematic of a line source of radioactive material that emits S_L gamma rays/s per cm for (a) an "infinitely long" ($\ell \cong \infty$) length that produces a flux at P and a finite length that is either (b) the sum of two right-angle segments, one of which may be zero, or (c) the difference between two right-angle segments.

$$d\phi_\ell = \frac{S_L}{4\pi r^2} d\ell$$

which can be integrated by substituting $\ell^2 + x^2$ for r^2. For an "infinite line" source the limits of integration are $-\infty$ and $+\infty$ and the flux at P is

$$\phi_{\ell,\infty}(\gamma/cm^2 \cdot s) = \frac{S_L}{4x}$$

The photon flux for a "finite line" source is best obtained by integration of individual right angle segments with P located perpendicular to each segment as shown in Figure 7-5b. By breaking the line into segments from 0 to ℓ_1 and 0 to ℓ_2, the integration is performed over each segment from $\ell = 0$ to $\ell = \ell_i$ to yield

$$\phi_\ell = \frac{S_L}{4\pi x}\left[\tan^{-1}\left[\frac{\ell_i}{x}\right]\right]$$

where x is the perpendicular distance from P to the line source. For the two segmented line in Figure 7-5b

$$\phi_\ell = \frac{S_L}{4\pi x}\left[\tan^{-1}\left[\frac{\ell_2}{x}\right] + \tan^{-1}\left[\frac{\ell_1}{x}\right]\right]$$

where the $\tan^{-1}$ solutions are expressed in radians. For the geometry shown in Figure 7-5c the solution is such that the second term is $-\tan^{-1}\ell_1/x$ since in this case ℓ_1 is a fictitious source and thus must be subtracted, or

$$\phi_\ell = \frac{S_L}{4\pi x}\left[\tan^{-1}\left[\frac{\ell_2}{x}\right] - \tan^{-1}\left[\frac{\ell_1}{x}\right]\right]$$

A very useful general rule for line sources is that the fluence or fluence rate from line sources of photon emitters varies as $1/x$.

Example 7-7: (a) What is the gamma flux at a point one meter (3.291 ft) from a 100-ft pipe and 30 ft from one end if the lineal gamma emission rate is $10\,\gamma/cm \cdot s$, and (b) how much does the flux change if the pipe is assumed to be an infinite line source?
Solution: (a) The gamma flux at 30 ft from one end is

$$\phi_{\ell,\infty} = \frac{10\gamma/cm \cdot s}{4\pi(100\,cm)}\left[\tan^{-1}\left(\frac{70\,ft}{3.281\,ft}\right) + \tan^{-1}\left(\frac{30\,ft}{3.281\,ft}\right)\right]$$
$$= 7.96 \times 10^{-3}[87.32° + 83.76°]$$

and since one radian $= 57.3°$

$$\phi_\ell = 7.96 \times 10^{-3} \times 2.986\,\text{radians} = 2.38 \times 10^{-2}\gamma/\text{cm}^2 \cdot \text{s}$$

(b) If an infinite line source is assumed that has the same lineal emission rate, the flux is

$$\phi_{\ell,\infty} = \frac{S_L}{4x} = \frac{10\gamma/\text{cm} \cdot \text{s}}{4 \times 100\,\text{cm}} = 2.5 \times 10^{-2}\gamma/\text{cm}^2 \cdot \text{s}$$

which conservatively overestimates the flux by about 5%.

Shielding of a radioactive line source is often done by placing a sheet of metal close to it or constructing an annular ring around the lineal source. Interposing shielding material between the line source and the receptor point of interest not only attenuates the photons emitted but introduces a scattered component that must be accounted for by an appropriate buildup factor. **An approximation** (emphasis added) to the shielded photon flux at a point P away from the lineal source can be made by assuming that all photons penetrate through the shield in a perpendicular direction even though the true direction from most points along the line will be along an angular (and longer) path through the shield. This approximation overestimates the flux and the resultant exposure rate by only a few percent and is reasonable to use in lieu of the more complex integration necessary to account for an angular distance that constantly varies.

Example 7-8: What would be the photon flux in Example 7-7 if the 100-ft line source were to be enclosed in a 2-cm thick annular shield constructed of lead? **Solution:** It is reasonable to assume that all photons are emitted perpendicular to the pipe which simplifies the consideration of buildup; therefore,

$$\phi_s = \phi_{us} B e^{-\mu x}$$

where B is determined from Table 7-4 to be 1.56 for 1 MeV photons and since $\mu = 0.8061\,\text{cm}^{-1}$ for 1 MeV photons in lead, $\mu x = 0.8061\,\text{cm}^{-1} \times 2\,\text{cm} = 1.61$; thus, the shielded flux is

$$\phi_s = (2.38 \times 10^{-2}\gamma/\text{cm}^2 \cdot \text{s})(1.56)(e^{-1.61})$$
$$= 7.4 \times 10^{-3}\gamma/\text{cm}^2 \cdot \text{s}$$

which can be used in the usual way to determine radiation exposure in air or radiation absorbed dose.

Disc and Planar Sources

Spill areas on floors and/or contaminated sites can produce a flux of gamma rays and an exposure field at points above them. These area sources can be modeled as a disc source made up of a series of annular rings as shown in

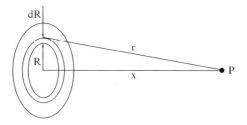

Figure 7-6. A disc source with a uniform emission rate of S_A photons/s per unit area produces a differential flux at point P at distance r from a thin ring of width dR and area $2\pi R dR$.

Figure 7-6. If the activity is uniformly spread over the area such that gamma rays are emitted isotropically as $S_A\gamma/cm^2 \cdot s$, the differential flux contributed by each ring at a point P a distance x away from the center of the disc is

$$d\phi = S_A\frac{2\pi R dR}{4\pi r^2} = S_A\frac{R dR}{2r^2}$$

The total unscattered flux at P is obtained by integrating over all annuli encompassed in the disc of radius R

$$\phi(P) = \frac{S_A}{2}\int_0^R \frac{R dR}{r^2}$$

and since $r^2 = x^2 + R^2$

$$\phi_u(P)_A = \frac{S_A}{4}\ln\left[1 + \frac{R^2}{x^2}\right]$$

This is the general solution for all values of x, which is the distance P from the center of a disc-shaped source of radius R. When the area source is very much larger than the distance x, as it typically is, the flux is

$$\phi_u(P)_A = \frac{S_A}{2}\ln\left[\frac{R}{x}\right], R \gg x$$

Therefore, for large-area sources the gamma flux (and radiation exposure) decreases as $1/x$.

Volume Sources

Volume sources such as large drums or tanks of radioactive material produce scattered photons due to self-absorption by the medium in which they are produced. Radiation exposure calculations for these various cylindrical and spherical geometries are fairly complex; however, good information can be obtained for such geometries by dividing them up into several subdivisions which are considered as approximate "point" sources and summing the contributions of each. Such calculations are generally conservative in that they tend to overestimate exposure, but considerable simplification of the calculations is obtained and errors in the estimates are not large. Calculations for numerous small volume elements increases the accuracy, and integration over all possible differential volume sources is of course the most precise.

Example 7-9: Estimate the exposure rate in air at 1 m. (a) from the center of a 10-cm diameter plastic pipe if the pipe is 1.5 m long and contains 5 Ci of ^{137}Cs in H_2O, and (b) with a 3-cm thick lead shield between the pipe and the receptor.

Solution: The 1.5 m pipe is divided into 5 segments of 30 cm, each of which contains 1 Ci as shown in Figure 7-7. The activity of each subvolume is assumed to be concentrated at a point in the center and the exposure at P is calculated for the "point sources" at 1, 2, and 3, and because of symmetry the exposures for 4 and 5 are the same as 1 and 2. The individual exposures are added to obtain the total. It is necessary to determine the distance between each "point source" and P, the angular thickness of water and Pb to be penetrated, the attenuation of flux by each of these thicknesses, and the buildup flux for each thickness. The results of these determinations, which are shown in Table 7-5, yield the following results:

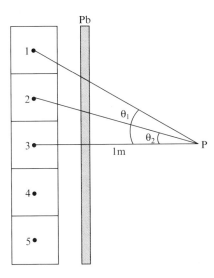

Figure 7-7. Volume source divided into five segments, each of which contains one Ci assumed to be concentrated at a "point" in the center of the segment.

Table 7-5. Parameters and Exposure Calculations for a 10-cm Diameter Pipe Modeled as Five One-curie "Point" Sources.

Point Source	Distance (cm)			μx		B(μx)		Attenuated Flux at P ($\gamma/cm^2 \cdot s$)			Exposure in air (mR/hr) x B	
	to P	in H₂O	in Air	for H₂O	for Pb	for H₂O	for Pb	by Air	by H₂O	by Pb	H₂O only	H₂O +Pb
1	116.6	5.83	3.50	0.503	4.51	1.58	1.76	1.84×10^5	1.11×10^5	2.02×10^3	224	4.53
2	104.4	5.22	3.13	0.450	4.03	1.56	1.71	2.29×10^5	1.46×10^5	4.07×10^3	290	8.88
3	100	5.00	3.00	0.631	3.87	1.55	1.70	2.5×10^5	1.62×10^5	5.21×10^3	321	11.30
4	104.4	5.22	3.13	0.450	4.03	1.56	1.71	2.29×10^5	1.46×10^5	4.07×10^3	290	8.88
5	116.6	5.83	3.50	0.503	4.51	1.58	1.76	1.84×10^5	1.11×10^5	2.02×10^3	224	4.53
										TOTAL	1,349	38.12

* Exposure rate in air for 0.662 MeV gamma rays is 1.276×10^{-6} R/hr · γ.

(a) exposure at P without the Pb shield is 1,350 mR/h due to photon absorption and the buildup effect in the H_2O solution.

(b) with the lead shield in place, the exposure rate is reduced to 38.1 mR/h.

Better accuracy and somewhat less conservatism (slightly lower total exposure) could be obtained by further subdividing the segments, perhaps into two lengthwise segments and then further dividing each of these into nine segments. And, alternatively, the activity in the volume could be separated into several (perhaps 3 for Example 7-9) line sources and the unscattered and scattered flux at P calculated for each and then summed. With appropriate positioning of these "line" sources, it is possible to consider the lengthwise distribution of the activity, the mean thickness of self absorption in the volume, and the effect of any shielding placed between the source and a receptor.

SUMMARY

Radiation shielding, which is a very complex discipline, is a form of radiation protection for many radiation sources and the many geometric configurations in which they may occur. Alpha particles and other light ions are easy to shield, usually by a sealer, since their relatively large mass and charge limit their range in most media to a few tens of microns. Shielding of beta particles from a source is also straightforward by choosing a thickness of medium that matches or exceeds the maximum range; however, since high-energy beta particles can produce bremsstrahlung, especially in high-Z materials, beta shields should use plastic, aluminum, or other low-Z material followed by lead or some other dense material to absorb any bremsstrahlung and characteristic x rays that are produced in the beta shield.

Photon shields, unlike those for charged particles, are governed by the exponential, or probabilistic, attenuation of electromagnetic radiation, and the flux of shielded photons is a complex mixture of scattered and unscattered photons characterized as "poor geometry." A calculated value of $I(x)$ based on the attenuation coefficient, μ, which is determined in "good geometry" conditions, will thus underestimate the number of photons reaching the receptor which implies that absorption is greater than what actually occurs. A buildup factor, $B > 1.0$, is used to correct "good geometry" calculations to more accurately reflect actual, or "poor geometry" conditions.

Many radiation sources can, with its ease and utility, be represented as a point or an approximate point source; however, many "real world" situations cannot, for example long pipes or tubes approximate a line source, a contaminated area that is representative of a disc or infinite planar source, and various volume sources. Fortunately, various practical calculations, some of which are fairly complex, can be used to determine the photon flux, which can then be applied in the usual way to calculate radiation exposure. Such calculations are generally conservative in that they tend to overestimate exposure, but considerable simplification of the calculations is obtained and errors in the estimates are relatively small.

Acknowledgments

Many of the data resources in this chapter came from the very helpful people at the National Institute of Standards and Technology, in particular Dr. John H. Hubbell and his colleagues M.J. Berger and S. M. Seltzer.

ADDITIONAL RESOURCES

1   Hubbell, J.H., and Seltzer, S.M., Tables of X-Ray Attenuation Coefficients and Mass Absorption Coefficients 1 keV to 20 MeV for Elements Z = 1 to 92 and 48 Additional Substances of dosimetric Interest, NISTIR 5632, National Institute of Standards and Technology, Gaithersbuy, MD, 1995.

2   Lamarsh, J.R., *Introduction to Nuclear Engineering*, 2nd ed., Chapters 5 and 9 (Addition-Wesley, 1983).

3   Morgan, K.Z., and Turner, J.E., *Principles of Radiation Dosimetry* (New York: Wiley, 1967).

4   NCRP Report No. 51, Radiation Protection Design Guidelines for 0.1–100 MeV Paticle Accelerator Facilities, National Council on Radiation Protection and Measuremens, Bethesda, MD, March 1977.

PROBLEMS

1   By what fraction will 2 cm of aluminum reduce a narrow beam of 1.0 MeV photons?

2   What thickness of iron is required to attenuate a narrow beam of 500 keV photons to one half of the original number?

3   Calculate the thickness of lead shielding needed to reduce the exposure rate 2 m from a 1 Ci point source of ^{137}Cs to 1.0 mR/h if scattered photons are not considered, i.e., without buildup.

4   Recalculate the exposure rate for the shield design in problem 3 when the buildup of scattered photons is considered.

5   How thick must a spherical lead container be in order to reduce the exposure rate 1 m from 0.1 Ci source of ^{24}Na to 2.0 mR/h?

6   An ion exchange column 9 m tall and 0.2 m in diameter contains radioactive materials that emit 5.5×10^6 1.0 MeV gammas per cm of length. Ignoring air attenuation and scatter from the floor and walls, calculate the un-shielded dose rate at a point 6 m away and 1 m above the floor.

7 A solution of ^{131}I containing an activity of 10^7t/s is spilled on an approximate circular area 1 m in diameter. Determine the dose rate 1 m above the center of the contaminated area due to the two principal photon emissions.

8 Design a shield for a "point source" of ^{90}Sr–^{90}Y that has a ^{90}Sr activity of 10^9t/s such that the surface dose rate will be 20 mrads per hour due to beta particle emissions.

9 What would be the dose rate due to photons 1 m from the source in problem 8 if the shield is made of lead?

10 A shallow circular impoundment roughly 10 m in diameter was drained exposing a thin sediment layer containing ^{137}Cs which dried. If the estimated activity of the thin sediment layer is 1 mCi/cm^2, what is the gamma dose rate 1 m above the center?

8

MEASUREMENT OF RADIATION

"When you can measure what you are speaking about and express it in numbers, you know something about it; when you cannot . . . your knowledge is of a meager and unsatisfactory kind."

– Lord Kelvin, 1889

The interactions of various types of radiation in matter provide mechanisms for measuring the amount of radiation emitted by a source, and with careful techniques the identity of the radiation source or radionuclide producing it. Radiation instruments include those that only detect, as well as those that provide details for quantification of radioactivity, exposure, and the absorbed dose. Detectors used for such purposes can be roughly divided into two categories: gas-filled chambers and crystalline solids, each of which is based on the liberation of electrons in a medium and the collection and processing of the ions by electronic means. These include portable survey instruments for field use, fixed laboratory instruments that require preparation of samples, and "fixed" monitors to read continuous levels in areas.

GAS-FILLED DETECTORS

One of the simplest detection methods is to enclose a gas inside a chamber and to collect and process the ions produced by radiation interactions in the gas, which can be air or a specially selected mixture to enhance interactions. An electrostatic field is established between the wall of the chamber and a positive electrode located on the axis of the chamber, and insulated from it as illustrated in Figure 8-1. The gas is usually at a pressure of one atmosphere or less, but can be pressurized to enhance interactions. When radiation is absorbed by the gas contained in the chamber, ion pairs (a positive ion and an electron) are produced, which constitutes a charge. An external circuit collects the charge produced and amplifies and records it.

The collection of ion pairs in a gas-filled detector is a function of the applied voltage as shown in Figure 8-2 for equal numbers of beta particles (lower curve) and alpha particles (upper curve). If there is no voltage across the chamber, the ion pairs will recombine, and no charge will flow in the external circuit for either type of radiation. As the voltage is increased, say to a few volts, some ion pairs will still recombine but others will flow to the electrodes and be collected. At a voltage (V) of perhaps 10 volts or more, recombination becomes negligible, and all of the electrons produced by ionization will reach the central electrode. As V is increased to several tens of volts the number of ion pairs collected is independent of the applied voltage, and the curve will remain horizontal as long as the radiation source produces ionizing radiation at a steady rate. If 10 ion pairs are formed initially, the response will be steady as shown by the lower curve; however, if the source strength or ionizing rate (e.g., an alpha emitter) is greater by a factor of say, 10, 100 ion pairs are formed, and the response will be 10 times greater as shown in the upper curve. These curves, labeled N_α and N_β in Figure 8-2, are parallel to each other and the current collected is directly related to the ionization produced by the incoming radiation. This region is called the **Ionization Region**; the current pulse measured is directly related to the amount of ionization produced, which will be quite different for equal source strengths of alpha particles, beta particles, and x or gamma rays due to differences in the number of ion pairs produced by each as they interact in the medium.

A **Proportional Region** is produced by increasing the voltage on a gas-filled detector above the ionization region, which in turn causes the electrons released

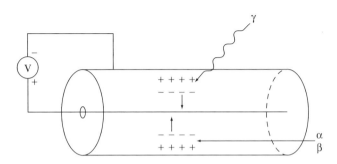

Figure 8-1. Schematic of a gas-filled detector operated with a varying voltage applied between the chamber wall (cathode) and a central collecting electrode (the anode).

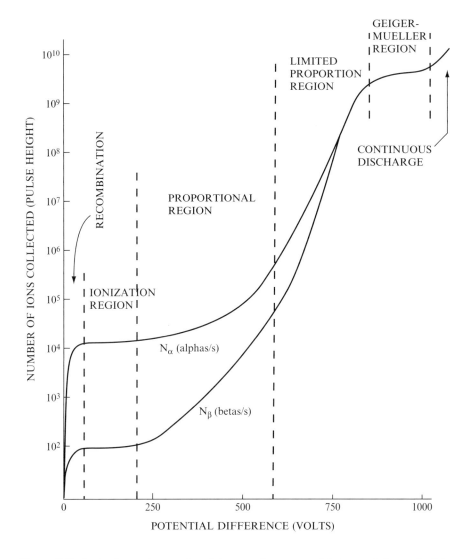

Figure 8-2. Ionization rates produced by equal sources of beta particles and alpha particles in a gas-filled detector operated in different voltage ranges.

by the primary ionizations to acquire enough energy to produce additional ionizations as they collide with the gas molecules in the chamber. The number of electrons collected increases roughly exponentially with V because each initial electron is accelerated to produce a small "avalanche" of electrons; hence the number of ion pairs collected is proportional to the initial ionization. Increasing V to several hundred volts or so increases the gas multiplication effect very rapidly, and as more electrons produce avalanches, the latter begin to interact with one another creating a region of **Limited Proportionality**. Increasing the voltage still further causes each ionizing event to create an avalanche of ions and the charge collected becomes independent of the ionization initiating it. Curves N_α and N_β not only become identical but form a plateau as voltage is increased; this plateau forms the **Geiger–Mueller Region** where each detected radiation produces a pulse of equal size. Further increases in the voltage above the plateau produces a region of continuous discharge in the presence of radiation.

Three radiation detector types have been developed based on the regions of applied voltage illustrated in Figure 8-2: the ionization chamber (based on the ionization region), the proportional counter (based on the proportional region), and the Geiger–Mueller counter (based on the G–M region). The ions produced in response to the incoming radiation, either multiplied in number or not, are collected to produce a voltage pulse which may be as small as 10 microvolts; these are in turn amplified to 5 to 10 volts and fed to a galvanometer or a pulse counter (or scaler) so that their rates can be registered.

The **Ionization Chamber**, operated at voltages in the ionization region, is characterized by complete collection, without gas amplification, of all the electrons liberated by the passage of the ionizing radiation, either in the form of particles or photons. A typical ionization chamber consists of a cylinder that forms the outer electrode, with a center electrode that is insulated from it, but parallel-plate designs are also used. The applied voltage is selected to assure collection of all the ions formed; i.e., it is high enough to prevent recombination of the ions produced but still on the plateau where current amplification does not occur. The current pulse is proportional to the number of ionizing events produced, and the collected charge is a direct measure of radiation exposure.

The **Proportional Counter**, which is operated at voltages in the proportional region, is characterized by gas multiplication which produces a pulse proportional to the initial ionization. Since alpha particles are highly ionizing relative to beta particles, proportional counters are useful to both count the particles and to discriminate between them on the basis of the sizes of the pulse each produces. The proportional counter thus offers a particular advantage for measurements of alpha and beta radiation because it can be used in conjunction with electronics to sort the smaller beta pulses from the larger ones produced by alpha particles, which allows the activity of each to be measured in the presence of the other.

The **Geiger–Mueller (or G–M) Counter** is operated in the Geiger–Mueller region and is characterized by a plateau voltage which produces an avalanche of discharge throughout the counter for each ionizing radiation that enters the chamber. This avalanche of charge produces a pulse, the size of which is independent of the initial ionization; therefore, the G–M counter is especially useful for counting lightly ionizing radiations such as beta particles or gamma rays and is specially designed to take advantage of this effect. It is difficult to make tubes with windows thin enough for alpha particles to penetrate into the gas chamber, and because of this limitation G–M counters are used mainly for the more penetrating beta and gamma radiations, although a thin end-window is often provided for detection of energetic particles. A GM tube usually consists of a fine wire electrode (e.g., tungsten) mounted along the axis of a tube containing a mixture of 90% argon and 10% ethyl alcohol (for quenching) at a fraction of atmospheric pressure. A potential difference of 800 to 2,000 volts (nominally 900 V) is applied to make the tube negative with respect to the wire. And, because of their mode of operation, G–M counters cannot be used to identify the type of radiation being detected nor its energy.

CRYSTALLINE DETECTORS AND SPECTROMETERS

Various crystals and solid state detectors can be used with electronic instrumentation to quantitate the number of photons of a given energy that activate the detector and the rate at which such events occur. Some of the more important detectors are scintillation crystals which have a response that is proportional to photon energy (i.e., as a spectrometer), solid state detectors for various radiations, and thermoluminescent phosphors to store energy deposited by radiation.

A scintillating crystal such as sodium iodide with a thallium additive [NaI(Tl)] will produce a "flash" of light proportional to the energy deposited by a photon that interacts in the crystal. The light (also a photon with a wavelength in the visible region of the electromagnetic spectrum) is reflected onto a photocathode connected to a photomultiplier (PM) tube. When the light photons strike the light-sensitive cathode, a few electrons (Figure 8-3) proportional to the energy of the absorbed photon are emitted. These are accelerated by a potential difference across the first dynode which emits approximately four secondary electrons for each incident electron. A series of 10 dynodes gives an amplification factor of 4^{10}, or approximately a million. Each pulse leaving the 10th dynode is proportional to the amount of energy absorbed, which in turn is proportional to the energy of the photon striking the crystal. These pulses are sorted according to their size and stored in corresponding channels of a multichannel analyzer as counts, allowing the source to be identified by its unique gamma energy.

Sodium iodide crystals have a high efficiency for detecting gamma rays (because of their density and the large Z of iodine), and their pulse resolving time of about $0.25\,\mu$sec permits their use for high count rates. Plastic scintillators have even shorter resolving times (several nsec), but their detection efficiency for photons is low and the light output with energy is not linear, resulting in spectra with poor resolution.

Semiconducting Detectors

The deposition of energy in a semiconducting material such as intrinsically pure germanium excites electrons from filled valence bands to conduction bands

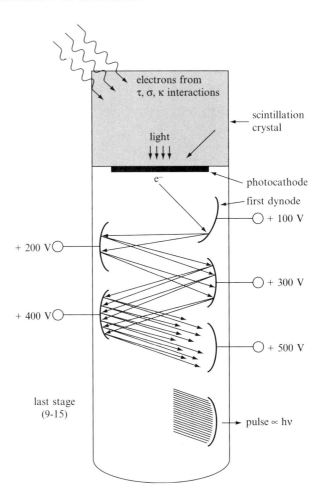

Figure 8-3. Scintillation detector with associated photomultiplier tube.

producing pairs of conduction electrons and electron vacancies, or holes. A bias voltage is applied across the semiconductor which causes these charge carriers to move, producing a current pulse. The energy needed to produce an electron-hole pair in a semiconductor is typically about 1 eV, which is considerably less than that required to produce ionizations in a scintillator; therefore, a relatively large number of charge carriers is produced for each photon absorbed. Consequently, the statistical fluctuations in the number of atoms excited or ionized is much less for a semiconductor detector, and when used with a multichannel analyzer, very sharp peaks (high resolution) are obtained which allows energies to be determined very accurately. This technology has advanced to a stage that high efficiency germanium detectors are routinely produced which provide excellent photon resolution (an example of a spectrum obtained with such a detector is shown in Figure 8-10).

Semiconductor detectors have several advantages over scintillators. Although they are not as efficient as NaI (Tl) detectors, they readily detect photons because of their density and produce fast pulses (typically a few nanoseconds) with excellent resolution. Germanium and lithium-drifted germanium [Ge(Li)] detectors must be operated at liquid nitrogen temperatures to reduce thermal noise, which can be a disadvantage. Ge(Li) (sometimes called "jelly") detectors will quickly deteriorate if not stored at liquid nitrogen temperature due to unwanted drifting of the Li into the germanium matrix, and for this reason Ge(Li) detectors have been supplanted by intrinsically pure germanium detectors which can be stored at room temperature between uses; however, they must be cooled during use.

Other semiconductor detectors are made of silicon with lithium additives to detect electrons and alpha particles, and when coupled with pulse height analyzers, they are used for spectral analysis. These detectors also provide excellent resolution and can be stored at room temperatures because the mobility of Li is less in Si than it is in Ge.

PORTABLE FIELD INSTRUMENTS

Portable G–M detectors, proportional counters, and ion chambers make field measurements convenient and reliable, though with differing degrees of sensitivity compared to laboratory instruments.

Geiger Counters, especially those equipped with pancake probes as shown in Figure 8-4, are very useful for general surveys of personnel contamination, area contamination, and the presence of external radiation fields. When used for such surveys, the G–M probe should be moved slowly to allow time for the instrument to respond, and a check source should be measured before use to assure that the instrument is functioning properly.

High-exposure fields can saturate G–M counters; therefore, care should be taken to assure that Geiger counter instruments used for such measurements are of the *nonsaturating* type. If not, GM instruments can read near zero when placed in a high-radiation field; care needs to be exercised to recognize such circumstances, and when they exist, an ion chamber should be used.

Ion Chambers are the instrument of choice for measuring exposure and are frequently used for measurements of x rays and gamma ray fields. These instruments have a very flat energy response, and although they are less sensitive than the Geiger counter, they respond correctly in much higher fields. A typical ionization chamber instrument encloses a chamber of about $200 \, \text{cm}^3$ of ambient air that is sealed at one end with aluminized mylar with a density thickness of $7 \, \text{mg/cm}^2$. A plastic or metal shield is provided to protect the mylar window, but it can be removed to detect beta particles as well as gamma radiation (Figure 8-5). If both are present, readings are taken with and without

PORTABLE FIELD INSTRUMENTS

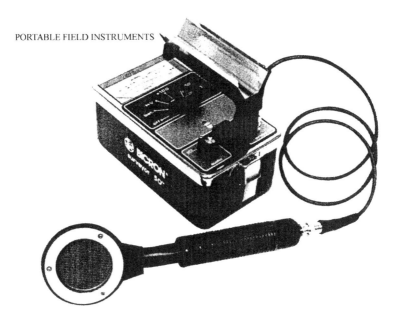

Figure 8-4. A Geiger Counter equipped with a "thin window" pancake probe.

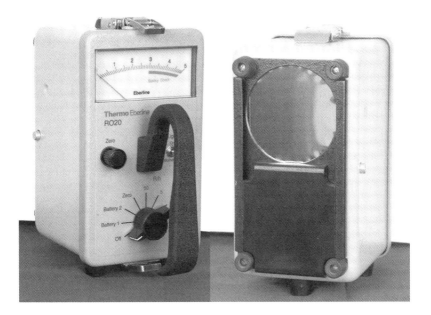

Figure 8-5. A portable ion chamber and the thin window uncovered.

the beta shield in place, and the two readings are subtracted to determine the beta component.

Micro-rem Meters are typically about 10 times more sensitive than commercial Geiger counters because they use a scintillation crystal to increase sensitivity. The counter consists of a suitable phosphor which is optically coupled to a photomultiplier tube, which in turn is connected to an electronic circuit. If the counter is used simply to detect, the output circuit often consists of a battery-operated power supply, an amplifier and pulse shaper, and a rate meter. If the device is used for energy analysis, the output circuit includes a pulse-height analyzer and a scaler and can be operated as a single or multichannel analyzer.

Alpha Radiation Monitoring is typically done with an alpha-proportional counter or an alpha-scintillation counter. Portable alpha-proportional counters are of two types: one uses air at ambient pressure as the counting gas while the other makes use of propane gas which is supplied by an external cylinder. Portable alpha-proportional counters require special discriminator circuits, very stable high-voltage supplies, and, in some cases, very sensitive amplifiers; their main advantage is the ability to discriminate between radiation types. An air-proportional counter does not require the gas cylinder and associated plumbing necessary for a propane-proportional counter; consequently, it is lighter and less cumbersome to set up. Both must have an extremely thin window in order for the alpha particles to penetrate to the sensitive region of the detector, and the counting rates that need to be measured are generally well below the normal background rate in a G–M counter. Alpha particles produce a rather large electrical pulse in a proportional counter and good discrimination is possible against beta and gamma-ray interference, although some interference could be caused by fast neutrons, if present, by elastic scattering reactions with counting gas molecules. Interference by either can be detected quite easily by moving the alpha probe about 10 cm away from the surface being monitored; if the counts cease, then true alpha contamination is being detected.

Alpha-scintillation detectors use a silver-activated zinc sulfide [AnS(Ag)] phosphor that is quite sensitive to alpha particles. The alpha-scintillation counter is less rugged than either the air- or propane-proportional counters because of the fragile photomultiplier assembly, but it has good detection efficiency, ranking between the air-proportional counter and the propane-proportional counter.

Beta Radiation Surveys are made with thin-window Geiger counters which have good sensitivity for detecting beta contamination; however, an ion chamber should be used if the beta dose rate is required. Mixed beta/gamma fields can be monitored by covering the window with a shield (see Figure 8-5) to detect gamma radiation only and removing it to detect both.

Removable Radioactive Surface Contamination is monitored by the smear or wipe test. The technique involves swiping a surface area of about $100 \, cm^2$ by a cloth, paper, plastic, foam, or fiberglass disk. These smears are in turn counted for alpha and/or beta contamination in a gas-flow proportional counter; gamma contamination is measured with a scintillation or semiconductor detector.

Low-energy beta emitters such as tritium or ^{14}C are counted in liquid scintillation counters which can be used to count other beta emitters and, with appropriate adjustment, alpha emitters as well.

PERSONNEL DOSIMETERS

Personnel dosimeters are usually one of two types: film badges which contain photographic film or thermoluminescent dosimeters (or TLDs). Both are used to measure the radiation dose received by persons over a period of time and to provide a record of the exposure received, typically for a month or a quarter. For short-term monitoring of work, a self-reading pocket dosimeter or an electronic dosimeter is worn; the latter can be set to indicate an alert or warning level of exposure.

Film Badges contain photographic dental x-ray film packets inserted into a "clip-on" holder with thin-metal shields in several places (called windows) to distinguish beta rays from gamma radiation. An unshielded area (the "open window") gives the total dose from beta and gamma rays and this can be apportioned by the amount of film density behind other windows shielded by metal strips. Two separate pieces of film differing in sensitivity by about a factor of 100 are often used. The "low-range film" (highest sensitivity) uses a double emulsion and covers 10 mR to 10 R and the "high-range film" covers 1 R to 1,000 R for accident readings. At photon energies above 200 keV or so, most film has a response that is independent of energy; however, at low photon energies, the film may overrespond by as much as 2,000 to 4,000%.

Film badges continue to be used because they are relatively inexpensive, the processed film provides a permanent record, and their accuracy satisfactorily meets accreditation requirements, especially for x and gamma radiation. With proper calibration and appropriate corrections for energy response, the film badge can be used to reliably report exposures from about 10 mR to about 1000 R. It can also be used to measure neutrons if the film is impregnated or covered with a material such as lithium or boron which has a high probability for neutron absorption reactions.

Thermoluminescence Dosimeters (TLDs), that contain lithium fluoride (LiF) and calcium fluoride (CaF_2) phosphors have been used as a dosimetry technique since the 1970s or so. As indicated in Figure 8-6, exposure to radiation causes orbital electrons in the phosphor to be excited to the conduction band from which they fall into one of the isolated levels provided by impurities in the crystal such that they are "trapped." The trapped electrons will remain so until energy is supplied (usually by heat) to free them. Heating the crystal elevates the trapped electrons back to the conduction band, and when they return to a valence "hole," a photon of visible light is emitted. The total light emitted is a measure of the number of trapped electrons and therefore of the total absorbed

hν

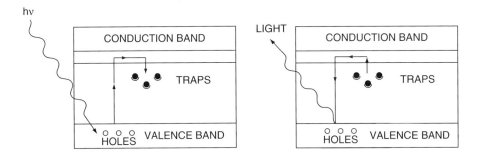

Figure 8-6. Energy level diagram of a TLD crystal. Absorption of radiation excites electrons from the valence level to electron traps provided by added impurities which can be de-excited by heating; this causes the emission of light photons that are a measure of the amount of radiation absorbed by the TLD.

radiation even after months of storage. These crystals are called thermoluminescent dosimeters because thermal heating (thermo-) produces luminescence emitted by the crystal that is proportional to the radiant energy absorbed. TLDs are generally more accurate than film badges, are less subject to fading, are usable over a much larger range of radiation levels, can be used over and over again, and processing does not require film development but can be done by a one-step electronic reader.

A TLD reader consists of a support for the phosphor, a heater to raise the temperature, a photomultiplier tube to measure the light output, and some type of meter or digital readout to display the information. The TLD reader also resets the phosphor by heating it to a high temperature to release all trapped electrons, a process called *annealing*. After proper annealing, the phosphor has the same sensitivity as previously and can be reused. This feature can also be a disadvantage because once the phosphor is annealed the dose information is permanently destroyed. Even though the TLD phosphor itself is "reset" by the reading cycle, the glow curve from the TLD reader can be stored electronically to provide a permanent record if required.

Pocket Dosimeters are often used in conjunction with a TLD dosimeter or film badge. It is a small electroscope, about the size and shape of a fountain pen, and is usually fitted with an eye piece and a calibrated scale so that an individual user can read the amount of exposure received. The dosimeter is charged before use, and exposure to radiation causes it to lose charge proportional to the radiation exposure received. Since any leakage of charge produces a reading, good insulation of the electrode is needed. The response of the dosimeter is only linear in the region of the calibrated scale; therefore, exposures should not be estimated if the device reads slightly above the full-scale reading.

GAMMA SPECTROSCOPY

Since photon interactions are enhanced in high-Z materials, photon detectors are generally made of germanium ($Z = 32$) or sodium iodide ($Z = 53$) with a thallium impurity [NaI(Tl)] to increase sensitivity. The most distinctive features of gamma-ray spectra can be roughly divided as those for energies below the threshold for pair production ($h\nu \leq 1.022\,\text{MeV}$) and those above it.

Gamma-ray Spectra: $h\nu \leq 1.022\,\text{MeV}$

Light scintillations produced in a crystal of NaI(Tl) are proportional to the absorbed photon energy and are due solely to ionizations and excitations produced by the electrons that are released when photons interact in the crystal. Being charged particles, virtually all of the electrons so released are absorbed in the crystal to produce an output pulse that corresponds to the photon energy absorbed, thus **counting** the photon interaction and yielding its energy. A typical NaI(Tl) detector/spectrometer is shown in Figure 8-7. The detection of photons in NaI(Tl) is determined by the attenuation coefficient which, as shown in Figure 8-8, is dominated by photoelectric effect interactions with iodine atoms ($Z = 53$) for photon energies up to about 250 keV (some inter-actions occur in sodium, but with $Z = 11$ these are considerably less probable).

Electrons ejected by photoelectric absorption have an energy equal to the photon energy minus 33.3 keV supplied to overcome the binding energy of K-shell electrons in iodine atoms. The orbital vacancy thus produced will be promptly filled followed by characteristic iodine x rays which are very soft and are highly likely to be absorbed in the crystal by other photoelectric inter-actions. If all of these processes (absorption of the photoelectron and the

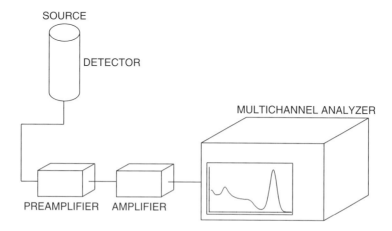

Figure 8-7. Schematic of a NaI(Tl) photon detector/spectrometer.

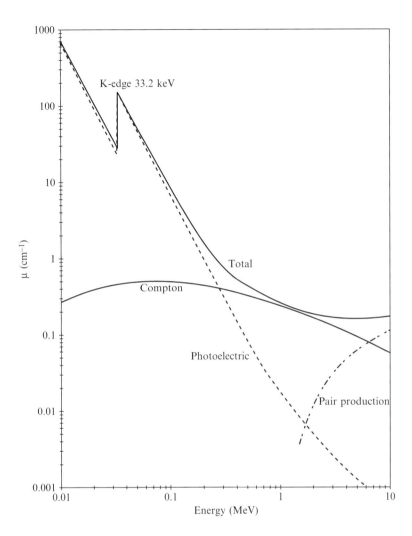

Figure 8-8. Linear absorption coefficients for photons in sodium iodide.

subsequent photoelectron liberated in the absorption of the K_α x rays) take place within the luminous lifetime of the phosphor, the visible light produced will correspond to that of the original photon energy, $h\nu$. This in fact occurs for a large fraction of absorbed photons (determined by crystal size and efficiency) and yields a full photopeak as illustrated in Figure 8-9 for ^{137}Cs.

The "photopeak" is the most prominent feature of the gamma-ray spectrum. Other prominent features of gamma-ray spectra are the Compton continuum, the Compton edge, and a backscatter peak, as denoted in Figure 8-9. Since the number of electrons produced in the photocathode of the PM tube is relatively

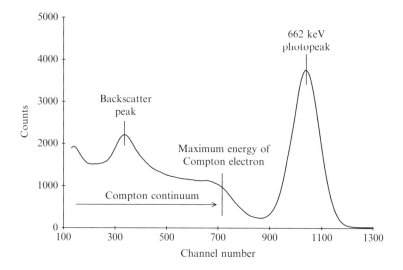

Figure 8-9. Gamma-ray spectrum of 662-keV gamma rays of ^{137}Cs measured with a sodium iodide crystal.

small, the peaks are subject to statistical fluctuations which are nearly normally distributed and serve to broaden the peak (Figure 8-9). Because of this fluctuation, the energy resolution is usually measured as the full width at half-maximum (FWHM) of the photopeak and is expressed as a percentage of the photon energy (e.g., 9% for the 662-keV photopeak in Figure 8-9).

Compton interactions, which are prominent in NaI(Tl) above about 250 keV (Figure 8-9), yield multistage processes each of which can enter the PM tube and be recorded as separate events or summed in various ways that appear in the gamma spectrum. The Compton electron will be completely absorbed, and if the Compton-scattered photon is also absorbed in the crystal along with it, the light output will be due to the energy dissipated in both events and will produce a pulse that corresponds to the total energy of the photon. This pulse will appear in the photopeak, an event that is quite likely if the crystal is fairly large since many of the scattered photons will be of lower energy and photoelectric absorption is favorable. If, however, the Compton-scattered photon escapes from the crystal, the light output will correspond only to the energy transferred to the Compton electron. These electrons form a continuum of light scintillations and output pulses that are registered as a continuum of energies, called the Compton continuum, as shown in Figure 8-9. The energies of the Compton electrons are dependent on the scattering angle or

$$E_{ce} = h\nu \frac{\alpha(1 - \cos\theta)}{1 + \alpha(1 - \cos\theta)}$$

where $\alpha = h\nu/m_oc^2$. These energies range from zero up to the maximum that the photon in question can transfer to an electron in a Compton interaction which occurs when the photon is scattered backward towards the source or when $\theta = 180°$. The Compton continuum thus extends up to the maximum energy of the Compton electrons, at which point it drops off sharply to form the "Compton edge."

A "backscatter peak" with an energy of about 0.25 MeV is also produced (Fig. 8-9) because some of the primary photons from a source miss the detector and produce Compton-scattered photons due to interactions in the shield and other materials. Because they must undergo large-angle scattering to reach the detector, the energy of these Compton-backscattered photons approaches $m_oc^2/2$ or about 0.25 MeV, or slightly less. These lower-energy backscattered photons are very likely to be detected separately by photoelectric interactions to produce a small "backscatter peak" near the low-energy end of the spectrum at 0.25 MeV or so.

Photon interactions in gamma detectors thus yield multistage processes each of which can produce light scintillations that enter the photomultplier tube either as separate events or summed in various combinations to produce features of the gamma spectrum. Although the events for any individual photon will be random, the average effect of a large number of photon events for a given source/detector produce predictable and reproducible spectra so that photon sources can be identified and quantified.

Gamma-ray Spectra: h$\nu \geq$ 1.022 MeV

More complicated gamma-ray spectra occur for photon energies ≥ 1.022 MeV because pair-production interactions also occur as illustrated in the gamma spectrum of ^{24}Na, (Figure 8-10), which was measured with an intrinsically pure germanium crystal with excellent resolution. Sodium-24 emits two high-energy gamma rays at 1.369 MeV and 2.754 MeV (see Appendix C), both of which produce the usual photopeaks and others as well due to pair production interactions. The total kinetic energy shared between the positron and electron pair ($h\nu - 1.022$ MeV) produced in pair production interactions will be absorbed in the crystal followed by positron annihilation. If both of the annihilation photons are absorbed in the crystal, the light energy produced adds to that of the absorbed electron pair and the pulse produced will correspond to the full energy of the original photon and be registered in the full photopeak. If, however, both of the annihilation photons escape from the crystal, a peak will be seen at an energy of $h\nu - 1.022$ MeV. If only one of the 0.511-MeV photons escapes and the other is absorbed, a peak will occur at an energy of $h\nu - 0.511$ MeV. The peak labeled 2.75 − 0.511 MeV is known as the single escape peak and the one at 2.75 − 1.022 MeV corresponds to the double escape of both of the annihilation photons from the crystal. Pair-production interactions also produce a strong peak at 0.511 MeV. This peak occurs because many of the photons emitted by the source will miss the crystal entirely and be

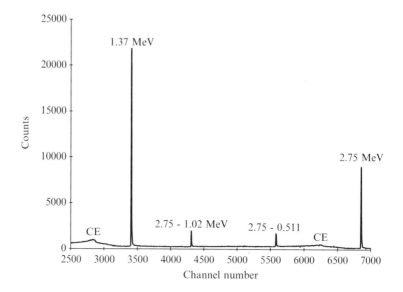

Figure 8-10. Pulse height spectrum of ^{24}Na measured with a germanium detector showing the excellent resolution obtained with such a detector.

absorbed in the shield around the crystal, and due to the 180° separation between the two annihilation photons thus produced, only one will strike the crystal and be detected.

Sum peaks are another feature of gamma-ray spectra (see Figure 8-11). These are produced when two or more photons are emitted in quick succession such that they are absorbed within the luminous lifetime of the crystal (i.e., simultaneously, or very nearly so). When this happens they are recorded as a single event but at an energy equal to the sum of the two energies. Sum peaks are often observed for high activity sources that emit two or more photons per transformation, or for photon sources with $h\nu \geq 1.022$ MeV which yield annihilation photons; these peaks will be much smaller than the photopeaks since they depend upon simultaneous detection of two emission events in coincidence, which is much less probable.

Gamma-ray Spectroscopy of Positron Emitters

Positron emitters produce gamma spectra with distinct features, and if they also emit gamma rays, other complexities, as shown in Figure 8-11 for the spectrum for ^{22}Na, a positron emitter that also emits a 1.275 MeV gamma ray. The 1.275-MeV

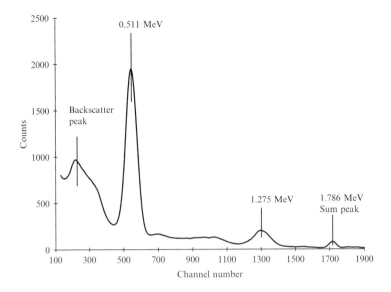

Figure 8-11. Gamma-ray spectrum of ^{22}Na measured with a NaI(Tl) detector. The spectrum shows the typical 0.511-MeV annihilation peak of a positron emitter, photopeak produced by the 1.275-MeV gamma ray, two sum peaks, the Compton continuum, and backscatter and escape peaks.

gamma ray of ^{22}Na is absorbed by photoelectric, Compton, and pair-production interactions, and if the sequence of electron-producing events for each interaction all occur within the luminous lifetime of the crystal, the pulses will produce a photopeak at 1.275 MeV. If these events are detected simultaneously with an annihilation photon, a small-sum peak will be produced at 1.786 MeV due to the reduced probability of both sequences occurring simultaneously. This sum peak is evident in the spectrum shown in Figure 8-11, in addition to the strong peak at 0.511 MeV produced when ^{22}Na positrons are annihilated and the full energy photopeak produced by the 1.275-MeV gamma rays.

A positron emitter such as ^{22}Na will always show a peak at 0.511 MeV due to the annihilation of the positron which occurs within the source itself, outside the crystal, or at the surface of the crystal where the positrons are absorbed. Since the annihilation photons are emitted 180° from each other, only one of them is likely to be absorbed in the crystal and thus will be detected as a separate event; however, if two of them are detected simultaneously, a sum peak at 1.022 MeV will be produced. The Compton continuum also occurs below the 0.511-MeV and 1.275-MeV photopeaks, as does the usual backscatter peak.

LABORATORY INSTRUMENTS

Liquid Scintillation Analysis

Liquid scintillation analysis or counting (LSA or LSC) of samples is achieved by mixing the radioactive sample into a liquid scintillant made up of chemicals that produce visible light when radiation is absorbed. Vials that transmit light are used and the light flashes, which are proportional to the energy of the radiation emitted (and absorbed), are measured by one or more PM tubes as shown in Figure 8-12. Liquid scintillation solutions contain low-Z materials (typically $Z = 6$ to 8), and consequently have relatively low counting efficiency for x and γ rays above 40 keV or so. On the other hand, good efficiency is obtained for low-energy x and γ rays and especially so for most beta particles because of their short range in liquids. The liquid scintillation counter is often the best and perhaps the only practical detector for measuring low-energy beta emitters such as ^{3}H and ^{14}C. It is also a useful instrument for measuring beta-emitting radionuclides on wipe and leak test samples because most LSCs are equipped with automatic sample changers and computerized readouts.

The detectors used in an LSC are placed in a darkened chamber with a tight lid made of materials that exclude ultraviolet radiation so that only the light produced by the sample is observed. Optical reflectors around the counting vials direct the emitted light produced in the liquid scintillant to two PM tubes placed 180° apart with outputs routed to a coincidence circuit (Figure 8-12).

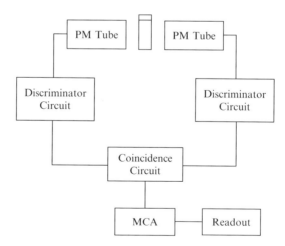

Figure 8-12. Schematic of a Liquid Scintillation Counter (LSC) in which two photo-multiplier (PM) tubes measure light produced by absorption of emitted radiations in the liquid scintillation sample. The signals from each of the PM tubes pass through discriminators and are registered as a function of energy if they arrive in coincidence; if not, the signal is presumed to be noise and is rejected.

Since the light emitted from a radiation event will be reflected into both PM tubes, two pulses are produced in coincidence and the event will be detected and registered; thermal noise that occurs randomly in one of the PM tubes will produce only a single pulse that is rejected by the circuitry.

The signal from the PM tubes is proportional to the light energy collected which in turn is directly proportional to the radiation energy absorbed; thus the pulses that are produced can be sorted to produce a spectrum. Most modern LSCs have several thousand channels and the data can be accumulated and displayed electronically as well as plotted. With appropriate calibration the endpoint beta energy, or $E_{\beta\,max}$, can be determined and this is often a good determinant of the radionuclide. Another technique is to compare the spectrum obtained with reference spectra obtained with pure standards, many of which have unique shapes. For example, a full spectrum of ^{137}Cs will show the sharp conversion electron peak above the continuous spectrum of beta particles (see Figure 3-19) and a ^{60}Co spectrum will show a unique feature at the upper energies due to absorption of its gamma rays.

LSC Beta Spectrocopy can also be used to identify and quantify two or more beta emitters in a sample in the presence of each other. The spectrum obtained is the sum of both as shown in Figure 8-13 for tritium and ^{14}C, and the counts from each beta emitter can be accumulated by setting appropriate energy windows. This approach requires counting of a standard of each radionuclide to set each energy window, determining the counting efficiency in each set window, and determining the fractional overlap of each into the other window so that it may be subtracted as shown in Example 8-1.

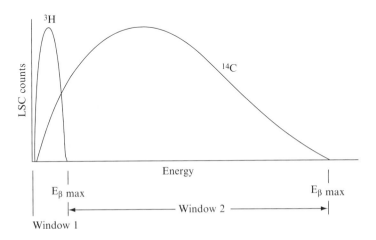

Figure 8-13. Typical beta spectra of ^{3}H and ^{14}C obtained by liquid scintillation counting.

Example 8-1: A mixed sample containing ^{3}H and ^{14}C is counted in an LSC system with two window settings with upper level discriminator settings just above 18.6 keV for ^{3}H beta particles and 156 keV for ^{14}C beta particles. The measured counting efficiencies are 36% for the ^{3}H window and 80% for ^{14}C, and it is determined that 8% of the counts produced by the continuous beta spectrum of ^{14}C will be registered in the ^{3}H window. A sample that contains both ^{3}H and ^{14}C is counted with a net (minus background) count rate of 2,000 cpm in the ^{3}H window and 6,000 cpm in the ^{14}C window. Determine the activity of each in the sample.

Solution: First, the number of counts in the ^{3}H window due to the spectral distribution of ^{14}C counts is subtracted to obtain the number of counts due to ^{3}H alone.

$$^3\text{H (net cpm)} = 2,000 - (0.08)6,000 = 1,520 \, \text{cpm}$$

And since ^{3}H beta particles are not energetic enough to exceed the lower setting of the ^{14}C window, the net count rate for ^{14}C is 6,000 cpm. The activity of each is

$$^3\text{H(tpm)} = \frac{1520}{0.36} = 4,222 \, t/m$$

$$^{14}\text{C(tpm)} = \frac{6,000}{0.80} = 7,500 \, t/m$$

The spectral analysis method shown in Example 8-1 can be developed and used, with appropriate standards and technique, for various other beta-emitting radionuclides.

Sample Quenching, which is the term applied to any process which reduces the emitted light output, is a major consideration in liquid scintillation counting. There are two general sub-classes – chemical and optical quenching. The effect of sample quench, as shown in Figure 8-14, is to shift the beta spectrum toward lower energies and thus reduce the overall number of counts observed. This problem is dealt with by measuring the amount of quench and applying a "quench correction" to the counter results. Sample quench can be minimized by using small sample volumes and assuring good mixing of the sample in the liquid scintillant.

The amount of quench in each sample is usually measured by briefly exposing it to a high-activity radiation standard (e.g., ^{133}Ba) located next to the sample chamber. The measured light output produced by such exposure is calculated as a "quench number" which has been given various designations by equipment manufacturers. A standard of the radionuclide of interest is mixed with distilled water, added to the LSC cocktail, and counted. A chemical contaminant such as carbon tetrachloride is then added to the "known" in the LSC cocktail in increasing amounts and the quench number and count rate are recorded and plotted to produce the quench curve which is in turn used to

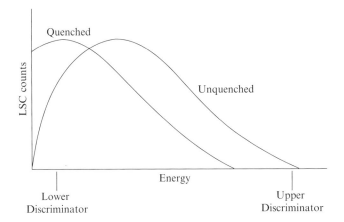

Figure 8-14. The effect of sample quench in liquid scintillation counting is to lower the number of counts observed in a set window.

adjust the count rate for each sample to the true count rate.. Modern LSC instruments can store quench correction curves and such corrections can be made electronically to increase the number of recorded counts to those that would have been recorded if quench losses had not occurred.

Proportional Counters

Gas proportional counters take advantage of the amplification obtained when radiation is absorbed in the counting gas to provide good sensitivity as well as to discriminate charged particles on the basis of the proportional pulse sizes produced; e.g., beta emitters from alpha emitters. Because of these factors and their relatively poor sensitivity to x and gamma radiation, proportional counters are used primarily for counting charged particles, principally beta particles or alpha particles or one in the presence of the other. A hemispherical counting chamber (Figure 8-15) with a central electrode, or anode, usually in the form of a small wire loop, is used that encloses an optimal counting gas in intimate contact with the sample to be counted. The electric field changes rapidly with distance in the immediate vicinity of the anode wire, and most of the amplification occurs in its vicinity. In this configuration (typically referred to as a "2π chamber" or "2π geometry"), the counter may be "windowless," or more typically, it will have a thin window such that the sample to be counted is placed just outside the window. Both designs have advantages and disadvantages in terms of sensitivity and control of contamination.

Proportional counters have good counting efficiency and can be used at very high counting rates because the negative ions, which are barely influenced by the slower-moving positive ions, have to move only a few mean-free-paths in the intense field region of the counter to be collected. When operated in the

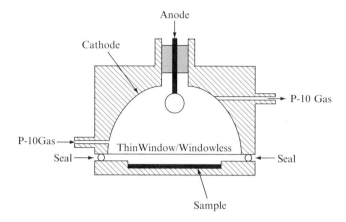

Figure 8-15. A 2π proportional counter with a window; if used without a window in either 2π or 4π geometry, it is necessary to provide a mechanism to reseal the chamber after samples are inserted and to provide rapid flushing with the counting gas to replace any outside air.

proportional region, secondary electrons are formed in the immediate vicinity of the primaries produced by the incoming radiation as it is absorbed, and with a typical gas amplification of 10^3 the absorption of a highly ionizing alpha particle might lead to a pulse containing 10^8 ions while a pulse initiated by a beta particle or a gamma ray would contain around 10^5 ions. These pulses are in turn amplified to a desired output level, but the relative pulse sizes remain the same and are sorted accordingly. The smaller beta-particle pulses are not registered by the alpha-particle circuitry; however, when the counter is operated in the beta-particle region, both are registered. The sizes of the pulses are significantly different and can be sorted electronically for a separate alpha count in the alpha region where beta particles are excluded and substracted from the beta-plus-alpha count. A discriminator can also be set to reject all pulses below some desired level, allowing alpha particles to be registered separately even in the presence of beta and gamma radiation. These characteristics lead to different degrees of amplification vs detector voltage for alpha particles and beta particles, producing two distinct "plateaus" for counter operation as shown in Figure 8-16. The counter can be operated to just detect alpha particles or at a higher voltage to detect both alpha particles and beta particles. Electronic circuitry can be used to register one or the other or both.

Various gases and mixtures are suitable for proportional counting but a mixture of 90% argon and 10% methane, called P-10 gas, is often used. Argon has good performance because of its high density but it has long-lived excited states which may trigger spurious discharges; therefore, methane is added to counteract these. The gas is usually introduced at atmospheric pressure because

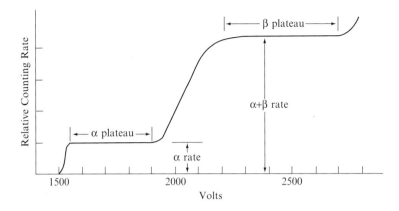

Figure 8-16. Response of a proportional counter vs applied voltage showing a voltage plateau for counting energetic alpha particles and a much higher voltage plateau for beta particles.

the voltages, which range from 1,000 to 5,000 volts, present no technical problems and there are advantages to a counter that can be readily opened for the insertion of samples, a necessity for windowless counters. If a windowless chamber is used, it is necessary to provide a mechanism to reseal the chamber after samples are inserted. This is usually done by causing the sample tray to press against an o-ring or similar sealing device when the sample holder brings the sample into the detector volume (or under it for a thin-window detector). It is also necessary to flush the windowless chamber with counting gas to assure that no outside air remains in the chamber. A detector with a window does not require flushing since a steady gas flow is maintained; however, the window must be quite thin to minimize absorption of alpha and beta particles.

End-Window G–M Counters

A laboratory version of the Geiger–Mueller counter is used with a high-quality thin-window G–M tube to count various samples such as smears or for determining beta-particle absorption. Although newer devices are replacing it because of its inability to distinguish between radiations, the end-window G–M counter was one of the first practical detectors used in radiation protection and it is still used for routine counting. The counter is enclosed in a shield to both reduce background radiation and provide physical protection of the G–M tube itself. Samples are placed just below the end window and the output routed to a scaler unit, often with a timer, to record the counts. The operating voltage is set in the G–M plateau region (see Figure 8-2) and the counter provides good sensitivity for routine counting of many types of samples. It is usually calibrated for efficiency with a medium energy beta source, or a standard of the radio-nuclide that is routinely counted.

Surface Barrier Detectors

Specially constructed silicon surface barrier semiconductors are used for alpha and beta spectroscopy. A pure silicon crystal will normally have an equal number of electrons and holes, but impurities can be added to create an excess number of electrons (an *n*-region) or an excess number of holes (a *p*-region). Silicon (and germanium) is in group IV of the periodic table. If atoms of group V, each of which has five valence electrons, are added, four of the five electrons in each of the added atoms will be shared by silicon atoms to form a covalent bond. The fifth electron from the impurity is thus an excess electron and is free to move about in the crystal and to participate in the flow of electric current. Similarly, adding an impurity from group III with three valence electrons creates bonds with missing electrons, or holes, or a *p*-type crystal.

Surface barrier detectors are constructed to provide a junction between an *n*-type and a *p*-type crystal. When a voltage bias is applied to a silicon crystal layered in such a way as to create an *n-p* junction, the excess electrons are swept in one direction and the "holes" in the opposite direction. This creates a depletion layer between the two that is nonconducting. If, however, an alpha (or beta) particle is absorbed in the depletion layer, it creates numerous charge pairs by ionizing silicon atoms and these migrate quickly to create a pulse. The size of the pulse is directly proportional to the energy deposited and can be processed electronically to provide a spectrum for identification and quantification of the source. A large number of charge units is produced because only 1 eV or so is required to ionize silicon; thus excellent resolution is obtained because the statistical fluctuation of the collected ions is minimal.

The depletion layer can be constructed just thick enough to equal the maximum range of the particles to be detected and thus minimize interference by other types of radiation. A detector with a very thin depletion layer is used for alpha spectroscopy, and these detectors have essentially zero background since the probability of photon interactions is minimal. When used for alpha spectroscopy, silicon surface barrier detectors provide good energy resolution because alpha particles are emitted monoenergetically. Any self-absorption of the alpha particles in the source will degrade their energy before they reach the detector, and only the energy actually deposited will be recorded; therefore, sharp peaks will only be attained if the alpha source is deposited in a thin layer to minimize self-absorption.

Beta spectroscopy is achieved by increasing the depletion layer , and both alpha and beta particles will be detected with such a detector. Beta particles, which are less subject to self-absorption, will be detected and displayed as a continuous energy spectrum up to $E_{\beta Max}$ because of the mode of beta transformation. Since no peaks will be formed, except perhaps for conversion electrons associated with gamma emission (e.g., ^{137}Cs), beta spectroscopy with silicon detectors requires analysis of the entire spectrum as in liquid scintillation analysis.

STATISTICS OF RADIATION MEASUREMENTS

All measurements of radiation must recognize and account for the random statistical behavior of radioactive atoms; thus rather straightforward statistical techniques are applied to several important circumstances:

- specifying the amount of uncertainty at a given level of confidence for a measurement of radioactivity;
- determining whether or not a sample actually contains activity, especially if the measured activity level in the sample is very close to the natural background; and
- checking whether a counting instrument is functioning properly by comparing the statistically predicted variance of the sample counts to that obtained experimentally.

Nature of Counting Distributions

The transformation of radioactive atoms is subject to the laws of chance and each atom of a given radionuclide has the same probability for transformation in an interval of time regardless of its past history. The random nature of radioactivity is demonstrated by making successive measurements of a steady radiation source in a fixed geometry with a stable counting system. When this is done, it is found that the source does not consistently produce the same count but that the counts vary from one identical counting interval to another; i.e., they are distributed in frequency. Figure 8-17, which is a scatter plot of one thousand one-minute counts of a sample, illustrates important features of such data sets. First, the one-minute counts cluster around a central value which is most likely the true count rate of the sample; and second, the values around the central value form a fairly symmetrical distribution pattern. If one additional count is made on the sample, it is highly likely that it would be one of the values obtained before; however, it is also possible, but not as likely, that the new value would be outside the distribution of the data. These features are described with statistical models using three terms: the mean, the variance, and the standard deviation of the data.

When a radioactive sample is measured, the objective is to arrive at a value as close to the true transformation rate, or activity, of the sample as possible. It is a fact of nature that the true value cannot be determined exactly unless an infinite number of measurements is made, which is not practical. It is possible, however, due to the random nature of radioactive transformation to use one or several measurements to describe the measurement and its reliability using the features of the binomial, Poisson, and Normal (or Gaussian) distributions.

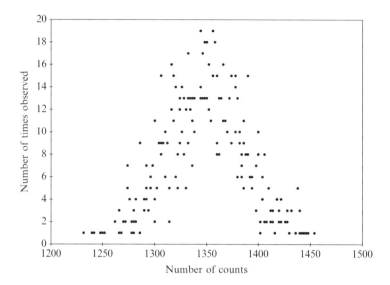

Figure 8-17. A plot of the number of times (i.e., the frequency) a given count was observed when a sample was counted repeatedly for a one-minute period.

The binomial distribution is the most general and exact description of radioactive transformation; however, it is difficult to use for more than a few events. The Poisson distribution, which is derived from the binomial distribution function for low-probability events ($p \ll 1$), is broadly applicable to radiation events because it is characterized by two properties of a data set: the arithmetic mean $\bar{x}$ and its standard deviation which is obtained as $\sqrt{x}$. A data set of 17 or more Poisson-distributed events (e.g., detector counts) can be represented as a Normal distribution where only one parameter, the arithmetic mean $\bar{x}$, needs to be known. If, however, the number of counts falls to a very small value (less than 17), the distribution no longer follows the normal curve and the statement of confidence limits must be based on the Poisson distribution.

The standard deviation of a normal distribution can be readily obtained once the best estimate of the mean is made; it is simply the square root of the estimated mean, $\bar{x}$. Because the normal distribution curve is symmetrical about the mean, multiples of σ provide a set of points for plotting the probability distribution of radiation events, e.g., sample counts, etc.

Checkpoints

Before proceeding on to practical applications of statistics to common radiation protection circumstances, the following points need to be emphasized:

- Radioactivity measurements simply record the number of random events (transformations) that occur in a given time interval, and these events are truly random with very small probabilities of occurrence. For this reason, they are statistical in nature and statistical models are used to characterize them.

- Since radioactivity events are binary (they either occur or they don't), their predictability is governed by the broadly applicable binomial probability distribution function, but this is generally too cumbersome to use for more than a few counts. Counts are discrete events each with a probability of $\ll 1$, and since the Poisson distribution is for discrete events only, it is an accurate and practical simplification of the binomial probability distribution.

- The normal (or Gaussian) distribution is a continuous function that is symmetrical about the mean. It is easier to use, plot, and describe than either the binomial or Poisson distributions, which is the main reason it is used to describe radioactivity measurements.

- If the number of events (e.g., counts) is larger than 17 to 20, the normal distribution provides essentially the same statistical results for the set of events as does the more accurate Poisson distribution. Both distributions are characterized by the mean (estimated as $\bar{x}$ from a single measurement or from n measurements) from which the standard deviation can be calculated as $\sqrt{\bar{x}}$.

Uncertainty in a Single Measurement

A single observation of a radioactive sample is often made, which is the best estimate of the mean transformation rate of the sample since no other value is in hand. The true mean is probably different from this measurement, but confidence limits can be specified, based on the normal distribution, within which the true value lies at a stated probability. In effect, a normal curve is drawn around the estimated mean based on the square root of the measured counts. This can be shown to be a reasonable practice using the data in Table 8-1. If only one count is selected randomly from the table, say 108, the standard deviation is $\sqrt{108} = 10.4$. This single measurement would be reported as 108 ± 10.4 counts, which is equivalent to stating that the "true" activity of the source has a 68.3% probability (or 1σ) of being between 97.6 and 118.4 counts in one minute. The mean count rate for the 10 separate observations listed in Table 8-1, is 99 counts per minute which falls within the range. Although the single measurement and its reported uncertainty is a reasonable estimate, the uncertainty interval for the single measurement is much larger than the one obtained from the 10 observations (or alternately a 10-minute count) where the mean is 99 ± 3.1 cpm. Larger count times or more observations tighten the estimate of the true activity and narrow the uncertainty interval specified for the measurement.

Table 8-1. Ten One-minute Counts of a Long-lived Sample Using the Same Equipment/Sample Setup.

Counts in one minute	
89	108
120	85
94	83
110	101
105	95

Statistical models apply only to an observed number of counts; therefore, a calculation of the standard deviation of a radioactivity measurement as equal to the square root of $\bar{x}$ can be made only if x represents a recorded number of events. The standard deviation or any multiples of it cannot be associated with the square root of any quantity that is not a directly measured number of counts.

PROPAGATION OF ERROR

The aggregate uncertainly of two or more values $A \pm \sigma_A$ and $B \pm \sigma_B$, each with its own uncertainty value, can be determined by propagation of error as follows:

- for addition or subtraction of $(A \pm \sigma_A)$ and $(B \pm \sigma_B)$, the results and the aggregate standard deviations are

$$(A \pm B) \pm \sqrt{\sigma_A^2 + \sigma_B^2}$$

- for multiplication or division of $(A \pm \sigma_A)$ and $(B \pm \sigma_B)$, the results and the aggregate standard deviations are

$$AB \pm AB\sqrt{(\sigma_A/A)^2 + (\sigma_B/B)^2}$$

or for division of the two quantities

$$\frac{A}{B} \pm \frac{A}{B}\sqrt{(\sigma_A/A)^2 + (\sigma_B/B)^2}$$

Multiplying or dividing total counts or a count rate by a constant is not propagated by a statistical operation but is simply altered by the constant applied. A common example of this is conversion of an observed number of counts to a count rate r by dividing by the count time, t, or

$$r = \frac{C_{s+b}}{t}$$

The standard deviation of a counting rate r can be calculated one of two ways:

$$\sigma = \frac{\sqrt{C_{s+b}}}{t}$$

or

$$\sigma = \sqrt{\frac{r}{t}}$$

where r = counting rate, C_{s+b} = gross sample counts including background, and t = sample counting time.

Example 8-2: A long-lived sample was counted for 10 minutes and produced a total of 87,775 counts. What is the counting rate and its associated standard deviation?

Solution: The counting rate r is

$$r = \frac{87,775}{10} = 8,778 \, \text{cpm}$$

and the standard deviation of the count rate is

$$\sigma = \frac{\sqrt{87,775}}{10} = 30 \, \text{cpm}$$

or, alternately,

$$\sigma = \sqrt{\frac{8,778}{10}} = 30 \, \text{cpm}$$

and the counting rate with an uncertainty of 1σ would be expressed as $8,778 \pm 30 \, \text{cpm}$.

Statistical Subtraction of a Background Count

When a measurement includes a significant contribution from the radiation background, the net count above background is determined by subtracting the two and the uncertainty is determined by propagation of the error associated with each measured quantity. The resulting net count has a standard deviation associated with it that is greater than that of either the sample or the background alone. The net count rate, R_{net}, is calculated as

$$R_{net} = \frac{C_{s+b}}{t_{s+b}} - \frac{B}{t_b}$$

and the standard deviation of the net count rate is

$$\sigma_{net} = \sqrt{\left(\frac{C_{s+b}}{t_{s+b}^2} + \frac{B}{t_b^2}\right)} \qquad (8\text{-}1)$$

where R_{net} = net count rate (cpm), C_{s+b} = gross count of sample plus background, B = background count, t_{s+b} = gross sample counting time, t_b = background counting time, and σ_{net} = standard deviation of the net count rate.

If the counting time for both the sample and the background is the same, this simplifies to

$$R = \frac{C_{s+b} - B}{t} \quad \text{and} \quad \sigma = \frac{\sqrt{C_{s+b} + B}}{t}$$

which is actually a specialized application of error propagation, as shown in Examples 8-3 and 8-4.

Example 8-3: A sample counted for 10 minutes yields 3300 counts. A 1-minute background measurement yields 45 counts. Find the net counting rate and the standard deviation.
Solution:

$$\text{Net count rate} = \frac{3300}{10} - \frac{45}{1} = 285\,\text{cpm}$$

$$\sigma_{net} = \sqrt{\left(\frac{3300}{10^2} + \frac{45}{1^2}\right)} = 8.83\,\text{cpm}$$

or $R_{net} = 285 \pm 8.83\,\text{cpm}$

Example 8-4: What is the net count and its standard deviation for a sample if the sample count is 400 ± 20 and the background count is 64 ± 18?

Solution:

$$\text{Sample net count} = (A - B) \pm \sqrt{\sigma_A^2 + \sigma_B^2}$$

$$= (400 - 64) \pm \sqrt{(20)^2 + (18)^2}$$

$$= 336 \pm 27 \text{ counts}$$

These examples are computed with uncertainties based on "1-sigma" (68.3%) confidence intervals. If a 95% confidence interval is desired for the measurement in Example 8-4, it would be computed as A − B with 1.96σ for each value and reported as 400 ± 39 and 64 ± 35. The result and the aggregate uncertainty at the 95% confidence level would be $(400 - 64) \pm 53$ counts or 336 ± 53 counts. The uncertainty of 1.96σ can be calculated for A and B separately and aggregated as shown in Example 8-4 or applied to the result (i.e., $1.96 \times 27 = 53$ counts).

Error Propagation of Several Uncertain Parameters

Laboratory results are often calculated from a number of parameters, more than one of which may have a specified uncertainty. For example, a sample count is often divided by the detector efficiency and the chemical/physical recovery of the procedure to determine the activity in a collected sample, i.e., before it is processed. Detector efficiency is determined by counting a known source and dividing the measured count rate by the disintegration rate which requires a statement of uncertainty. Similarly, the chemical yield is usually determined by processing a tracer or several known spikes, determining the mean of the measured values, and calculating their uncertainties. Count times can be reasonably assumed to be constant because electronic timers are very accurate, and geometry factors can also be considered constant if samples are held in a stable configuration.

The reported uncertainty in a measured value is obtained by propagating the relative standard errors (i.e., one sigma) of each of the uncertainties according to the rule for dividing or multiplying values with attendant uncertainties. The uncertainty in a calculated value, u, is

$$\sigma_u = u\left[\left(\frac{\sigma_a}{a}\right)^2 + \left(\frac{\sigma_b}{b}\right)^2 + \left(\frac{\sigma_c}{c}\right)^2 + \ldots\right]^{\frac{1}{2}}$$

where u is the calculated value and σ_a, σ_b, ... are the relative standard errors of each of the parameters used to calculate u. Typical cases are shown in Examples 8-5 and 8-6.

Example 8-5: A counting standard with a transformation rate of 1000 ± 30 dpm is measured with a counting system that yields a count rate of 200 ± 10 cpm. Calculate (a) the efficiency of the counting system, and (b) its uncertainty.

Solution: (a) The efficiency is

$$\epsilon = \frac{200 \text{ min}^{-1}}{1000 \text{ min}^{-1}} = 0.2 \text{ or } 20\%$$

(b) the standard deviation of the efficiency, is calculated as

$$\sigma_\epsilon = 0.2\sqrt{\left(\frac{30}{1000}\right)^2 + \left(\frac{10}{200}\right)^2} = 0.058 = 5.8\%$$

and the recorded efficiency with a one-sigma confidence interval would be $20 \pm 5.8\%$ or 0.2 ± 0.058.

Example 8-6: A one-gram sample is processed in the laboratory and a count rate of 30 ± 5 cpm was obtained by a measurement system with a precise timer and a fixed geometry that has a counting efficiency of $20 \pm 2\%$. The chemical yield was determined by processing several known spikes and was recorded as $70 \pm 5\%$. Calculate (a) the activity concentration, and (b) the associated uncertainty in the concentration.

Solution: (a) The concentration of activity in the sample is

$$\text{Conc (dpm/g)} = \frac{30 \text{ cpm}}{0.2 \text{ cpm/dpm} \times 0.7 \times 1 \text{ g}}$$
$$= 214 \text{ dpm/g}$$

(b) the uncertainty for the measurement is

$$\sigma_{\text{Conc}} = 214\sqrt{\left(\frac{5}{30}\right)^2 + \left(\frac{0.02}{0.2}\right)^2 + \left(\frac{0.05}{0.7}\right)^2}$$
$$= 214\sqrt{0.0278 + 0.01 + 0.0051}$$
$$= 214\sqrt{0.0429} = 44 \text{ dpm/g}$$

The result would be reported as 214 ± 44 dpm/g at a one-sigma confidence interval; i.e., the true concentration has a 68.3% probability of being between 170 and 258 dpm/g. The 95% confidence interval would be 1.96×44 dpm/g and the result would be reported as 214 ± 86 dpm/g.

MINIMUM DETECTION LEVELS

A repeat measurement of a radioactive sample will almost always produce a different value. It is tempting to take advantage of this statistical fact to obtain a desired result (either lower or higher) but this practice is not in accord with the laws of statistics. A common question in many measurement procedures is whether the observed counts are due to just background radiation or whether the sample actually contains radioactivity above background. Two concepts related to such circumstances are the critical level, L_c, and the lower limit of detection, or LLD. Both are related to the smallest amount of sample activity that will yield a net count for which there is a predetermined level of confidence that activity is present above background. Each is related only to the observed counts registered by the counting instrument; i.e., neither is dependent on other factors involved in the measurement method or on the sample characteristics. Detector efficiency, chemical recovery, and the measured volume or mass also have an uncertainty, and each of these should be propagated through the calculation of the measured activity of the sample, but they do not influence L_c or LLD which is based solely on the recorded events (e.g., a number of counts measured by the counting system).

The Critical Level is a value that is used to determine whether a sample contains radioactivity above a true blank; it is defined in terms of the distribution of a zero net count as $L_c = k_\alpha \sigma_b \sqrt{2}$ for equal count times for the sample and a blank. The value of k_α is selected such that a false positive would be determined only at a specified value of p, usually $p = 0.05$, so that a conclusion from the measurement that activity is present when in fact it is not would occur only 5% of the time. The calculated value of L_c is thus a net count above that of the background blank, as illustrated in Figure 8-18, and the probability of making a Type I (or α) false negative determination is limited to a specific value, also typically $p = 0.05$, or $k = 1.645$; therefore,

$$L_c = 2.33\sigma_b = 2.33\sqrt{B} = 2.33\sigma_b$$

at a confidence level of 95%. If, however, the measured count of a sample minus that from a true blank has a net value of L_c, the net value will also follow a normal distribution (Figure 8-19), and concluding that a net count of L_c is radioactive has a 50–50 chance of being less than L_c, which is not sufficient confidence to conclude that the sample is in fact radioactive. The additional confidence needed is provided by the concept of LLD which is related to L_c as shown in Figure 8-19.

The Lower Limit of Detection (LLD) is, as shown in Figure 8-19, defined in terms of the critical level as

$$LLD = L_c + k_\beta \sigma_d$$

where k_β is the one-sided confidence factor and σ_d is the standard deviation of a net count rate equal to LLD. The LLD is based on a willingness to accept being wrong but only at a specified level, typically 5% of the time; i.e., one has 95% confidence that a measured activity level that is at or above the LLD represents a true presence of radioactivity above the system background. If k_β is chosen so that 95% of the measurements of a true mean net count equal to LLD is detected (i.e., $k_\beta = 1.65$), then LLD can be used as a practical detection limit.

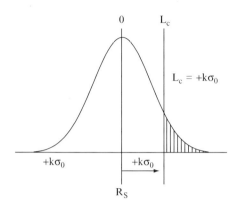

Figure 8-18. The Critical Level, L_c, is that point on the normal distribution curve of counts for a true zero net count above which there is a specified level of confidence that a true mean count of zero would be falsely recorded as positive; i.e., activity is recorded as present when in fact it is not.

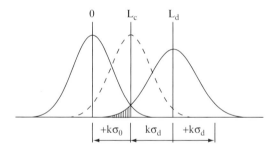

Figure 8-19. The lower limit of detection, LLD, is the smallest amount of net activity above background that will be registered as positive with a given level of confidence, typically with a false positive level of $p = 0.05$, or $k_\beta = 1.645$.

It is chosen in such a way that the distribution of the net count about the mean yields a Type II (or β) probability of concluding that activity is present when in fact it is not, and that it is limited to a predetermined value, typically $p = 0.05$. The value of k_β is chosen so that β is the one-sided confidence interval since wrong determinations relative to the distribution of the net mean of zero are the only ones of interest.

The general relationship for LLD, assuming that $k_\alpha = k_\beta = k$, in counts per unit time is

$$\text{LLD} = \frac{k^2}{t_{s+b}} + 2k\left[\frac{R_b}{t_b}\left(1 + \frac{t_b}{t_{s+b}}\right)\right]^{1/2}$$

Table 8-2. Equations for Calculating the Lower Limit of Detection (LLD) for $k = k_\alpha = k_\beta$ and Various Conditions of Use.

Prerequisites for use	Equation*
General equation for LLD as a count rate for unequal count times and any value of k	$LLD = \frac{k^2}{t_{s+b}} + 2k\left[\frac{R_b}{t_b}\left(1 + \frac{t_b}{t_{s+b}}\right)\right]^{1/2}$
LLD as a count rate for equal count times for sample and background and for any value of k	$LLD = \frac{k^2}{t} + 2k\left(2\frac{R_b}{t}\right)^{1/2}$
LLD as a count rate for equal count times for background and sample and $k = 1.645$ (95% confidence level)	$LLD = \frac{2.706}{t} + 4.653\sqrt{\frac{R_b}{t}}$
LLD in counts for a background B (cts) and $t \gg 2.706$	$LLD = 4.653\sqrt{B} = 4.653\sigma_b$

* R_b in these equations is the background count rate in counts per unit time t and σ_b is the standard deviation of the background counts B.

The specific cases that are derivable from this expression and the conditions for the use of each are summarized in Table 8-2. The latter two expressions, for which it is assumed that $k_\alpha = k_\beta = 1.645$, are frequently used for measurements of radioactivity at a 95% confidence interval. The calculated result for each states that for a measured net activity level that is equal to (or above) LLD there is only a 5% chance of concluding that activity is not present when in fact it is. If, however, a different level of confidence is desired, then it is necessary to use the more general equation, as shown in Example 8-7. Values of k that correspond to various confidence intervals are

False Positive Probability	k_α
0.010	2.326
0.025	1.96
0.050	1.645
0.100	1.282

Example 8-7: What is the LLD for a detection system designed for two-hour measurements of samples if the system records 50 background counts in two hours at: (a) the 95% confidence level; and (b) at a 90% confidence level?

Solution: (a) For 95% confidence $k = 1.645$, and

$$LLD = \frac{(1.645)^2}{120 \text{ min}} + 2(1.645)\left[2\left(\frac{0.42 \text{ c/m}}{120 \text{ min}}\right)\right]^{1/2} = 0.02 + 0.27 = 0.29 \text{ cpm}$$

(b) For 90% confidence, $k = 1.282$ and

$$LLD = \frac{(1.282)^2}{120 \text{ min}} + 2(1.282)\left[2\left(\frac{0.42 \text{ c/m}}{120 \text{ min}}\right)\right]^{1/2} = 0.014 + 0.215 = 0.23 \text{ cpm}$$

Example 8-8: A single measurement of a sample yields a gross count of 465 cpm. If the counting system has a background of 400 ± 10 cpm, should the measurement be reported as being radioactive?

Solution: The Critical Level, L_c, and the LLD, both determined at the 95% confidence level are

$$L_c = 2.33\sigma_b = 2.33 \times 10 = 23.3 \text{ cpm}$$

$$LLD = 2.71 + 4.653 \times 10 = 49.2 \text{ cpm}$$

Since the net count rate of $465 - 400 = 65$ cpm exceeds the calculated value of both L_c and LLD for the system, the sample is clearly radioactive. The LLD value provides a considerably higher level of assurance that normal statistical variations in the net count do not lead to a false conclusion, and because of this a measured net count at LLD (or above) is used in regulatory decisions.

The **Minimum Detectable Concentration (MDConc.)** is a level (not a limit) derived from the LLD. It is an *a priori* level of radioactivity in a sample that is practically achievable by the overall method of measurement at a selected level of confidence, usually 95%. The MDConc. considers not only the instrument characteristics (background and efficiency), but all other factors and conditions which influence the measurement. These include sample size, counting time, self-absorption and decay corrections, chemical yield, and any other factors that influence the determination of the amount of radioactivity in a sample. It cannot serve as a detection limit per se, because any change in measurement conditions or factors will influence its value. It establishes that some minimum overall measurement conditions are met, and is often used for regulatory purposes; when so used it is commonly referred to as the minimum detectable activity or MDA.

The MDConc. (or alternatively the MDA) for a measurement system, derived from the LLD and expressed as pCi per unit volume or weight of the sample, is

$$MDConc.(pCi/unit) = \frac{LLD}{(E \cdot Y)(\text{volume or weight})(2.22)t}$$

where
MDConc. = minimum detectable concentration of activity in a sample,

$$LLD = 2.71 + 4.653\sqrt{B}$$

B = counts due to background

E – counting efficiency for a given detector and geometry

Y = chemical recovery

2.22 = dpm/pCi

t = count time for B (minutes)

One use of LLD in radiation protection is determining whether certain materials meet regulatory requirements. For example, various facilities consolidate liquids that may contain radioactive material and then determine whether it can be handled without regard to its radioactivity. The regulatory basis for this decision is the LLD of the detector/sampling system, as long as the activity meets other environmental and public health regulations. This situation is illustrated in Example 8-9.

Example 8-9: Floor drains are consolidated at a nuclear power plant. Two liters of background water are counted in a Marinelli beaker on a germanium detector that has an efficiency of 8% for ^{137}Cs and produces 2000 counts in 2 hours. (a) What concentration of ^{137}Cs in the floor drain sump could be disposed without regard to its radioactivity; (b) What concentration would be allowed if the background water were measured under the same conditions for just 30 minutes yielding 500 counts?

Solution: (a) The limiting concentration, for regulatory purposes, is based on the LLD of the detector system, which for a 95% confidence level is

$$\text{LLD} = 2.71 + 4.653\sqrt{2000} = 211 \text{ counts}$$

Therefore, any 2 liter sump sample for which the recorded counts is 2211 or more (2000+LLD) in a two-hour measurement would require additional consideration. This would correspond to a measured MDConc. in sump water of

$$\text{MDConc.} = \frac{(2000 + 211)\,\text{cts}}{0.08\,\text{c/d} \times 120\,\text{min} \times 2\,\text{L} \times 2.22\,\text{dpm/pCi}} = 51.9\,\text{pCi/L}$$

(b) Similarly, if the background water yielded 500 counts for a 30-minute measurement,

$$\text{MDConc.} = 56.9\,\text{pCi/L}$$

These two results indicate that a 10% greater concentration could escape regulation if the routine method of measuring sump water were based on a 30-minute count of sample and background rather than 120 minutes. Conversely, if increased sensitivity were the main goal, the longer count would be chosen.

THE CHI-SQUARE TEST OF A DETECTOR SYSTEM

The operation of a counting system can be tested by comparing the distribution of a series of repeat measurements with the expected distribution. One measure of performance is the Poisson index of dispersion, or the chi-square statistic, defined as

$$\chi^2 = \sum_{i=1}^{n} \frac{(x_i - \bar{x})^2}{\bar{x}}$$

Chi square is plotted as a function of the degrees of freedom F where $F = n - 1$ (Figure 8-20). The system is expected to work normally if the probability of χ^2 lies between 0.1 and 0.9; and alternatively its performance is suspect if the p-value of χ^2 lies outside this range. If the measured value of χ^2 is < 0.1, the replicate counts are fluctuating more than would be expected for a Poisson distribution, and a stability problem in the electronics of the system is likely. If the measured value of χ^2 is > 0.9 (e.g., $p = 0.99$), the replicate counts are not fluctuating as would be expected in a Poisson distribution, and it is expected that the counting system is strongly biased.

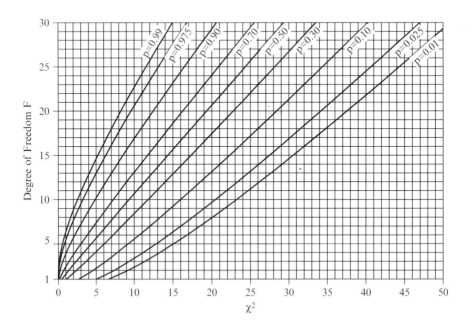

Figure 8-20. Integrals of the chi-square distribution.

Example 8-10: Five 1-minute counts of a given sample had values of 11, 9, 11, 13, and 9 for which the mean $\bar{x} = 10.6$. Does the system follow the Poisson distribution?

Solution: The χ^2 value for the five counts is

$$\chi^2 = \frac{\sum (x_i - \bar{x})^2}{\bar{x}} = \frac{11.2}{10.6} = 1.06$$

which from Figure 8-20 corresponds to $p = 0.9$ at 4 degrees of freedom. The data look somewhat more uniform than expected, but they follow the Poisson distribution and the counting system can be judged to be functioning properly.

SUMMARY

The interactions of various types of radiation in matter provide mechanisms for measuring the amount of radiation emitted and absorbed by a source(s), and with careful techniques the identity of the radiation source or radionuclide producing it. Radiation instruments include portable survey instruments that are designed to detect radiation and measure exposure or absorbed dose, and laboratory instruments that allow precise quantitation and identification of the radiation source. Various detectors are used in them, and these can be roughly divided into two categories: gas-filled chambers and crystalline materials. Each of these devices is based on the liberation of electrons in a medium and the collection and processing of the ions by electronic means.

The transformation of any given radioactive atom is subject to the laws of chance; therefore, the science of statistics is used to specify the uncertainty in a measurement of radioactivity at a given level of confidence, determining whether a sample actually contains radioactivity, and verifying whether a counting instrument is functioning properly. Such determinations are based on the Poisson distribution law, but, fortunately the Normal distribution can be used to represent the Poisson-distributed events if n, the number of counts, is greater than 16. Both normal and Poisson statistical models apply only to an observed number of counts and the standard deviation is equal to the square root of $\bar{x}$.

When a measurement includes a significant contribution from the radiation background, the background value must be subtracted and the total uncertainty in the net result is obtained by propagation of the error associated with each. The resulting net count has a standard deviation associated with it that is greater than that of either the sample or the background alone.

Acknowledgments

Professor James E. Watson of the University of North Carolina provided valuable information on the statistics of radiation measurements.

ADDITIONAL RESOURCES

1   Knoll, G. F., *Nuclear Radiation Detection* 3rd ed., (New York: Wiley Interscience, 2001).
2   Turner, J. E., *Atoms, Radiation, and Radiation Protection*, 2nd ed., (New York: Wiley Interscience, 1995).

PROBLEMS

1   K- and L-shell binding energies for cesium are 28 keV and 5 keV, respectively. What are the kinetic energies of photoelectrons released from the K and L shells when 40 keV photons interact in cesium?

2   Calculate the number of electrons entering the first stage of a photomultiplier tube from the interaction of a 0.46 MeV photon in a 12% efficient NaI(Tl) crystal.

3   Calculate the energy of the Compton edge for 1 MeV photons.

4   For ^{88}Y photons of energy, $h\nu = 1.836$ MeV, incident on an intrinsically pure germanium detector/spectrometer, calculate the energies of the "single escape" and "double escape" peaks.

5   A biomedical research laboratory uses ^{14}C and ^{32}P in tracer experiments. Devise a method for setting up a liquid scintillation counter to quantitate each radionuclide separately and in the presence of each other.

6   An LSC is calibrated such that a low-energy window has a counting efficiency of 30% for ^{3}H and a higher-energy window has a counting efficiency of 72% for ^{32}P. When a ^{32}P standard is counted, it is noted that 12% of the counts in window 2 are recorded in window 1. A mixed sample containing ^{3}H and ^{32}P yields 3,800 counts in window 1 and 5,800 counts in window 2. Determine the activity of each radionuclide.

7   What relative meaning does the term "quench" have in regard to a proportional counter, a G–M counter, and a liquid scintillation counter?

8   A single count of a radioactive sample yields 100 counts in 2 minutes. What is the uncertainty in the count and the count rate for a one-sigma confidence interval and a 95% confidence interval.

9   Compare the 95% uncertainties of a 2-minute measurement yielding 100 counts (see problem 8) and another 2-minute count that gives 1000 counts.

10   A single one-minute count of a radioactive ^{137}Cs source is 2000 counts. What is the expected range of values within which another count could be expected to appear with the same counting conditions: (a) at a 68.3% confidence level, and (b) at 95% confidence?

11   If the counter setup in problem 10 has a measured background rate of 100 cpm (measured during a counting time of one minute), calculate the net count rate and its uncertainty: (a) at a 68.3% confidence level, and (b) at 95% confidence. What percent errors do these results have?

12   A long-lived radioactive sample produces 1100 counts in 20 minutes on a detector system with measured background of 900 counts in 30 minutes. What is the net sample count rate and its standard deviation?

13   A sample was measured and reported at the 95% confidence level as 65.2 ± 3.5 cpm. The background was measured at 14.7 ± 4.2 cpm. Determine the net count rate for the sample and its uncertainty at the 95% confidence level.

14   An environmental sample yielded 530 counts in 10 minutes and a 30-minute background for the detector registered 50 counts per minute. Determine whether this difference is significant at the 95% confidence level; i.e., whether the sample contains radioactivity.

9

INTERNAL RADIATION DOSIMETRY

"Madame Curie died of aplastic pernicious anemia . . . the bone marrow did not react, probably because it had been injured by a long accumulation of radiations."
— *Dr. Tobé, July 4, 1934*

There are a number of circumstances in radiation protection or medical uses of radioactive material that require a determination of the radiation dose from radionuclides deposited in the body. Inhalation or ingestion of radionuclides by workers can occur either routinely or due to incidents to produce an internal dose to various organs and tissues, and similar exposures of the population can occur due to intakes of air, water, or food that contain radioactivity due to environmental releases. Radionuclides are also introduced into the body, in diagnosis or treatment of disease, by direct injection, a puncture wound, or skin absorption. Submersion in an atmosphere containing a noble gas such as xenon, krypton, or radon may also irradiate internal organs, but these are typically treated as an external, rather than internal, exposure.

ABSORBED DOSE IN TISSUE

Internal radiation doses cannot be measured; they must be calculated based on an estimated/measured intake, a measured/estimated quantity in an organ, or an amount eliminated from the body. Such calculations begin with the definition of absorbed dose, which is energy (joules or ergs) deposited per unit mass, and several key assumptions. First, it is assumed that the deposited radionuclide (expressed as a transformation rate, q, per unit time) is uniformly distributed throughout the tissue mass of a source organ. Second, the radionuclide emits energy while in a source organ S that is absorbed in the organ itself or a separate target tissue T which is characterized by an absorbed fraction, AF (T ← S). If the deposited radionuclide is a pure alpha or beta emitter, the source organ will be the only target tissue and all of the energy emitted will be deposited in it; i.e., AF (T ← S) = 1.0. For x and gamma rays, AF (T ← S) will generally be less than 1.0 and will vary considerably depending on the photon energy and the masses of the source organ and target tissues. And since many radionuclides produce more than one form of radiation emission per transformation, it is necessary to account for the energy absorbed due to each emitted radiation and its fraction Y_i of the number of transformations that occur.

If the transformation rate $q(t)$ of a radionuclide uniformly deposited in a tissue mass, m_T, is in t/s, the energy deposition rate is

$$\dot{D}(ergs/g \cdot s) = 1.6022 \times 10^{-6} q(t) \frac{\sum Y_i \bar{E}_i AF(T \leftarrow S)_i}{m_T}$$

or in Joules/kg · s

$$\dot{D}(Joules/kg \cdot s) = 1.6022 \times 10^{-10} q(t) \frac{\sum Y_i \bar{E}_i AF(T \leftarrow S)_i}{m_T}$$

where

    $q(t)$ = activity (t/s) of the radionuclide in the tissue at any time t
    Y_i = fractional yield per transformation of each radiation emitted
    $\bar{E}_i$ = average energy (MeV) of each emitted radiation
    AF(T ← S)$_i$ = the fraction of energy emitted by a source tissue, S, that is absorbed in a target tissue, T; and
    m_T = mass (g) of target tissue, T.

These expressions yield the "instantaneous" dose rate adjusted for the particular energy units of interest, and each is a direct function of the amount of activity, $q(t)$, that exists at any time t in a source organ.

Example 9-1: Calculate the instantaneous dose rate (rads/h) for $1\ \mu Ci$ of ^{32}P uniformly deposited in the liver ($m_T = 1800\ g$).
Solution: One $\mu Ci = 3.7 \times 10^4$ t/s and since ^{32}P is a pure beta emitter ($Y_i = 1.0$) with $E_{\beta,\ avg} = 0.695\ MeV/t$, it is reasonable to assume that all of the emitted beta energy is uniformly absorbed in the liver ($m_T = 1800\ g$); therefore,

$$\dot{D}(Liv \leftarrow Liv) = 1.6022 \times 10^{-6} \times 3.7 \times 10^4\ t/s \times \frac{1.0 \times 0.695\ MeV/t \times 1.0}{1800\ g}$$

$$= 2.29 \times 10^{-5}\ erg/g \cdot s$$

$$= 8.24 \times 10^{-4}\ rad/h$$

ACCUMULATED DOSE

Although the "instantaneous" dose rate is of some interest, it is the accumulation of dose delivered over a period of time (usually the total dose) that is
of most importance. All of the parameters that determine the "instantaneous"
dose rate are constant for a given exposure situation except $q(t)$ which varies
with time after deposition in an organ due to the combined effects of radioactive transformation (physical loss) and biological turnover, which
steadily decrease the number of atoms, as shown schematically in Figure 9-1.
The rate constants, λ_r (radioactive transformation) and λ_b (biological removal), are additive, thus their sum is an "effective" removal constant, λ_{eff},
or

$$\lambda_{eff} = \lambda_r + \lambda_b$$

And, since $\lambda_i = \ln 2 / T_i$, an effective half-life (T_e) can be determined from the
radiological (T_r) and biological (T_b) half-lives as

$$T_e = \frac{T_r T_b}{T_r + T_b}$$

The radiological half life, T_r, is a well-known quantity for each radionuclide;
however, T_b is characteristic of the radionuclide, its physical/chemical form,
and the physiological dynamics of the organ in which it is deposited. Values
of T_b for various tissues have been determined experimentally for a number of
compounds and these are listed in various compilations, the most useful
of which are metabolic models. Values of T_b and T_r for selected radionuclides
are listed in Table 9-6 (see below).

Since the only factor in the instantaneous dose equation that changes with
time is the activity, $q(t)$, the dose rate will diminish with time as shown in Figure
9-2 as a function of $q(t)$ which is in turn a function of q_0, the activity initially
deposited, and the effective removal constant, λ_{eff}, or

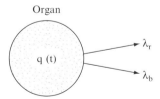

Figure 9-1. A radionuclide consisting of N atoms with activity $q(t) = \lambda N$ is assumed to
be distributed uniformly in an organ mass from which radioactive atoms are cleared by
both radioactive transformation (λ_r) and biological removal (λ_b).

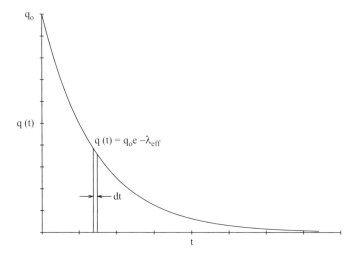

Figure 9-2. Variation of the "instantaneous dose rate" with time after deposition in a tissue from which it is removed by radioactive transformation and biological processes, or with an effective removal constant $\lambda_{\text{eff}} = \lambda_r + \lambda_b$.

$$q(t) = q_0 e^{-\lambda_{\text{eff}} t}$$

The Accumulated Radiation Dose is obtained by integrating the "instantaneous" dose rate equation from the time of initial deposition ($t = 0$) in the tissue to a later time t. If the initial activity deposited in a source organ S is q_0, the total energy deposition (ergs/g) in a target tissue T is

$$D_{\text{TOT}} = 1.6022 \times 10^{-6} \frac{\sum Y_i \bar{E}_i \, \text{AF}(T \leftarrow S)}{m_T} q_0 \int_0^t e^{-\lambda_{\text{eff}} t}$$

which can be integrated directly. Since $\lambda_{\text{eff}} = \ln 2 / T_e$, the total energy deposition is

$$D_{\text{TOT}}(\text{ergs/g}) = 1.6022 \times 10^{-6} \left[\frac{q_0}{\lambda_e} \left(1 - \exp\left(-\frac{\ln 2}{T_e} t \right) \right) \right] \times \left[\frac{\sum Y_i E_i \text{AF}(T \leftarrow S)_i}{m_T} \right]$$

where the first term in brackets represents the total number of transformations that occur for a time t following the initial deposition of an activity q_0; the second term is the energy deposition (or dose) rate (ergs/g · t) in the tissue mass m_T; and the constant accounts for the units. D_{TOT} can also be expressed as J/kg by multiplying by 10^{-7} J/erg and 10^3 g/kg.

In SI Units, the total dose, D_{TOT}, is defined as the committed dose equivalent, $H_{50, T}$, in units of Sievert (Sv) delivered over a 50-y period following the initial deposition of an activity q_0 in t/s or Bq, or

$$H_{50,\mathrm{T}}(\mathrm{Sv}) = 1.6 \times 10^{-10} \left[\frac{q_0}{\lambda_e} \left(1 - \exp\left(-\frac{\ln 2}{T_e} t\right) \right) \right] \times \left[\frac{\sum Y_i \bar{E}_i \mathrm{AF}(\mathrm{T} \leftarrow \mathrm{S})_i Q_i}{m_\mathrm{T}} \right]$$

where the first term in brackets is defined as U_s, the total number of transformations that occur due to an initially deposited activity q_0 (Bq), the second (or energy deposition) term in brackets is defined as the Specific Effective Energy (SEE) with units of MeV/g·t adjusted by the quality factor Q_i for each form of emitted radiation, and the constant 1.6×10^{-10} converts MeV/g·t to Joules/kg to yield Sieverts (Sv), the committed dose equivalent (CDE) for an internal emitter in SI units, or

$$H_{50,\mathrm{T}}(\mathrm{Sv}) - 1.6 \times 10^{-10} U_s \times \mathrm{SEE}\,(\mathrm{T} \leftarrow \mathrm{S})$$

U_s is calculated from an initially deposited activity of q_0(Bq) where T_e is in seconds; and since $\lambda_e = \ln 2/T_e$, it is

$$U_s(\#t) = 1.443 q_0 T_e \left(1 - \exp\left(-\frac{\ln 2}{T_e} t\right) \right)$$

The exposure period for a radionuclide deposited in a tissue is usually assumed to be 50 years, or a working lifetime, which is much longer than biological clearance for most radionuclides, thus the total number of transformations that occur in an organ due to an uptake, q_0 (t/s), reduces to

$$U_s(\#t) = 1.443 q_0 T_e$$

with an error of less than 1% for exposure times that are 7 to 10 times the effective half-life, T_e. This simplified expression of the total number of transformations can be used for internal dose calculations if, and only if, the integrated period of exposure is 7–10 times greater than the effective half-life, T_e, which is the case for most radionuclides in tissue. Possible exceptions are long-lived radionuclides such as $^{90}\mathrm{Sr}$, $^{226}\mathrm{Ra}$, and long-lived isotopes of thorium and the transuranic elements that are deposited in bone and lymph tissues that have very slow biological clearance. In these cases it is necessary to integrate the transformation rate over the period of interest, usually a working lifetime of 50 years.

INTERNAL DOSE—MEDICAL USES

Medical applications usually determine q_0 in μCi and T_e in hours, and the cumulative activity has units of μCi · h where one μCi · h $= 1.332 \times 10^8$t. The Medical Internal Radiation Dose (MIRD) Committee of the Society of Nuclear Medicine has incorporated these parameters into a straightforward schema as

$$D_{\mathrm{TOT}} = \tilde{A} \cdot \mathrm{S}(\mathrm{T} \leftarrow \mathrm{S})$$

where $\tilde{A}$ is the cumulative activity in $\mu Ci \cdot h$ due to an initial deposition of q_0 μCi in a source tissue and S (T $\leftarrow$ S) incorporates the energy deposition parameters and conversion constants to yield rads/$\mu Ci \cdot h$ for each deposited radionuclide. The S-value is fixed for a given radionuclide and source-target; therefore, use of the MIRD schema is basically a process of determining $\tilde{A}$ due to a deposited radionuclide and use of the respective S-value table to determine dose(s) to a relevant tissue(s).

Checkpoints

All internal radiation dose determinations involve three quantities, which vary only to account for the units of dose (rads or Sieverts):

1. the total number of transformations that occur following the deposition of an initial activity q_0 in a source organ;
2. the amount of energy (MeV) deposited per gram of target tissue for each transformation that occurs in a source organ; and
3. a constant to adjust units: for D_{TOT} in ergs/g it is 1.6022×10^{-6}; in SI units of J/kg (or Sv) it is rounded to 1.6×10^{-10} J/MeV $\cdot$ kg.

FACTORS IN THE INTERNAL DOSE EQUATION

The internal dose equation can be generalized in terms of the total number of transformations that occur in a source organ, which is a function of q_0 and λ_{eff}, and three key factors: m_T, $\sum Y_i \bar{E}_i$, and AF(T $\leftarrow$ S)$_i$, which are constants for any given radionuclide and a selected source-target organ pair.

The Initial Activity q_0 in a source organ is commonly derived from stated conditions, measured by whole-body or tissue counting, or estimated from the amount entering the body based on established models and uptake fractions.

The Target Tissue Mass, m_T *(in grams)* is used in the calculation when actual values are available; otherwise the values in Table 9-1, which are based on a 70-kg reference person, are used.

The Energy Emission per Transformation $\sum Y_i E_i$, is combined with the absorbed fraction AF (T $\leftarrow$ S) to obtain the energy deposition in a target tissue T for each transformation that occurs in a source organ S. Energy emission and deposition from pure beta or alpha emitters that have a single energy are straightforward, but many radionuclides undergo transformation by emitting one or more particles and one or more photons. Whereas the emission energies of alpha particles and photons are discrete, the average energy, $\bar{E}_\beta$, for beta radiation and positron emission is dependent upon the shape of the beta-particle energy spectrum and weighted average energy values (as tabulated in

Table 9-1. Tissue Masses (g) for a Reference 70-kg Adult Male (Unique Source Organs are Denoted as (S), and Specific Target Tissues, which are Usually Irradiated by Source Organ Contents, are Denoted as (T)).

Tissue	Mass (g)
Ovaries	11
Testes	35
Muscle	28,000
Red marrow	1,500
Lungs	1,000
Thyroid	20
Stomach contents (S)	250
Small intestine contents (S)	400
Upper large intestine contents (S)	220
Lower large intestine contents (S)	135
Stomach wall (T)	150
Small intestine wall (T)	640
Upper large intestine wall (T)	210
Lower large intestine wall (T)	160
Kidneys	310
Liver	1,800
Pancreas	100
Cortical bone	4,000
Trabecular bone	1,000
Bone surfaces (T)	120
Skin	2,600
Spleen	180
Adrenals	14
Thymus	20
Uterus	80
Bladder contents (S)	200
Bladder wall (T)	45
Total body	70,000

Appendix C) should be used in dose calculations; when weighted values of $\bar{E}_\beta$ are not available it is common to use the approximation $\bar{E}_\beta = 0.33 E_{\beta\,max}$.

The Absorbed Fraction, AF (T ← S), is a dimensionless quantity to account for the fraction of the energy emitted by a source organ S that is actually absorbed in a target tissue T, which may be the source organ itself. For most practical purposes it is 1.0 for alpha particles and electrons (beta particles and positrons), and the source organ as well as the target organ will be the same (the range of a 5 MeV alpha particle is about 50 microns in tissue and the range of most beta particles in tissue is at most a few mm). On the other hand, complete absorption of all emitted photons in a source organ is unlikely (possible exceptions are those with very low energies) and the absorbed fraction, AF (T ← S), accounts for the energy deposited when photons from a source organ S irradiate a target organ some distance away; it is a function of the energy of the radiation and the size, shape, and relative locations of the source organ(s) and the target organ(s) as illustrated in Figure 9-3. Tables 9-2, 9-3, and 9-4 list absorbed fractions for photon energies between 0.02 and 4 MeV for the lungs, liver, and whole body as source organs. Similar data have been compiled by the MIRD Committee for up to 25 source/target-organ pairs and are available at www.nndc.bnl.gov. Specific absorbed fractions, which already include tissue masses based on a reference person, are also available and are easier to use in detailed internal dose calculations.

The Dose Reciprocity Theorem provides a mechanism for determining the absorbed energy per gram in a tissue that is not listed as a target organ but may be listed as a source organ or vice versa. Under the dose reciprocity theorem the energy absorbed per gram is the same for radiation travelling from T to S as it is for radiation travelling from S to T. Therefore, if AF (T ← S) is known, then AF (S ← T) can be obtained from the dose reciprocity theorem by adjusting the known quantity to account for the different masses; i.e.,

$$AF(S \leftarrow T)/m_S = AF(T \leftarrow S)/m_T$$

Therefore,

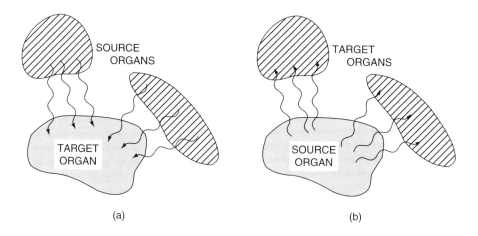

Figure 9-3. A radionuclide may be deposited such that (a) more than one source organ produces absorption of energy in a target tissue or (b) a source organ can produce energy absorption in more than one target tissue.

Table 9-2. Absorbed Fractions, AF (T ← Lu), for Photons Emitted from Radionuclides Deposited in the Lungs.

Target Organ	Photon Energy, E (MeV)									
	0.020	0.030	0.050	0.100	0.200	0.500	1.000	1.500	2.000	4.000
Adrenals	—	1.59E−04	1.55E−04	1.64E−04	1.03E−04	1.57E−04	1.25E−04	1.07E−04	1.85E−04	1.22E−04
Bladder	—	—	—	4.79E−05	1.45E−04	1.61E−04	1.58E−04	1.77E−04	3.25E−04	2.95E−04
GI (Stom)	8.75E−04	2.63E−03	3.62E−03	2.63E−03	2.50E−03	2.28E−03	1.84E−03	1.90E−03	2.02E−03	1.68E−03
GI (SI)	—	1.70E−04	8.79E−04	1.37E−03	1.60E−03	1.72E−03	1.95E−03	2.47E−03	1.90E−03	2.03E−03
GI (ULI)	—	—	2.94E−04	4.46E−04	4.53E−04	4.32E−04	4.01E−04	6.26E−04	4.34E−04	5.17E−04
GI (LLI)	—	—	—	5.17E−05	9.73E−05	6.22E−05	1.43E−04	1.25E−04	1.46E−04	2.09E−04
Heart	9.83E−03	2.28E−02	2.03E−02	1.39E−02	1.30E−02	1.11E−02	1.02E−02	8.98E−03	9.40E−03	7.18E−03
Kidneys	—	3.26E−04	9.23E−04	9.01E−04	9.34E−04	1.16E−03	8.14E−04	9.15E−04	9.39E−04	6.32E−04
Liver	1.42E−02	2.58E−02	2.39E−02	1.73E−02	1.55E−02	1.54E−02	1.38E−02	1.33E−02	1.17E−02	1.06E−02
Lungs	4.75E−01	2.31E−01	8.92E−02	4.93E−02	4.98E−02	5.14E−02	4.52E−02	4.13E−02	3.82E−02	3.14E−02
Marrow	1.33E−02	3.86E−02	4.66E−02	2.54E−02	1.61E−02	1.33E−02	1.25E−02	1.14E−02	1.13E−02	9.80E−03
Pancreas	—	3.79E−04	5.73E−04	6.07E−04	4.30E−04	4.62E−04	5.71E−04	5.69E−04	6.47E−04	2.40E−04
Skel (Rib)	3.12E−02	7.01E−02	5.77E−02	2.43E−02	1.45E−02	1.19E−02	1.10E−02	1.04E−02	1.01E−02	8.58E−03
Skel (Pelvis)	—	—	1.96E−04	4.07E−04	3.48E−04	4.34E−04	4.72E−04	3.46E−04	4.98E−04	6.37E−04
Skel (Spine)	1.28E−03	1.48E−02	3.55E−02	2.35E−02	1.53E−02	1.21E−02	1.07E−02	1.02E−02	9.91E−03	8.27E−03
Skel (Total)	3.31E−02	9.52E−02	1.13E−02	6.11E−02	3.89E−02	3.24E−02	3.07E−02	2.78E−02	2.78E−02	2.42E−02
Skull	—	2.64E−04	1.37E−03	1.20E−03	9.70E−04	1.50E−03	1.37E−03	1.23E−03	1.58E−03	1.07E−03
Skin	1.76E−03	5.73E−03	5.99E−03	5.41E−03	5.47E−03	6.37E−03	6.12E−03	6.68E−03	6.10E−03	4.71E−03
Spleen	8.91E−04	1.82E−03	2.05E−03	1.46E−03	1.46E−03	1.19E−03	1.13E−03	1.08E−03	1.03E−03	8.54E−04
Thyroid	—	—	3.83E−05	3.29E−05	5.39E−05	6.49E−05	6.33E−05	—	9.76E−05	—
Uterus	—	—	—	1.53E−05	1.57E−05	2.25E−05	—	—	—	—
Total body	9.81E−01	8.78E−01	6.10E−01	3.92E−01	3.59E−01	3.54E−01	3.29E−01	3.01E−01	2.88E−01	2.41E−01

Source: Society of Nuclear Medicine, August 1969.

Table 9-3. Absorbed Fractions, AF (T ← Liv), for Photons Emitted from Radionuclides Deposited in the Liver.

Target Organ	Photon Energy, E (MeV)									
	0.020	0.030	0.050	0.100	0.200	0.500	1.000	1.500	2.000	4.000
Adrenals	1.83E−04	4.40E−04	3.92E−04	2.70E−04	2.37E−04	1.98E−04	2.21E−04	2.33E−04	3.27E−04	2.65E−04
Bladder	—	—	1.69E−04	2.75E−04	3.89E−04	3.58E−04	5.21E−04	5.15E−04	7.35E−04	4.74E−04
GI (Stom)	1.71E−04	1.51E−03	3.60E−03	3.00E−03	2.71E−03	2.80E−03	2.40E−03	2.54E−03	2.10E−03	1.50E−03
GI (SI)	1.17E−04	4.93E−03	1.08E−02	1.09E−02	1.00E−02	9.17E−03	9.25E−03	9.04E−03	7.91E−03	6.46E−03
GI (ULI)	9.27E−04	3.18E−03	4.48E−03	4.01E−03	3.87E−03	3.64E−03	3.20E−03	2.87E−03	3.40E−03	2.32E−03
GI (LLI)	—	—	6.13E−05	1.49E−04	2.11E−04	3.91E−04	2.68E−04	2.86E−04	3.34E−04	3.91E−04
Heart	1.32E−03	5.31E−03	7.62E−03	6.74E−03	5.70E−03	5.73E−03	5.01E−03	4.98E−03	3.92E−03	0.37E−02
Kidneys	1.06E−03	4.37E−03	5.66E−03	4.37E−03	3.90E−03	3.86E−03	3.35E−03	2.94E−03	3.01E−03	2.56E−03
Liver	7.84E−01	5.43E−01	2.78E−01	1.65E−01	1.58E−01	1.57E−01	1.44E−01	1.32E−01	1.22E−01	1.01E−01
Lungs	8.59E−02	1.65E−03	1.47E−02	1.01E−02	9.23E−03	8.38E−03	8.25E−03	6.96E−03	6.01E−03	5.68E−03
Marrow	8.19E−03	2.28E−02	3.25E−02	2.06E−02	1.33E−02	1.07E−02	9.35E−03	9.24E−03	8.88E−03	7.20E−03
Pancreas	1.86E−04	1.07E−03	1.30E−03	1.05E−02	1.02E−03	8.22E−04	8.64E−04	5.87E−04	7.65E−04	6.12E−04
Sk. (Rib)	1.83E−02	4.02E−02	3.66E−02	1.81E−02	1.11E−02	8.67E−03	7.60E−03	7.48E−03	6.35E−03	5.27E−03
Sk. (Pelvis)	—	5.23E−04	2.65E−03	3.08E−03	2.16E−03	1.82E−03	1.71E−03	1.81E−03	2.07E−03	1.57E−03
Sk. (Spine)	3.93E−04	5.66E−03	2.17E−02	1.67E−02	1.08E−02	0.857E−02	6.94E−03	6.70E−03	7.01E−03	5.21E−03
Sk. (Skull)	—	—	—	6.29E−05	1.40E−04	1.87E−04	2.62E−04	3.60E−04	4.20E−04	3.70E−04
Skel (Total)	2.09E−02	5.87E−02	8.03E−02	4.98E−02	3.24E−02	2.60E−02	2.31E−02	2.29E−02	2.17E−02	1.79E−02
Skin	1.36E−03	4.68E−03	5.58E−03	4.99E−03	5.07E−03	5.61E−03	5.81E−03	5.67E−03	5.60E−03	5.43E−03
Spleen	—	6.17E−05	5.33E−04	6.06E−04	6.45E−04	6.19E−04	6.33E−04	3.96E−04	3.36E−04	5.03E−04
Thyroid	—	—	—	—	—	—	—	—	2.38E−05	—
Uterus	—	—	5.64E−05	1.15E−04	1.36E−04	1.30E−04	1.27E−04	7.29E−05	9.20E−05	—
Total body	9.84E−01	9.05E−01	6.61E−01	4.54E−01	4.15E−01	4.07E−01	3.81E−01	3.55E−01	3.36E−01	2.82E−01

Source: Society of Nuclear Medicine, August 1969.

Table 9-4. Absorbed Fractions, AF (T ← TB), for Photons Emitted from Radionuclides Deposited in the Whole Body.

Target Organ	Photon Energy, E (MeV)									
	0.020	0.030	0.050	0.100	0.200	0.500	1.000	1.500	2.000	4.000
Adrenals	1.75E-04	2.09E-04	1.31E-04	1.01E-04	3.52E-05	1.38E-04	1.00E-04	1.07E-04	1.14E-04	—
Bladder	6.83E-03	6.25E-03	4.45E-03	3.25E-03	3.27E-03	3.41E-03	2.74E-03	2.91E-03	2.31E-03	1.47E-03
GI (Stom)	5.73E-03	5.60E-03	3.91E-03	2.73E-03	2.18E-03	2.58E-03	1.81E-03	1.99E-03	2.12E-03	1.19E-03
GI (SI)	2.34E-02	2.09E-02	1.63E-02	1.20E-02	1.06E-02	1.14E-02	1.09E-02	9.15E-03	8.20E-03	4.09E-03
GI (ULI)	6.47E-03	5.33E-03	3.74E-03	2.62E-03	2.56E-03	3.06E-03	2.28E-03	2.09E-03	1.97E-03	1.60E-03
GI (LLI)	4.57E-03	2.85E-03	2.56E-03	1.87E-03	1.51E-03	1.84E-03	1.78E-03	1.81E-03	1.57E-03	6.73E-04
Heart	7.69E-03	6.35E-03	4.69E-03	4.20E-03	3.37E-03	3.72E-03	3.01E-03	3.45E-03	3.12E-03	1.45E-03
Kidneys	4.12E-03	3.38E-03	2.33E-03	1.83E-03	1.71E-03	1.42E-03	1.61E-03	1.52E-03	1.54E-03	9.04E-04
Liver	2.49E-02	2.21E-02	1.54E-02	1.20E-02	1.11E-02	1.01E-02	8.96E-03	9.12E-03	8.47E-03	5.60E-03
Lungs	1.38E-02	1.22E-02	8.08E-03	5.51E-03	5.07E-03	4.96E-03	4.66E-03	4.66E-03	4.27E-03	5.68E-03
Marrow	—	7.40E-02	6.13E-02	3.29E-02	2.21E-02	1.94E-02	1.82E-02	1.64E-02	1.56E-02	9.69E-03
Pancreas	8.28E-04	7.80E-04	5.67E-04	4.49E-04	4.44E-04	3.82E-04	5.34E-04	3.48E-04	3.58E-04	1.42E-04
Skel (Rib)	2.47E-02	2.63E-02	1.76E-02	7.64E-03	5.05E-03	4.35E-03	4.21E-03	4.05E-03	3.50E-03	3.38E-03
Skel (Pelvis)	1.63E-02	2.20E-02	1.99E-02	1.03E-02	6.68E-03	5.69E-03	5.62E-03	5.11E-03	4.22E-03	2.56E-03
Skel (Spine)	2.34E-02	2.53E-02	2.29E-02	1.44E-02	9.10E-03	7.63E-03	7.51E-03	6.10E-03	6.06E-03	3.41E-03
Skel (Total)	1.67E-01	1.88E-01	1.53E-01	8.10E-02	5.50E-02	4.88E-02	4.56E-02	4.13E-02	3.96E-02	2.52E-02
Skull	1.23E-02	1.28E-02	7.22E-03	3.13E-03	2.77E-03	3.04E-03	2.80E-03	2.54E-03	2.92E-03	2.24E-03
Skin	1.69E-02	1.16E-02	7.58E-03	5.85E-03	6.77E-03	7.57E-03	7.45E-03	7.59E-03	6.64E-03	1.23E-02
Spleen	2.42E-03	2.23E-03	1.49E-03	1.11E-03	7.98E-04	1.16E-03	9.14E-04	9.03E-04	7.40E-04	3.68E-04
Thyroid	6.02E-05	1.11E-04	1.14E-04	8.73E-05	4.18E-05	—	—	—	8.10E-05	—
Uterus	—	—	1.22E-03	9.24E-04	4.08E-04	4.73E-04	5.17E-04	3.23E-04	3.64E-04	2.38E-04
Total body	8.92E-01	7.74E-01	5.48E-01	3.70E-01	3.38E-01	3.40E-01	3.21E-01	3.02E-01	2.84E-01	2.40E-01

Source: Society of Nuclear Medicine, August 1969.

$$AF(S \leftarrow T) = m_S/m_T \; AF(T \leftarrow S)$$

as illustrated in Example 9-2.

Example 9-2: Use the dose reciprocity theorem and the data in Table 9-2 to determine the absorbed fraction of 1 MeV photons emitted from the thyroid and absorbed in the lung.
Solution: From Table 9-2, AF (Thy $\leftarrow$ Lung) = 0.0000633, and from Table 9-1 the masses of the lung and the thyroid are 1,000 g and 20 g, respectively; therefore,

$$AF(Lung \leftarrow Thy) = 0.0000633 \times [1000\,g/20\,g]$$
$$= 0.003165$$

Energy Absorption in a Source Organ is often the largest component of internal dose produced by a radionuclide. The dose coefficient (ergs/g · t or J/kg · t) associated with the source organ can be calculated by absorbing all of the particle energy in the source organ mass and adding that due to the fractional absorption of the photon(s). Absorbed fractions of photon energies produced in tissues that are both the source and target are listed in Table 9-5; e.g., thyroid to thyroid, whole body to whole body, etc.

DEPOSITION AND CLEARANCE DATA

Calculations of internal dose require information on the fractional uptake of a radionuclide from the blood stream to a source organ and the pattern and rate(s) of clearance from it. Representative data for selected radioelements are summarized in Table 9-6, which also shows that deposition can occur in various tissues and clearance can be rather complex.

Multi-Compartment Clearance/Retention occurs when a given radionuclide is eliminated from an organ or tissue at different rates, suggesting two or more biological compartments of retention with different effective removal half-lives (see Table 9-6). In such cases, it is necessary to calculate the number of transformations (or cumulative activity) that occur in each compartment separately and to sum them to determine the total dose delivered to the tissue. For example, it has been established from biological studies that 10% of cesium that is deposited in the whole body is cleared from a sub-compartment with a biological half-life of 2 days and the remaining 90% is cleared with a biological half-life of 110 days (see Table 9-6). The amount, $q(t)$, of cesium in the total body at any time after a deposition of q_0 is called the retention function, and is expressed for cesium as

$$q(t) = 0.1q_0 e^{-(\ln 2/2\,d)t} + 0.9q_0 e^{-(\ln 2/110\,d)t}$$

Table 9-5. Absorbed Fractions, AF (S ← S), for Photon Energies Released in a Source Organ and Absorbed Therein.

Source/Target Tissue	Photon Energy (MeV)									
	0.020	0.030	0.050	0.100	0.200	0.500	1.000	1.500	2.000	4.000
Bladder	7.45E−01	4.64E−01	2.01E−01	1.17E−01	1.16E−01	1.16E−01	1.07E−01	1.00E−01	9.10E−02	7.22E−02
Brain	8.03E−01	5.55E−01	2.68E−01	1.61E−01	1.59E−01	1.60E−01	1.49E−01	1.36E−01	1.26E−01	1.00E−01
GI (Stom)	7.05E−01	4.14E−01	1.76E−01	1.01E−01	1.01E−01	1.01E−01	9.32E−02	8.46E−02	7.92E−02	6.43E−02
GI (SI)	7.56E−01	5.17E−01	2.64E−01	1.59E−01	1.54E−01	1.50E−01	1.40E−01	1.25E−01	1.17E−01	9.36E−02
GI (ULI)	6.25E−01	3.35E−01	1.34E−01	7.83E−02	8.04E−02	8.11E−02	7.49E−02	7.00E−02	6.38E−02	4.89E−02
GI (LLI)	5.56E−01	2.76E−01	1.05E−01	6.24E−02	6.31E−02	6.49E−02	6.01E−02	5.61E−02	4.93E−02	4.04E−02
Kidneys	5.82E−01	2.98E−01	1.12E−01	6.61E−02	6.76E−02	7.30E−02	6.70E−02	6.00E−02	5.59E−02	4.49E−02
Liver	7.84E−01	5.43E−01	2.78E−01	1.65E−01	1.58E−01	1.57E−01	1.44E−01	1.32E−01	1.22E−01	1.01E−01
Lungs	4.75E−01	2.31E−01	8.92E−02	4.93E−02	4.98E−02	5.14E−02	4.52E−02	4.13E−02	3.82E−02	3.14E−02
Ovaries	2.72E−01	9.84E−02	2.94E−02	1.74E−02	1.96E−02	2.29E−02	1.99E−02	1.76E−02	1.70E−02	1.37E−02
Pancreas	4.42E−01	1.95E−01	6.76E−02	3.83E−02	4.18E−02	4.37E−02	4.01E−02	3.74E−02	3.41E−02	2.67E−02
Skeleton	8.30E−01	6.81E−01	4.00E−01	1.73E−01	1.23E−01	1.18E−01	1.10E−01	1.02E−01	9.40E−02	7.97E−02
Spleen	6.25E−01	3.31E−01	1.28E−01	7.09E−02	7.35E−02	7.69E−02	6.99E−02	6.50E−02	5.99E−02	4.83E−02
Testes	4.69E−01	2.08E−01	7.03E−02	3.99E−02	4.33E−02	4.46E−02	4.12E−02	3.83E−02	3.54E−02	2.89E−02
Thyroid	3.66E−01	1.49E−01	4.80E−02	2.78E−02	3.06E−02	3.19E−02	2.95E−02	2.76E−02	2.54E−02	2.00E−02
Total Body	8.92E−01	7.74E−01	5.48E−01	3.70E−01	3.38E−01	3.40E−01	3.21E−01	3.02E−01	2.84E−01	2.40E−01

Table 9-6. Distribution of Radioelements in Adults from the Transfer Compartment (Blood) to Tissue in the Body, the Percentage (%) Clearance, Subcompartments of Each Tissue, and the Biological Half-life (T_b) for Each.

Element (f_1) Nuclide ($T_{1/2}$)	Transfer from Blood Tissue	%	Clearance from Tissue Compartment	%[a]	T_b
Carbon ($f_1 = 1.0$) ^{11}C (20.3 m) ^{14}C (5715 y)	Tot. Body	100 %	Tot. Body	100 %	40 d
Phosphorus ($f_1 = 0.8$) 32P (14.1 d) ^{33}P (25.34 d)	Tot. Body Bone	70 % 30 %	T. Body Exc. Tot. Body A Tot. Body B Bone	21.5 % 21.5 % 57 % 100 %	0.5 d 2 d 19 d ∞
Sulphur ($f_1 = 0.8$) ^{35}S (87.44 d)	Tot. Body	100 %	T. Body Exc. Tot. Body A Tot. Body B	80 % 15 % 5 %	0.3 d 20 d 200 d
Manganese ($f_1 = 0.1$) ^{54}Mn (312.12 d) ^{56}Mn (2.5785 h)	Bone Liver Other Tiss.	35 % 25 % 40 %	Bone Liver A Liver B Other Tiss. A Other Tiss. B	100 % 10 % 15 % 20 % 20 %	40 d 4 d 40 d 4 d 40 d
Cobalt ($f_1 = 0.1$) ^{57}Co (270.9 d) ^{58}Co (70.8 d) ^{60}Co (5.27 y)	Excretion Tot. Body Liver	50 % 45 % 5 %	Excretion Tot. Body A Tot. Body B Tot. Body C Liver A Liver B Liver C	100 % 60 % 20 % 20 % 60 % 20 % 20 %	0.5 d 6 d 60 d 800 d 6 d 600 d 800 d
Nickel ($f_1 = 0.05$) ^{59}Ni (75.000 y)	Kidneys Other Tiss.	2 % 30 %	Kidneys Other Tiss.	100 % 100 %	0.2 d 1200 d
Zinc (f1 = 0.5) ^{65}Zn (245 d)	Tot. Body	100 %	Skeleton A Skeleton B Tot. Body A Tot. Body B	19.5 % 0.5 % 24 % 56 %	400 d 10,000 d 20 d 400 d
Selenium ($f_1 = 0.8$) ^{75}Se (119.8 d)	Liver Kidneys Spleen Pancreas Testes Ovaries Other Tiss.	25 % 10 % 1 % 0.5 % 0.1 % 0.02 % 63.4 %	All Tiss. A All Tiss. B All Tiss. C	10 % 40 % 50 %	3 d 30 d 200 d
Strontium ($f_1 = 0.3$) ^{89}Sr (50.5 d) ^{90}Sr (29.1 y)	Soft Tiss.	100 %	S. Tiss. A S. Tiss. B S. Tiss. C	80 % 15 % 5 %	2 d 30 d 200 d

Table 9-6. (*continued*)

Element (f_1) Nuclide ($T_{1/2}$)	Transfer from Blood Tissue	%	Clearance from Tissue Compartment	%[a]	T_b
Zirconium ($f_1 = 0.01$)	Skeleton	50 %	Skeleton	100 %	10,000 d
^{95}Zr (64.02 d)	Other Tiss.	50 %	Other Tiss.	100 %	7 d
Molybdenum ($f_1 = 1.0$)	Skeleton	10 %	Skeleton	100 %	10,000 d
^{99}Mo (66 h)	Liver	25 %	Other Tiss. A	10 %	1 d
	Kidney	5 %	Other Tiss. B	90 %	50 d
	Other Tissue	60 %			
Technetium ($f_1 = 0.5$)	Thyroid	4 %	Thyroid	100 %	0.5 d
^{99m}Tc (6.0 h)	Stom. Wall	10 %	Other Tiss. A	75 %	1.6 d
^{99}Tc (2.1E5 y)	Liver	3 %	Other Tiss. B	20 %	3.7 d
	Tot. Body	83 %	Other Tiss. C	5 %	22 d
Ruthenium ($f_1 = 0.05$)	Tot. Body	100 %	T. Body Exc.	15 %	0.3 d
^{103}Ru (39.28 d)			Tot. Body A	35 %	8 d
^{106}Ru (368.2 d)			Tot. Body B	30 %	35 d
			Tot. Body C	20 %	1,000 d
Silver ($f_1 = 0.05$)	Liver	50 %	Liver A	10 %	3.5 d
^{110m}Ag (249.9 d)	Other Tiss.	50 %	Liver B	80 %	50 d
			Liver C	10 %	500 d
			Other Tiss. A	10 %	3.5 d
			Other Tiss. B	80 %	50 d
			Other Tiss. C	10 %	500 d
Antimony ($f_1 = 0.1$)	Excretion	20 %	All Tiss. A	85 %	5 d
^{124}Sb (60.2 d)	Skeleton	40 %	All Tiss. B	10 %	100 d
^{125}Sb (2.77 y)	Liver	5 %	All Tiss. C	5 %	5,000 d
	Other Tiss.	35 %			
Iodine ($f_1 = 1.0$)	Thyroid	30 %	Thyroid	100 %	80 d
^{125}I (60 d)					
^{129}I (1.57E7 y)					
^{131}I (8.02 d)					
Cesium ($f_1 = 1.0$)	Tot. Body	100 %	Tot. Body A	10 %	2 d
^{134}Cs (2.065 y)			Tot. Body B	90 %	110 d
^{137}Cs (30.07 y)					
Cerium ($f_1 = 0.0005$)	Skeleton	30 %	All Tissues		3,500 d
^{141}Ce (32.5 d)	Liver	50 %			
^{144}Ce (284.3 d)	Other Tiss.	20 %			

[a]Listed value represents percent clearance of amount deposited in the listed tissue, e.g., for ^{32}P 70% of that entering the blood (transfer compartment) distributes to the whole body and 21.5%, 21.5%, and 57% of that is cleared with half lives of 0.5 d, 2 d, and 19 d, respectively, and the remaining 30% goes to bone where all of it (100%) is retained with an infinite half life.

where, as indicated by the first term, $0.1q_0$ is cleared with a biological half-life of 2 days, and as shown by the second term, $0.9q_0$ is cleared with a biological half-life of 110 days.

Retention Functions are provided in various compilations for stable elements as sums of exponential functions (e.g., cesium with two compartments); i.e., the clearance half-lives are due to biological removal only (e.g., all cesium, whether radioactive or not, clears the same way). It is necessary, therefore, to multiply through by $e^{-\lambda_r t}$ to account for radioactive removal as well as biological turnover for each compartment of retention/clearance; however, for long-lived radionuclides such as ^{137}Cs ($T_{1/2} = 30.04$ y) the biological half-life for each retention compartment is, for all practical purposes, the effective half-life.

The number of transformations due to deposition of an initial activity q_0 is obtained by integrating each term of the retention function over the period of interest (usually 50 y). For example, for ^{137}Cs distributed in the total body

$$U_s(\#t) = \frac{0.1q_0}{\ln 2/2d}(1 - e^{-\frac{\ln 2}{2d}t}) + \frac{0.9q_0}{\ln 2/110d}(1 - e^{-\frac{\ln 2}{110d}t})$$

which for $t \geq 7 - 10$ effective half lives for each compartment simplifies to

$$U_s(\#t) = (0.1q_0)(1.443 \times 2d) + (0.9q_0)(1.443 \times 110d) \times 86,400\,\text{s/d}$$

if q_0 is in units of t/s. The parameters in the internal dose equation and their use are shown in Examples 9-3 and 9-4.

Example 9-3: Twenty μCi of ^{32}P ($T_{1/2} = 14.3$d) is injected into the blood stream of a reference adult. Calculate (a) the absorbed dose constant (erg/g · t) for ^{32}P irradiating the whole body uniformly, (b) the total number of transformations that occur in the total body as a source organ, and (c) the total body dose.
Solution: (a) The energy deposition rate per transformation is

$$\dot{D}(\text{TB} \leftarrow \text{TB}) = 1.6022 \times 10^{-6}\,\text{erg/MeV}\,\frac{1.0 \times 0.695\,\text{MeV/t} \times 1.0}{70,000\,\text{g}}$$

$$= 1.59 \times 10^{-11}\,(\text{erg/g} \cdot \text{t})$$

(b) As shown in Table 9-6 ^{32}P is metabolized in such a way that 40% is uniformly distributed throughout the whole body where it is cleared with a biological half-life of 19 days; thus, the total number of transformations that occur in the total body from an injection of 20 μCi is determined from the initial activity distributed in the total body ($q_0 = 0.40 \times 20\,\mu Ci = 8\,\mu Ci$) and the effective half-life of ^{32}P in the total body, or

$$T_{eff} = \frac{19d \times 14.3d}{19d + 14.3d} = 8.16d = 196h$$

And since T_{eff} is quite short compared to 50 y,

$$\#t = 1.443 \times (196h \times 3600s/h) \times (8\,\mu Ci \times 3.7 \times 10^4 t/s \cdot \mu Ci)$$
$$= 3.014 \times 10^{11} t$$

(c) The total dose to the whole body is obtained by multiplying the absorbed dose constant (ergs/g · t) and the total number of transformations, or

$$D_{TOT}(TB \leftarrow TB) = (1.59 \times 10^{-11} erg/g \cdot t) \times 3.014 \times 10^{11} t$$
$$= 4.79\,erg/g$$
$$= 47.9\,mrads$$

Example 9-4: For 2 μCi of ^{137}Cs deposited in the total body, determine (a) the total number of transformations that occur over a 50-y working career; (b) the energy deposition constant (MeV/g · t); and (c) the total body dose.
Solution: (a) From Table 9-6, cesium is distributed uniformly in the body, but is cleared from two compartments with biological half-lives of 2 d and 110 d, respectively, which are, because $T_r = 30.04$ y, also the effective half-lives. And since the exposure time t is long compared to the effective half-lives, the number of transformations in the total body due to retention in the two compartments is

$$U_s = 1.443 \times (2\,\mu Ci \times 3.7 \times 10^4 t/s \cdot \mu Ci) \times [(0.1 \times 2d) + (0.9 \times 110d)] \times 86,400\,s/d$$
$$= 9.15 \times 10^{11} t$$

(b) The decay scheme (App. C) for ^{137}Cs lists two beta emissions ($\bar{E}_\beta = 0.1734\,MeV$ @ 94.4% and 0.4163 MeV @ 5.6%), two primary conversion electron transitions ($E = 0.6242\,MeV$ @ 7.66% and 0.6557 MeV @ 1.12%), and one gamma emission of 0.662 MeV at 85.1% (*Note:* other lesser emissions are neglected). The energy deposition constant for the non-penetrating particle emissions, where AF $(T \leftarrow S) = 1.0$, is

$$\sum \frac{Y_i \bar{E}_i \times 1.0}{70,000\,g} = \frac{1}{70,000}[(0.1734 \times 0.944) + (0.4163 \times 0.056) + (0.6242 \times 0.076) + (0.6557 \times 0.0112)]$$
$$= 3.4594 \times 10^{-6}\,MeV/g \cdot t$$

and for the penetrating γ energy where AF (T.Body $\leftarrow$ T.Body) = 0.334 (Table 9-5)

$$= \frac{0.662 \times 0.851 \times 0.334}{70,000\,\text{g}} = 2.688 \times 10^{-6}\,\text{MeV/g} \cdot \text{t}$$

for a total energy deposition per transformation of $6.1474 \times 10^{-6}\,\text{MeV/g} \cdot \text{t}$
(c) The whole body dose is

$$H_{50,\text{T}} = 1.6 \times 10^{-6}\,\text{ergs/MeV} \times 6.1474 \times 10^{-6}\,\text{MeV/g} \cdot \text{t} \times 9.15 \times 10^{11}\text{t}$$
$$= 9.0122\,\text{erg/g} = 0.09\,\text{rads} = 90\,\text{mrads}$$

INTERNAL RADIATION DOSE STANDARDS

Internal radiation dose and intake are directly related, thus, controls for internal radiation dose have been based on establishing an annual limit of intake (ALI) for each particular radionuclide. Such limits, which have varied over the years, are now based on an explicit consideration of risk; previous standards for internal dose (since the 1940s) were to a large extent also based on risk, however, it was not explicitly stated, at least not numerically. Risk has thus become the primary protection standard, and dose, whether received internally or externally or both, is a physical surrogate, or secondary standard, for meeting the risk standard.

Internal dose limits were, prior to 1977, devised for a critical organ which, depending on its sensitivity, was limited to 15 rads/y for most tissues – a factor of 3 greater than the total body dose limit. For special tissues, such as the thyroid, the skin, and the gonads, the limits were 30, 30, and 5 rads/y, respectively, and as much as 50 rads/y to skeletal tissue for bone seekers. Dose limits for specific tissues were allowed to be higher than those for the total body because they posed an overall lesser risk to the total organism, and perhaps more significantly organ doses were considered singly and separately from external dose on the presumption that one or the other would dominate in an occupational setting.

The ICRP, in 1977, estimated that a whole body dose of 1 rem (0.01 Sv) to 10^6 exposed persons would produce a lifetime risk of 165 serious health effects based on a linear-quadratic extrapolation model of dose versus risk. These effects were observed to occur in six major tissue groups plus a varying number of other tissues (referred to as the remainder), and would produce a total body risk that could be considered acceptable if the total effective dose equivalent (TEDE) to each individual was $\leq 0.05\,\text{Sv}$ (5 rem) in each year of exposure. The ICRP schema also limited the 50-year committed dose for any tissue to 50 rem $(0.5\,Sv)/y$ to preclude nonstochastic effects. Both stochastic and nonstochastic dose limits presume that no external or other doses occur.

The ICRP revisited the risk data in 1990 (ICRP Report 60) and concluded that a linear nonthreshold model was a more appropriate descriptor of risk than the linear quadratic model used in its 1977 analysis. Although the number of health effects in the study populations was essentially the same, this change in the extrapolation model (see Chapter 6) yielded a risk factor of about 5×10^{-4}/rem, or about a factor of 3 higher than the previously estimated value of 1.65×10^{-4}/rem. Rather than raise the level of risk that had been so arduously justified in 1977, the ICRP chose to recommend an overall lower (by a factor of about 3) standard of 10 rem for 5-year periods of exposure not to exceed 5 rems in any year. Two major effects of the 1990 recommendations are (1) they limit the average annual dose to 2 rems, and (2) they affirm the previous level of 0.05 Sv (5 rem) in one year, but restrict its application such that a 5-year average would be $\leq 0.1\,Sv$ (or 10 rem).

Radiation protection standards in the US are also now based on an assessment of risk in accordance with Federal Radiation Guidance recommended by EPA and approved by the President in 1987. EPA analyses of the risk data have consistently used the linear nonthreshold model, which yields a risk coefficient on the order of 5×10^{-4}/rem, and US standards have remained at

$$\text{TEDE} \leq 5 \ \text{rem}(0.05\,Sv)/y$$

which limits radiation risk regardless of whether the dose was received internally, externally, or the sum of both. And, consistent with ICRP recommendations, the annual committed dose to any individual tissue is also limited to 50 rems (0.5 Sv) to preclude nonstochastic effects.

Effective Dose Equivalent

Since the basic protection standards were derived from risk data based on whole body doses, a mechanism is necessary to adjust doses to internal organs so that the risk incurred is equivalent to that of the whole body irradiated uniformly; i.e., an effective whole body dose. A relative weighting factor, w_T, is thus assigned to each of the major tissues at risk and the aggregate (or whole body) risk is obtained by summing these weighted doses. The 1977 schema assigned values of w_T for six major tissues and the next five highest (the remainder); however, by 1990 sufficient risk data were available to derive w_T values (in ICRP 60) for several additional tissues. Values of w_T for tissues at risk are

	w_T (ICRP 26)	w_T (ICRP 60)
Gonads	0.25	0.20
Breast	0.15	0.05
Colon	–	0.12
Red Marrow	0.12	0.12
Lung	0.12	0.12
Stomach	–	0.12
Ur. Bladder	–	0.05
Liver	–	0.05
Esophagus	–	0.05
Thyroid	0.03	0.05
Bone surfaces	0.03	0.01
Skin	–	0.01
Remainder ($w_T \times 5$)	0.06×5	0.01×5
TOTAL	1.00	1.00

Values of w_T can thus be used to convert internal doses to organs and tissues to an effective whole body dose; e.g., a dose to the lung of 10 rads (0.1 Sv) multiplied by the weighing factor of 0.12 for lung yields 1.2 rads, which has a risk equivalent to that of the whole body irradiated uniformly, and under the risk-based schema is considered as a committed effective dose equivalent (CEDE) of 1.2 rads (or 1.2 rem if $Q = 1.0$). And since the risk-based TEDE standard is 5 rem/y (0.05 Sv/y) the sum of the weighted CEDE to all affected tissues due to an intake is limited to

$$\sum w_T H_{50,T} \leq 5 \text{ rems}(0.05 \text{ Sv}) \text{ CEDE/y}$$

where $H_{50,T}$ is the 50-year committed dose equivalent to each target tissue T due to the intake of a radionuclide.

Two tertiary requirements are used to ensure that internal doses do not exceed the TEDE limit: (1) an annual limit on intake (ALI) and, (2) a derived air concentration (DAC) which, as the name implies, is derived from the ALI and is the concentration of a radionuclide in air which, if breathed by a reference person, will result in an intake just equal to the ALI. The ALI for each radionuclide taken into the body is calculated from the weighted sum of tissue doses as

$$\text{ALI(Bq)} = \frac{0.05 \text{ Sv}}{\sum w_T H_{50,T} \text{Sv/Bq of intake}}$$

where 0.05 Sv (5 rem) is the limiting CEDE for stochastic risks (for nonstochastic risks the ALI would be based on 0.5 Sv, or 50 rem to the maximum tissue exposed). The DAC is calculated as

$$\text{DAC (Bq/m}^3) = \frac{\text{ALI (Bq)}}{2.4 \times 10^3 \, \text{m}^3}$$

where $2.4 \times 10^3 \, \text{m}^3$ is the amount of air breathed by a reference person during light activity (20 L/min) for a working year of 2000 hours, and the ALI is the most restrictive value based on limiting either stochastic or nonstochastic effects.

US Regulations for Internal Emitters currently include ALIs and DACs based on a TEDE of 5 rem/y (see Table 9-8). The ICRP has also used its recommended 2 rem/y average value to develop new calculations of ALIs and DACs which are provided in ICRP Report 61.

BIOKINETICS AND DOSE FOR RADIONUCLIDE INTAKES

Inhalation or ingestion of radionuclides introduces complexity to internal dose determinations because radionuclides must first pass through the respiratory system and/or the gastrointestinal (GI) tract before entering the bloodstream from which uptake to an organ occurs. These circumstances are quite different from those where the amount of a radionuclide in the organ of interest is known or can be reasonably determined (e.g., in medical uses); thus, it is necessary to consider the mechanisms by which they are deposited and the physiological mechanisms by which they are translocated to other tissues which become source organs. It is also necessary to determine radiation doses to the lung and the various segments of the GI tract as the radioactive material passes through them, as well as doses to tissues where they subsequently deposit and those that surround them.

Biokinetic Models—Risk-Based Internal Dosimetry

Fairly extensive models have been developed for calculating doses due to inhaled and ingested radionuclides, in particular a respiratory deposition/clearance model and a GI tract model which are shown together in Figure 9-4. Both assume an initial deposition of radioactive material that is translocated to various tissues to produce a tissue dose according to the chemical and physical characteristics of the radionuclide. A special model has also been developed for bone dosimetry because of the unique distribution and deposition of bone-seeking radionuclides.

Inhaled radionuclides are first deposited in various lung compartments (shown in Figure 9-4) followed by clearance to other tissues in the body either by direct absorption to the bloodstream or clearance to and absorption from the GI tract. Radioactive material translocated from the lung to the GI tract clears the same as ingested material; therefore, both are shown in Figure 9-4 as interconnected routes. A radionuclide entering the stomach can produce a radiation dose to the stomach wall while it clears to the small intestine where it then produces a dose to the small intestine wall. And, depending on its solubility (denoted by f_1), the radionuclide can be absorbed into the blood stream and translocated to other organs and tissues, producing a dose to them. The remaining portion passes to the upper and lower large intestine where it delivers a dose to the ULI and LLI walls as the material is translocated for eventual excretion. The GI tract is thus an important route for translocation of both inhaled and ingested radioactive material to other tissues and organs of the body.

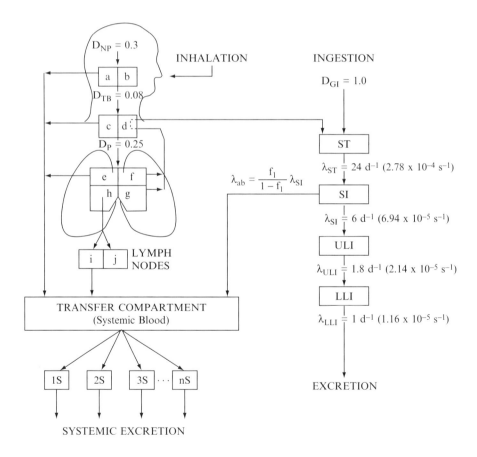

Figure 9-4. Deposition and clearance routes of inhaled and ingested materials according to ICRP 30 schema.

Radiation Doses Due to Inhaled Radionuclides

Current US regulations for controlling internal radiation dose are based on a respiratory model (ICRP 30) that assigns the deposition of inhaled radioactive materials to four major regions: (1) the nasal-pharangeal (or NP) region, (2) the trachea and bronchial tree (TB), (3) the pulmonary parenchyma (P), and (4) the lymphatic (L) system which is considered separately as a tissue compartment connected to the pulmonary region. The deposited material is assigned to one of three broad clearance classes based on its chemical and physical characteristics:

- Class D (daily) particles are quite soluble and have a half- time in the lung of less than about 10 days.
- Class W (weekly) particles are slightly soluble and have a half-time in the lung of 10 to 100 days (or several weeks).
- Class Y (yearly) particles are largely insoluble and can remain in the lung with half-times greater than 100 days (i.e., on the order of years).

Each region of the ICRP 30 lung model has two or more subcompartments (see Figure 9-4) each of which has unique deposition fractions and clearance routes and rates depending on the clearance class (D, W, Y) of the element (see Table 9-7). Material deposited in each sub-region of the lung is cleared as follows:

- Material deposited in compartments a, c, and e is absorbed directly into the transfer compartment, or the systemic blood.
- Material deposited in compartments b, d, f, and g is transported by mucociliary action to the oropharynx where it is swallowed, thus entering the stomach where it is translocated through the rest of the GI tract where, depending on its solubility, it can be absorbed into the systemic blood through the small intestine wall or, if not absorbed, translocated through the upper and lower large intestines.
- Material deposited in compartment h is translocated via phagocytosis and absorption to the pulmonary lymph nodes (region L) which contains two compartments, i and j. Material transferred to compartment i is slowly translocated to body fluids, and some class Y aerosols (i.e., highly insoluble particulate matter) may be translocated to compartment j where they are assumed to be retained indefinitely.

Radioactive material that is cleared to the systemic blood supply (mathematically labeled the transfer compartment) can be taken up in various amounts by systemic organs and tissues, can circulate in the intercellular fluid, or be excreted directly. The time-dependent rate of translocation and retention in the lung, the blood, and various interconnected tissues can be modeled as a system of first-order kinetics to determine the number of transformations that occur in each region.

Table 9-7. Deposition Fractions(F), Clearance Half-lives (in Days), and Clearance Constants (s⁻¹) for 1 μm AMAD Aerosols in Each Subregion of the Lung.

Region/Compartment		Class D			Class W			Class Y		
		F	T (day)	$\lambda(s^{-1})$	F	T (day)	$\lambda(s^{-1})$	F	T (day)	$\lambda(s^{-1})$
Nasal-Pharynx	a	0.50	0.01	8.02E−04	0.10	0.01	8.02E−04	0.01	0.01	8.02E−04
($D_{NP} = 0.30$)	b	0.50	0.01	8.02E−04	0.90	0.40	2.01E−05	0.99	0.40	2.01E−05
Trachea-Bronchi	c	0.95	0.01	8.02E−04	0.50	0.01	8.02E−04	0.01	0.01	8.02E−04
($D_{TB} = 0.08$)	d	0.05	0.20	4.01E−05	0.50	0.20	4.01E−05	0.99	0.20	4.01E−05
Pulmonary	e	0.80	0.50	1.60E−05	0.15	50	1.60E−07	0.05	500	1.60E−08
($D_P = 0.25$)	f	n.a.	n.a.	n.a.	0.40	1.0	8.02E−06	0.40	1.0	8.02E−06
	g	n.a.	n.a.	n.a.	0.40	50	1.60E−07	0.40	500	1.60E−08
	h	0.20	0.50	1.60E−05	0.05	50	1.60E−07	0.15	500	1.60E−08
Lymph	i	1.0	0.50	1.60E−05	1.0	50	1.60E−07	0.90	1000	8.02E−09
	j	n.a.	n.a.	n.a.	n.a.	n.a.	n.a.	0.10	∞	n.a.

The clearance class of each radiocompound and relevant metabolic data are listed in the various parts and supplements of ICRP Publication 30. The 1996 (or ICRP 66) model has more regions to account for complexities learned since 1966 (see below); however, it also designates clearance of radiocompounds in three classes that are similar to those in the ICRP 30 model: fast (F), medium (M), and slow (S) and a class (V) for vapors and gases.

Lung Deposition of Radionuclides is determined by the particle size of the inhaled aerosols expressed as the activity median aerodynamic diameter (AMAD). Determinations of lung deposition and clearance have been normalized for $1\,\mu m$ AMAD particles, and the deposition fractions for the naso–pharyngeal, the trachea–bronchial, and the pulmonary regions are $D_{NP} = 0.30$, $D_{TB} = 0.08$, and $D_P = 0.25$, respectively, which totals 0.63, indicating that the remaining fraction (0.37 for $1.0\,\mu m$ AMAD particles) is exhaled. Particles different from 1 mm AMAD particles will be uniquely deposited in the NP, TB, and P regions of the lung, but once deposited will be cleared according to the basic model. Unique deposition fractions can be determined from the deposition model (see Figure 9-5) for particles ranging from 0.1 to 20 μm (particles greater than $20\,\mu m$ are assumed to deposit only in the NP region). The deposition fractions for each clearance class are listed in Table 9-7 for each compartment of the lung for $1.0\,\mu m$ AMAD (activity median aerodynamic diameter) aerosols along with the corresponding clearance half-lives (in days) and clearance constants (s^{-1}).

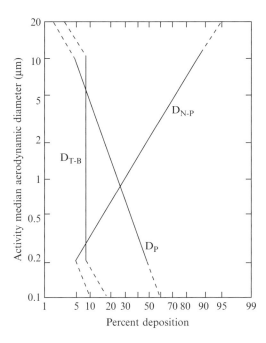

Figure 9-5. Percent deposition of different AMAD aerosols in the naso–pharangeal (NP), trachea–bronchial (TB), and pulmonary (P) regions of the lung of an adult male.

Inhalation Dose Factors (Sv/Bq of intake) corresponding to inhalation of nominal 1 μm AMAD particles have been calculated for radionuclides in one or more clearance classes; values of these (in CDE) for selected radionuclides are listed in Table 9-8 for the six major tissue groups which have specific weighting factors, plus up to five remainder tissues, each of which has a weighting factor of 0.06. These CDE values for the specific tissues and the "remainder" tissue(s) have been multiplied by their respective weighting factors, w_T, and summed to obtain the committed effective dose equivalent (CEDE) per unit intake (Sv/Bq). If the limiting CEDE is the effective dose (0.05 Sv or 5 rem), which it quite often is, it is shown in bold lettering; if however, the controlling effect is due to nonstochastic effects (i.e., 0.5 Sv or 50 rem) in one of the tissues, then the dose per unit intake is shown in bold lettering for the particular tissue that is limiting.

Table 9-8 also lists the annual limit on intake (ALI) for each radionuclide which is calculated by dividing the effective dose per unit of intake into 0.05 Sv (5 rem) if the effective dose is controlling, or 0.5 Sv (50 rem) if the nonstochastic CDE for a particular tissue is limiting (see Example 9-5). And, once the ALI is known, the corresponding derived air concentration (also listed as DAC in Table 9-8) can be obtained by dividing the ALI by the nominal annual breathing rate of 2.4×10^3 m^3 for a reference worker engaged in light activity.

Example 9-5: From the data in Table 9-8, calculate the ALI and DAC for: (a) class Y ^{60}Co, and (b) class D ^{131}I.

Solution: (a) As shown in Table 9-8, the limiting dose (shown by the bold lettering) for ^{60}Co is the effective dose which is 5.91×10^{-8} Sv/Bq; thus, the ALI is based on stochastic effects and is

$$ALI_S = \frac{0.05\ Sv}{5.91 \times 10^{-8}\ Sv/Bq} = 8.46 \times 10^5 Bq$$

The derived air concentration (DAC) to preclude inhalation of one ALI in 2000 work hours in a year is

$$DAC\ (Bq/m^3) = \frac{ALI}{2.4 \times 10^3\ m^3} = 3.53 \times 10^2\ Bq/m^3$$

(b) For ^{131}I, Table 9-8 shows the thyroid dose per Bq in bold lettering as 2.92×10^{-7} Sv/Bq; thus, the ALI_{NS} is based on a capping dose of 0.5 Sv to the thyroid to preclude nonstochastic effects, or

$$ALI_{NS}(Bq) = \frac{0.5\ Sv}{2.92 \times 10^{-7}\ Sv/Bq} = 1.7 \times 10^6\ Bq$$

The data in Table 9-8 can also be used to calculate the stochastic ALI for ^{131}I; this is based on the listed effective dose of 8.89×10^{-9} Sv/Bq, or

$$^{131}I\ ALI_S = \frac{0.05\ Sv}{8.89 \times 10^{-9}\ Sv/Bq} = 5.6 \times 10^6\ Bq$$

Table 9-8. Dose Conversion Factors for Inhalaton of One Bq of Different Classes (D,W,Y) of Selected Radionuclides.

Nuclide	Class	f_1	Gonads	Breast	Lung	R Marrow	B Surface	Thyroid	Remainder	Effective	ALI (Bq)	DAC (Bq/m³)
H-3	V	1.0	1.73E−11	1.73E−11	1.73E−11	1.73E−11	1.73E−11	1.73E−11	1.73E−11	**1.73E−11**	2.89E+09	1.20E+06
Be-7	Y	0.005	3.17E−11	3.82E−11	3.73E−10	3.99E−11	2.98E−11	3.10E−11	7.23E−11	**8.67E−11**	5.77E+08	2.40E+05
C−11		1.0	3.41E−12	2.98E−12	3.09E−12	3.18E−12	3.03E−12	2.97E−12	3.54E−12	**3.29E−12**	1.52E+10	6.33E+06
C−14		1.0	5.64E−10	5.64E−10	5.64E−10	5.64E−10	5.64E−10	5.64E−10	5.64E−10	**5.64E−10**	8.87E+07	3.69E+04
F−18	D	1.0	2.17E−12	3.88E−12	1.09E−10	2.76E−11	2.79E−11	3.47E−10	1.37E−11	**2.26E−11**	2.21E+09	9.22E+05
Na-22	D	1.0	1.77E−09	1.65E−09	2.47E−09	2.73E−09	3.51E−10	1.60E−09	2.00E−09	**2.07E−09**	2.42E+07	1.01E+04
Na-24	D	1.0	1.78E−10	1.61E−10	1.25E−09	2.13E−10	2.58E−10	1.53E−10	2.35E−10	**3.27E−10**	1.53E+08	6.37E+04
P-32	W	0.8	3.37E−10	3.37E−10	2.56E−10	4.17E−09	4.05E−09	3.37E−10	1.18E−09	**4.19E−09**	1.19E+07	4.97E+03
S-35	W	0.1	4.54E−11	4.54E−11	5.07E−09	4.54E−11	4.54E−11	4.54E−11	1.15E−10	**6.69E−10**	7.47E+07	3.11E+04
Cl-36	W	1.0	5.04E−10	5.04E−10	4.56E−08	5.04E−10	5.04E−10	5.04E−10	5.36E−10	**5.93E−09**	8.43E+06	3.51E+03
K-40	D	1.0	3.19E−09	3.08E−09	4.66E−09	3.10E−09	3.07E−09	3.06E−09	3.21E−09	**3.34E−09**	1.50E+07	6.24E+03
Ca-45	W	0.3	4.49E−11	4.49E−11	9.67E−09	2.92E−09	4.39E−09	4.49E−11	4.27E−10	**1.79E−09**	2.79E+07	1.16E+04
Cr-51	Y	0.1	2.21E−11	1.50E−11	3.77E−10	1.87E−11	1.50E−11	1.00E−11	4.93E−11	**7.08E−11**	7.06E+08	2.94E+05
Mn-54	W	0.1	7.09E−10	8.59E−10	6.66E−09	1.10E−10	1.25E−09	7.40E−10	1.72E−09	**1.81E−09**	2.76E+07	1.15E+04
Mn-56	D	0.1	9.46E−12	7.79E−12	5.37E−10	1.02E−11	8.23E−12	6.18E−12	6.50E−11	**8.91E−11**	5.61E+08	2.34E+05
Fe-55	D	0.1	5.23E−10	5.09E−10	5.19E−10	5.17E−10	5.14E−10	5.42E−10	1.21E−10	**7.26E−10**	6.89E+07	2.87E+04
Fe-59	D	0.1	3.32E−09	3.01E−09	3.50E−09	3.18E−09	2.91E−09	2.95E−09	5.81E−09	**4.00E−09**	1.25E+07	5.21E+03
Co-57	Y	0.05	1.24E−10	3.75E−10	1.69E−08	5.88E−10	4.52E−10	2.71E−10	8.22E−10	**2.45E−09**	2.04E+07	8.50E+03
Co-58	Y	0.05	6.17E−10	9.37E−10	1.60E−08	9.23E−10	6.93E−10	8.72E−10	1.89E−09	**2.94E−09**	1.70E+07	7.09E+03
Co-60	Y	0.05	4.76E−09	1.84E−08	3.45E−07	1.72E−08	1.35E−08	1.62E−08	3.60E−08	**5.91E−08**	8.46E+05	3.53E+02
Ni-59	D	0.05	3.59E−10	3.46E−10	3.59E−10	3.54E−10	3.51E−10	3.77E−10	3.63E−10	**3.58E−10**	1.40E+08	5.82E+04
Ni-63	D	0.05	8.22E−10	8.22E−10	8.74E−10	8.22E−10	8.22E−10	8.22E−10	8.59E−10	**8.39E−10**	5.96E+07	2.48E+04
Ni-65	D	0.05	8.46E−12	6.48E−12	3.11E−10	6.70E−12	5.79E−12	5.54E−12	7.98E−11	**6.55E−11**	7.63E+08	3.18E+05
Cu-64	Y	0.5	1.24E−11	6.38E−12	3.50E−10	7.12E−12	5.29E−12	4.98E−12	9.20E−11	**7.48E−11**	6.68E+08	2.79E+05
Zn-65	Y	0.5	2.03E−09	3.08E−09	2.10E−08	3.62E−09	3.36E−09	3.02E−09	4.66E−09	**5.51E−09**	9.07E+06	3.78E+03

(continues)

Table 9-8. (*continued*)

Nuclide	Class	f_1	Gonads	Breast	Lung	R Marrow	B Surface	Thyroid	Remainder	Effective	ALI (Bq)	DAC (Bq/m³)
Ga-67	Y	0.001	6.12E-11	1.83E-11	5.37E-10	3.71E-11	7.84E-11	9.43E-12	2.04E-10	**1.51E-10**	3.31E+08	1.38E+05
Ga-68	D	0.001	5.49E-12	4.40E-12	1.88E-10	5.71E-12	4.91E-12	3.77E-12	3.95E-11	**3.74E-11**	1.34E+09	5.57E+05
Ge-67	D	1.0	1.22E-12	1.63E-12	1.01E-10	1.66E-12	1.49E-12	1.47E-12	1.16E-11	**1.64E-11**	3.05E+09	1.27E+06
Ge-68	W	1.0	2.16E-10	7.50E-10	1.11E-07	7.17E-10	5.94E-10	6.90E-10	1.43E-09	**1.40E-08**	3.57E+06	1.49E+03
As-76	W	0.5	7.54E-11	5.33E-11	5.02E-09	5.59E-11	4.90E-11	4.80E-11	1.24E-09	**1.01E-09**	4.95E+07	2.06E+04
Br-82	W	1.0	1.69E-10	2.10E-10	1.68E-10	2.18E-10	1.92E-10	2.06E-10	3.31E-10	**4.13E-10**	1.21E+08	5.04E+04
Rb-86	D	1.0	1.34E-09	1.33E-09	3.30E-09	2.32E-09	4.27E-09	1.33E-09	1.38E-09	**1.79E-09**	2.79E+07	1.16E+04
Sr-85	Y	0.3	3.34E-10	4.65E-10	7.15E-09	4.65E-10	3.50E-10	3.85E-10	9.05E-10	**1.36E-09**	3.68E+07	1.53E+04
Sr-89	Y	0.3	7.95E-12	7.96E-12	8.35E-08	1.07E-10	1.59E-10	7.96E-12	3.97E-09	**1.12E-08**	4.46E+06	1.86E+03
Sr-90	Y	0.3	2.69E-10	2.69E-10	2.86E-06	3.28E-08	7.09E-08	2.69E-10	5.73E-09	**3.51E-07**	1.42E+05	5.94E+01
Zr-95	W	0.02	8.40E-10	9.32E-10	1.86E-08	3.24E-09	2.17E-08	7.82E-10	2.13E-09	**4.29E-09**	1.17E+07	4.86E+03
Nb-94	Y	0.01	4.42E-09	2.24E-08	7.48E-07	2.26E-08	1.97E-08	2.22E-08	4.45E-08	**1.12E-07**	4.46E+05	1.86E+02
Nb-95	Y	0.01	4.32E-10	4.07E-10	8.32E-09	4.42E-10	5.13E-10	3.58E-10	1.07E-09	**1.57E-09**	3.18E+07	1.33E+04
Mo-99	Y	0.8	9.51E-11	2.75E-11	4.29E-09	5.24E-11	4.13E-11	1.52E-11	1.74E-09	**1.07E-09**	4.67E+07	1.95E+04
Tc-99	W	0.8	3.99E-11	3.99E-11	1.67E-08	3.99E-11	3.99E-11	1.07E-09	6.26E-10	**2.25E-09**	2.22E+07	9.26E+03
Tc-99m	D	0.8	2.77E-12	2.15E-12	2.28E-11	3.36E-12	2.62E-12	5.01E-11	1.02E-11	**8.80E-12**	5.68E+09	2.37E+06
Ru-103	Y	0.05	3.07E-10	3.11E-10	1.56E-08	3.19E-10	2.37E-10	2.57E-10	1.25E-09	**2.42E-09**	2.07E+07	8.61E+03
Ru-106	Y	0.05	1.30E-09	1.78E-09	1.04E-06	1.76E-09	1.61E-09	1.72E-09	1.20E-08	**1.29E-07**	3.88E+05	1.61E+02
Ag-110m	Y	0.05	2.43E-09	7.10E-09	1.20E-07	6.74E-09	5.19E-09	6.39E-09	1.51E-08	**2.17E-08**	2.30E+06	9.60E+02
Cd-109	D	0.05	2.71E-09	2.97E-09	3.34E-09	3.45E-09	3.14E-09	2.66E-09	9.59E-08	**3.09E-08**	1.62E+06	6.74E+02
									3.95E-07	**Kidneys**		5.29E+02
In-115	Y	0.02	1.17E-07	1.17E-07	1.17E-07	3.67E-06	1.89E-06	1.17E-07	1.49E-06	**1.01E-06**	4.95E+04	2.06E+01
Sn-113	W	0.02	3.16E-10	2.99E-10	1.84E-08	7.71E-10	1.32E-09	2.27E-10	1.38E-09	**2.88E-09**	1.74E+07	7.23E+03
Sb-124	W	0.01	1.04E-09	8.94E-10	4.14E-08	1.09E-09	1.24E-09	6.74E-09	4.18E-09	**6.80E-09**	7.35E+06	3.06E+03
Sb-125	W	0.01	3.60E-10	4.16E-10	2.17E-08	5.35E-10	9.78E-10	3.24E-10	1.45E-09	**3.30E-09**	1.52E+07	6.31E+03
I-123	D	1.0	2.89E-12	4.87E-12	6.57E-11	5.97E-12	5.18E-12	**2.25E-09**	7.89E-12	8.01E-11	**2.22E+08**	9.25E+04
I-125	D	1.0	1.84E-11	9.25E-11	1.19E-10	4.41E-11	4.27E-11	**2.16E-07**	3.33E-11	6.53E-09	**2.31E+06**	9.63E+02
I-129	D	1.0	8.69E-11	2.09E-10	3.14E-10	1.40E-10	1.38E-10	**1.56E-06**	1.18E-10	4.69E-08	**3.21E+05**	1.34E+02
I-130	D	1.0	2.81E-11	4.87E-11	6.03E-10	4.55E-11	4.03E-11	**1.99E-08**	8.02E-11	7.14E-10	**2.51E+07**	1.05E+04

I-131	D	1.0	2.53E-11	7.88E-11	6.57E-10	5.73E-11	6.26E-11	8.03E-11	**2.92E-07**	**8.89E-09**	**1.71E+06**	7.13E+02
Cs-134	D	1.0	1.30E-08	1.08E-08	1.18E-08	1.10E-08	1.18E-08	1.39E-08	1.11E-08	**1.25E-08**	4.00E+06	1.67E+03
Cs-137	D	1.0	8.76E-09	7.84E-09	8.82E-09	7.94E-09	8.30E-09	9.12E-09	7.93E-09	**8.63E-09**	5.79E+06	2.41E+03
Cs-138	D	1.0	3.28E-12	4.02E-12	1.59E-10	3.55E-12	3.95E-12	2.06E-11	3.57E-12	**2.74E-11**	1.82E+09	7.60E+05
Ba-133	D	0.1	1.07E-09	1.10E-09	1.29E-09	9.51E-09	6.56E-09	1.41E-09	9.99E-10	**2.11E-09**	2.37E+07	9.87E+03
Ba-140	D	0.1	4.30E-10	2.87E-10	1.66E-09	2.41E-09	1.29E-09	1.41E-09	2.56E-10	**1.01E-09**	4.95E+07	2.06E+04
La-140	W	0.001	4.54E-10	1.45E-10	4.21E-10	1.41E-10	2.14E-10	2.12E-11	6.87E-11	**1.31E-09**	3.82E+07	1.59E+04
Ce-141	Y	0.0003	5.54E-11	4.46E-11	1.67E-08	2.54E-10	8.96E-11	1.26E-09	2.55E-11	**2.42E-09**	2.07E+07	8.61E+03
Ce-144	Y	0.0003	2.39E-10	3.48E-10	7.91E-07	4.72E-09	2.88E-08	1.91E-08	2.92E-10	**1.01E-07**	4.95E+05	2.06E+02
Pr-144	Y	0.0003	2.41E-15	1.05E-14	9.40E-11	1.47E-14	1.38E-14	1.40E-12	8.47E-15	**1.17E-11**	4.27E+09	1.78E+06
Nd-147	Y	0.0003	8.41E-11	3.45E-11	1.06E-08	3.26E-10	9.19E-11	1.76E-09	1.82E-11	**1.85E-09**	2.70E+07	1.13E+04
Pm-147	W	0.0003	8.25E-15	3.60E-14	7.74E-08	2.01E-08	1.61E-09	1.56E-09	1.98E-14	**1.06E-08**	4.72E+06	1.97E+03
Eu-152	W	0.001	1.31E-08	1.74E-08	5.76E-08	2.40E-07	7.91E-08	9.99E-08	8.25E-09	**5.97E-08**	8.38E+05	3.49E+02
Eu-154	W	0.001	1.17E-08	1.55E-08	7.92E-08	5.23E-07	1.06E-07	1.13E-07	7.14E-09	**7.73E-08**	6.47E+05	2.70E+02
Gd-148	D	0.0003	0.00E+01	0.00E+01	1.98E-07	**1.76E-03**	1.41E-04	6.45E-05	0.00E+01	**8.91E-05**	**2.84E+02**	1.18E-01
Gd-152	D	0.0003	0.00E+01	0.00E+01	1.33E-07	**1.30E-03**	1.04E-04	4.77E-05	0.00E+01	**6.58E-05**	**3.85E+02**	1.60E-01
Lu-176	D	0.0003	3.86E-09	1.10E-08	9.99E-07	1.19E-06	1.21E-07	2.10E-08	8.24E-09	**1.79E-07**	2.79E+05	1.16E+02
Lu-177	W	0.0003	1.93E-11	5.79E-12	3.33E-09	1.03E-10	1.82E-11	8.42E-10	2.47E-12	**6.63E-10**	7.54E+07	3.14E+04
Ta-182	Y	0.001	8.99E-10	1.79E-10	8.28E-08	1.51E-09	1.92E-09	4.37E-09	1.53E-09	**1.21E-08**	4.13E+06	1.72E+03
W-188	D	0.3	7.97E-12	4.88E-12	1.36E-09	1.65E-09	5.54E-10	2.75E-09	2.72E-12	**1.11E-09**	4.50E+07	1.88E+04
Re-186	W	0.8	4.53E-11	4.48E-11	4.42E-09	4.59E-11	4.72E-11	8.10E-10	2.19E-09	**8.64E-10**	5.79E+07	2.41E+04
Ir-192	Y	0.01	6.08E-10	8.63E-10	5.24E-08	7.00E-10	9.38E-10	2.94E-09	6.51E-10	**7.61E-09**	6.57E+06	2.74E+03
Pt-193	D	0.01	1.43E-11	1.33E-11	3.86E-11	1.35E-11	1.36E-11	1.62E-10	1.42E-11	**6.14E-11**	8.14E+08	3.39E+05
Pt-193m	D	0.01	3.78E-11	3.55E-11	4.09E-10	3.70E-11	3.86E-11	5.55E-10	3.50E-11	**2.37E-10**	2.11E+08	8.79E+04
Pt-195m	D	0.01	6.80E-11	5.42E-11	5.65E-10	6.29E-11	7.24E-11	7.48E-10	5.04E-11	**3.29E-10**	1.52E+08	6.33E+04
Pt-197	D	0.01	1.64E-11	1.45E-11	4.55E-10	1.50E-11	1.59E-11	2.97E-10	1.41E-11	**1.53E-10**	3.27E+08	1.36E+05
Pt-197m	D	0.01	3.24E-12	2.80E-12	1.40E-10	2.92E-12	3.28E-12	4.83E-11	2.62E-23	**3.31E-11**	1.51E+09	6.29E+05
Pt-199	D	0.01	1.09E-12	9.87E-13	6.61E-11	9.66E-13	1.07E-12	1.26E-11	9.03E-13	**1.23E-11**	4.07E+09	1.69E+06
Au-195	Y	0.1	7.67E-11	2.29E-10	2.65E-08	3.39E-10	4.35E-10	6.63E-10	1.08E-10	**3.50E-09**	1.43E+07	5.95E+03
Au-198	Y	0.1	1.40E-10	4.16E-11	3.51E-09	3.19E-11	5.40E-11	1.39E-09	2.37E-11	**8.87E-10**	5.64E+07	2.35E+04
Hg-197	W	0.02	3.38E-11	1.20E-11	7.21E-10	1.62E-11	2.65E-11	2.84E-10	6.98E-12	**1.86E-10**	2.69E+08	1.12E+05

(continues)

Table 9-8. (*continued*)

Nuclide	Class	f_1	Gonads	Breast	Lung	R Marrow	B Surface	Thyroid	Remainder	Effective	ALI (Bq)	DAC (Bq/m³)
Hg-197	W	vapor	3.15E−11	3.47E−11	1.12E−09	5.34E−11	4.70E−11	3.09E−11	1.18E−10	**1.92E−10**	2.60E+08	1.09E+05
Hg-203	W	0.02	2.74E−10	2.14E−10	8.78E−09	2.63E−10	2.09E−10	1.77E−10	1.19E−09	**1.55E−09**	3.23E+07	1.34E+04
Hg-203	W	vapor	8.65E−10	7.90E−10	3.32E−09	9.45E−10	8.49E−10	7.32E−10	2.77E−09	**1.73E−09**	2.89E+07	1.20E+04
Tl-201	D	1.0	3.66E−11	3.32E−11	1.69E−10	5.37E−11	4.77E−11	3.14E−11	6.71E−11	**6.34E−11**	7.89E+08	3.29E+05
Tl-204	D	1.0	4.14E−10	4.14E−10	1.13E−09	4.15E−10	4.15E−10	4.14E−10	9.14E−10	**6.50E−10**	7.69E+07	3.21E+04
Pb-210	D	0.2	3.18E−07	3.18E−07	3.18E−07	3.75E−06	**5.47E−05**	3.18E−07	4.69E−06	3.67E−06	**9.14E+03**	3.81E−00
Bi-210	D	0.05	6.47E−11	6.47E−11	4.26E−07	6.47E−11	6.47E−11	6.47E−11	5.66E−09	**5.29E−08**	9.45E+05	3.94E+02
Po-210	D	0.1	4.04E−07	4.04E−07	7.29E−07	4.04E−07	4.04E−07	4.04E−07	7.40E−06	**2.54E−06**	1.97E+04	8.20E−00
Ra-226	W	0.2	1.02E−07	1.02E−07	1.61E−05	6.64E−07	7.59E−06	1.02E−07	1.07E−07	**2.32E−06**	2.16E+04	8.98E−00
Ra-227	W	0.2	2.27E−12	2.36E−12	3.32E−10	4.81E−11	**9.59E−10**	2.30E−12	4.60E−12	7.68E−11	**5.21E+08**	2.17E+05
Ra-228	W	0.2	1.83E−07	1.84E−07	7.22E−06	7.38E−07	6.51E−06	1.83E−07	1.87E−07	**1.29E−04**	3.88E+02	1.61E−01
Th-230	W	0.0002	4.08E−07	4.08E−07	1.61E−05	1.73E−04	**2.16E−03**	4.08E+07	1.05E−06	8.80E−05	**2.31E+02**	9.63E−02
Th-232	W	0.0002	7.62E−07	7.72E−07	1.44E−05	8.93E−04	**1.11E−02**	7.44E−07	1.87E−06	4.43E−04	**4.50E+01**	1.88E−02
Pa-231	W	0.001	6.90E−09	8.79E−09	1.72E−05	6.97E−04	**8.70E−03**	7.64E−09	2.87E−07	3.47E−04	**5.75E+01**	2.40E−02
U-235	Y	0.002	2.84E−09	5.37E−09	2.76E−04	7.15E−08	1.05E−06	4.11E−09	1.02E−07	**3.32E−05**	1.51E+03	6.28E−01
U-238	Y	0.002	2.42E−09	2.91E−09	2.66E−04	6.88E−08	1.01E−06	2.73E−09	9.61E−08	**3.20E−05**	1.56E+03	6.51E−01
Np-237	W	0.001	2.96E−05	1.69E−08	1.61E−05	2.62E−04	**3.27E−03**	1.34E−08	2.34E−05	1.46E−04	**1.53E+02**	6.38E−02
Pu-238	Y	0.00001	1.04E−05	4.40E−10	3.20E−04	5.80E−05	7.25E−04	3.86E−10	2.74E−05	**7.79E−05**	6.42E+02	2.67E−01
Pu-239	Y	0.001	3.18E−05	9.22E−10	1.73E−05	1.69E−04	**2.11E−03**	9.03E−10	7.56r-5	1.16E−04	**2.37E+02**	9.88E−02
Pu-241	W	0.001	6.82E−07	3.06E−11	7.42E−09	3.36E−06	**4.20E−05**	1.24E−11	1.31E−06	2.23E−06	**1.19E+02**	4.96E−02
Pu-242	W	0.001	3.02E−05	9.45E−10	1.64E−05	1.61E−04	**2.01E−03**	8.79E−10	7.18E−05	1.11E−04	**2.49E+02**	1.04E−01
Am-241	W	0.001	3.25E−05	2.67E−09	1.84E−05	1.74E−04	**2.17E−03**	1.60E−09	7.82E−05	1.20E−04	**2.30E+02**	9.58E−02
Cm-244	W	0.001	1.59E−05	1.04E−09	1.93E−05	9.38E−05	**1.17E−03**	1.01E−09	4.78E−05	6.70E−05	**4.27E+02**	1.78E−01
Cf-252	W	0.001	5.43E−06	6.56E−08	3.74E−05	5.50E−05	**6.86E−04**	3.38E−08	1.33E−05	3.70E−05	**7.29E+02**	3.04E−01

Adapted from Federal Guidance Report no. 11, EPA 520/11-88−020 1988.

Obviously, the nonstochastic ALI is the more limiting condition and is the one chosen to limit intakes of ^{131}I; the DAC is derived from it as

$$DAC = \frac{1.7 \times 10^6 \text{ Bq}}{2.4 \times 10^3 \text{ m}^3} = 7.1 \times 10^2 \text{ Bq/m}^3$$

Radiation Doses due to Ingested Radionuclides

A dynamic GI tract model, as illustrated in Figure 9-6 and based on linear first-order kinetics, has been used to determine translocation and retention of radioactive material deposited in the stomach (ST) which in turn passes through the small intestine (SI) and the upper (ULI) and lower (LLI) large intestines. The GI tract model includes absorption through the small intestine wall into the blood stream where it can be taken up by organs and tissues, circulated in intercellular fluids, or excreted directly.

Ingestion Dose Factors (Sv/Bq of intake) are provided in Table 9-9 for selected radionuclides that are ingested into the body along with the

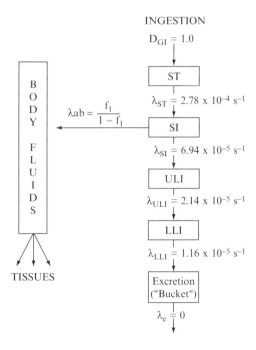

Figure 9-6. Model of the gastrointestinal (GI) tract used for determining radiation dose to segments of the GI tract and for dynamic translocation of ingested radionuclides to other tissues.

Table 9-9. Committed Dose Equivalent Conversion Factors (Sv/Bq) and Annual Limits on Intake (Bq) for Ingestion of Selected Radionuclides. (ALIs Marked with * are Based on Limiting Nonstochastic Risks.)

Nuclide	f_1	Gonads	Breast	Lung	R. Marrow	B. Surface	Thyroid	Remainder	Effective	ALI (Bq)
H-3	1.0	1.73E−11	1.73E−11	1.73E−11	1.73E−11	1.73E−11	1.73E−11	1.73E−11	**1.73E−11**	2.89E+09
Be-7	0.005	5.67E−11	6.97E−12	1.41E−12	1.23E−11	5.03E−12	6.08E−13	5.83E−11	**3.45E−11**	1.45E+09
C−11	1.0	3.41E−12	2.98E−12	3.09E−12	3.18E−12	3.03E−12	2.97E−12	3.54E−12	**3.29E−12**	1.52E+10
C−14	1.0	5.64E−10	5.64E−10	5.64E−10	5.64E−10	5.64E−10	5.64E−10	5.64E−10	**5.64E−10**	8.87E+07
F−18	1.0	4.97E−12	6.36E−12	6.54E−12	5.94E−11	6.02E−11	4.52E−12	7.03E−11	**3.31E−11**	1.51E+09
								2.87E−10	**ST wall**	**1.74E+09***
Na-22	1.0	2.81E−09	2.58E−09	2.51E−09	4.29E−09	5.54E−09	2.50E−09	3.18E−09	**3.10E−09**	1.61E+07
Na-24	1.0	3.43E−10	2.71E−10	2.60E−10	3.74E−10	4.68E−10	2.60E−10	5.31E−10	**3.84E−10**	1.30E+08
P-32	0.8	6.55E−10	6.55E−10	6.55E−10	8.09E−09	7.87E−09	6.55E−10	2.67E−09	**2.37E−09**	2.11E+07
S-35	0.1	9.53E−12	9.53E−12	9.53E−12	9.53E−12	9.53E−12	9.53E−12	6.39E−10	**1.98E−10**	2.53E+08
								2.23E−09	**LLI wall**	**2.24E+08***
Cl-36	1.0	7.99E−10	7.99E−10	7.99E−10	7.99E−10	7.99E−10	7.99E−10	8.61E−10	**8.18E−10**	6.11E+07
K-40	1.0	5.07E−09	4.89E−09	4.85E−09	4.91E−09	4.88E−09	4.85E−09	5.18E−09	**5.02E−09**	9.96E+06
Ca-45	0.3	5.36E−11	5.36E−11	5.36E−11	3.47E−09	5.23E−09	5.36E−11	8.40E−10	**8.55E−10**	5.85E+07
Cr-51	0.1	4.00E−11	7.51E−12	4.38E−12	1.25E−11	7.86E−12	3.71E−12	8.75E−11	**3.98E−11**	1.26E+09
Mn-54	0.1	9.48E−10	2.77E−10	2.29E−10	4.89E−10	5.71E−10	1.33E−10	1.21E−09	**7.48E−10**	6.68E+07
Mn-56	0.1	8.53E−11	1.76E−11	8.00E−12	2.43E−11	1.06E−11	2.40E−12	7.84E−10	**2.64E−10**	1.89E+08
Fe-55	0.1	1.07E−10	1.04E−10	1.02E−10	1.05E−10	1.05E−10	1.10E−10	3.00E−10	**1.64E−10**	3.05E+08
Fe-59	0.1	1.66E−09	7.37E−10	6.35E−10	8.45E−10	6.61E−10	6.03E−10	3.56E−09	**1.81E−09**	2.76E+07
Co-57	0.05	1.83E−10	4.10E−11	2.89E−11	8.84E−11	4.92E−11	1.93E−11	4.42E−10	**2.01E−10**	2.49E+08
Co-58	0.05	1.04E−09	1.79E−10	8.53E−11	2.60E−10	1.25E−10	6.31E−11	1.58E−09	**8.09E−10**	6.18E+07
Co-60	0.05	3.19E−09	1.10E−09	8.77E−10	1.32E−09	9.39E−10	7.88E−10	4.97E−09	**2.77E−09**	1.81E+07
Ni-59	0.05	3.83E−11	3.58E−11	3.50E−11	3.66E−11	3.62E−11	3.90E−11	1.03E−10	**5.67E−11**	8.82E+08

Nuclide	f_1									
Ni-63	0.05	8.50E−11	8.50E−11	8.50E−11	8.50E−11	8.50E−11	8.50E−11	3.20E−10	**1.56E−10**	3.21E+08
Ni-65	0.05	2.43E−11	5.63E−12	2.75E−12	7.26E−12	2.89E−12	6.79E−13	5.32E−10	**1.68E−10**	2.98E+08
Cu-64	0.5	4.78E−11	1.59E−11	1.28E−11	1.94E−11	1.39E−11	1.13E−11	3.57E−10	**1.26E−10**	3.97E+08
Zn-65	0.5	3.56E−09	3.28E−09	3.08E−09	4.50E−09	4.50E−09	3.21E−09	4.59E−09	**3.90E−09**	1.28E+07
Ga-67	0.001	1.58E−10	1.70E−11	2.38E−11	4.14E−11	1.40E−11	2.43E−13	5.49E−10	**2.12E−10**	2.36E+08
Ga-68	0.001	1.95E−11	4.56E−12	2.79E−12	5.81E−12	2.18E−12	2.61E−13	2.86E−10	**9.24E−11**	5.41E+08
Ge-67	1.0	2.71E−12	3.25E−12	3.64E−12	3.09E−12	2.63E−12	2.17E−12	1.10E−10	**3.52E−11**	1.42E+09
								5.09E−10	**ST wall**	9.82E+08*
Ge-68	1.0	2.42E−10	2.23E−10	2.28E−10	2.33E−10	2.25E−10	2.22E−10	4.22E−10	**2.89E−10**	1.73E+08
As-76	0.5	2.16E−10	1.09E−10	9.83E−11	1.20E−10	1.02E−10	9.35E−11	4.35E−09	**1.41E−09**	3.55E+07
Br-82	1.0	4.48E−10	3.81E−10	3.84E−10	4.14E−10	3.80E−10	3.83E−10	5.80E−10	**4.62E−10**	1.08E+08
Rb-86	1.0	2.15E−09	2.14E−09	2.14E−09	3.72E−09	6.86E−09	2.14E−09	2.33E−09	**2.53E−09**	1.98E+07
Sr-85	0.3	6.25E−10	2.53E−10	2.06E−10	5.97E−10	6.06E−10	2.05E−10	7.31E−10	**5.34E−10**	9.36E+07
Sr-89	0.3	2.40E−10	2.40E−10	2.40E−10	3.23E−09	4.81E−09	2.40E−10	6.11E−10	**2.50E−09**	2.00E+07
Sr-90	0.3	1.51E−09	1.51E−09	1.51E−09	1.94E−07	**4.19E−07**	1.51E−09	6.14E−09	**3.85E−08**	1.19E+06*
Zr-95	0.02	8.16E−10	1.05E−10	2.34E−11	2.14E−10	4.86E−10	8.27E−12	2.53E−09	**1.02E−09**	4.90E+07
Nb-94	0.01	1.80E−09	3.47E−10	1.72E−10	7.39E−10	7.65E−10	1.23E−10	4.30E−09	**1.93E−09**	2.59E+07
Nb-95	0.01	8.05E−10	1.07E−10	2.74E−11	1.99E−10	2.94E−10	1.18E−11	1.47E−09	**6.95E−10**	7.19E+07
Mo-99	0.8	2.21E−10	1.83E−10	1.93E−10	5.33E−10	7.69E−10	1.64E−10	2.08E−09	**8.22E−10**	6.08E+07
Tc-99	0.8	6.04E−11	6.04E−11	6.04E−11	6.04E−11	6.04E−11	1.62E−09	1.02E−09	**3.95E−10**	1.27E+08
Tc-99m	0.8	9.75E−12	3.57E−12	3.14E−12	6.29E−12	4.06E−12	8.46E−11	3.34E−11	**1.68E−11**	2.98E+09
Ru-103	0.05	5.72E−10	1.20E−10	7.31E−11	1.66E−10	9.63E−11	6.25E−11	2.10E−09	**8.24E−10**	6.07E+07
Ru-106	0.05	1.64E−09	1.44E−09	1.42E−09	1.46E−09	1.43E−09	1.41E−09	2.11E−08	**7.40E−09**	6.76E+06
Ag-110m	0.05	2.99E−09	7.51E−10	8.30E−10	9.42E−10	4.93E−10	1.81E−10	6.08E−09	**2.92E−09**	1.71E+07
Cd-109	0.05	3.46E−10	3.10E−10	3.17E−10	3.70E−10	3.28E−10	2.75E−10	1.10E−08	**3.55E−09**	1.22E+07*
								4.08E−08	**Kidneys**	

(continues)

Table 9-9. (*continued*)

Nuclide	f_1	Gonads	Breast	Lung	R. Marrow	B. Surface	Thyroid	Remainder	Effective	ALI (Bq)
In−115	0.02	4.86E−09	4.86E−09	4.86E−09	1.53E−07	7.91E−08	4.86E−09	6.41E−08	**4.26E−08**	1.17E+06
Sn−113	0.02	3.88E−10	5.68E−10	2.54E−11	1.78E−10	2.34E−10	2.16E−11	2.32E−09	**8.33E−10**	6.00E+07
								7.91E−09	**LLI wall**	6.32E+07*
Sb−124	0.01	1.78E−09	2.30E−10	5.40E−11	3.81E−10	1.89E−10	1.76E−11	7.34E−09	**2.74E−09**	1.82E+07
Sb−125	0.01	5.27E−10	6.22E−11	1.36E−11	1.21E−10	9.05E−11	5.58E−12	1.99E−09	**7.57E−10**	6.61E+07
I−123	1.0	5.61E−12	7.23E−12	6.66E−12	8.68E−12	7.65E−12	**4.42E−09**	2.01E−11	1.43E−10	1.13E+08*
I−125	1.0	2.93E−11	1.45E−10	4.08E−10	6.82E−11	6.63E−11	**3.44E−07**	5.80E−11	1.04E−08	1.45E+06*
I−129	1.0	1.38E−10	3.31E−10	1.65E−10	2.21E−10	2.17E−10	**2.48E−06**	1.99E−10	7.46E−08	2.02E+05*
I−130	1.0	5.52E−11	7.32E−11	7.18E−11	6.74E−11	6.12E−11	**3.94E−08**	1.97E−10	1.28E−09	1.27E+07*
I−131	1.0	4.07E−11	1.21E−10	1.02E−10	9.44E−11	8.72E−11	**4.76E−07**	1.57E−10	1.44E−08	1.05E+06*
Cs−134	1.0	2.06E−08	1.72E−08	1.76E−08	1.87E−08	1.74E−08	1.76E−08	2.21E−08	**1.98E−08**	2.53E+06
Cs−137	1.0	1.39E−08	1.24E−08	1.27E−08	1.32E−08	1.26E−08	1.26E−08	1.45E−08	**1.35E−08**	3.70E+06
Cs−138	1.0	8.00E−12	8.00E−12	8.53E−12	7.37E−12	6.47E−12	5.73E−12	1.57E−10	5.25E−11	9.52E+08
								7.01E−10	**ST wall**	7.13E+08*
Ba−133	0.1	7.33E−10	2.73E−10	2.19E−10	1.46E−09	1.97E−09	2.03E−10	1.43E−09	**9.19E−10**	5.44E+07
Ba−140	0.1	9.96E−10	1.59E−10	6.63E−11	4.39E−10	5.53E−10	5.25E−11	7.37E−09	2.56E−09	1.95E+07
								2.64E−08	**LLI wall**	1.89E+07*
La−140	0.001	1.34E−09	1.80E−10	4.01E−11	2.81E−10	9.77E−11	6.40E−12	6.26E−09	**2.28E−09**	2.19E+07
Ce−141	3E−04	1.08E−10	1.11E−11	1.43E−12	3.39E−11	2.30E−11	1.80E−13	2.50E−09	7.83E−10	6.39E+07
								8.64E−09	**LLI wall**	5.79E+07*
Ce−144	3E−04	6.98E−11	1.22E−11	6.52E−12	8.92E−11	1.28E−10	5.15E−12	1.88E−08	5.68E−09	8.80E+06
								6.64E−08	**LLI wall**	7.53E+06*
Pr−144	3E−04	7.38E−14	3.38E−14	3.15E−14	3.22E−14	1.52E−14	3.59E−15	1.05E−10	3.15E−11	1.59E+09*
								4.09E−10	**ST wall**	1.22E+09*

Nd-147	3E−04	1.79E−10	1.87E−11	2.44E−12	5.05E−11	2.22E−11	2.64E−13	3.76E−09	1.18E−09	4.24E+07*
								1.28E−08	**LLI wall**	3.91E+07*
Pm-147	3E−04	6.86E−15	7.45E−16	1.96E−16	2.09E−11	2.61E−10	3.12E−17	9.08E−10	2.83E−10	1.77E+08
								3.17E−09	**LLI wall**	1.58E+08*
Eu-152	0.001	1.33E−09	2.85E−10	2.40E−10	9.19E−10	2.09E−09	6.66E−11	3.92E−09	**1.75E−09**	2.86E+07
Eu-154	0.001	1.37E−09	2.79E−10	2.16E−10	1.15E−09	4.46E−09	5.71E−11	6.32E−09	**2.58E−09**	1.94E+07
Gd-148	3E−04	0.00E+00	0.00E+00	0.00E+00	8.90E−08	**1.11E−06**	0.00E+00	4.98E−08	5.89E−08	4.50E+05*
Gd-152	3E−04	0.00E+00	0.00E+00	0.00E+00	6.57E−08	**8.21E−07**	0.00E+00	3.62E−08	4.34E−08	6.09E+05*
W-188	0.3	3.31E−11	5.57E−12	2.76E−12	3.25E−10	9.52E−10	1.42E−12	6.78E−09	2.11E−09	2.37E+07
								2.33E−08	**LLI wall**	2.15E+07*
Re-186	0.8	1.00E−10	9.54E−11	9.53E−11	9.89E−11	9.69E−11	4.79E−09	1.95E−09	**7.95E−10**	6.29E+07
Ir-192	0.01	1.03E−09	1.51E−10	6.54E−11	2.54E−10	1.11E−10	3.78E−11	4.08E−09	**1.55E−09**	3.23E+07
Pt-193	0.01	2.02E−12	2.95E−13	2.73E−13	3.35E−13	2.94E−13	2.97E−13	1.05E−10	3.21E−11	1.56E+09
								3.60E−10	**LLI wall**	1.39E+09*
Au-195	0.1	1.33E−10	1.98E−11	9.07E−12	6.29E−11	2.56E−11	7.43E−12	8.05E−10	**2.87E−10**	1.74E+08
Au-198	0.1	3.43E−10	5.51E−11	2.44E−11	8.57E−11	4.06E−11	1.85E−11	3.44E−09	**1.14E−09**	4.39E+07
Hg-197	0.02	8.26E−11	9.43E−12	2.15E−12	3.53E−11	1.14E−11	1.27E−12	7.72E−10	**2.59E−10**	1.93E+08
Hg-203	0.02	3.30E−10	5.41E−11	2.69E−11	9.38E−11	4.61E−11	2.15E−11	1.71E−09	**6.21E−10**	8.05E+07
Tl-201	1.0	6.19E−11	5.41E−11	5.78E−11	8.71E−11	7.76E−11	5.19E−11	1.21E−10	**8.11E−11**	6.17E+08
Tl-204	1.0	6.57E−10	6.57E−10	6.57E−10	6.59E−10	6.59E−10	6.57E−10	1.49E−09	**9.08E−10**	5.51E+07
Pb-210	0.2	1.25E−07	1.25E−07	1.25E−07	1.48E−06	**2.16E−05**	1.25E−07	1.85E−06	1.45E−06	2.31E+04*
Bi-210	0.05	1.97E−11	1.97E−11	1.97E−11	1.97E−11	1.97E−11	1.97E−11	5.72E−09	**1.73E−09**	2.89E+07
Po-210	0.1	8.23E−08	8.23E−08	8.23E−08	8.23E−08	8.23E−08	8.23E−08	1.52E−06	**5.14E−07**	9.73E+04
Ra-226	0.2	9.16E−08	9.17E−08	9.16E−08	5.98E−07	**6.83E−06**	9.15E−08	1.03E−07	3.58E−07	7.32E+04*

(continues)

Table 9-9. (*continued*)

Nuclide	f_1	Gonads	Breast	Lung	R. Marrow	B. Surface	Thyroid	Remainder	Effective	ALI (Bq)
Ra-227	0.2	3.65E−12	2.31E−12	2.16E−12	4.30E−11	**8.54E−10**	1.84E−12	9.56E−11	6.10E−11	5.85E+08*
Ra-228	0.2	1.58E−07	1.57E−07	1.57E−07	6.53E−07	**5.82E−06**	1.57E−07	1.63E−07	3.88E−07	8.59E+05*
Th-230	2E−04	6.82E−10	6.80E−10	6.80E−10	2.89E−07	**3.60E−06**	6.80E−10	1.54E−08	1.48E−07	1.39E+05*
Th-232	2E−04	1.25E−09	1.26E−09	1.25E−09	1.48E−06	**1.85E−05**	1.21E−09	1.47E−08	7.38E−07	2.70E+04*
Pa-231	0.001	1.21E−10	7.81E−11	6.80E−11	5.78E−06	**7.22E−05**	6.33E−11	1.71E−08	2.86E−06	6.93E+03*
U-235	0.002	3.34E−10	1.21E−10	1.01E−10	2.78E−09	4.20E−08	9.82E−11	1.84E−08	**7.22E−09**	6.93E+06
U-238	0.002	1.02E−10	9.33E−11	9.22E−11	2.72E−09	4.04E−08	9.20E−11	1.61E−08	**6.42E−09**	7.79E+06
Np-237	0.001	2.46E−07	1.45E−10	1.53E−10	2.18E−06	**2.72E−05**	1.10E−10	2.10E−07	1.20E−06	1.84E+04*
Pu-238	1E−05	2.33E−09	1.80E−13	8.64E−14	1.27E−08	**1.58E−07**	7.99E−14	2.18E−08	1.34E−08	3.16E+06*
Pu-239	0.001	2.64E−07	7.69E−12	7.74E−12	1.41E−06	**1.76E−05**	7.49E−12	6.43E−07	9.56E−07	2.84E+04*
Pu-241	0.001	5.66E−09	2.52E−13	4.45E−13	2.78E−08	**3.48E−07**	1.01E−13	1.10E−08	1.85E−08	1.44E+06*
Pu-242	0.001	2.51E−07	8.00E−12	7.88E−12	1.34E−06	**1.67E−05**	7.29E−12	6.10E−07	9.08E−07	2.99E+04*
Am-241	0.001	2.70E−07	2.62E−11	3.36E−11	1.45E−06	**1.81E−05**	1.32E−11	6.66E−07	9.84E−07	2.76E+04*
Cm-244	0.001	1.33E−07	8.82E−12	8.81E−12	7.82E−07	**9.77E−06**	8.44E−12	4.15E−07	5.45E−07	5.12E+04*
Cf-252	0.001	5.39E−08	1.49E−09	4.67E−10	4.69E−07	**5.84E−06**	2.68E−10	1.58E−07	2.93E−07	8.56E+04*

Adapted from Federal Guidance Report no. 11, EPA 520/11-88-020 1988.

corresponding ALI values. No concentration values have been calculated since it is doubtful that workers would consume contaminated water, and certainly not continuously over a year. Such a limiting concentration for a radionuclide in water might be more appropriate for the general public; however, national and international standards organizations do not recommend the use of these derived values for the general public since they are based on a reference male adult worker and not a typical member of the public. Regardless of these cautions, various groups, and in particular the US Nuclear Regulatory Commission, have used inhalation and ingestion ALIs, appropriately adjusted, to derive concentration values in air and water for the general public (see Table 11-2).

Example 9-6: A Norwegian farmer consumed milk from his dairy herd over several days following a release of ^{131}I from a nearby facility. If the ^{131}I intake was 10 μCi (3.7 × 10^5 Bq), estimate: (a) the committed dose to the thyroid, and (b) the committed effective whole body dose.
Solution: (a) From Table 9-9, the CDE to thyroid from ^{131}I ingestion is 4.76 × 10^{-7} Sv/Bq; thus, the thyroid dose is

$$\text{CDE} = (4.76 \times 10^{-7}\,\text{Sv/Bq})(3.7 \times 10^5\,\text{Bq}) = 0.176\,\text{Sv (17.6 rem)}$$

(b) The committed effective dose equivalent (CEDE) due to the thyroid uptake can be determined by use of the weighting factor 0.03 (p. 294) which yields

$$\text{CEDE} = w_T H_{50,\,T}$$
$$= 0.03 \times 0.176\,\text{Sv} = 5.28 \times 10^{-3}\,\text{Sv (0.53 rem)}$$

or, alternatively, from Table 9-9 the effective dose coefficient is 1.44 × 10^{-8} Sv/Bq; thus, the effective whole body dose for the intake of 3.7 × 10^5 Bq(10 μCi) is

$$\text{CEDE} = (1.44 \times 10^{-8}\,\text{Sv/Bq})(3.7 \times 10^5\,\text{Bq}) = 5.3 \times 10^{-3}\,\text{Sv (0.53 rem)}$$

Note: The CEDE dose factor incorporates the weighing factor w_T for all tissues which is dominated by $w_T = 0.03$ for thyroid.

Submersion Dose

Exposure of persons immersed in a cloud of radioactive material (submersion) can also be considered a form of internal radiation exposure; however, in most cases a total body external exposure is received due to the concentration of the radionuclide in the air surrounding the person. Noble gases and tritium are the most important radionuclides for submersion exposure, and the air concentration is the limiting condition to preclude exceeding the TEDE limit. Values of dose equivalent rates (Sv/h) per unit air concentration (Bq/m^3) are provided in Table 9-10. The effective whole body dose is usually limiting but the

Table 9-10. Exposure-to-dose Conversion Factors for Submersion Exposure to Airborne Concentrations of Tritium and Noble Gases.

Nuclide	Dose Equivalent Rates per Unit Air Concentration (Sv/hr per Bq/m³)							
	Gonad	Breast	Lung	R. Marrow	B Surface	Thyroid	Remainder	Effective
H-3			9.90×10^{-15}					1.19×10^{-15}
Ar-41	1.90×10^{-10}	2.32×10^{-10}	2.20×10^{-10}	2.28×10^{-10}	2.47×10^{-10}	2.07×10^{-10}	2.24×10^{-10}	2.17×10^{-10}
Kr-85	5.18×10^{-13}	4.52×10^{-13}	4.31×10^{-13}	5.75×10^{-13}	6.15×10^{-13}	2.50×10^{-13}	4.20×10^{-13}	4.70×10^{-13}
							$\mathbf{4.66 \times 10^{-11}}$	Skin
Kr-85 m	3.35×10^{-11}	2.66×10^{-11}	2.57×10^{-11}	4.43×10^{-11}	4.72×10^{-11}	2.95×10^{-11}	2.25×10^{-11}	2.98×10^{-11}
Kr-87	1.26×10^{-10}	1.48×10^{-10}	1.41×10^{-10}	1.52×10^{-10}	1.67×10^{-10}	1.42×10^{-10}	1.46×10^{-10}	1.42×10^{-10}
Kr-88	3.48×10^{-10}	3.65×10^{-10}	3.49×10^{-10}	3.48×10^{-10}	3.85×10^{-10}	3.74×10^{-10}	3.72×10^{-10}	3.60×10^{-10}
Xe-133	6.30×10^{-12}	5.62×10^{-12}	4.84×10^{-12}	1.08×10^{-11}	1.18×10^{-11}	7.12×10^{-12}	4.03×10^{-12}	6.07×10^{-12}
Xe-133 m	6.80×10^{-12}	4.88×10^{-12}	4.33×10^{-12}	7.37×10^{-12}	7.95×10^{-12}	4.89×10^{-12}	3.84×10^{-12}	5.38×10^{-12}
Xe-135	5.63×10^{-11}	4.21×10^{-11}	4.07×10^{-11}	6.16×10^{-11}	6.59×10^{-11}	3.80×10^{-11}	3.67×10^{-11}	4.68×10^{-11}
Xe-135 m	8.27×10^{-11}	7.32×10^{-11}	7.04×10^{-11}	8.62×10^{-11}	9.21×10^{-11}	3.32×10^{-11}	7.04×10^{-11}	7.53×10^{-11}
Xe-138	1.65×10^{-10}	2.06×10^{-10}	1.98×10^{-10}	2.11×10^{-10}	2.31×10^{-10}	1.91×10^{-10}	1.95×10^{-10}	1.92×10^{-10}

nonstochastic effect for skin or the lens of the eye may be the dominant effect in some cases; when the limiting concentration is based on nonstochastic effects, the limiting tissue is also listed in Table 9-10 for the particular gaseous radionuclide (see Example 9-7).

Example 9-7: A worker is exposed to an atmosphere of ^{85}Kr for 2000 h/y. What is the limiting concentration of ^{85}Kr to preclude exceeding radiation protection limits?

Solution: As shown in bold lettering in Table 9-10, the limiting dose rate for ^{85}Kr is 4.66×10^{-11} Sv/h per Bq/m^3 of ^{85}Kr which is based on precluding nonstochastic effects in skin. The limiting air concentration that corresponds to a skin dose of 0.5 Sv (50 rem) in 2000 h of exposure (a nominal working year) is

$$0.5\,\text{Sv} = \chi(\text{Bq/m}^3) \times 4.66 \times 10^{-11}\,\text{Sv/h per Bq/m}^3 \times 2000\,\text{h}$$

$$\chi = 5.36 \times 10^6\,\text{Bq/m}^3$$

which, because of the 2000 h exposure period, is also the derived air concentration (DAC) for assuring that an occupational dose to skin does not exceed the 50 rem nonstochastic limit.

OPERATIONAL DETERMINATIONS OF INTERNAL DOSE

A straightforward operational practice for assigning internal radiation doses to persons is to determine the intake, which is then used with the data in Tables 9-8 and 9-9 to determine the committed dose equivalent (CDE) to individual tissues or the committed effective dose equivalent (or CEDE). For airborne radioactivity, the inhalation intake (or alternatively the dose due to submersion) can be determined by measuring the air concentration and combining it with a standard breathing rate and the time of exposure. The ingestion intake due to radioactivity in water or food can be determined by similar means.

Example 9-8: A person ingests food and water over a period of several weeks for an estimated intake of 3.7×10^5 Bq (10 μCi) of ^{60}Co. Determine the committed dose equivalent (CDE) to the lung and the committed effective dose equivalent (CEDE) to the whole body.

Solution: Since the intake period is relatively short, it is reasonable to consider it as an acute intake. From Table 9-9, the CDE to lung due to ingested ^{60}Co is 8.77×10^{-10} Sv/Bq and the total body effective dose is 2.77×10^{-9} Sv/Bq; therefore

$$\text{CDE(lung)} = 8.77 \times 10^{-10}\,\text{Sv/Bq} \times 3.7 \times 10^5\,\text{Bq}$$

$$= 3.25 \times 10^{-4}\,\text{Sv}(32.5\,\text{mrem})$$

and for the total body

$$\text{CEDE(TB)} = 2.77 \times 10^{-9}\,\text{Sv/Bq} \times 3.7 \times 10^5\,\text{Bq}$$
$$= 1.02 \times 10^{-3}\,\text{Sv}(102.5\,\text{mrem})$$

BIOASSAY DETERMINATION OF INTAKE

If it is not possible to determine the radionuclide intake by physical measurement of airborne levels or its concentration in ingested food, water, or contamination, it may be necessary to measure the person after-the-fact, perhaps following an incident or spill. This process is known as bioassay and a measurement of the activity A_i of a radionuclide in a given tissue (e.g., the thyroid), the whole body, or in urine or feces is related to the intake by

$$I = A_i/r_i$$

where r_i is the time-dependent fraction retained (referred to as the intake retention fraction) in a given tissue or excretion compartment at the time of the measurement of A_i. And, since the committed effective dose is also directly related to the intake, bioassay measurements can be used to determine the dose received if the intake retention fraction is known (see Examples 9-9 and 9-10).

Intake Retention Fractions, r_i have been calculated for all internally deposited radionuclides (NUREG-4884), selected values of which are listed in Table 9-11 at various times (in days) post intake by inhalation and ingestion. The values of r_i for inhalation in Table 9-11 are based on 1 μm AMAD particles and reference person biology, and were chosen for radiation protection purposes for the most restrictive clearance class; values for other compounds can be obtained from NUREG-4884. Similarly, the values for ingestion were chosen for the most restrictive compounds, typically those with the greatest amount of absorption (based on the f_1 value) through the small intestine wall.

Example 9-9: Estimate: (a) the intake for a worker who inhaled class D ^{131}I in a contaminated atmosphere if the measured thyroid burden one day later is 3.7×10^4 Bq; (b) the thyroid dose due to the intake; and (c) the effective whole body dose due to the intake.
Solution: (a) From Table 9-11 the fraction of ^{131}I retained in the thyroid after one day is 0.133; thus, the intake is

$$I = \frac{3.7 \times 10^4\,\text{Bq}}{0.133} = 2.8 \times 10^5\,\text{Bq}$$

which is 14% of the nonstochastic ALI (2×10^6 Bq) for class D ^{131}I.
(b) From Table 9-9, the committed dose equivalent coefficient for thyroid is 2.92×10^{-7} Sv/Bq; thus, the thyroid dose is

Table 9-11. Intake Retention Fractions (r_i) in Major Tissues and/or 24-h Excreta at Days Post-intake by Inhalation and/or Ingestion of One Activity Unit of Selected Radionuclides.

Days	Tritium Oxide 24-h Urine	C-14 Oxide 24-h Excr.	P-32 Class W 24-h Urine	P-32 Ingestion 24-h Urine	S-35 Class W 24-h Urine	S-35 Ingestion W. Body	Ca-45 Class W 24-h Urine	Cr-51 Class Y W. Body	Cr-51 Ingestion W. Body	Mn-54 Class W W. Body	Fe-59 Class W W. Body	Fe-59 Ingestion W. Body
0.5	—	—	—	—	—	5.33E−01	—	6.23E−01	9.09E−01	6.31E−01	6.29E−01	9.25E−01
1	3.85E−02	1.64E−04	3.32E−02	1.12E−01	1.73E−01	3.34E−01	5.00E−03	5.70E−01	6.94E−01	5.93E−01	5.94E−01	7.24E−01
3	3.51E−02	1.58E−04	1.92E−02	2.73E−02	2.64E−02	1.67E−01	3.26E−03	2.87E−01	1.60E−01	3.57E−01	3.58E−01	2.04E−01
5	3.05E−02	1.53E−04	9.10E−03	1.37E−02	6.09E−03	1.39E−01	2.75E−03	1.82E−01	6.00E−01	2.58E−01	2.64E−01	1.07E−01
7	2.66E−02	1.48E−04	5.59E−03	8.90E−03	3.30E−03	1.27E−01	2.34E−03	1.52E−01	4.10E−02	2.27E−01	2.35E−01	9.14E−02
10	2.16E−02	1.41E−04	3.38E−03	5.53E−03	2.41E−03	1.15E−01	1.84E−03	1.33E−01	3.18E−01	2.08E−01	2.17E−01	8.54E−02
20	1.08E−02	1.19E−04	1.28E−03	2.02E−03	1.72E−03	8.52E−02	8.71E−04	9.88E−02	1.82E−02	1.75E−01	1.77E−01	7.27E−02
30	5.37E−03	1.04E−04	5.94E−04	8.61E−04	1.29E−03	6.48E−02	4.70E−04	7.51E−02	1.20E−03	1.51E−01	1.46E−01	6.20E−02
50	1.34E−03	7.24E−05	1.36E−04	1.62E−04	7.52E−04	4.08E−02	2.13E−04	4.40E−02	6.07E−03	1.13E−01	9.95E−02	4.51E−02
70	3.34E−04	5.19E−05	3.31E−05	3.12E−05	4.55E−04	2.85E−02	1.42E−04	2.60E−02	3.20E−03	8.45E−02	6.88E−02	3.28E−02
100	4.16E−05	3.15E−05	4.45E−06	2.93E−06	2.25E−04	1.92E−02	9.45E−05	1.18E−02	1.25E−03	5.48E−02	4.03E−02	2.04E−02
300	4.01E−08	1.13E−06	—	—	3.91E−06	3.34E−03	1.12E−05	6.31E−05	3.26E−06	3.01E−03	1.49E−03	8.45E−04
500	—	—	—	—	1.84E−07	6.39E−04	2.37E−06	3.47E−07	1.48E−08	1.59E−04	6.12E−05	3.51E−05
700	—	—	—	—	2.64E−08	1.22E−04	7.02E−07	—	—	7.97E−06	2.54E−06	1.45E−06

(continues)

Table 9-11. (*continued*)

	Co-60		Zn-65		Sr-90		I-131				Cs-137	
	Class Y	Ingestion	Class Y	Ingestion	Class D	Ingestion	Class D		Ingestion		Class D	Ingestion
Days	W. Body	W. Body	W. Body	W. Body	24-h Urine	24-h urine	Thyroid	24-h Urine	Thyroid	24-h urine	W. Body	W. Body
0.5	6.30E−01	9.21E−01	6.33E−01	9.50E−01	—	—	9.59E−02	—	2.06E−01	—	6.31E−01	9.83E−01
1	5.83E−01	7.06E−01	6.05E−01	8.31E−01	8.57E−02	5.85E−02	1.33E−01	3.04E−01	2.54E−01	5.79E−01	6.22E−01	9.66E−01
3	2.97E−01	1.42E−01	4.54E−01	5.41E−01	3.69E−02	1.94E−02	1.42E−01	1.62E−02	2.27E−01	2.33E−03	5.92E−01	9.19E−01
5	1.90E−01	3.54E−02	3.96E−01	4.80E−01	2.45E−02	1.30E−02	1.20E−01	1.31E−03	1.88E−01	2.43E−04	5.72E−01	8.90E−01
7	1.65E−01	1.87E−02	3.78E−01	4.62E−01	1.71E−02	9.15E−03	9.95E−02	2.53E−04	1.55E−01	2.81E−04	5.58E−01	8.70E−01
10	1.57E−01	1.43E−02	3.66E−01	4.46E−01	1.04E−02	5.60E−03	7.51E−02	1.88E−04	1.17E−01	2.99E−04	5.43E−01	8.48E−01
20	1.52E−01	1.03E−02	3.41E−01	4.04E−01	2.26E−03	1.23E−03	2.95E−02	1.27E−04	4.61E−01	2.00E−04	5.08E−01	7.93E−01
30	1.49E−01	8.80E−03	3.19E−01	3.70E−01	6.21E−04	3.41E−04	1.16E−02	6.25E−03	1.82E−01	9.80E−05	4.76E−01	7.44E−01
50	1.44E−01	7.51E−03	2.85E−01	3.21E−01	1.94E−04	1.10E−04	1.82E−03	1.16E−05	2.86E−03	1.81E−05	4.19E−01	6.55E−01
70	1.40E−01	6.77E−03	2.59E−01	2.85E−01	1.54E−04	8.74E−05	2.87E−04	1.93E−06	4.52E−04	3.02E−06	3.69E−01	5.77E−01
100	1.34E−01	5.95E−03	2.27E−01	2.43E−01	1.26E−04	7.15E−05	1.80E−05	1.24E−07	2.85E−05	1.93E−07	3.05E−01	4.76E−01
300	1.02E−01	3.60E−03	1.00E−01	9.64E−02	5.04E−05	2.86E−05	—	—	—	—	8.55E−01	1.33E−01
500	7.82E−02	2.72E−03	4.51E−02	3.86E−02	2.86E−05	1.62E−05	—	—	—	—	2.39E−02	3.74E−02
700	6.05E−02	2.12E−03	2.04E−02	1.55E−02	1.86E−05	1.06E−05	—	—	—	—	6.70E−03	1.05E−02

(*continues*)

Table 9-11. (*continued*)

Days	Hg (org.) Ingestion W. Body	Tl-204 Class D 24-h Urine	Class Y 24-h Feces	Ce-144 Ingestion W. Body	Ingestion 24-h Feces	Class Y Lungs	Ir-192 Ingestion W. Body	Ingestion 24-h Feces
0.5	9.96 E−01	—	—	9.27 E−01	—	2.46E−01	9.23 E−01	—
1	9.92 E−01	3.20E−02	5.20E−02	7.12 E−01	2.82 E−01	2.11E−01	7.07 E−01	2.82 E−01
3	9.76 E−01	3.71E−02	1.30E−01	1.28 E−01	1.95 E−01	1.60E−01	1.31 E−01	1.87 E−01
5	9.60 E−01	3.31E−02	3.60E−02	1.83 E−02	3.10 E−02	1.46E−01	2.42 E−02	2.89 E−02
7	9.44 E−01	2.89E−02	7.99E−03	2.74 E−03	4.31 E−03	1.40E−01	8.99 E−03	3.94 E−03
10	9.21 E−01	2.34E−02	9.62E−04	4.13 E−04	2.14 E−04	1.35E−01	6.37 E−03	1.92 E−04
20	8.49 E−01	1.17E−02	1.29E−04	2.85 E−04	—	1.21E−01	5.19 E−03	—
30	7.83 E−01	5.80E−03	1.24E−04	2.77 E−04	—	1.09E−01	4.45 E−03	—
50	6.66 E−01	1.44E−03	1.14E−04	2.63 E−04	—	8.89E−02	3.39 E−03	—
70	5.68 E−01	3.55E−04	1.06E−04	2.49 E−04	—	7.22E−02	2.61 E−03	—
100	4.49 E−01	4.38E−05	9.46E−05	2.30 E−04	—	5.29E−02	1.78 E−03	—
300	1.20 E−01	—	4.40E−05	1.36 E−04	—	6.64E−03	1.60 E−04	—
500	6.08 E−02	—	2.05E−05	8.03 E−05	—	8.39E−04	1.05 E−05	—
700	4.98 E−02	—	9.53E−06	4.74 E−05	—	1.07E−04	8.06 E−07	—

Days	Th-232 Class Y		U-238 (Insol.) Class Y		Ingestion	Pu-239 Class Y		Ingestion	Am-241 Class W	
	Lungs	24-h Feces	Lungs	24-h Feces	24-h Feces	Lungs	24-h Feces	24-h Feces	Lungs	24-h Feces
0.5	2.47E−01	—	2.47E−01	—	—	2.47E−01	—	—	2.40E−01	—
1	2.13E−01	5.21E−02	2.13E−01	3.20E−02	2.76E−01	2.13E−01	5.21E−02	2.82E−01	2.11E−01	4.21E−02
3	1.65E−01	1.31E−01	1.65E−01	1.31E−01	1.84E−01	1.65E−01	1.31E−01	1.97E−01	1.60E−01	1.14E−01
5	1.53E−01	3.64E−02	1.53E−01	3.63E−02	2.89E−02	1.53E−01	3.64E−02	3.14E−02	1.45E−01	3.41E−02
7	1.49E−01	8.13E−03	1.49E−01	8.11E−03	4.00E−03	1.49E−01	8.13E−03	4.38E−03	1.38E−01	8.78E−03
10	1.48E−01	9.85E−04	1.48E−01	9.83E−04	2.00E−04	1.48E−01	9.87E−04	2.20E−04	1.32E−01	2.07E−03
20	1.46E−01	1.36E−04	1.46E−01	1.35E−04		1.46E−01	1.36E−04		1.16E−01	1.08E−03
30	1.45E−01	1.33E−04	1.45E−01	1.33E−04		1.45E−01	1.34E−04		1.02E−01	9.44E−04
50	1.42E−01	1.29E−04	1.42E−01	1.29E−04		1.42E−01	1.30E−04		7.93E−02	7.15E−04
70	1.39E−01	1.24E−04	1.39E−01	1.26E−04		1.39E−01	1.26E−04		6.14E−02	5.42E−04
100	1.35E−01	1.21E−04	1.35E−01	1.20E−04		1.35E−01	1.21E−04		4.18E−02	3.58E−04
300	1.10E−01	9.15E−05	1.10E−01	9.13E−05		1.10E−01	9.21E−05		3.15E−03	2.24E−05
500	9.06E−02	6.94E−05	9.06E−02	6.92E−05		9.06E−02	7.00E−05		2.31E−04	1.40E−06
700	7.49E−02	5.26E−05	7.49E−02	5.24E−05		7.49E−02	5.33E−05		1.65E−05	8.70E−08

$$H_{50,\,T}(\text{Thyroid}) = 2.92 \times 10^{-7}\,\text{Sv/Bq} \times 2.8 \times 10^5\,\text{Bq} = 0.082\,\text{Sv}(8.2\,\text{rem})$$

(c) The committed effective dose equivalent (CEDE) is obtained from $w_T = 0.03$ for thyroid, or

$$\text{CEDE} = 8.2\,\text{rem} \times 0.03 = 0.246\,\text{rem}\ (246\,\text{mrem})$$

Alternatively, the dose coefficient in Table 9-9 can be used, thus

$$\text{CEDE} = 8.89 \times 10^{-9}\,\text{Sv/Bq} \times 2.8 \times 10^5\,\text{Bq} = 2.49\,\text{mSv}\ (249\,\text{mrem})$$

which is a good example of the significance of weighting each tissue dose to determine the total body risk.

Example 9-10: Estimate: (a) the intake for a worker observed on an annual screening to have a positive whole body count of 520 Bq of ^{137}Cs; and (b) the committed effective dose equivalent.
Solution: (a) Since the intake time is unknown, it is conservatively assumed that it occurred by inhalation one year previously or right after the last whole-body count. From Table 9-11, r_i for class D ^{137}Cs (the most common class) at 365 days post intake is, by interpolation, 5.93×10^{-2}; therefore, the estimated intake of ^{137}Cs one year earlier is

$$I = \frac{520\,\text{Bq}}{5.93 \times 10^{-2}} = 8770\,\text{Bq}$$

(b) The committed effective dose equivalent (CEDE) for this intake can be estimated from the data in Table 9-9, which lists the effective dose coefficient as 8.63×10^{-9} Sv/Bq of intake; thus, the 50-year committed effective dose equivalent is

$$\begin{aligned} H_{50,\,T} &= 8770\,\text{Bq} \times 8.63 \times 10^{-9}\,\text{Sv/Bq} \\ &= 7.6 \times 10^{-5}\,\text{Sv}\ (7.6\,\text{mrem}) \end{aligned}$$

Alternatively, the dose can be calculated from the ALI for ^{137}Cs (6×10^6) which corresponds to a CEDE of 0.05 Sv, or 1.2×10^8 Bq/Sv; therefore,

$$H_{50,\,T} = \frac{8770\,\text{Bq}}{1.2 \times 10^8\,\text{Bq/Sv}} = 7.31 \times 10^{-5}\,\text{Sv}\ (7.3\,\text{mrem})$$

where the difference is due to rounding the calculated ALI to one significant figure.

Weighted Estimate of Intake

Sometimes a series of measurements will be made at different times post intake, and it is useful to weight these measurements by a least-squares fit to the measured data to obtain the intake I. A commonly used formulation, of the several that are available for this purpose, is

$$I = \frac{\sum_i^n r_i A_i}{\sum_i^n r_i^2}$$

where A_i is the measured activity at each respective time post intake and r_i is the retention fraction for each value of t.

Example 9-11: A nuclear medicine scientist working with ^{131}I to tag a radio-pharmaceutical furnished, as required by procedure, a routine 24-h urine sample following the radio-labeling procedure. The result was positive, but 18 days had elapsed by the time it was reported. A thyroid counter measurement determined a thyroid burden of 4.81×10^5 Bq of ^{131}I on the 18th day, and a followup measurement on the 20th day yielded 3.9×10^5 Bq. What is the best estimate of the person's intake?

Solution: As shown in Table 9-11, the thyroid r_i for 18 days post inhalation is 3.55×10^{-2} and at 20 days is 2.95×10^{-2}. The weighted least-squares fit of the data is

$$I = \frac{\sum r_i A_i}{\sum r_i^2}$$

$$I(Bq) = \frac{1.7076 \times 10^4 + 1.150 \times 10^4}{1.26 \times 10^{-3} + 8.7 \times 10^{-4}} = 1.34 \times 10^7 \text{ Bq}$$

TRITIUM—A SPECIAL CASE

Persons can be exposed to tritium either as elemental ^3H or ingestion of tritiated water in the form of HTO or tritium oxide, ^3H_2O. Exposure to elemental ^3H deposits a minimal amount in the body water due to inhalation and very little is absorbed through the skin; the remainder, which is difficult to know exactly, will be exhaled where it may become oxidized and be available for other exposure. The form of tritium can change rapidly, thus it is a perplexing radionuclide due to such variability.

Most radiation protection determinations for tritium are based on the oxide form since it is incorporated directly into body water to produce a whole body exposure. For a person exposed to airborne tritium oxide, an amount equal to 50% of the inhaled intake is assumed also to be absorbed through skin; therefore, a person in a tritium atmosphere that inhales 200 Bq of ^3H_2O will also absorb 100 Bq through the skin for a total intake to the body water of 300 Bq.

Tritium entering the body distributes rapidly and uniformly in 42 L of body water where it is uniformly distributed in and irradiates 63 kg of soft tissue. The daily intake of fluids by a reference male adult is 3 L/d, and an equal amount is

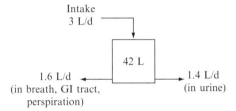

Figure 9-7. Fluid balance in the total body of a reference adult male represented by a tidal pool of 42 L in which a daily intake of 3 L/d replenishes average daily losses of 1.4 L/d in urine and 1.6 L/d through the GI tract, breath moisture, and perspiration.

eliminated each day (see Figure 9-7) to maintain a tidal volume of 42 L of water in the body; thus, the biological removal coefficient for water in the body is

$$\lambda_b = \frac{3\,L/d}{42\,L} = 7.14 \times 10^{-2} \text{ d}^{-1}$$

which is also the biological removal constant for tritium since it is uniformly distributed in body water. The biological half-life is thus

$$T_b = \frac{\ln 2}{\lambda_b} = 9.7\text{d}$$

or about 10 days. For an acute intake of tritium, its initial concentration in body water is

$$C_0 = \frac{I}{42\,L}$$

and as shown schematically in Figure 9-7, the concentration of tritium in body water at any time t following an acute intake, I, is

$$C(t) = \frac{I}{42\,L}\,e^{-(\lambda_b + \lambda_r)t}$$

where $\lambda_b + \lambda_r$ is the effective removal constant λ_{eff} which is effectively λ_b since the radiological half-life of ^{3}H is 12.3 years; therefore

$$C(t) = C_0 e^{-\lambda_b t} = C_0 e^{\frac{-\ln 2}{T_b}t}$$

And, since tritium (as HTO) spreads uniformly and quickly throughout the body water, this is also the concentration in urine at a time t following an accute intake.

The Dose due to an Intake of Tritium is, as described previously, the product of a unit-conversion constant, the total number of transformations that are associated with an initial deposition of activity q_0, and the absorbed energy per transformation. In SI units

$$D_{TOT}(J/kg) = 1.6022 \times 10^{-10} \frac{q_0}{\lambda_e}\left(1 - \exp\left(-\frac{\ln 2}{T_e}t\right)\right)\left[\frac{\sum Y_i \bar{E}_i AF(T \leftarrow S)_i Q_i}{m_T}\right]$$

where the first term in brackets is the total number of transformations, U_s, that occur due to an intake of q_0 (Bq), or $1.443 q_0$ (Bq)T_b since $T_e = T_b = 9.7\,d$ (due to the long radiological half-life of tritium), or for each Bq of a tritium intake

$$U_s(\#t) = 1.443 \times 1\,Bq \times 1t/s \cdot Bq \times 9.7\,d \times 86,400\,s/d$$
$$= 1.21 \times 10^6 t$$

Since ^{3}H is a pure beta emitter Y_i, Q_i, and AF $(T \leftarrow S) = 1.0$ and the average energy per transformation is 5.865×10^{-3} MeV/t; thus the Specific Effective Energy (SEE) for tritium uniformly distributed in 63 kg of soft tissue is

$$SEE(TB \leftarrow TB) = \frac{\sum Y_i \bar{E}_i \times AF(TB \leftarrow TB)_i Q_i}{m_T(g)}$$
$$= 5.685 \times 10^{-3} \frac{MeV/t \times 1 \times 1}{63,000\,g}$$
$$= 9 \times 10^{-8}\,MeV/g \cdot t$$

The committed effective dose equivalent, $H_{50,T}$, due to an intake of 1 Bq of ^{3}H is

$$H_{50,T}(TB \leftarrow TB) = 1.6 \times 10^{-10} \times 1.21 \times 10^6\,t \times 9 \times 10^{-8}\,MeV/g \cdot t$$
$$= 1.74 \times 10^{-11}\,Sv/Bq \text{ of intake}$$

Example 9-12: If the whole body dose limit for a reference person is 0.05 Sv (5 rem) per year, what is the maximum intake of tritium (i.e., the ALI) that could occur without exceeding the dose limit?
Solution: The committed effective dose equivalent for 1 Bq of tritium in oxide form (HTO) is 1.74×10^{-11} Sv/Bq, and the intake that corresponds to the stochastic limit of 0.05 Sv in one year is

$$I = \frac{0.05\,Sv/y}{1.74 \times 10^{-11}\,Sv/Bq} = 2.9 \times 10^9\,Bq/y$$

which is usually expressed to one significant figure, thus the annual limit on intake is

$$ALI_s(^3H) = 3 \times 10^9 \text{ Bq}$$

The Derived Air Concentration (or DAC) for limiting exposure to an air concentration of ^3H_2O must recognize that 1/3 of the body burden of ^3H_2O will enter through the skin; i.e., a person in a tritium atmosphere that inhales 200 Bq of ^3H_2O will also absorb 100 Bq into the body water by absorption through the skin such that only 2/3 of the ALI can be by inhalation, or

$$\text{DAC} (^3H) = \frac{2/3 \text{ ALI(Bq)}}{2.4 \times 10^3 \text{ m}^3 \text{ of air}} = 8 \times 10^5 \text{ Bq/m}^3$$

Bioassay of Tritium

A bioassay measurement of the tritium concentration in urine can be used to determine the intake of tritium because tritium in oxide form mixes rapidly in the body water. Since λ_r is very small relative to λ_b, the tritium concentration in body fluids, including urine, at any time, t, following an acute intake, I, is

$$C_u(t) = \frac{I}{42 \text{ L}} e^{-(\lambda_b + \lambda_r)t} = C_0 e^{-\lambda_b t}$$

where $C_u(t)$ is the tritium concentration (Bq/L) in urine at any time t post intake and C_0 is the initial concentration due to distribution of the intake in 42 L of body water. C_0 (Bq/L) is back calculated from the measured value of $C_u(t)$ at t and this value is multiplied by 42 liters of body water to determine the intake which is then used in the dose calculation. A 24-hour measurement of tritium in urine can also be used with the r_i values in Table 9-11 for a time post intake to determine tritium intake; however, the listed r_i values are for reference person which assumes a T_b of 9.7 days which is generally conservative.

A person-specific biological half-life of tritium elimination can in many cases be determined from several excretion measurements, and a calculation based on this person-specific value provides the most accurate determination of intake and dose, as illustrated in Example 9-13 and problems 9-18 and 9-20.

Example 9-13: An adult male worker was put on a high liquid intake (cheap beer?) and subjected to intense physical exercise following an acute inhalation intake of HTO. Tritium concentrations in urine were measured during this period and were found to fit an exponential function yielding a biological (actually effective) half-life of 5.96 days. Extrapolation of the urine concentration data back to the time of intake for this clearance rate yielded an initial concentration, C_0, of 9×10^7 Bq/L. If the worker was otherwise typical of a reference person, what was his dose?

Solution: The worker's intake was

$$\text{Intake} = 9 \times 10^7 \text{ Bq/L} \times 42 \text{ L} = 3.78 \times 10^9 \text{ Bq}$$

and based on the measured biological half-life of 5.96 d, the number of transformations is

$$U_s = 1.443 \times 3.78 \times 10^9 \text{ t/s}(5.96 \text{ d})(86,400 \text{ s/d})$$
$$= 2.81 \times 10^{15}t.$$

And since the specific effective energy (SEE) for tritium in body water is 9×10^{-8} MeV/g · t (see above) the CEDE is

$$H_{50,\text{T}} = 1.6 \times 10^{-10} U_s \times \text{SEE}$$

$$H_{50,\text{T}} = 1.6 \times 10^{-10}(2.81 \times 10^{15} \text{ t})(9 \times 10^{-8} \text{ MeV/g · t}) = 0.04 \text{ Sv}(4 \text{ rem})$$

which is close enough to the limiting CEDE for the whole body of 0.05 Sv (5 rem) that an overexposure would be assigned if the measured effective half-life had not been used; i.e., the high liquid intake and exercise regimen was to the benefit of the worker (see problem 9-18).

SUMMARY

Internal radiation doses are calculated as the product of three quantities: (1) the number of transformations produced by a radionuclide in a source organ at a given time period, which is a function of the activity of the deposited radionuclide and the time it remains in the source organ, (2) the energy deposition per nuclear transformation which is a constant for a specific radionuclide and a source-target organ pair, and (3) a constant to adjust the units. Because energy deposition is a constant for a specific radionuclide and source–target organ pair, calculation of internal dose thus becomes an exercise in determining how much of a radionuclide is deposited in a source organ and the subsequent number of transformations associated with the deposited activity. Both are very much determined by human biology, and since this occurs with varying degrees of complexity, determining internal dose can also be complex. This is very much the case in occupational/ environmental settings because radionuclides must first pass through the respiratory system and/or the gastrointestinal (GI) tract before entering the bloodstream from which uptake to an organ occurs.

Radiation doses occur to source organs, other surrounding target tissues, and to the lung and the various segments of the GI tract as the radioactive material passes through them. Fairly extensive biokinetic models have been developed for calculating doses due to intakes by inhalation and ingestion of radionuclides, in particular a respiratory deposition/clearance model and a GI tract model. Both assume an initial deposition of radioactive material that is, depending on chemical and physical characteristics, translocated to various tissues to produce a tissue dose. Controls for internal radiation dose are based on establishing an annual limit of intake (ALI) for each particular radionuclide such that a dose limit is not exceeded. Such limits, which have varied over the years, are now based on an explicit consideration of risk, and the resultant values are currently known as risk-based standards.

The intake of a radionuclide by inhalation or ingestion is directly related to the internal radiation dose, which provides a straightforward method for controlling and assigning internal radiation doses to persons. Radionuclide intakes can be determined from measured radionuclide concentrations in inhaled or ingested food, water, or contamination. If, however, such measurements are not available it may be necessary to perform bioassay measurements of the person after the fact, perhaps following an incident or spill. The measured amount in the whole body, a specific organ or tissue, and/or excreta is directly proportional to the intake; thus, bioassay measurements can, with appropriate adjustment, be used to determine the intake of the radionuclide, and thus the dose received.

Acknowledgments

Patricia Ellis compiled many of the data tables in this chapter from complex data sets with patience and diligence.

ADDITIONAL RESOURCES

1 Publications of the International Commission on Radiological Protection, in particular ICRP Reports 26, 30 (all parts with supplements), 60, 61, 66, 71, and 72.
2 EPA Federal Guidance Report 11.
3 Lessard, E.T. et. al., "Interpretation of Bioassay Measurements," US Nuclear Regulatory Commission NUREG/CR-4884, July 1987.

PROBLEMS

1 Determine: (a) the effective half-life of ^{131}I ($T_{1/2} = 8.02$ d) in the thyroid if its biological half-life is 80 days; and (b) explain the magnitude of the calculated value.

2   A researcher working with $^{14}CO_2$ was found through routine urine analysis to have a body burden of 20 μCi of ^{14}C ($\bar{E} = 0.049$ MeV; $T_b = 40$ d): (a) what dose rate (rads/h) would be delivered to soft tissue (m = 63,000 g) from this body level; and (b) if removed from the work, what total dose would be delivered due to the amount of ^{14}C accumulated in the body?

3   Determine: (a) the instantaneous dose rate (rads/h) due to the principal beta emission ($\bar{E}_\beta = 0.192$ MeV) of ^{131}I which occurs in 89.4% of transformations for 1 mCi of ^{131}I($T_r = 8.02$ d and $T_b = 80$ d) deposited in the 12 g thyroid gland of a 10-year old person; and (b) the total accumulated dose delivered due to this emission.

4   Use the dose reciprocity theorem to determine the absorbed fraction AF (Liver $\leftarrow$ Kidney) for 0.5 MeV photons.

5   Calculate: (a) the photon dose to the lung and the total body from 1 μCi of ^{60}Co deposited in the lung and cleared with a biological half-life of 60 days, and (b) the photon dose to the lung and the total body for 1 μCi of ^{60}Co deposited uniformly in the total body that clears with a biological half-life of 800 days.

6   Inorganic compounds of metallic mercury are presumed from metabolic studies to distribute uniformly after inhalation or ingestion with 8% going to the liver ($f = 0.08$) and 92% to the total body ($f = 0.92$). Biological clearance of mercury in each of these tissues occurs by two compartments, 95% with a biological half-life of 40 d and the remaining 5% with a biological half-life of 10,000 d. For an intake of 1 Bq of ^{203}Hg($T_{1/2} = 46.61$ d), determine: (a) how many transformations occur in the liver; and (b) in the total body.

7   For $^{137}Cs - {}^{137m}Ba$ distributed uniformly in the total body, the specific effective energy is 6.31×10^{-6} MeV/g $\cdot$ t. If an initially measured amount in the body of an adult is 4×10^5 Bq, determine: (a) the total number of transformations based on the retention function for ^{137}Cs; and (b) the 50-year integrated whole body dose.

8   For the first 8 months of a calendar year, an adult worker inhales 2×10^6 Bq of ^{137}Cs, 1×10^6 Bq of ^{131}I, and 6×10^5 Bq of ^{60}Co. The appropriate ALIs are 6×10^6, 2×10^6, and 6×10^6 Bq, respectively. (a) If the worker is reassigned to an area where only external gamma exposure is possible, how much exposure (if any) could he receive without exceeding the stochastic risk limit?; (b) if the worker were exposed to ^{131}I aerosols instead of external γ, how many Bq could he take in without exceeding ICRP 30 recommended limits?

9   For the ICRP 30 derived DAC, (Table 9-8), how many DAC-h could the worker in Example 9-9 be exposed to without exceeding a stochastic dose limit of 0.05 Sv?

10 A lung burden of 2.45×10^7 Bq of ^{51}Cr is measured in an adult male worker at 4 days post intake of class Y ^{51}Cr aerosols. Estimate: (a) the

initial intake; and (b) the committed effective dose equivalent to the worker.

11 Use the data in Table 9-8 to determine the stochastic ALI and DAC for inhalation of: (a) 1 μm AMAD aerosols of ^{59}Ni; and (b) ^{125}I.

12 Use the data in Table 9-9 to determine the ALI for ingestion of: (a) ^{32}P; and (b) ^{35}S.

13 A routine *annual* whole body measurement of a worker identified a whole body accumulation of 190 Bq of ^{60}Co assumed to be due to exposure to Class D 1 μm AMAD aerosols. Determine: (a) a best estimate of the intake; (b) the CEDE based on the ^{60}Co ALI (6×10^6 Bq); (c) the CEDE using the data in Table 9-2 (from Federal Guidance Report No. 11), and (d) compare the results.

14 If in the absence of a whole body counter it was necessary to address the conditions for the previous problem (# 13) by a 24-h urine sample, and the LLD for ^{60}Co for such a measurement was 10 Bq, what would be the minumum intake/dose that could be assigned based on urine sampling? Fecal sampling?

15 A 24-h urine sample collected 7 days after a potential exposure to ^{14}C-labeled compounds had a measured activity of 9×10^6 Bq: (a) estimate the intake, and the CEDE; (b) determine whether ingestion or inhalation is the most restrictive; and (c) explain why.

16 The following measurements of daily urine following an acute uptake of ^{131}I by inhalation yielded 1440, 970, 380, and 300 Bq at 5, 7, 10, and 20 d, respectively. Determine the best estimate of the intake using the weighted least-squares fit of the urine data.

17 If the worker in Problem 16 had thyroid burdens of 2.2×10^5 Bq, 1.2×10^5 Bq and 7×10^4 Bq at 5, 10, and 20 d post intake, estimate the intake by a weighted least-squares fit of the data. Based on results and the listings in Table 9-5, which approach would be preferable and why?

18 What would be the estimated dose to the worker in Example 9-13 due to the intake of tritium if he had not been on the high liquid diet and exercise regimen?

19 Assuming reference person radiobiology, determine the initial intake of tritium for a person whose urine concentration was $2 \times 10^{-4} \mu$Ci/L 6 d post intake.

20 Urine samples from an adult male worker who was placed on a high liquid diet after exposure to airborne tritium (assumed to be in oxide form) were 65, 40 and 22 Bq/L at 2, 5, and 9 d, after exposure: (a) plot the data and determine the effective half-life for elimination; (b) estimate the initial intake of HTO; and (c) determine the whole body dose due to the intake.

10

RADIATION PROTECTION STANDARDS

"Those who love sausage and the law should never watch either being made."
– Unknown

Setting radiation protection standards for workers and the public has been a major technical and policy challenge since the discovery of x rays and radio-activity (in 1895 and 1896, respectively). and the subsequent discovery of biological effects in humans exposed to them. These effects, the benefits of various activities that produce radiation, and the physics of radiation emission and dose have been considered by the major governmental and advisory groups that have developed and recommended protection standards. The first group to do so was the medical profession, especially radiologists, who were at greatest risk but who also had an interest in assuring that controls did not unduly restrict medical uses of x rays and radium. Professional and governmental organizations have developed radiation protection standards within this frame-work over the past several decades, and they continue to be challenged to balance radiation exposure and public health goals, worker protection, energy supply, national defense, etc.

EVOLUTION OF RADIATION PROTECTION STANDARDS

The major trends and events that have shaped radiation protection policies and standards, as summarized in Table 10-1, consist of several important stages:

- **Discovery of Radiation Injury** encompassed a period of roughly 50 years following the discovery of x rays and radioactivity, a period in which biological effects were directly observable in radiologists and radium workers and controls were based on a "Tolerance Level."
- **Bioeffects Studies** that demonstrated (a) that genetic mutations occurred in Drosophila (fruit flies) at a rate that was linear with the x-ray dose which provided an early basis for a non-threshold policy for radiation risk; and (b) a statistically significant increase in human cancers of all types in Japanese bomb survivors, but in sharp contrast to the fruit fly experiments, no detectable genetic effects.
- **A Major Shift of Controls from Medical Uses to Nuclear Activities** during and after World War II due to the nuclear weapons complex, nuclear electricity generation, and expanded uses of radioisotopes with a significant increase in the numbers of persons potentially exposed.
- **Standards Based on an Explicit Statement of Risk**, made possible with human data on bioeffects, to limit radiation risks to levels comparable to those in other safe industries.
- **Dilution of AE Act Authority and Public Controversy** in the 1960s, 70s, and subsequent years that led to creation of the Federal Radiation Council in 1959; a reorganization of the AEC in 1974 into the Department of Energy (DOE) and the US Nuclear Regulatory Commission (NRC) to separate development and regulation of nuclear energy; abolishment of the Joint Committee on Atomic Energy and dispersal of its sole oversight to other congressional committees; transfer of environmental radiation standards and Federal radiation guidance authorities to EPA from the AE Act; and establishment of new EPA authorities for radioactive materials under RCRA, CERCLA, the Clean Air Act, and other environmental statutes.

Historical Trends in Radiation Policies

In a period of roughly 50 years following the discovery of x rays and radioactivity, biological effects were directly observable in radiologists and radium workers. These effects were various cancers, callousses and burns on the hands and, most notably, skin erythema. By the 1920s it was clear that radiologists were experiencing a distinctly observable increase in premature deaths, and excess bone cancers were occurring in radium dial painters who were mostly young women. Development of the hot cathode Coolidge x-ray tube and shielded x-ray tubes considerably reduced exposures and observable effects; however, it was common to give x-ray technologists 4–6 weeks of vacation for recovery from fatigue and other effects, mostly blood changes.

A qualitative unit called the skin erythema dose (SED) was commonly used to characterize radiation exposure, although Rollins (1903) used darkening of x-ray film as a rough measure of exposure (the first "film badge"). The SED was defined as the amount of radiation that produces a well-defined reddening

Table 10-1. Evolution of Radiation Protection Concepts and Standards.

1895–1925: Discovery of Radiation Injury; Minimal Controls

- Use of x rays and radium in diagnosis and treatment; high exposures to radiologists (200 deaths by 1930), technicians, and radium dial painters; erythema common; 198 cancers, 54 deaths by 1914, including Edison's assistant in 1904.
- Coolidge x-ray tube (1913) reduced exposure; Rollins used film as a primitive monitor.

1925–1945: Bioeffects Studies; Protection Organizations; Atomic Energy

- Muller discovers radiation mutations in Drosophila (1927); genetic effects dominate protection policy; study of bone cancers in Ra workers.
- Mutscheller and Sievert, acting independently, recommend (1925) x-ray controls at ~1/10 SED or ~0.2 R/day (~70 R/y); ICR International X-ray and Radium Committee (1928) and its US counterpart (1929) adopt it as a "tolerance level" (1931 and 34); US (1935) reduces it to 0.1 R/d and adds 5 R/d for hands of Ra workers.
- The Roentgen is defined in 1925 by ICRU and adopted in 1937.
- Nuclear fission increased sources and persons exposed; controls based on NCRP values of 0.1 R/d (~35 R/y); 0.1 µg of Ra or equivalent used for weapons materials.

1945–1960: Shift from Medicine to Fission and its Byproducts

- Sweeping government authority by Atomic Energy Act of 1946 to develop atomic energy in interest of "the common defense and health and safety"; 10 CFR 20 regulations (1954) based on NCRP and ICRP.
- Atomic Bomb Casualty Commission (now RERF) established in 1947 as joint US/Japan effort to treat Japanese survivors and develop data on radiation effects.
- "Maximum permissible dose" replaces "tolerance level" which had implied no effects; NCRP (1949) and ICRP (1950) adopt 0.3 R/wk (~15 rem/y) based on feasibility and more persons exposed.
- ICRP (in 1956), to avoid technical overruns, adopts 3 rems/qtr not to exceed an accumulated dose of $5(N - 18)$ rems; organ limits of 15 rem/y; public dose ≤ 0.5 rem/y.
- US establishes (1960) cabinet-level Federal Radiation Council; it restates ICRP guides $[5(N - 18)]$ as values not to be exceeded without careful consideration; public exposure limited to 0.17 rem/y to a suitable sample to assure no individual exceeds 0.5 rem/y.

1960–1980: Public Controversy Influences Federal Standards and Programs

- Nuclear power growth; public concerns led to Appendix I to 10 CFR 50 (AEC 1971) which stated 5 mrem/y for as low as reasonably achievable (ALARA) operation of nuclear plants.
- National Academy of Science (NAS) concludes (1972) that cancer is major bioeffect.
- EPA established (1970) with AE Act authority for environmental standards and FRC function; new authorities from RCRA (1976), CERCLA (1980), Clean Air Act Amendments (1977), uranium mill tailings radiation control act (UMTRCA, 1978), Radon Control Act.
- AEC divided (1974) into energy/weapons agency (ERDA, now DOE) and US Nuclear Regulatory Commission; Joint Committee on Atomic Energy oversight given to other committees.
- EPA's uranium fuel cycle standards (1977) set 25 mrem/y for routine operations and limited releases per GW(e) for Kr-85, TRUs, and I-129; implemented by NRC.
- Risk-based standards developed (ICRP 1977) corresponding to 5 rem/y for total (external plus internal) dose; President issues EPA's Federal guidance for medical x-rays (1977) and workers (1987).

(continues)

Table 10-1. (*Continued*)

1980–2000: Risk-Based Policies; Contaminated Sites; ~50 years of Risk Data

- BEIR III (1980) used linear quadratic risk model yielding 110 effects/10^6 person-rem; BEIR V (1990) used linear non-threshold model and DDREF to yield ~ 500 effects/10^6 person-rems. EPA leaves standards intact.
- 10 CFR 20 revised (1/1/94); ALARA programs required (previously recommended); internal and external doses must be summed if either >10% of limit.
- Below Regulatory Concern (BRC) of 10 mrem/y proposed by NRC in July 1990; withdrawn in 1992.
- EPA regulatory activities (Superfund, Clean Air Act, RCRA, high-level and low-level wastes) dominated by $10^{-4} - 10^{-6}$ lifetime risk (corresponds to ~15 mrem/y).

---

of the skin. The SED was obviously a very qualitative measure for radiation control that varied by the type and rate of radiation exposure and was of little use for workers in the radium-dial painting industry which flourished during World War I. Activities during World War I also increased exposures to people from expanded uses of x rays on the battlefield (including M. Curie who saved countless lives by operating x-ray equipment in the field at considerable exposure to herself) and for detecting flaws in materials, but because of the war no real progress was made in formulating the standards needed.

The various attempts at radiation protection in these formative years were thus essentially trial-and-error processes, complicated to a degree by the lack of measurement methods, reference units, and maintenance of medical discretion. The International College of Radiology (ICR) formed the International X-ray and Radium Protection Committee (1928), predecessor of the International Commission on Radiological Protection (ICRP), because of increases in premature deaths in radiologists and fatal bone cancers in radium-dial painters. The US formed the Advisory Committee on X-Ray and Radium Protection (1929), which later became the National Council on Radiation Protection and Measurements (NCRP). These bodies had two primary goals: (1) a unit to assess exposure; and (2) establishing a "tolerance level" below which it could be presumed that effects would not occur. The Roentgen (R), which was based on measured ionization in air, was established for quantitation of exposure to x and gamma rays by the International Commission on Radiological Units and Measurement (ICRU) which was formed in 1927. The R was formally adopted in 1937 and continues to serve as a practical and accurate measurement of radiation exposure for radiation control.

The "tolerance dose" concept was recommended by Dr. R. Sievert in 1925 to obviate the occurrence of radiation effects; it was proposed as one-tenth of the SED, or about 0.2 R/d, and was adopted by the ICRP in 1928. The tolerance dose concept suggested a threshold below which complete recovery from any radiation effects, if they occurred, was possible, and in fact erythema would disappear (or not appear) when personnel exposure was limited to one-tenth of the SED. The US Advisory Committee on X-ray and Radium Protection adopted the same value in 1929 but reduced it in 1935 to 0.1 R/d (the first factor of two) and included a recommendation of 5 R/d to the hands for radium workers, the first quantitative standards for such work. These medically oriented and industrial activities were the main focus of radiation protection policies up to the 1940s, and during and after World War II they formed the basis for controlling exposure due to new materials associated with nuclear energy uses, in particular an expanding nuclear weapons complex.

Nuclear Activities   during and after World War II dominated radiation controls. The US Atomic Energy Commission (AEC) was established in the US by the Atomic Energy Act of 1946 which made all uses of nuclear energy the exclusive province of the US government primarily to assure civilian control over the development and control of atomic energy and its byproducts for "the common defense and in the interest of public health and safety." The AEC chose to carry out its broad mandate for "health and safety" by incorporating the recommendations of both ICRP and NCRP into its own facilities and in its regulatory programs.

The NCRP in 1949 again reduced its recommended limits for external whole body exposure from 0.1 R/d to 0.3 rem/wk (a second factor of two) because the number of persons who could potentially be exposed was increasing substantially, making it important to preclude genetic risks in this larger population. For similar considerations, the NCRP also introduced the concept that any radiation exposure should be maintained "as low as practicable" below the recommended "maximum permissible dose" (MPD). Both of these new terms were chosen to supplant the concept of a "tolerance dose" with a policy that any exposure carried a risk of effects and should be avoided by practical measures below a ceiling that was chosen to be protective of a given individual by limiting his/her direct risk, but which would also limit genetic risks to the total population.

The Atomic Energy (AE) Act was amended in 1954 to allow nongovernmental uses of AE Act materials as part of President Eisenhower's initiative for peaceful uses of atomic energy in research, medicine, and nuclear electricity generation. These uses were to be conducted under AEC licenses which in turn required a set of federal regulations that were issued in 1954 as Part 20 of Title 10 of the US Code of Federal Regulations (10 CFR 20). Thus, radiation standards and the first federal regulations that implemented them (10 CFR 20) were firmly rooted in two policies: "maximum permissible dose" and a concept of keeping exposure "as low as practicable," both of which were based on a non-threshold model of radiation risk.

It was also recognized that intakes of radioactive materials were likely due to expanded nuclear energy activities and research and development using radioactive materials, and that these intakes posed internal radiation doses and potential effects. Consequently, the NCRP also recommended for the first time (in 1953) maximum permissible organ burdens for a limited number of radionuclides to control internal dose to a *critical organ* defined as the organ which received the greatest damage (and usually the highest dose) from an internally deposited radionuclide. These organ burdens were in turn used to establish maximum permissible concentrations (MPCs) of radionuclides in air and water as a practical means of limiting radiation dose due to intakes of radionuclides.

The NCRP lowered its recommended MPD again in 1956 to 5 rem/y (a factor of 3), and the AEC dutifully revised its regulations (in 1959) to incorporate the reduced limits. The primary reason for lowering the MPD was that it was practicable to achieve it, and therefore prudent. The new limits were stated as an accumulated dose of $\leq 5(N - 18)$, where N is a person's age with the provision that the dose in any calendar quarter not exceed 3 rem in order to provide regulatory flexibility to deal with technical overruns which were not viewed as serious since the new limits were based on practicability and prudence. Even though the average annual dose limit was 5 rem, the $5(N - 18)$ provision made it technically possible for a previously unexposed worker to receive 12 rems in a year, a result that was not originally intended and unfortunately was sometimes abused. Doses to most organs were limited to 15 rem/y with exceptions for thyroid and skin of 30 rem/y and 5 rem/y to the gonads. And since the general public, a more diverse group that did not benefit from a paycheck, could be exposed due to some activities, the concept was introduced that their exposure be limited to one-tenth of the radiation-worker standard, or 0.5 rem/y (further reduced to 0.1 rem/y in 1976 because no necessary activity required such exposures).

Bioeffects Studies have greatly influenced radiation standards. H. J. Muller's findings of genetic mutations in Drosophila in 1927 and the extensive and long-term studies of health effects in Japanese bomb survivors have been incorporated into radiation protection policies and standards. Muller's studies provided the basis for a non-threshold policy for genetic risks, and data from the Japanese atomic bomb survivors demonstrated that cancers of all types, rather than genetic effects, were the dominant risk of radiation exposure, risks that continue to be evaluated as occurring without threshold. The Japanese survivor data are of inestimable value for assessing the delayed effects of radiation since they are based on direct observations of effects in humans. Even though the circumstances were unfortunate, it is a credit to humankind and the Japanese in particular who formed the Atomic Bomb Casualty Commission and its successor, the Radiation Effects Research Foundation (RERF). These organizations, both of which have been jointly sponsored by the Japanese and US governments, have provided compassionate and continuous treatment and followup of the survivors since 1947.

Radiation bioeffects data have been reviewed periodically by the Committee on the Biological Effects of Ionizing Radiation (BEIR) of the National Academy of Sciences (NAS) since 1956 in the US and the United Nations Scientific Committee on the Effects of Atomic Radiation (UNSCEAR) which published its first report in 1958. In 1972, the NAS BEIR Committee concluded that cancer, not genetic risk, was the major effect of radiation exposure, and radiation-control efforts and protection standards have focused on limiting radiogenic cancer risk ever since. Most professional groups and federal standards agencies have consistently used this basic policy. Although various models have been devised by the various BEIR and UNSCEAR committees to fit the data in terms of risk and dose (e.g., the linear quadratic and the linear models), all have assumed some effect at any dose.

Current radiation protection policy, as stated in radiation protection standards and regulations, is based on prudent public health. Although it is beyond the ability of epidemiologic science to demonstrate effects at recommended protection levels, it is assumed that the mechanisms that produce observable effects at high exposures (demonstrable at greater than 20 rads or so) also occur at lower exposure levels, and the inability to observe them directly is due to their low probability. Since the manifestation of an effect is probabilistic (stochastic), it is prudent to presume that the same mechanisms and effects occur in a population of persons exposed at lower doses and dose rates and to derive protection standards on the basis of calculated (nonmeasurable) potential risks. This has become the basis for various extrapolation models (see Figure 6-6), each of which extrapolates data from high exposures to a level of zero additional dose from a given source above background. Once this presumption was adopted, all radiation protection decisions became a tradeoff between the projected risk of exposure and the practicability of changing the conditions and levels of exposure, be it medical uses, nuclear energy, research, or national defense.

The non-threshold concept has also formed the basis of recommendations by NCRP, ICRP, and standards-setting agencies to maintain actual exposures "as low as practicable." The concept was incorporated for the first time into regulations issued by the US Atomic Energy Commission (AEC) in 1972 for design objectives for nuclear power reactors. The AEC, forerunner of the current US Nuclear Regulatory Commission (NRC), stated the concept as one of "as low as reasonably achievable" (ALARA) to allow consideration of costs (the word "practicable" is defined as capable of being done; i.e., regardless of cost).

Standards Based on an Explicit Statement of Risk   were recommended by the ICRP in Publications 26 and 30 issued in 1977 and 1979, respectively. Although previous radiation standards were designed to control risks, the ICRP recommendations were the first to explicitly quantitate the number of excess effects that could be expected should a defined population receive doses allowed by the recommended standard. Recommendations were thus designed to limit the level of stochastic effects to those comparable to risks in other safe industries and to preclude nonstochastic (or threshold) effects altogether. The ICRP recommendations specifically state policies that exist to varying degrees in recommendations issued by NCRP in 1987 (NCRP-94) and in the most recent ICRP revision (ICRP-60):

- justification (no radiation exposure without benefit);
- optimization of any justified exposure (i.e., ALARA); and
- compliance with recommended dose limits.

The prevention of nonstochastic (or threshold) effects is widely accepted; however, the limitation of probabilistic stochastic effects (i.e., without threshold) is quite controversial because some groups refuse to accept any human-induced radiation exposure whereas others believe that current standards require excessive actions not justified by the potential risks (i.e., they are more protective than necessary).

The adoption of risk as the primary radiation protection standard means that radiation dose is an implementing standard which is achieved by an annual total effective dose equivalent (TEDE) limit of 5 rem (0.05 Sv) to the whole body irradiated uniformly. For non-uniform partial body irradiation or irradiation of various tissues by internal emitters, the risk-based whole-body limit is apportioned through the use of tissue weighting factors to ensure that stochastic effects do not exceed the risk associated with a TEDE of 5 rem (0.05 Sv) per year. Nonstochastic effects are precluded by applying a dose equivalent limit of 50 rem (0.5 Sv) in a year to all tissues except the lens of the eye, for which the recommendation is 15 rem (0.15 Sv) in a year.

Adoption of a risk-based limit also requires that all exposures be considered in limiting radiation risk; therefore, it became necessary (for the first time in radiation control) to sum both internal and external doses. Adoption of a risk-based standard also suggests that if risk data change, the dose limit must also change or a new basis for justifying risk associated with exposures corresponding to the standard needs to be stated. In 1990, the ICRP re-evaluated the Japanese radiation risk data based on a linear extrapolation model which increased its previous risk coefficient of 165 effects/10^6 person-rems (which was based on a linear quadratic model) to ~ 500 effects/10^6 person-rems. Thus ICRP revised its limits (in ICRP Report 60) to 10 rem in any 5-year period, not to exceed 5 rem in any given year, or a factor of about 3 to an annual average dose of 2 rem. The numbers of observed cancers had not changed; only the model for extrapolating risk data (from a linear quadratic model to a linear one). The effect of this 5-year average standard has been to lower long-term risk while validating continuation of the annual limit at 5 rem/y for justifiable exposure.

Dilution of AE Act Authority and Considerable Public Controversy over radiation standards has occurred since about 1970, a period in which the AEC was reorganized into the Department of Energy (DOE) and a separate regulatory agency (the US Nuclear Regulatory Commission), and EPA was given authorities for radiation standards under the AE Act, RCRA, CERCLA, the Clean Air Act, and other environmental statutes. These redirections of radiation control authorities reflect challenges of federal programs and regulations by various organizations, primarily through the Administrative Procedures Act, based on a premise that there is no safe level of radiation. Others believe just as strongly that the exposure guides that have been developed should be considered protective and "safe" and that this is a sufficient basis for radiation protection of workers and the public.

Current radiation protection guides and recommendations, and standards and regulations based on them, appear to be a compromise between opposing views on radiation risk. They not only state an upper limit of dose that is not to be exceeded without careful consideration, but two additional requirements: all exposures are to be justified and actual doses are to be maintained as far below the limit as is reasonably achievable (ALARA). As a result, AEC, NRC, DOE, and EPA have focused on optimization of radiation protection; i.e., maintaining actual exposures at ALARA levels and well within maximum limits.

CURRENT RADIATION PROTECTION STANDARDS

Radiation protection standards have remained virtually the same since the late 1950s although various interpretations of bioeffects data have been made, including refinements in the bases for choosing protection levels, most notably the incorporation of risk-based limits recommended by ICRP in 1977. The recommended risk-weighted dose limitation system is based on a weighted sum of all the dose equivalents delivered to all parts of the body, either externally or by internally deposited radionuclides. This weighted sum is defined as the effective dose equivalent, and accounts for variations in radiosensitivity of various organs and tissues and potential susceptibility to hereditary effects.

Current US Standards for Occupational Workers are based on federal radiation guidance recommended by EPA and approved by the President in 1987; these are:

Occupational Dose

- Stochastic effects $\leq$ 5 rem/y (0.05 Sv) effective whole body
- Nonstochastic effects $\leq$ 50 rem/y (0.5 Sv) to any tissue other than the eye
- Extremities $\leq$ 50 rem/y (0.5 Sv) hands/feet
- Lens of the eye $\leq$ 15 rem/y (0.15 Sv).

These standards are consistent with the 1977 recommendations of the ICRP and later the NCRP; they do not incorporate the 1990 recommendations of ICRP (ICPR-60) to limit occupational dose to 10 rems (0.1 Sv) in a 5-year period not to exceed 5 rem (0.05 Sv) in any given year. Unlike those of the ICRP, EPA's bioeffects analyses have always been based on the linear risk model which provided more consistent (i.e., minimally varying) levels of risk.

Current US Standards for the Public are still derived from the 1960 federal radiation guidance, which (as of 2001) has not been formally revised by EPA. The US Nuclear Regulatory Commission (NRC) has, however, incorporated NCRP and ICRP recommendations into its regulations (in 10 CFR 20) for members of the public, which are

Population Dose

- Long-term average ≤ 0.1 rem/y (1 mSv) effective whole body
- Temporary ≤ 0.5 rem/y (5 mSv) effective whole body.

The temporary limit of 500 mrem (5 mSv) in a given year can only be used by a licensee upon petition to and approval by NRC based on appropriate justification, and then only for a specified period. This approach, in effect, continues to legitimize the previous level of 500 mrem (5 mSv) per year (or 0.17 rem/y to a suitable sample of persons) which has been in effect for many years. NCRP and others state this as acceptable for temporary but justified situations; however, the long term average is to be maintained at 100 mrem (1 mSv) per year. In recognition that members of the public may receive exposure from more than one regulated activity, it is also general policy that activities that could expose the public to radiation through environmental releases, etc., be controlled to a fraction of the 0.1 rem/y value to assure that no individual would receive exposures in excess of the standard should more than one exposure circumstance be encountered.

These values exclude, as have all previous ones, exposures due to medical uses of radiation and the pre-existing background (no longer just natural levels). The values also do not state an inconsequential (or *de minimus*) dose level for occupational and public groups even though various efforts have been made to incorporate such a value(s) into national standards and regulations. Regulations that implement these numerical limits include procedural and other programmatic requirements that promote, without stating dose levels, justification of all radiation exposures and optimization of actual ones.

As Low as Reasonably Achievable (ALARA)   is now a stated requirement in US radiation standards whereas it was previously recommended (and expected) as part of an effective program. In general, efforts to increase the level of radiation protection implies a cost for society in terms of work, time, materials, and even other risks, and it is not clear if these costs have always been considered explicitly. For a practice to be ALARA, an appropriate weighing of the collective dose (the sum of the dose delivered to all exposed individuals) against the effort to change it should be considered, any determination to use or not use a control should be justified, and caution should be exercised to avoid merely shifting exposures from one group to another.

The ALARA policy has been interpreted by many regulated groups, perhaps for political, legal, or public-relations reasons, as a dose reduction requirement; it is, however, a statement that exposure(s) be balanced in terms of risk and practical means (cost) of changing it. A strict application of this approach means that an existing practice could justifiably be relaxed because the change in risk is not worth the effort to achieve it.

In sum, national and international standards have incorporated the widely accepted, three-pronged policy of justification (no exposure without benefit), optimization (ALARA within exposure limits), and limitation (control of exposures below recommended limits). The ALARA requirement, in particular, is consistent with good radiation protection policies: improving radiation measurements; training personnel; providing support personnel, equipment and facilities; balancing the cost and feasibility of achieving a given protection level; and weighing the impact of different controls and practices on various activities.

ALARA-BASED STANDARDS FOR SPECIFIC CIRCUMSTANCES

The major standards-setting and regulatory bodies have, in adherence to the non-threshold presumption, chosen limits that themselves are as low as reasonably achievable (ALARA) based on a broad assessment of the activity (e.g., normal operation of nuclear power reactors). Regulations promulgated for such activities are in essence a statement that the day-to-day practices are justified; i.e., that actual exposures (and risks) are balanced against available means (cost) of changing them. And since there is a further requirement to optimize conditions of exposure, practices conducted under ALARA-based limits also assure that actual exposures are well below the limits and at a level that is no higher than necessary. Such ALARA-based standards and regulations have been developed for the design of nuclear power reactors, uranium fuel cycle operations, public drinking water supplies, emissions of radionuclides as hazardous air pollutants, radon guides, and operations associated with low-level and high-level radioactive wastes (but not their disposal).

It is important to emphasize that the numerical statements in ALARA-based standards are applicable only to the specific activities for which they were developed (see below) since they are based on a trade off of risks and costs that are generally not applicable to other circumstances. None the less, many people interpret any numerical radiation standard as a statement of an upper level of hazard and apply it broadly. While such interpretations are understandable, differing ALARA-based standards are not inconsistent with other differing requirements accepted by society; for example, speed limits are lower in neighborhoods and school zones for good reasons even though much higher values are permitted on major highways. Despite the confusion that may be associated with different numerical requirements for specific activities, it is useful to state optimal levels of control for sources/situations where the availability and costs of controls can be evaluated.

Design Objectives for Nuclear Power Plants (Appendix I, 10 CFR 50)

The first formal requirements based on the ALARA concept were issued by the US AEC in 1973 as Appendix I to regulations in 10CFR50 for the design and operation of commercial nuclear power plants. The amendment was in large part a response to intense public opposition to nuclear power reactors, but was also influenced to a large degree by ICRP and NCRP radiation protection philosophy to maintain radiation exposures as far below stated limits as practicable.

After careful examination of the effectiveness of effluent control systems and their cost and feasibility, a number of design objectives and operational requirements were promulgated that would limit public radiation exposures within 50 miles of such facilities to

   $\leq$ 10 mrem/y for gamma radiation due to air releases;
   $\leq$ 20 mrem/y from beta radiation due to air releases;
   $\leq$ 3 mrem/y due to releases in liquid effluents; or
   $\leq$ 5 mrem/y to the whole body or 15 mrem to any organ via all pathways.

An important additional requirement was to apply all available control systems up to a cost/risk value of $1,000 per person-rem to the total body or $1,000 per thyroid-rem. These individual dose limits have been met by all nuclear power stations; however, since the annual doses cannot be measured directly, compliance is based on effluent measurements and use of calculational models for environmental pathways (see Chapter 12). Control systems have also been applied to meet the $/person-rem criteria, and in many cases well in excess of the criterion.

Uranium Fuel Cycle Standards—40 CFR 190

The U.S. Environmental Protection Agency, acting under AE Act authority transferred to it by Reorganization Plan No. 3 of 1970, issued ALARA-based standards for all operations of the uranium fuel cycle used for domestic production of electricity. These standards, which were based on a review of radiation risks for each operation in the fuel cycle and control costs, were designed as cost-effective limits stated in terms of exposure of actual members of the general public. Two unique features of the standards were (1) to limit the releases of specified long-lived radionuclides (I-129, Kr-85, and TRUs with half-lives greater than 5 years) based on the amount (in Gwe-y) of electricity generated; and (2) to allow the numerical limits to be exceeded for justified conditions (e.g., a sustainable power supply) if an analysis is performed and made a matter of public record and operations are returned to levels in accord with the standards as soon as feasible.

The numerical standards in 40 CFR 190 were stated for any member of the public (an actual person) due to operations in the uranium fuel cycle as a whole as

   $\leq$ 25 mrem/y to the whole body;
   $\leq$ 75 mrem/y to the thyroid;
   $\leq$ 25 mrem/y to any given organ;
   $\leq$ 50,000 Ci for ^{85}Kr per GW(e) $\cdot$ y of electricity;
   $\leq$ 5 mCi ^{129}I per GW(e) $\cdot$ y of electricity; and
   $\leq$ 0.5 mCi per GW(e) $\cdot$ y of alpha-emitting TRUs with half-lives greater than 5 years.

Although the 25 mrem standard might appear to be inconsistent with NRC's design objective of 5 mrem/y for reactors, the 40 CFR 190 standard is for operations of all facilities in the cycle, including reactors. It thus provides a practicable standard within which NRC can regulate power plants with a zone of flexibility. Even though the 40 CFR 190 standard was specified for just UFC operations, it has also been applied broadly to other situations; e.g., waste disposal operations.

STANDARDS FOR RADIONUCLIDES IN AIR—40 CFR 60

A National Emission Standard for Hazardous Air Pollutants (NESHAP) has been issued by EPA for radionuclide emissions to ambient air at 10 mrem/y. This NESHAP standard, which corresponds to a lifetime excess cancer risk of approximately 2×10^{-4}, presumes that more stringent standards would produce only marginal risk reductions, and the costs associated with a lower ambient standard would be high and may not be feasible using available technology.

RADIONUCLIDES IN DRINKING WATER

The Safe Drinking Water Act (1973) required EPA to set standards for hazardous substances in public drinking water sources, including radionuclides and U and Ra in particular. A public drinking water supply is one with 25 or more connections, which was intended to exempt those who obtained water from privately owned wells but not a municipality that used groundwater wells as a source of public drinking water. The limits for radionuclides in public drinking water supplies are

≤ 5 pCi/L for ^{226}Ra
≤ 15 pCi/L for gross alpha emitters
≤ 4 mrem/y for other radionuclides.

The 4 mrem/y standard represents an interesting bureaucratic tradeoff. A value of 5 mrem/y was proposed and supported by EPA's Offices of Radiation Programs and Drinking Water, respectively, as a cost-effective standard that could be met with available treatment systems; i.e., a prudent public health approach. During agency review, however, an administrative official recommended that the standard be 3 mrem/y to provide a safety cushion: a compromise value of 4 mrem/y was settled upon so that the standard could be issued. It is, however, an ALARA-based value despite many tendencies in recent years to use it as an upper limit of risk by those who do not know (or remember) its history.

Another aspect of the drinking water standard has developed recently for dissolved radon. EPA, adhering to a 10^{-4} to 10^{-6} risk range criterion that has evolved from the Superfund program, first proposed and has since adopted a maximum concentration limit (MCL) of 300 pCi/L of radon in drinking water. Although EPA science advisors pointed out that the significant radon risk is not through drinking water but airborne progeny in homes or other occupied structures, the agency took the position that it was required to regulate all materials in public drinking water, including natural constituents, and that its policy of limiting lifetime environmental and public health risks to 10^{-4} to 10^{-6} would make a level different from 300 pCi/L technically inconsistent with EPA policy (it is not required to regulate indoor radon even though lifetime risks associated with the EPA guideline are $1 - 3 \times 10^{-2}$, well above its preferred risk range).

RADON GUIDES

EPA, in response to the requirements of the Radon Control Act, developed guides for testing homes for radon and remediation of those that contain elevated radon levels. Radon is not regulated in the traditional sense because it is a naturally occurring element and regulatory programs typically do not apply to homeowners in natural settings. None the less, radon is of sufficient public health concern that guidelines are needed for testing and remediating homes, especially since high levels can occur in geographic areas where significant geological outcroppings containing uranium/radium occur, e.g., in the vicinity of the Reading Prong in Pennsylvania and phosphate deposits in Florida.

Lung cancer risks have been shown to be quite significant for uranium miners exposed to radon and its progeny, and other instances have also come to light. As shown in Table 13-3, lifetime exposure to indoor radon at 4 pCi/L in air for typical individuals is estimated to have a risk of $1 - 3 \times 10^{-2}$. However, since ambient outdoor levels are about 0.5–1.0 pCi/L and indoor levels average 1.0–2.0 pCi/L, EPA has determined, based on feasibility and practicality, that a screening guide of 4 pCi/L would identify most of the homes that could be practically remediated. The value is quite controversial primarily because it has been used as a go/no-go safety standard especially in real estate transactions.

The 4pCi/L radon guide is closely linked to an earlier EPA program of evaluating Florida homes located in reclaimed phosphate mining areas, most of which used slab-on-grade construction (i.e., without basements). A reasonable correlation was found between indoor radon levels and soil concentrations such that soils that contained < 5 pCi/g of Ra-226 generally yielded indoor radon levels in the homes that were less than 4 pCi/L. Evaluations of homes built on soils containing uranium mill tailings in Grand Junction, CO, also indicated that if external gamma readings were $\leq 20 \, \mu R/h$, the radiation exposure of occupants would be ≤ 500 mrem/y. The 5 pCi/g and 20 $\mu R/h$ values have been used as reference points for decisions on lands containing naturally occurring radioactive material (NORM) and have been incorporated in some regulations for related circumstances.

URANIUM MILL TAILINGS

Standards (10 CFR 192) developed under Title I of The Uranium Mill Tailings Radiation Control Act (UMTRCA) apply to inactive uranium milling sites and vicinity properties; however, they have also been used as applicable, relevant, and appropriate requirements (ARARs) for Superfund decisions on remediation of sites containing NORM materials. The UMTRCA standards limit the concentration of radium-226 and 228 within 15 cm of the surface to no more than 5 pCi/g above background levels and to an average of 15 pCi/g of radium below 15 cm. These values are supplemented by two additional requirements: (1) radon progeny concentrations for remediation designs are limited to 0.02 Working Levels (WLs) but justified exceptions are allowed up to 0.03 WL; and (2) gamma radiation levels are required to be $\leq 20\,\mu R/h$ above background.

The UMTRCA regulations also require site remediation measures that are designed to be effective for 200 to 1,000 years and which provide assurance that releases of radon-222 to the atmosphere from residual radioactive material will not exceed an average release rate of $20\,pCi/m^2/s$, or increase the annual average concentration of radon-222 in the air at or above any location outside the disposal site by more than 0.5 pCi/L (40CFR192.02). Computational models, theories, and prevalent expert judgment may be used to determine whether a control system design will satisfy the standard.

NATURALLY OCCURRING OR ACCELERATOR-PRODUCED RADIOACTIVE MATERIALS (NARM)

The AE Act does not apply to naturally occurring or accelerator-produced radioactive materials; thus, they are controlled by the states and other federal statutes. Important subsets of NARM are naturally occurring radioactive material (NORM) and technologically enhanced NORM (TENORM), both of which are generally large volumes of soil or debris containing elevated levels of naturally occurring radioactive materials from the mining and processing of mineral-bearing ores (e.g., uranium, phosphate, aluminum, copper, titanium, etc.). The huge piles of phosphogypsum wastes at phosphate plants are an example of such high-bulk technologically enhanced residues. The most usual constituent is radium which, in addition to potential gamma exposure, can produce elevated radon levels; it can also accumulate in scales on the inside of pipes used in oil and gas operations and in sludges, resins, or charcoal used to process groundwater for public and other uses. These process materials are usually disposed in local landfills which may, under some circumstances, be considered NORM sites.

NARM materials have an interesting control history since neither AEC nor NRC have had control over them, and state programs, which are based on broad public health authority, have been irregular and in some cases non-existent. Recent public concerns over low-level and high-level AE Act wastes have provided an impetus for states to deal with NARM materials presumably because they are radioactive and have the potential to expose members of the public to radiation. It was perhaps this perspective that caused Congress to pass UMTRCA and bring legacy waste from Manhattan Project sites under federal responsibility even though the naturally occurring materials they contained appeared outside the AE Act. Regulatory authority for NARM, NORM, and TENORM has been extended considerably by RCRA and CERCLA and the Superfund Act Reauthorization Amendments (SARA) of 1986; the Denver (CO) radium sites are a good example of actions taken by EPA and the states under Superfund (see below).

SITE CLEANUP STANDARDS/CRITERIA

Few radiation protection standards developed under current federal statutes expressly apply to the cleanup of radioactively contaminated sites although both ICRP and NCRP have addressed aspects of the issue. For radioactive residues from previous events, ICRP states that *the need for and extent of remedial action has to be judged by comparing the benefit of the reductions in dose with the detriment of the remedial work, including that due to doses incurred in the remedial work*. The NCRP has recommended [NCRP Report 116 (1993)] that all exposures of the public due to man-made sources be justified (no exposures without expectation of benefit) and optimized (ALARA). It also recommends that if operations at a single site could potentially expose members of the public to more than 25% of the limit, the operator should assure that the dose from all man-made sources not exceed 100 mrem (1 mSv) per year.

EPA regulatory programs have played a significant role in site cleanup practices since about 1970; these are carried out under authority contained in the AE Act; the Comprehensive Emergency Response, Compensation, and Liability Act (CERCLA); the Resource Conservation and Recovery Act (RCRA); the Safe Drinking Water Act; the Toxic Substances Control Act (TSCA); the Clean Air Act; the Uranium Mill Tailings Radiation Control Act (UMTRCA); and other statutes.

The "Superfund" Law or CERCLA (Comprehensive Emergency Response, Compensation, and Liability Act of 1980) is widely applicable to contaminated sites and several sites contaminated with radionuclides have been listed on the National Priorities List (NPL). Remediation of sites on the NPL is done in accordance with nine criteria contained in the National Oil and Hazardous Substances Contingency Plan (NCP) as required under CERCLA (40 CFR Part 300):

1. **Overall protection of human health and environment** by use of treatment, engineering controls, or institutional controls to provide adequate protection and reduction of risks through relevant exposure pathways.

2. **Compliance with applicable or relevant and appropriate requirements (ARARs)**, including substantive regulatory standards.

3. **Long-term effectiveness and permanence** to maintain reliable protection of human health and the environment over time once cleanup goals have been met.

4. **Reduction of toxicity, mobility, or volume of the contaminants**, with a statutory (CERCLA) preference for on-site remedies which include a treatment component and its anticipated performance.

5. **Short-term effectiveness and timeliness** in protecting human health and the environment and potential adverse impacts that may occur during the construction and implementation period.

6. **Implementability,** including the availability of necessary materials and services.

7. **Cost**, including capital construction as well as operation and maintenance.

8. **Community Acceptance.**

9. **State Acceptance.**

Criteria 1 and 2 are threshold criteria which must be attained by the selected remedial action; criteria 3, 4, 5, 6, and 7 are balancing criteria; and 8 and 9 are modifying criteria. Residual risk levels due to actions based on these criteria are therefore highly varied because each circumstance is so different in terms of sources, the availability and feasibility of controls, and practical matters, including costs.

The NRC also controls licensee operations (10 CFR 20) such that doses to a member of the public are ALARA within an upper limit of 100 mrem/y with justified short-term activities up to 500 mrem in a given year. It has proposed a decommissioning criterion of 25 millirem/y plus ALARA, but has no specific criteria on when residual contamination is ALARA. Although the NRC regulations for various pathways (groundwater, soil, and buildings) vary, the general practice has been and will likely continue to be to apply cost-effective methods to reduce doses to levels below the regulatory requirements; i.e. ALARA is, for all practical purposes, the NRC cleanup standard, and the numerical limits are upper levels not to be exceeded.

Institutional Controls such as deed restrictions on land and water use are not specifically addressed in the NCP; however, they may be used if more active measures are not feasible. Superfund-financed remedial actions also cannot begin without assurances from the state that it will ensure that institutional controls implemented as part of the remedial action are in place and reliable and will remain in place for at least 30 years after initiation of a remedy.

EPA has generally considered cleanup to be protective if it results in a lifetime excess cancer incidence risk range of between 10^{-4} and 10^{-6} (for carcinogens) and a hazard index of less than 1 for noncarcinogens. There is no single set of regulations and guidelines prescribing the cleanup of sites

Table 10-2. Action Levels and Bases for Denver Radium Site Remediation.

Contaminant	Action Level	Comments
^{226}Ra – surface soils ($<$ 5 cm)	5 pCi/g above background	Averaged over 100 m^2 (40 CFR 192)
^{226}Ra – subsurface (5–15 cm)	15 pCi/g above background	Averaged over 100 m^2 (40 CFR 192)
^{230}Th in subsurface soils	42 pCi/g	Limits ingrowth of ^{226}Ra to 15 pCi/g due to human enhancement; if ^{226}Ra is present, ^{230}Th limit is reduced
Gamma Radiation	20 μR/h above background	Based on 40 CFR 192, Subpart B for identifying areas of contamination.
Natural U	75 pCi/g	Limits U toxicity
Arsenic	160 mg/kg	Health-based values for a trespasser scenario to restrict future uses of surface areas.
Selenium	490 mg/kg	
Lead	540 mg/kg	

containing radioactive material, and ARARs commonly vary from site to site because they depend on a site-specific analysis. The UMTRCA cleanup standard is typically used as an ARAR (see Table 10-2 for Denver Radium Sites) for naturally occurring radioactive materials and the records of decision (RODs) do not discuss the rationales for the risk levels achieved; however, most sites have achieved a risk range of 10^{-2} to 10^{-4} and these have been deemed acceptable.

EPA has also proposed that site remediations should ensure that water that is a current or potential source of drinking water not exceed maximum concentration limits (MCLs) i.e. 4 mrem/y for beta particles, 15 pCi/L for gross alpha, 5 pCi/L for Radium-228, and 5 pCi/L for Radium-226. Revision of the MCLs for Radium-228 and Radium-226 to 20 pCi/L was proposed by EPA in 1991, but they have not yet been adopted. Although UMTRCA and, in some circumstances CERCLA, has been used for the cleanup of NPL facilities, exposures of the general public are limited to ALARA levels below a limit of 100 mrem/y, and pathway dose limits of 10 mrem/y for air, 4 mrem/y for drinking water, and 25 mrem/y for waste management are to be met. For contaminated sites/facilities that are not on the NPL, a corrective action is required pursuant to RCRA (at facilities with RCRA permitted units) or other applicable authorities. The RCRA program requires a corrective measures study, which is similar to a CERCLA remedial investigation/feasibility study (RI/FS), as a basis for the formal selection of a remedy.

The Denver Radium Sites, which included some 31 separate locations, occurred because no regulations were in place in the early years of the 20th century to control radium production that was done by the National Radium Institute (NRI) located in Denver, CO. All operations, except that of the Shattuck Chemical Co., ceased in the 1920s, and the radioactive tailings and unprocessed ore became fill and foundation material and were largely forgotten. The sites were rediscovered in the late 1970s and placed as a single site on the Superfund National Priorities List with 11 operable units for remedial investigation and feasibility studies (RIFs), public meetings, and records of decision. All of the sites, with the exception of Shattuck Chemical, were excavated and backfilled with replacement soil, and some 230,000 cubic yards of contaminated soil were disposed at the Envirocare facility in Clive, UT. Cleanup was based primarily on EPA's UMTRCA standards (40 CFR 192 – Subpart B) as applicable, relevant, and appropriate requirements (ARARs) as follows:

The remedy for the Shattuck Chemical site was onsite stabilization of 50,000 cubic yards with cement and fly ash, which was determined to be the most appropriate because of the statutory preference of CERCLA for onsite remedies, the statutory preference for remedies involving treatment, and the high costs for permanent offsite disposal of contaminated soil. Other contaminated material underneath roadways was left in place; however, the city of Denver was required to develop a health and safety protocol for street and utility workers who might need to dig in the area and to assess and properly dispose of any contaminated material at that time. Remediation of parts of the Denver Radium Sites thus relied on environmental stewardship as an effective measure and included provisions for institutional controls, maintenance, and monitoring. Also included are deed restrictions and annotations denoting that the property is a dedicated waste disposal site. The notations also restrict excavation, construction, groundwater use, and agricultural use, and provide for post-closure maintenance of the cap and cover and for groundwater monitoring to detect releases from the site.

The Record of Decision for the Shattuck Chemical unit stated that "... the location of the site in a metropolitan area should not dictate selection of an offsite remedy ..." The city of Denver sued EPA over the selected remedy in federal court as provided in CERCLA, but EPA's decision was upheld. The decision and remedy are, however, likely to be revisited (Sowinski, personal communication)—a circumstance, if it occurs, which has interesting policy overtones about radioactive waste materials in urban areas.

A *de facto* **EPA Cleanup Standard** of 15 mrem/y above background radiation levels has evolved from EPA remedial actions under CERCLA and RCRA, and the agency is considering it as a requirement for at least 1,000 years following the completion of cleanup activities. This *de facto* standard for unrestricted site use corresponds, by EPA's evaluation of risk data, to an excess lifetime fatal cancer risk of 3×10^{-4} and attempts to ensure that actual exposures are a small fraction of the broader 100 mrem guideline recommended by NCRP and ICRP for exposures of the public from all sources. It is consistent with an EPA policy that is apparent in all of its regulations to achieve a lifetime risk level of 10^{-4} or less for the maximum exposed individual.

EPA is also considering a requirement that a proposed remedy demonstrate that exposures of members of the public would be limited to 75 mrem/y even if all active or effective institutional controls fail, which is an additional requirement to assure that the effectiveness of controls is projected well into the future, especially for long-lived radionuclides. This additional requirement for site cleanups corresponds to a lifetime excess cancer risk of 1.4×10^{-3} if it should persist after controls fail (i.e., no additional action is taken by future people), but is still consistent with the 100 mrem/y limit recommended by NCRP, ICRP, and others. It was derived by subtracting from 100 mrem the 25 mrem allowed for uranium fuel cycle facilities because it is extremely unlikely that there would be several sources of man-made radiation within the vicinity of a single site. The requirement attempts to strike a balance between protecting the public, should institutional controls fail, and imposing additional standards in those cases when availability of institutional controls justify leaving higher concentrations of radioactive material at a given site, e.g., release of a site for industrial/commercial use. Without the provision, a site would be required to either be cleaned up to a level for unrestricted use or not to be released at all.

SUMMARY

Recent radiation protection guides for occupational exposures recommended by the ICRP and NCRP and incorporated in US standards and regulations have been based on an explicit quantitative level of acceptable risk that reflects a careful review of the major radiation bioeffects studies in humans (Ch 6) and, to some extent, similar data from animal studies. Control levels have been stated well below those for which effects in humans have been observed and at risk levels that are indistinguishable from the background incidence of cancer and genetic effects in exposed populations. National and international groups recommend that exposure of the general public should be limited to a total effective dose equivalent of 100 millirems/y; that no single source should use the entire guide; that exposures for specific sources be ALARA below the 100 mrem/y guide; and that temporary exposures of 500 mrem in a given year could be allowed, but only for circumstances justified to the controlling authority. These considerations have been adopted for a wide range of activities that have the potential to expose the public to radiation.

No specific criteria exist for cleanup of contaminated sites; thus, applicable, relevant, and appropriate requirements (ARARs) are often used for sites on the Superfund national priorities list (NPL). Cleanup remedies and corrective actions have been and currently are based on site-specific remedial investigation and feasibility studies (RIFS), although guidance is lacking on what constitutes a cost effective remedy. EPA generally attempts to achieve a 10^{-4} to 10^{-6} lifetime risk range in its generally applicable environmental standards and other regulations and programs, which has been interpreted as 15 millirem per year from all pathways for unrestricted site use. It also requires that drinking water MCLs be met and allows restricted use with active controls if a protection level of 75 millirem/y for future conditions can be assured even if controls fail.

PROBLEMS

1 Compare the carcinogenic risk for workers protected under guides issued in 1925 and those for current US workers.

2 Determine: (a) the difference in the radiation risk level associated with EPA's policy of 15 mrem/y for exposure of the general public vs the 25 mrem/y regulation used by NRC for nuclear facilities; (b) the net change in risk by reducing the dose level for a member of the public below the 100 mrem/y guide recommend by ICRP; and (c) EPA's rationale for its 15 mrem/y level and the role of cost.

3 For a group of 1000 persons: (a) determine the health-risk benefit of a 4 mrem/y standard vs the 5 mrem/y standard originally proposed; and (b) construct an argument based on risk/cost for or against the change.

4 A lifetime risk objective of 10^{-4} is evident in various radiation policies. What annual dose level corresponds to such a risk level and how is it affected by the assumed period of exposure?

5 Several radiation protection regulations are based on the ALARA concept. Explain the mechanism and rationale of balancing risks and costs to arrive at ALARA standards for specific circumstances.

11

RADIATION PROTECTION PROGRAMS

"She scorned the precautions she so severely imposed on her students: to manipulate tubes of radioactivity with pincers, never to touch unguarded tubes, to use leaden 'bucklers'...; for thirty-five years Mme. Curie had handled radium and breathed (its emanation)... with a slight deterioration in the blood, annoying and painful burns on the hands...."

– Eve Curie, 1937

Radiation protection programs are designed and executed to meet radiation protection standards and the regulations that implement them, both of which focus on limiting radiation risks. Such programs not only assure that radiation doses (and risks) are maintained within established limits, but are, as a prudent measure, conducted to maintain radiation exposures as low as reasonably achievable (i.e., ALARA) below regulatory limits. Radiation levels and exposures for most radiation activities are, as a consequence of this two-pronged philosophy, kept at a fraction, and many times a very small fraction, of regulatory limits.

Radiation-induced health effects have not been observed at or below current standards; therefore, many believe that the exposure guides that have been developed should be considered protective and "safe" and that this is a sufficient basis for radiation protection of workers and the public. This, however, is not the policy used in radiation protection programs. Rather, radiation protection regulations are treated as upper limits of dose that are not to be exceeded without careful consideration, and two additional requirements are expected to be achieved: all exposures are to be justified and actual doses are to be maintained as far below regulatory limits as reasonably achievable.

Many of the elements of radiation protection programs rely on the physics of radiation production, its behavior in media, and energy deposition interactions that provide the basis for radiation dose determination and its measurement. Most regulations only state what must be done, and the basis for a given requirement is not always clear. Whereas this circumstance satisfies a strict regulatory approach, understanding the performance goal can be quite helpful in formulating an optimal control program for a given set of activities.

REGULATORY AUTHORITY FOR RADIATION AND RADIOACTIVE MATERIALS

Prior to World War II, x rays and naturally occurring radioactive materials (e.g., radium) were the major sources of radiation exposure, and these were generally unregulated. Although public health agencies had broad authority for health, specific regulatory authority for x rays and naturally occurring radioactive materials was virtually nonexistent and standards that existed were based on recommendations developed by professional groups rooted largely in the medical profession. The International Commission on Radiological Protection (ICRP) and the National Council on Radiation Protection and Measurements (NCRP), its US counterpart, and their predecessor organizations based recommendations on good practice and professional judgment, and implementation was voluntary. This approach of professional practice was not inappropriate because exposures that occurred were to small groups of specialists in such professions.

The discovery of nuclear fission, the production and use of special nuclear materials, and the numerous radioactive byproducts of fission led to expansive and stringent programs in the US to control these materials. The Atomic Energy Acts of 1946 and 1954 and subsequent amendments gave sweeping powers to the Federal government to own and regulate these unique materials, commonly known as Atomic Energy Act (AE Act) materials. X rays and radium and other naturally occurring radioactive material, except uranium and thorium ores containing more than 0.05% uranium or thorium (i.e., source materials), are not regulated by the federal government under the AE Act, but are regulated to varying degrees by the states under broad authorities to protect public health. In general, no private person could possess or use AE Act materials until after the 1954 amendments, but even then only under conditions of a license issued pursuant to the AE Act.

An extraordinary aspect of radiation protection, in contrast to other potentially hazardous materials, is the dominant role exercised by the US Nuclear Regulatory Commission (NRC) through its licensing and inspection authority. These authorities were carried out from 1946 to 1974 by the US Atomic Energy Commission (AEC) which was reorganized in 1974 to create the NRC and a separate agency which became the US Department of Energy (DOE) in 1977. The Atomic Energy Act requires the control of AE Act materials in the interest of "public health and safety," but allowed the AEC (and now the NRC) broad discretion in doing so through its various licensing and inspection programs or through state agreements (authorized by the 1954 AE Act amendments) where a state assumes regulatory responsibilities for AE Act materials within its borders. States also control naturally occurring radioactive materials (NORM) and x rays if (as most do) they have state legislative authority to do so. Control of AE Act materials by other agencies is and has been essentially only an extension of NRC's broad responsibility for the materials. Perhaps because of this government-dominated beginning for AE Act materials, the NRC, even though it is a spinoff agency, conducts very strict licensing and inspection programs that almost begrudgingly allow users access to AE Act materials, even in relatively small amounts.

Agreement States, of which there are 31, must carry out a program "equivalent in effect" to that of the NRC which retains ultimate responsibility under the AE Act; in non-agreement states the NRC is solely responsible for control of AE Act materials. The choice to conduct or not conduct a state program for federally controlled AE Act materials varies widely; however, those that choose to do so appear to believe it is worth the investment of state resources to provide service to its citizens and to assure that state interests are considered in radiation-related activities. Nonetheless an agreement state program, though not necessarily identical, is in effect a federal program subject to initial and continuing approval by the NRC. Such agreements generally cover licensing and inspection programs for possession and use of byproduct AE Act materials; they do not cover regulation of nuclear reactors, high-level waste disposal, or complex nuclear facilities. Congress clearly preempted the use and control of Atomic Energy Act Materials to the Federal government in all respects, and the US Supreme Court has consistently set aside state attempts to exercise state control over nuclear reactors, disposal of high-level wastes, defense sites, etc.

The US Environmental Protection Agency (USEPA), which was formed in 1970, sets generally applicable environmental radiation standards and carries out federal radiation guidance through the President. In an effort to assure that all government programs used the same basic standards, the Federal Radiation Council (FRC) was established in 1959 to advise the President who in turn issued standards for use by all federal agencies in setting regulations and in their work with the states. These federal guidance responsibilities, which were encoded in the AE Act (42 USC 2021h), were transferred to EPA in 1970. Although EPA has no regulatory authority for AE Act materials, it has broad responsibility to assure that basic environmental standards and Federal guides are carried out in federal regulatory programs. It does, however, have ultimate authority for radionuclides regulated under the Clean Air and Clean Water Acts.

The US Department of Transportation regulates the transportation of radioactive materials under a special agreement with the NRC, which has the ultimate authority and responsibility for the activities. The NRC depends on the US Department of Transportation (DOT) regulatory programs to control the shipment of AE Act materials under DOT regulations that are compatible with those of the International Atomic Energy Agency. In general, DOT requirements are identical in effect to NRC requirements and pertain to quantity limits, radiation limits from packages and vehicles, packaging requirements for types and quantities of waste, and vehicle and operator requirements. The Interstate Commerce Commission, the US Coast Guard, the Federal Aviation Administration, the US Postal Service, and the DOE also have an interest in regulations for the shipment of radioactive substances.

The US Department of Energy conducts energy research and development functions begun initially by the AEC at national laboratories and nuclear weapons facilities operated by contractors. The DOE issues regulations which pertain to its own activities as well as those of its contractors, who are not required to have a license for AE Act materials. Those regulations that apply specifically to radiation protection are based upon federal radiation guidance recommended by the EPA and issued by the President, and generally incorporate the recommendations of the ICRP and NCRP. The DOE has an interesting role in that they must eventually take title, on behalf of the Federal Government, to sites containing radioactive waste once they close and institutional control periods have elapsed.

Regulatory Control of Atomic Energy Act Materials

A person must have a license to possess and use AE Act material, which is not the case for many other potentially hazardous materials or activities. The recipient or user of AE Act materials has been (and is) required to demonstrate in the license application the capability to meet the government's perspective of proper control, especially for workers and the general public. Furthermore, a licensee must assure that any transfer of the material occurs only to another licensee who has authority (specified in its license) to store or dispose of the material or subject it to other uses. This interconnected system of licensees subject to NRC regulations (or equivalent state regulations) is shown schematically in Figure 11-1; it is designed to assure strict control of AE Act materials, however used or transferred, and inspection programs are conducted to assure that each licensee meets NRC regulations and/or conditions specified in the license.

The NRC exercises its broad regulatory authority through various regulations published under Title 10 in the U.S. Code of Federal Regulations (10 CFR). The parts applicable to most users of radioactive materials are

Part 19   Rights of Radiation Workers (applies to all licensees)
Part 20   Radiation Protection Regulations (applies to all licensees)
Part 30   Radioactive Material Licenses (applies to most licensees)
Part 32   General and Special licenses by rule for various products, including exempt quantities and concentrations
Part 35   Licenses for medical uses of AE Act materials and associated radiations

A license from NRC is required for possession of AE Act Materials

Licensee (1)
. Conduct a control program
. Meet regulations for
  - workers
  - discharges to air or water

Licensee (2)
. Conduct a control program
. Meet regulations for
  - workers
  - discharges to air or water

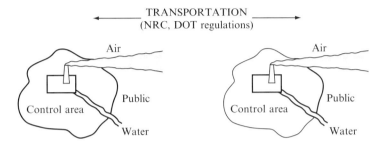

Figure 11-1. Conceptual framework for regulation of AE Act materials.

Part 50   Licensing requirements for nuclear reactors
Part 60   Licensing requirements for high-level radioactive wastes
Part 61   Licensing requirements for management and disposal of low-level radioactive wastes, including waste classification requirements
Part 71   Transportation regulations (keyed to and implemented by DOT regulations detailed in 49 CFR Parts 171 and 172)

Regulations in 10 CFR 30 and 10 CFR 35 are of particular relevance since they require that no person shall manufacture, produce, transfer, receive, acquire, own, possess, or use byproduct material (i.e., radioactive due to association with nuclear fission) except as authorized in a specific or general license issued by the NRC or an agreement state. The regulations in 10 CFR 30 prescribe requirements and provisions for uses of byproduct material, and 10 CFR 35 provides similar conditions for byproduct materials used in medicine including requirements for issuance of specific licenses authorizing the medical use of specified materials. The requirements and provisions of 10 CFR 30 and 35 are in addition to, and not a substitution for, other requirements and provisions under Title 10 of the Code of Federal Regulations, in particular Parts 19 and 20.

Radioactive material licenses are of two types: specific and broad scope. Each is obtained from the NRC by filing NRC Form 313 (see Figure 11-2), or from an agreement state. Information is required on the radionuclides to be obtained, the intended use, maximum quantities to be possessed at any one time, and a detailed description of the radiation safety program. Also required are listings, along with relevant training and experience, of the person(s)

NRC FORM 313	U. S. NUCLEAR REGULATORY COMMISSION	APPROVED BY OMB: NO. 3150-0120	EXPIRES:08/31/2002
(8-1999) 10 CFR 30, 32, 33 34, 35, 36, 39 and 40 **APPLICATION FOR MATERIAL LICENSE**		Estimated burden per response to comply with this mandatory information collection request 7.4 hours. Submittal of the application is necessary to determine that the applicant is qualified and that adequate procedures exist to protect the public health and safety. Send comments regarding burden estimate to the Records Management Branch (T-6 E6), U.S. Nuclear Regulatory Commission, Washington, DC 20555-0001, or by internet e-mail to bjs1@nrc.gov, and to the Desk Officer, Office of Information and Regulatory Affairs, NEOB-10202, (3150-0120), Office of Management and Budget, Washington, DC 20503. If a means used to impose an information collection does not display a currently valid OMB control number, NRC may not conduct or sponsor, and a person is not required to respond to, the information collection	

INSTRUCTIONS: SEE THE APPROPRIATE LICENSE APPLICATION GUIDE FOR DETAILED INSTRUCTIONS FOR COMPLETING APPLICATION. SEND TWO COPIES OF THE ENTIRE COMPLETED APPLICATION TO THE NRC OFFICE SPECIFIED BELOW.

APPLICATION FOR DISTRIBUTION OF EXEMPT PRODUCTS FILE APPLICATIONS WITH:

DIVISION OF INDUSTRIAL AND MEDICAL NUCLEAR SAFETY
OFFICE OF NUCLEAR MATERIALS SAFETY AND SAFEGUARDS
U.S. NUCLEAR REGULATORY COMMISSION
WASHINGTON, DC 20555-0001

ALL OTHER PERSONS FILE APPLICATIONS AS FOLLOWS:

IF YOU ARE LOCATED IN:

CONNECTICUT, DELAWARE, DISTRICT OF COLUMBIA, MAINE, MARYLAND, MASSACHUSETTS, NEW HAMPSHIRE, NEW JERSEY, NEW YORK, PENNSYLVANIA, RHODE ISLAND, OR VERMONT, SEND APPLICATIONS TO:

LICENSING ASSISTANT SECTION
NUCLEAR MATERIALS SAFETY BRANCH
U.S. NUCLEAR REGULATORY COMMISSION, REGION I
475 ALLENDALE ROAD
KING OF PRUSSIA, PA 19406-1415

ALABAMA, FLORIDA, GEORGIA, KENTUCKY, MISSISSIPPI, NORTH CAROLINA, PUERTO RICO, SOUTH CAROLINA, TENNESSEE, VIRGINIA, VIRGIN ISLANDS, OR WEST VIRGINIA, SEND APPLICATIONS TO:

SAM NUNN ATLANTA FEDERAL CENTER
U. S. NUCLEAR REGULATORY COMMISSION, REGION II
61 FORSYTH STREET, S.W., SUITE 23T85
ATLANTA, GEORGIA 30303-8931

IF YOU ARE LOCATED IN:

ILLINOIS, INDIANA, IOWA, MICHIGAN, MINNESOTA, MISSOURI, OHIO, OR WISCONSIN, SEND APPLICATIONS TO:

MATERIALS LICENSING SECTION
U.S. NUCLEAR REGULATORY COMMISSION, REGION III
801 WARRENVILLE RD
LISLE, IL 60532-4351

ALASKA, ARIZONA, ARKANSAS, CALIFORNIA, COLORADO, HAWAII, IDAHO, KANSAS, LOUISIANA, MONTANA, NEBRASKA, NEVADA, NEW MEXICO, NORTH DAKOTA, OKLAHOMA, OREGON, PACIFIC TRUST TERRITORIES, SOUTH DAKOTA, TEXAS, UTAH, WASHINGTON, OR WYOMING, SEND APPLICATIONS TO:

NUCLEAR MATERIALS LICENSING SECTION
U.S. NUCLEAR REGULATORY COMMISSION, REGION IV
611 RYAN PLAZA DRIVE, SUITE 400
ARLINGTON, TX 76011-8064

PERSONS LOCATED IN AGREEMENT STATES SEND APPLICATIONS TO THE U.S. NUCLEAR REGULATORY COMMISSION ONLY IF THEY WISH TO POSSESS AND USE LICENSED MATERIAL IN STATES SUBJECT TO U.S.NUCLEAR REGULATORY COMMISSION JURISDICTIONS.

1. THIS IS AN APPLICATION FOR *(Check appropriate item)* A. NEW LICENSE B. AMENDMENT TO LICENSE NUMBER _____ C. RENEWAL OF LICENSE NUMBER _____	2. NAME AND MAILING ADDRESS OF APPLICANT *(Include Zip code)*
3. ADDRESS(ES) WHERE LICENSED MATERIAL WILL BE USED OR POSSESSED	4. NAME OF PERSON TO BE CONTACTED ABOUT THIS APPLICATION TELEPHONE NUMBER

SUBMIT ITEMS 5 THROUGH 11 ON 8-1/2 X 11" PAPER. THE TYPE AND SCOPE OF INFORMATION TO BE PROVIDED IS DESCRIBED IN THE LICENSE APPLICATION GUIDE.

5. RADIOACTIVE MATERIAL a. Element and mass number; b. chemical and/or physical form; and c. maximum amount which will be possessed at any one time.	6. PURPOSE(S) FOR WHICH LICENSED MATERIAL WILL BE USED
7. INDIVIDUAL(S) RESPONSIBLE FOR RADIATION SAFETY PROGRAM AND THEIR TRAINING EXPERIENCE.	8. TRAINING FOR INDIVIDUALS WORKING IN OR FREQUENTING RESTRICTED AREAS
9. FACILITIES AND EQUIPMENT.	10. RADIATION SAFETY PROGRAM.
11. WASTE MANAGEMENT.	12. LICENSEE FEES *(See 10 CFR 170 and Section 170.31)* FEE CATEGORY / AMOUNT ENCLOSED $

13. CERTIFICATION. *(Must be completed by applicant)* THE APPLICANT UNDERSTANDS THAT ALL STATEMENTS AND REPRESENTATIONS MADE IN THIS APPLICATION ARE BINDING UPON THE APPLICANT.

THE APPLICANT AND ANY OFFICIAL EXECUTING THIS CERTIFICATION ON BEHALF OF THE APPLICANT, NAMED IN ITEM 2, CERTIFY THAT THIS APPLICATION IS PREPARED IN CONFORMITY WITH TITLE 10, CODE OF FEDERAL REGULATIONS, PARTS 30, 32, 33, 34, 35, 36, 39 AND 40, AND THAT ALL INFORMATION CONTAINED HEREIN IS TRUE AND CORRECT TO THE BEST OF THEIR KNOWLEDGE AND BELIEF.

WARNING: 18 U.S.C. SECTION 1001 ACT OF JUNE 25, 1948 62 STAT. 749 MAKES IT A CRIMINAL OFFENSE TO MAKE A WILLFULLY FALSE STATEMENT OR REPRESENTATION TO ANY DEPARTMENT OR AGENCY OF THE UNITED STATES AS TO ANY MATTER WITHIN ITS JURISDICTION.

CERTIFYING OFFICER -- TYPED/PRINTED NAME AND TITLE	SIGNATURE	DATE

FOR NRC USE ONLY

TYPE OF FEE	FEE LOG	FEE CATEGORY	AMOUNT RECEIVED	CHECK NUMBER	COMMENTS
			$		
APPROVED BY				DATE	

NRC FORM 313(8-1999)   PRINTED ON RECYCLED PAPAER

Figure 11-2. US NRC Form 313 for radioactive materials license application.

responsible for the radiation safety program and the individual users, radiation detection instruments available, personnel monitoring procedures to be used (including bioassay), laboratory facilities, handling equipment, and waste management procedures. The license authorizes possession and use only at the locations specified in the application, and radioactive material may not be transferred except as authorized by the regulations and the license. Compliance with the terms of the NRC regulations and the conditions of the license is the responsibility of the management of the institution.

The radiation safety program is implemented by a Radiation Safety Officer (RSO) whose qualifications must be assured in order to obtain and maintain the license. The duties of the RSO include preparing regulations, providing advice and training on matters of radiation protection, maintaining a system of accountability for all radioactive material from procurement to disposal, inspecting work spaces and handling procedures, determining personnel radiation exposures, monitoring environmental radiation levels, and instituting corrective action in the event of accidents or emergencies. The radiation safety program must include, to the extent practicable, procedures and engineering controls to ensure that doses to workers and members of the public are maintained ALARA.

Institutions engaged in research and development (R&D) or medical uses may obtain a license of broad scope (10 CFR 33) if requirements on staffing, experience, and control processes are met. A broad scope license does not name the individual users nor limit radionuclides to specific uses, but allows the institution to establish a Radiation Safety Committee to review applications and to authorize specified uses of radionuclides without special application to the regulatory agency. The Radiation Safety Committee is to include the Radiation Safety Officer, a representative of management, and persons trained and experienced in the safe use of radioactive materials. Users are obliged to ensure that individuals under their supervision have obtained the requisite training and indoctrination to ensure safe working habits, compliance with regulations, prevention of exposure to others or contamination of the surroundings, and that female workers receive specific instruction about prenatal exposure risks to the developing embryo and fetus. Failure to carry out these basic responsibilities with respect to supervision and training have been cited in radiation lawsuits as indicative of negligence. The ultimate responsibility for compliance rests with the institution's management which is also required to review the radiation protection program at least annually.

The NRC also issues a general license to any physician, clinical laboratory, or hospital for use of small quantities of specific radionuclides in pre-packaged units for *in vitro* clinical or laboratory tests (10 CFR 31.11). These and their maxima include ^{14}C, ^{75}Se, ^{125}I and ^{131}I (10 μCi each); ^{3}H (50 μCi); ^{59}Fe (20 μCi); and mock-^{125}I reference or calibration sources in units not exceeding 0.05 μCi^{129}I and 0.0005 μCi^{241}Am. The use of these materials is also exempt from the requirements of Part 20, except for mock-^{125}I (actually ^{129}I or ^{241}Am) sources, which are still subject to waste disposal restrictions (10 CFR 20.301).

Authorized users become directly responsible for complying with all regulations and for the safe use of radionuclides by other investigators or technicians under their direct supervision. The possession and use of radionuclides is limited to the quantities and for the purposes specified in the authorization.

RADIATION PROTECTION PROGRAMS/DOSE LIMITS

A major component of a radiation protection program is compliance with applicable regulations, and in fact many such programs are shaped by the regulations themselves. These are contained primarily in 10 CFR 19 and 20, first issued in 1954 after the Atomic Energy Act was amended to allow greater access to radioactive materials for research and industrial applications; these regulations have been revised and/or amended several times since 1954 with the most recent major revision taking effect on January 1, 1994. The purpose of the 1994 revision was to incorporate risk-based standards recommended by key national and international organizations, which represented a substantial change in the underlying philosophy for radiation protection regulations even though the control limits remained essentially the same.

The 10 CFR Part 19 Regulations establish the rights of workers to have information on the amount of radiation exposure they receive and associated risks. Licensees are required to document radiation exposures for each worker and to provide an annual report of accumulated dose whether regulated or not, and to provide interim data upon request. The regulations in Part 19 also establish requirements for notices, instructions, and reports by licensees to individuals participating in licensed activities, and options available to these individuals in connection with NRC inspections of licensees to ascertain compliance with regulations, orders, and licenses regarding radiological working conditions. The regulations in Part 19 also establish the rights of individuals during interviews and the responsibilities of the NRC with respect to workers during inspections and investigations.

The 10 CFR Part 20 Regulations outline the major requirements that a radiation protection program must meet. These requirements are detailed under the major sections of 10 CFR 20 which are listed in Table 11-1. They apply to all licensees and are designed to assure that a comprehensive program is conducted to protect workers and the public in accordance with radiation exposure standards for each, including several special categories. Licensees are expected to assure that users and supervised workers have basic training in radiological health and safety, radioactivity measurements and instruments, radiation calculations, and biological effects. These training requirements can generally be met for laboratory uses of radionuclides with about 20 hours of instruction.

Occupational Radiation Dose Limits

Licensees are required to conduct operations in restricted areas (i.e., under the control of the licensee) such that radiation doses to occupationally exposed persons (10 CFR 20.1201) meet the following limits:

- Stochastic effects $\leq$ 5 rem/y (0.05 Sv) Total Effective Dose Equivalent (TEDE)
- Nonstochastic $\leq$ 50 rem/y (0.5 Sv) to any tissue

Table 11-1. Major Sections of 10 CFR Part 20 Radiation Protection Regulations.

Subpart A	–	General Provisions (scope, definitions, units)
Subpart B	–	Radiation Protection Programs (license conditions, elements)
Subpart C	–	Occupational Dose Limits (limits for adults; summation of external and internal doses; airborne radioactive material; internal exposure; planned special exposures; dose limits for minors; dose to an embryo/fetus).
Subpart D	–	Radiation Dose Limits for Individual Members of the Public
Subpart E	–	[Reserved]
Subpart F	–	Surveys and Monitoring (for individuals; external and internal doses)
Subpart G	–	Control of Restricted Areas (high radiation areas; very high radiation areas)
Subpart H	–	Respiratory Protection (process or other engineering controls; requirements for and restrictions on uses of respiratory protection equipment).
Subpart I	–	Storage and Control of Licensed Material (security).
Subpart J	–	Precautionary Procedures (signs, posting, labeling, receipt of packages).
Subpart K	–	Waste Disposal (procedures, sewerage, incineration, ^{14}C and ^{3}H wastes).
Subpart L	–	Records (programs, surveys, doses, monitoring, wastes).
Subpart M	–	Reports (theft or loss, incidents, doses above limits).
Subpart N	–	Exemptions and Additional Requirements
Subpart O	–	Enforcement (violations, criminal penalties).

Appendix A: Protection factors for respirators
Appendix B: ALIs and DACs; concentrations in effluents and sewerage
Appendix C: Quantities requiring labeling.
Appendix D: US NRC Offices
Appendix E: [Reserved]
Appendix F: Waste disposal and manifest requirements.

- Extremities $\leq$ 50 rem/y (0.5 Sv) to hands and forearms or feet, ankles, and legs below the knee
- Lens of the eye $\leq$ 15 rem/y (0.15 Sv)

Summation of external and internal doses (10 CFR 20.1202) is required if both are subject to monitoring requirements [§§ 20.1502(a) and (b)], which "kick in" if doses could exceed 10% of limits; however, if the licensee is required to monitor only under one or the other [§ 20.1502(a) or § 20.1502(b)], then summation is not required. The summation requirement applies only to effective whole-body dose equivalents; the lens of the eye, the skin, and the extremities are not included in the summation but are subject to separate limits.

Radiation Dose Limits for the Public

Licensees are also required to conduct operations such that radiation doses in unrestricted areas (10 CFR 20.1301) are limited to

- Member of the public $\leq$ 0.1 rem/y (1 mSv/y) TEDE
- Dose rate $\leq$ 2 mrem in any one hour
- Special exposure $\leq$ 0.5 rem/y

If the licensee permits members of the public (i.e., they are not workers) to have access to controlled areas, the limits for members of the public (and not occupational limits) continue to apply to those individuals. Over and above these requirements, a licensee that operates a nuclear power plant or other facility in the uranium fuel cycle must also comply with EPA's generally applicable environmental radiation standards (40 CFR Part 190). Which limit doses to a member of the public to 25 mrem/y to the whole body, 75 mrem to the thyroid, and 25 mrem/yr to any other organ.

The limits for both occupational exposure and for members of the public are exclusive of the dose contributions from background radiation, medical uses of radiation and radioactive materials, and materials disposed into sanitary sewerage in accordance with § 20.2003.

COMPLIANCE WITH REGULATIONS

Compliance with radiation protection regulations for both workers and the public is based on radiation dose limits designed to limit radiation risk, the primary basis for radiation control. Doses are physical quantities that can be measured directly or calculated from other basic information, and demonstrating that the total effective dose equivalent (TEDE) to a person is below the dose limit is the overriding consideration in determining *de facto* compliance with regulations, although good practice and ALARA are expected as part of an approved radiation protection program.

External Radiation Dose Compliance is best demonstrated by measuring the delivered dose with film badges, TLDs, or direct reading or electronic dosimeters. Measurements with these devices are considered appropriate for the dose of record if they meet calibration and/or accreditation requirements for the types of radiation (e.g., photons, beta radiation, neutrons, etc.) to which they are exposed. Determination of the internal dose contribution to the total effective dose equivalent (TEDE) requires different methods since the dose delivered is internal to a person and cannot be measured directly. Because of this complexity, the regulations do not require monitoring to determine internal dose unless it is 10% or more of the dose limit. This 10% rule also applies to external dose measurement, but most facilities, as a legal or other precaution, require workers who could potentially be exposed to radiation to wear dosimeters. If, however, either form of exposure is monitored, it is to be recorded, and if both are monitored, even if one or the other or both is within the 10% exclusion, they are to be summed to record the TEDE received. As a practical matter, the 10% exclusion means that only one type of exposure needs to be monitored; only the more complex facilities will need to do both.

Internal Radiation Dose Determination is addressed in the regulations by indirect methods which are generally conservative. Annual Limits of Intake (ALI) of radionuclides that correspond to a committed effective whole-body dose equivalent (CEDE) of 5 rem/y (0.05 Sv/y) or a limiting tissue-specific non-stochastic dose of 50 rem/y (0.5 Sv/y) have been calculated and these are used to demonstrate compliance with applicable dose limits. Any intake of radioactive material in the workplace is usually by inhalation of airborne radioactive material since contaminated food or water is avoidable; therefore, air concentration values have been derived such that a reference adult working 2000 working hours per year would not exceed the ALI. These derived air concentration (DAC) values are obtained by dividing the ALI by 2.4×10^9 mL, the amount of air presumed to be inhaled in one year during light activity. Table 11-2 lists values of effective dose per unit of intake (rem/μCi) for radionuclides commonly encountered and the ALI (μCi), and the DAC (μCi /mL) that if met, will limit internal doses to workers in accordance with 10 CFR 20 requirements.

The ALIs and the corresponding DACs for radionuclides listed in Table 11-2 correspond to a committed effective dose equivalent of 5 rems (stochastic ALI), or a committed dose equivalent of 50 rems to an organ or tissue (non-stochastic ALI). The stochastic ALIs were derived based on ICRP weighting factors to result in a risk, due to irradiation of organs and tissues, comparable to the risk associated with a whole-body deep dose equivalent of 5 rems. The nonstochastic ALIs were derived to avoid nonstochastic effects such as prompt damage to tissue or reduction in organ function. When the derived ALI is based on the stochastic dose limit, only this value is listed in Table 11-2. If, however, it is based on the nonstochastic dose limit to an organ, the organ or tissue to which the limit applies is also shown (the stomach, small intestine, upper large intestine, and lower large intestine are treated as four separate organs) and the ALI corresponding to the stochastic limit is not shown; it can, however, be calculated by dividing 5000 mrem by the effective dose coefficient (mrem/μCi), also listed in Table 11-2 (see Example 11-2b).

A worker that inhales a radionuclide for 2000 h at a concentration equal to 1 DAC will receive an effective dose equivalent of 5 rem (or a 50-rem organ dose in the few instances when nonstochastic effects are limiting). If more than one radionuclide is inhaled over the course of a year, the number of DAC-hours of each is determined and compared to other modes of exposure via the "sum of fractions" technique (see Examples 11-1 and 11-2).

The data in Table 11-2 can be used to either specify the CEDE for known intakes of radionuclides or to determine whether the ALI has been met. A common technique is to determine the air concentration in a workspace and its fraction or multiple of the DAC and, based on the hours of exposure, to calculate the number of DAC-hours for each radionuclide inhaled. If the cumulative internal exposure is less than 2000 DAC-hours, the person is deemed to comply with the dose limit if no other external dose is received; if external dose is also recorded, then the number of DAC-hours allowed must be reduced accordingly (see Example 11-1).

Table 11-2. 10 CFR 20 Values of Effective Doses per μCi Due to Inhalation and Ingestion of Selected Radionuclides and the Corresponding ALIs, DACs, and Discharge Limits to Air, Water, and Sewage.

Nuclide	Clearance Data Class	f_1	Compounds	Inhalation Effective (mrem/μCi)	ALI (μCi)	DAC (μCi/mL)	Ingestion Effective (mrem/μCi)	ALI (μCi)	Discharge Limits (μCi/mL) Air	Water	Sewer
H-3	V	1.0	HTO	6.40E−02	8E+04	2E−05	6.40E−02	8E+04	1E−07	1E−03	1E−02
Be-7	Y	0.005	Ox, halides, NO	3.21E−01	2E+04	9E−06	1.28E−01	4E+04	3E−08	6E−04	6E−03
C-11		1.0	CO_2	1.22E−02	4E+05	2E−04	1.22E−02	4E+05	6E−07	6E−03	6E−02
C-14		1.0	CO_2	2.09E+00	2E+03	1E−06	2.09E+00	2E+03	3E−09	3E−05	3E−04
F-18	D	1.0	Fluorides (Fl)	8.36E−02	7E+04	3E−05	1.22E−01	5E+04 ST wall	1E−07	7E−04	7E−03
Na-22	D	1.0	all	7.66E+00	6E+02	3E−07	1.15E+01	4E+02	9E−10	6E−06	6E−05
Na-24	D	1.0	all	1.21E+00	5E+03	2E−06	1.42E+00	4E+03	7E−09	5E−05	5E−04
P-32	W	0.8	metal phosphates	1.55E+01	4E+02	2E−07	8.77E+00	6E+02	1E−09	9E−06	9E−05
S-35	W	0.8	elemental, sulfides	2.48E+00	2E+03	9E−07	7.33E−01	6E+03	2E−08	1E−04	1E−03
Cl-36	W	1.0	H, Li, Ha, K, Rb, Cs	2.19E+01	2E+02	1E−07	3.03E+00	2E+03	3E−09	2E−05	2E−04
K-40	D	1.0	all	1.24E+01	4E+02	2E−07	1.86E+01	3E+02	6E−10	4E−06	4E−05
Ca-45	W	0.3	all	6.62E+00	8E+02	4E−07	3.16E+00	2E+03	1E−09	2E−05	2E−04
Cr-51	Y	0.1	Ox, HO	2.62E−01	2E+04	8E−06	1.47E−01	4E+04	6E−08	5E−04	5E−03
Mn-54	W	0.1	Ox, HO, halides, NO	6.70E+00	8E+02	3E−07	2.77E+00	2E+03	1E−09	3E−05	3E−04
Mn-56	D	0.1	Ox, HO, halides, NO	3.30E−01	2E+04	6E−06	9.77E−01	5E+03	2E−08	7E−05	7E−04
Fe-55	D	0.1	all but Ox, OH, halides	2.69E+00	2E+03	8E−07	6.07E−01	9E+03	3E−09	1E−04	1E−03
Fe-59	D	0.1	all but Ox, OH, halides	1.48E+01	3E+02	1E−07	6.70E+00	8E+02	5E−10	1E−05	1E−04
Co-57	Y	0.05	Ox, OH, NO, halides	9.07E+00	7E+02	3E−07	7.44E−01	4E+03	9E−10	6E−05	6E−04
Co-58	Y	0.05	Ox, OH, NO, halides	1.09E+01	7E+02	3E−07	2.99E+00	1E+03	1E−09	2E−05	2E−04
Co-60	Y	0.05	Ox, OH, NO, halides	2.19E+02	3E+01	1E−08	1.02E+01	2E+02	5E−11	3E−06	3E−05
Ni-59	D	0.05	all but Ox, OH, CO	1.32E+00	4E+03	2E−06	2.10E−01	2E+04	5E−09	3E−04	3E−03

Nuclide	Class	Chemical form							
Ni-63	D 0.05	all but Ox, OH, CO	3.10E+00	2E+03	7E-07	5.77E-01	9E+03	2E-09	1E-04	1E-03
Ni-65	D 0.05	all but Ox, OH, CO	2.42E-01	2E+04	1E-05	6.22E-01	8E+03	3E-08	1E-04	1E-03
Cu-64	Y 0.5	Ox, OH	2.77E-01	2E+04	9E-06	4.66E-01	1E+04	3E-08	2E-04	2E-03
Zn-65	Y 0.5	all	2.04E+01	3E+02	1E-07	1.44E+01	4E+02	4E-10	5E-06	5E-05
Ga-67	W 0.001	Ox, OH, CO, O, halides	5.59E-01	1E+04	4E-06	7.84E-01	7E+03	1E-08	1E-04	1E-03
Ga-68	D 0.00	Ox, OH, CO, O, halides	1.38E-01	4E+04	2E-05	3.42E-01	2E+04	6E-08	2E-04	2E-03
Ge-67	D 1.0	all but Ox, SO, halides	6.07E-02	9E+04	4E-05	1.30E-01	3E+04	1E-07	6E-04	6E-03
							ST wall			
Ge-68	W 1.0	all but Ox, SO, halides	5.18E+01	1E+02	4E-08	1.07E+00	5E+04	1E-09	6E-05	6E-04
As-76	W 0.5	all	3.74E+00	1E+03	6E-07	5.22E+00	1E+03	2E-09	1E-05	1E-04
Br-82	W 1.0	all but H, U, Na, K, Cs	1.53E+00	4E+03	2E-06	1.71E+00	3E+03	5E-09	4E-05	4E-04
Rb-86	D 1.0	all	6.62E+00	8E+02	3E-07	9.36E+00	5E+02	1E-09	7E-06	7E-05
Sr-85	Y 0.01	all insoluble	5.03E+00	2E+03	6E-07	1.98E+00	3E+03	2E-09	4E-05	4E-04
Sr-89	Y 0.01	all insoluble	4.14E+01	1E+02	6E-08	9.25E+00	5E+02	2E-10	8E-06	8E-05
Sr-90	Y 0.01	all insoluble	1.30E+03	4E+00	2E-09	1.42E+02	4E+01	6E-12	5E-07	5E-06
Zr-95	W 0.002	Ox, OH, NO, halides	1.59E+01	3E+02	1E-07	3.77E+00	1E+03	5E-10	2E-05	2E-04
Nb-94	Y 0.01	Ox, OH	4.14E+02	2E+01	6E-09	7.14E+00	9E+02	2E-11	1E-05	1E-04
Nb-95	Y 0.01	Ox, OH	5.81E+00	1E+03	5E-07	2.57E+00	2E+03	2E-09	3E-05	3E-04
Mo-99	Y 0.05	Ox, OH, MoS$_2$	3.96E+00	1E+03	6E-07	3.04E+00	2E+03	2E-09	2E-05	2E-04
Tc-99	W 0.8	Ox, OH, NO, halides	8.33E+00	7E+02	3E-07	1.46E+00	4E+03	9E-10	6E-05	6E-04
Tc-99m	D 0.8	all other	3.26E-02	5E+05	2E-05	6.22E-02	8E+04	2E-07	1E-03	1E-02
Ru-103	Y 0.05	Ox, OH	8.95E+00	6E+02	3E-07	3.05E+00	2E+03	9E-10	3E-05	3E-04
Ru-106	Y 0.05	Ox, OH	4.77E+02	1E+01	5E-09	2.74E+01	2E+02	2E-11	3E-06	3E-05
Ag-110m	Y 0.05	Ox, OH	8.03E+01	9E+01	4E-07	1.08E+01	5E+02	1E-10	6E-06	6E-05
Cd-109	D 0.05	all but SO, NO, halides	1.14E+02	4E+01	1E-08	1.31E+01	4E+02	7E-11	6E-06	6E-05
				Kidneys			Kidneys			
In-115	D 0.02	all but Ox, OH, NO, halides	3.74E+03	1E+00	6E-10	1.58E+02	4E+01	2E-12	5E-07	5E-06
Sn-113	W 0.02	Ox, SO, OH, NO, halides	1.07E+01	5E+02	2E-07	3.08E+00	2E+03	8E-10	3E-05	3E-04
							LLI wall			

(continues)

Table 11-2. (*continued*)

Nuclide	Clearance Data Class f_1	Compounds	Inhalation Effective (mrem/μCi)	Inhalation ALI (μCi)	DAC (μCi/mL)	Ingestion Effective (mrem/μCi)	Ingestion ALI (μCi)	Discharge Limits (μCi/mL) Air	Water	Sewer
Sb-122	W 0.01	Ox, OH, SO, NO, halides	5.14E+00	1E+03	4E-07	7.29E+00	7E+02	2E-09	1E-05	1E-04
Sb-124	W 0.01	Ox, OH, SO, NO, halides	2.52E+01	2E+02	1E-07	1.01E+01	5E+02	3E-10	7E-06	7E-05
Sb-125	W 0.01	Ox, OH, SO, NO, halides	1.22E+01	5E+02	2E-07	2.80E+00	2E+03	7E-10	3E-05	3E-04
I-123	D 1.0	all	2.96E-01	6E+03	3E-06	5.29E-01	3E+03	2E-08	1E-04	1E-03
I-125	D 1.0	all	2.42E+01	6E+01	3E-08	3.85E+01	4E+02	3E-10	2E-06	2E-05
I-129	D 1.0	all	1.74E+02	9E+00	4E-09	2.76E+02	5E+00	4E-11	2E-07	2E-06
I-130	D 1.0	all	2.64E+00	7E+02	3E-07	4.74E+00	4E+02	3E-09	2E-05	2E-04
I-131	D 1.0	all	3.29E+01	5E+01	2E-08	5.33E+01	3E+01	2E-10	1E-06	1E-05
I-133	D 1.0	all	5.85E+00	3E+02	1E-07	1.04E+01	1E+02	1E-09	7E-06	7E-05
			(All iodines values based on thyroid)							
Cs-134	D 1.0	all	4.63E+01	1E+02	4E-08	7.33E+01	7E+01	2E-10	9E-07	9E-06
Cs-137	D 1.0	all	3.19E+01	2E+02	6E-08	5.00E+01	1E+02	2E-10	1E-06	1E-05
Cs-138	D 1.0	all	1.01E-01	6E+04	2E-05	1.94E-01	2E+04 ST wall	8E-08	4E-04	4E-03
Ba-133	D 0.1	all	7.81E+00	7E+02	3E-07	3.40E+00	2E+03 LLI wall	9E-10	2E-05	2E-04
Ba-140	D 0.1	all	4.85E+00	1E+03	6E-07	9.47E+00	5E+02 LLI wall	2E-09	8E-06	8E-05
La-140	W 0.001	Ox, OH	8.95E+00	1E+03	5E-07	8.44E+00	6E+02	2E-09	9E-06	9E-05
Ce-141	Y 0.0003	Ox, OH, Fl	3.74E+02	6E+02	2E-07	2.90E+00	2E+03 LLI wall	8E-10	3E-05	3E-04
Ce-144	Y 0.0003	Ox, OH, Fl	4.33E-02	1E+01	6E-09	2.10E+01	2E+02 LLI wall	2E-11	3E-06	3E-05

Pr-144	Y 0.0003	Ox, OH, CO, Fl	6.85E+00	1E+05	5E-05	1.17E-01	3E+04 ST wall	2E-07	6E-04	6E-03
Nd-147	Y 0.0003	Ox, OH, CO, Fl	3.92E+01	8E+02	4E-07	4.37E+00	1E+03 LLI wall	1E-09	2E-05	2E-04
Pm-147	Y 0.0003	Ox, OH, CO, Fl	1.10E+01	5E+03	2E-02	1.05E+00	4E+03 LLI wall	2E-10	7E-05	7E-04
Eu-152	W 0.001	all	2.86E+02	2E+01	1E-08	6.48E+00	8E+02	3E-11	1E-05	1E-04
Eu-154	W 0.001	all	3.30E+05	2E-02	8E-09	9.55E+00	5E+02	3E-11	7E-06	7E-05
Gd-148	D 0.0003	all but Ox, OH, Fl	2.43E+05	8E-03 B. surf.	3E-12	2.18E+02	1E+01 B. surf.	2E-14	3E-07	3E-06
Gd-152	D 0.0003	all but Ox, OH, Fl	2.38E+01	1E-02 B. surf.	4E-12	1.61E+02	2E+01 B. surf.	3E-14	4E-07	4E-06
W-188	D 0.3	all	5.14E+00	1E+03	5E-07	7.81E+00	4E+02 LLI wall	2E-09	7E-06	7E-05
Re-186	W 0.8	Ox, OH, NO	3.61E+01	2E+03	7E-07	2.94E+00	2E+03	2E-09	3E-05	3E-04
Ir-192	Y 0.01	Ox, OH	3.85E+02	2E+02	9E-08	5.74E+00	9E+02	3E-10	1E-05	1E-04
Pt-193	D 0.01	all	8.77E-01	2E+04	1E-05	1.19E-01	4E+04 LLI wall	3E-08	6E-04	6E-03
Au-195	Y 0.1	Ox, OH	3.28E+00	4E+02	2E-07	1.06E+00	5E+03	6E-10	7E-05	7E-04
Au-198	Y 0.1	Ox, OH	4.85E+00	2E+03	7E-07	4.22E+00	1E+03	2E-09	2E-05	2E-04
Hg-197	W 0.02	Ox, OH, NO, SO, halides	6.88E-01	9E+03	4E-06	9.58E-01	6E+03	1E-08	8E-05	8E-04
Hg-203	W 0.02	Ox, OH, NO, SO, halides	5.74E+00	1E+03	5E-07	2.30E+00	2E+03	2E-09	3E-05	3E-04
Tl-201	D 1.0	all	2.35E-01	2E+04	9E-06	3E-01	2E+04	3E-08	2E-04	2E-03
Tl-204	D 1.0	all	2.41E+00	2E+03	9E-07	3.36E+00	2E+03	3E-09	2E-05	2E-04
Pb-210	D 0.2	all	1.36E+04	2E+01 B. surf.	1E-10	5.37E+03	6E-01 B. surf.	6E-13	1E-08	1E-07

(continues)

Table 11-2. (*continued*)

Nuclide	Clearance Data Class f_1	Compounds	Inhalation Effective (mrem/μCi)	ALI (μCi)	DAC (μCi/mL)	Ingestion Effective (mrem/μCi)	ALI (μCi)	Discharge Limits (μCi/mL) Air	Water	Sewer
Bi-210	W 0.05	all but NO	1.96E+02	3E+01	1E-07	6.40E+00	8E+02 Kidneys	4E-11	1E-05	1E-04
Po-210	D 0.1	all but Ox, OH, NO	9.40E+03	6E-01	3E-10	1.90E+03	3E+00	9E-13	4E-08	4E-07
Ra-226	W 0.2	all	8.58E+03	6E-01	3E-10	1.32E+03	2E+00	9E-13	6E-08	6E-07
Ra-227	W 0.2	all	2.84E-01	1E+04 B. surf.	6E-06	2.26E-01	2E+04 B. surf.	3E-08	3E-04	3E-03
Ra-228	W 0.2	all	4.77E+05	1E+00	5E-10	1.44E+03	4E+00	2E-12	6E-08	6E-07
Th-230	W 0.0002	all but Ox, OH	3.26E+05	6E-03 B. surf.	3E-12	5.48E+02	4E+00 B. surf.	2E-14	1E-07	1E-06
Th-232	W 0.0002	Ox, OH	1.64E+06	1E+03 B. surf.	5E-13	2.73E+03	7E+01 B. surf.	4E-15	3E-08	3E-07
Pa-231	W 0.001	all but Ox, OH	1.28E+06	2E+03 B. surf.	6E-13	1.06E+04	2E-01 B. surf.	6E-15	6E-09	6E-08
U-235	Y 0.001	UO_2, U_3, O_8	1.23E+05	4E-02	2E-11	2.67E+01	1E+01 B. surf.	6E-14	3E-07	3E-06
U-238	Y 0.001	UO_2, U_3, O_8	1.18E+05	4E-02	2E-11	2.38E+01	1E+01 B. surf.	6E-14	3E-07	3E-06
Np-237	W 0.001	all	5.40E+05	4E-03 B. surf.	2E-12	4.44E+03	5E-01 B. surf.	1E-14	2E-08	2E-07

Pu-238	Y 0.001	PuO$_2$	2.88E+05	2E-02	8E-12	4.96E+01	9E+01 B. surf.	2E-14	2E-08	2E-07
Pu-239	W 0.001	all but PuO$_2$	4.29E+05	6E-03 B. surf.	3E-12	3.54E+03	8E-01 B. surf.	2E-14	2E-08	2E-07
Pu-241	W 0.001	all but PuO$_2$	8.25E+03	3E-01 B. surf.	1E-10	6.85E+01	4E+01 B. surf.	8E-13	1E-06	1E-05
Pu-242	W 0.001	all but PuO$_2$	4.11E+05	7E-03	3E-12	3.36E+03	1E+00 B. surf.	2E-14	2E-08	2E-07
Am-241	W 0.001	all	4.44E+05	6E-03 B. surf.	3E-12	3.64E+03	8E-01 B. surf.	2E-14	2E-08	2E-07
Cm-244	W 0.001	all	2.48E+05	1E-02 B. surf.	5E-12	2.02E+03	1E+00 B. surf.	3E-14	3E-08	3E-07
Cf-252	W 0.001	all but Ox, OH	1.37E+05	2E-02 B. surf.	8E-12	1.08E+03	2E+00 B. surf.	5E-14	7E-08	7E-07

Note: when a tissue is listed, the respective ALI and the corresponding DAC, if applicable, in the line just above it is based on the nonstochastic dose limit of 50 rem to the listed tissue.
Adapted from 10CFR20, Appendix B.

Example 11-1: A worker received 1 rem of exposure and was then reassigned to work with ^{14}C compounds to avoid further external dose; however, airborne ^{14}C levels are present in the work area. How many DAC-hours could the worker be exposed to and not exceed occupational dose limits?

Solution: The sum of fractions of each mode of exposure must be ≤ 1.0, and since the external dose is 1/5 of the dose limit of 5 rem TEDE, an additional exposure of 4/5 is possible (assuming ALARA considerations are met); thus 4/5 of 2000 DAC-hours or 1600 DAC-hours could be received.

If the ALI is based on nonstochastic effects, then its use in limiting intakes will ensure that nonstochastic effects are avoided, and since the nonstochastic value is a capping dose that is more restrictive, the risk of stochastic effects will also be limited. If, in a particular situation involving a radionuclide for which the nonstochastic ALI is limiting, and the use of that nonstochastic ALI is considered unduly conservative, the regulations allow the stochastic ALI to be used in determining the committed effective dose equivalent, provided that the 50-rem dose equivalent limit for any organ or tissue is not exceeded by the sum of the external deep dose equivalent plus the internal committed dose to that organ (not the effective dose). These considerations are illustrated in Example 11-2.

Example 11-2: A worker received in the course of a year 2 rem of external gamma dose and is exposed to $4 \times 10^{-8} \mu Ci/mL$ of ^{137}Cs aerosols for 1200 h and $4 \times 10^{-8} \mu Ci/mL$ of airborne ^{131}I for 250 h: (a) determine whether the person exceeded the regulations; and (b) determine the TEDE.

Solution: (a) Internal dose due to inhalation exposure is best determined in terms of DAC-hours; thus, from Table 11-2 the ^{137}Cs exposure concentration is 2/3 of the DAC and that for ^{131}I is 2 times the DAC, for a total DAC-hrs of

$$\begin{aligned} DAC \cdot h &= (2/3 \ DAC \times 1200 \ hr) + (2 \ DAC \times 250 \ h) \\ &= 800 \ DAC \cdot h(^{137}Cs) + 500 \ DAC \cdot h(^{131}I) \\ &= 1300 \ DAC \cdot h \end{aligned}$$

The external dose and the inhalation exposures are considered together by the sum of fractions

$$\frac{2 \ rem}{5 \ rem} + \frac{1300 \ DAC \cdot h}{2000 \ DAC \cdot h} = 1.05$$

which exceeds 1.0, and by strict application of the regulatory schema the worker is overexposed.

(b) If, however, the ^{131}I intake is considered in terms of stochastic effects using the effective dose coefficient of $3.29 \times 10^{1} \ mrem/\mu Ci (3.29 \times 10^{-2} \ rem/\mu Ci)$ from Table 11-2, the intake is

$$4 \times 10^{-8} \frac{\mu Ci}{mL} \times 2 \times 10^4 \frac{mL}{min} \times 60 \frac{min}{h} \times 250 \, h = 12 \mu Ci$$

and the TEDE is

$$TEDE = 2 \, rem + \left(\frac{800}{2000} \times 5 \, rem \right) + \left(12 \mu Ci \times 3.29 \times 10^{-2} \frac{rem}{\mu Ci} \right) = 4.4 \, rem$$

which is below the TEDE limit of 5 rem, the overriding consideration in controlling worker exposure. When this approach is used, it is necessary to assure that the nonstochastic limit is not exceeded, which is the case in this example since 500 DAC-hrs due to ^{131}I exposure is well below the limit of 2000 DAC-hrs based on the nonstochastic limiting dose of 50 rem to the thyroid.

Intakes of Radioactive Material can be monitored by determining concentrations of radionuclides in air, water, or food, or by bioassay techniques designed to detect intakes of radioactive material through inhalation, ingestion, or skin penetration/absorption. Bioassay requirements are not specifically stated in NRC regulations but are often stated as a license condition. The regulations do, however, require assessment of the committed effective dose equivalent for persons who receive, in one year, an intake in excess of 10% of the applicable Annual Limits on Intake, and for minors and declared pregnant women who may receive a committed effective dose equivalent in excess of 50 mrem (0.5 mSv) in any one year. Such assessments usually rely on whole-body counting, organ measurements, or bioassay of excreta. Measured values can be used with the data in Table 9-11 to determine the intake and in turn the dose to tissues and/or the effective whole-body dose.

Compliance for Members of the Public is demonstrated by measurement or calculation of the total effective dose equivalent to the individual(s) likely to receive the highest dose, or by demonstrating that the annual average concentrations of radioactive material released in gaseous and liquid effluents at the boundary of the unrestricted area do not exceed the values for selected radionuclides in Table 11-2 (adapted from Table 2 of Appendix B of 10 CFR 20). Inhalation or ingestion of radionuclides at the concentrations listed in Table 11-2 continuously over the course of a year would produce a total effective dose equivalent of 50 millirem (or 0.5 mSv) to an average member of the population. The regulations also require that the dose to an individual continuously present in an unrestricted area not exceed 2 millirem (0.002 rem or 0.02 mSv) in any one hour and 50 millirem (or 0.5 mSv) in a year, which is a factor of 2 below the standard of 100 mrem/y for a member of the public.

Nonstochastic limits were not considered in deriving the air and water effluent concentration limits because nonstochastic effects are presumed not to occur at the dose levels established for individual members of the public; therefore, only the stochastic ALI was used in deriving the corresponding effluent limits. The effluent concentration values in Table 11-2 may, upon approval by NRC, be adjusted for the actual physical and chemical characteristics of the effluents (e.g., aerosol size distribution, solubility, density, radioactive decay equilibrium, chemical form).

Airborne Radioactivity in Unrestricted Areas is also controlled by DACs. These are also listed in Table 11-2 and are based on continuous occupancy (168 hours per week, including sleeping hours). These values are based on an annual effective dose equivalent of 100 mrem for members of the public as the primary dose limit, reduced by an additional factor of 2 to reflect the potential exposure of children in releases to the environment. The corresponding limit of 50 mrem/y is a reduction from the previous value of 500 mrem/y; thus, the DACs for unrestricted areas have been reduced considerably in recent years (since 1994) for most of the radionuclides (iodine-125 is a notable exception). The air concentration values were derived by dividing the occupational stochastic inhalation ALI by 2.4×10^9 mL to obtain the inhalation DAC, and then dividing by a factor of 50 to relate the 5-rem annual occupational dose limit to the 0.1-rem limit for members of the public, a factor of 3 to adjust for the differences in exposure time and inhalation rate for workers versus members of the public, and a factor of 2 to adjust the occupational values (derived for adults) to a value of 50 mrem so that they are applicable to other age groups. The H_2O concentrations are also based on the occupational stochastic annual limit on intake (ALI) for ingestion divided by the annual water intake of 7.3×10^5 mL by a reference adult and adjusted by the same factors of 50 and 2 for members of the public.

The discharge limits for sewer are monthly average concentrations for release to sanitary sewers (as provided in § 20.2003); the concentration values are a factor of 10 greater than the H_2O discharge concentrations, which takes into consideration dilution within the system and processing prior to any potential public consumption.

The regulations also allow, upon application to and approval by NRC, temporary doses up to 0.5 rem (5 mSv) in a year. The need for the exception and the expected duration of operations in excess of the 0.1 rem/y limit must be demonstrated as well as the program to assess and control dose to ALARA levels within the limit of 0.5 rem/y (5 mSv/y). The NRC may also allow discharges of higher concentrations in air or H_2O if it is not likely that individuals would be exposed to levels in excess of applicable radiation protection guides, but any such acceptance is based on the condition that the user must first take every reasonable measure to keep releases of radioactivity in effluents as low as reasonably achievable.

Good Practice Requirements for Radiation Safety

Various precautionary procedures are also expected in an effective radiation control program. Storage (20.1801) and control (20.1802) of licensed material are considered to be particularly important, and licensees are required to assure the security of stored material and the control of material not in storage. Other requirements designed to preclude inadvertent exposure of personnel are caution signs (20.1901), posting requirements (20.1902), labeling of containers (20.1904), and procedures (20.1906) for receiving and opening packages.

Package Receipt and Opening procedures are detailed in 10 CFR 20.1906 and consist of timely inspection, monitoring, and security. Packages should be monitored within 3 hours after receipt if received during normal working hours, or within 3 hours of the beginning of the next working day if received after hours. Packages should be received at a designated central location where they are inspected, monitored, and logged by trained personnel and the material secured until transferred to the authorized user. If removable beta–gamma radioactive contamination in excess of 2,000 dis/min per 100 cm^2 of package surface is found on the external surfaces of the package, or if the radiation levels exceed 20 mrem/h at the surface of the package or 10 mrem/h at 1 meter from the surface, the licensee must immediately notify the final delivering carrier and the appropriate NRC Inspection and Enforcement office.

Storage and Control of Radionuclides (20.1801 and 20.1802) are required to preclude unauthorized removal or access either by placing them in a locked location or under the control of a responsible individual(s). The radiation level at 30 cm (1 ft) from the surface of packages should be less than 5 mrem/h, which may require adding shielding. If the level at 30 cm (1 ft) could exceed 100 mrem in 1 h, the area must be treated as a high-radiation area and be posted accordingly and equipped with visible or audible alarm signals.

Caution Signs are essential because individuals might otherwise be unaware of the presence of potential radiation exposure. The three-bladed radioactive caution symbol (magenta or purple on yellow background) is designed to be universally recognizable; i.e., similar in concept to the stop sign used in traffic control. On the other hand, unnecessary use of caution signs can defeat their purpose when workers know hazards are minimal; i.e., don't be cautious with caution signs.

Posting of Areas is required (20.1902) where significant levels of radiation or radioactivity may be present. These and the conditions for their use are

- "CAUTION—RADIATION AREA" for areas accessible to personnel in which a major portion of the body could receive in any 1 h a dose of 5 mrem or in any 5 consecutive days a dose in excess of 100 mrem; these limits are applicable at 30 cm (1 ft) from the object containing the source.

- "CAUTION—RADIOACTIVE MATERIAL" for areas or rooms where radioactive material is used or stored or on any container in which radioactive material is transported or stored. When containers are used for storage, the labels must also state the quantities and kinds of radioactive materials in the container and the date measured.

- "AIRBORNE RADIOACTIVITY AREA" when airborne radioactivity exists in concentrations such that an individual could receive in the number of hours spent in the area during one week an intake of 0.6 percent of an ALI or 12 DAC-hrs.

- "CAUTION—HIGH RADIATION AREA" is required if the radiation dose to a major portion of the body of a person in an area could exceed 100 mrem in any 1 h at a distance of 30 cm from the object containing the source.
- "GRAVE DANGER—VERY HIGH RADIATION AREA" is required if the radiation dose rate from a source could exceed 500 rads in any 1 hour at a distance of 30 cm.

Posting of high radiation areas or very high radiation areas is not sufficient for prevention of unauthorized access to areas where exposure can be quite significant. One or more of the following methods must also be used if a person is able to proceed past a designated point:

- a control device (e.g., a shuttertrip) to reduce the radiation level to < 0.1 rem in 1 h at 30 cm;
- a control device energizing a conspicuous visible or audible alarm; and/or
- locked entryways with positive control over each individual entry

Continuous direct or electronic surveillance may be substituted provided it is demonstrated to be capable of preventing unauthorized entry.

Licensees are also required to post a special form (NRC-3) notifying employees of the regulations and their rights to information on their exposure to radiation and inspections. They must also inform workers by posted notices of the availability of copies of 10 CFR 19 and 10 CFR 20 regulations, the license and any amendments, the operating procedures applicable to licensed activities, and notices of violations involving radiological working conditions (10 CFR 19.11).

Radioactive Waste Disposal

Disposal of waste or unwanted radioactive or other material is authorized by one of four methods: (1) transfer to a licensed person; (2) decay in storage; (3) release in effluents within authorized limits, or (4) as otherwise authorized by NRC. Radionuclides with half-lives less than 70 days can be stored for 10 half-lives, after which they can be disposed of as nonradioactive wastes. Many licensees will, however, transfer these wastes to a licensed commercial waste company rather than bother with the requirements for storage and security. Records are required for the disposal of all radioactive wastes.

Unrestricted disposal of ^{3}H and ^{14}C in biomedical waste is allowed without regard for their radioactivity if the concentration is less than 0.05 μCi/g. Dispersible liquids below such a concentration may be discharged to the sanitary sewage and animals may be incinerated or frozen and disposed in municipal landfills. Assurance must be provided, however, that the disposed material will not be used as food or animal feed.

Disposal of Liquid Wastes to Sewerage Systems is permitted (10 CFR 20.2003) if the concentrations are controlled to the limits shown in Table 11-2 for selected radionuclides, which are 10 times the concentration limits for ingestion of radioactive material in water by members of the public. The material must be dispersible or soluble in water, and if several radionuclides are being discharged, a sum of fractions must be ≤ 1.0 for daily and monthly concentrations. Maximum annual limits are 5 Ci for ^{3}H, 1 Ci for ^{14}C, and 1 Ci for all other radioactive material combined. Excreta from individuals undergoing medical diagnosis or therapy with radioactive material is exempt from these limitations.

Solid Low-Level Radioactive Wastes should be placed in covered metal cans (the type equipped with a foot-operated lid is convenient for small volumes) that are easily distinguishable from cans used for ordinary trash, and they should display a "radioactive materials" label to prevent accidental disposal of radioactive materials into the regular trash. A sturdy plastic bag should be used as a liner since this will likely be the container for transfer to the collection point for consolidation and disposal. Hypodermic needles and other sharp objects should be wrapped so that they do not pierce through the liner because even a mere scratch can result in serious infection or disease. Materials contaminated with radioiodine should be enclosed in two bags before discarding, and animal carcasses are best stored in a freezer prior to final disposal.

Records and Reports

Records and reports are required by the regulations in order to document compliance with the elements of the control program. These are almost always reviewed for compliance and sufficiency when the license is inspected. Records are required of the elements of and any changes in the radiation protection program; doses to workers, including any special exposures; doses to individual members of the public; surveys; amounts of radionuclides disposed as waste; instrument calibrations; and testing of entry control devices for "high" and "very high radiation areas". Dose monitoring records for individuals, when required by 20.1202 and 20.1502, include

- deep dose and shallow dose equivalents (i.e., to skin and extremities);
- intakes of radionuclides and the committed effective dose equivalent (CEDE), including specific information used to calculate internal dose;
- the total dose to the most irradiated organ; and
- the total effective dose equivalent (TEDE) from all modes of exposure.

Records (20.2110) may be the original, a reproduced copy, microform, or stored on electronic media. Licensees are required (20.2102) to maintain records of the radiation protection program for the life of the license. Records of audits and other reviews must be retained for 3 years. Records of doses to members of the public and supporting documents (e.g., effluent data) are to be retained until the license is terminated.

Notification of State and Federal Authorities is required for significant upset conditions (20.2202). Immediate notification of NRC is required of any event (generally at 5 times normal occupational limits) which has caused or threatens to cause individual doses in excess of 25 rems TEDE, 75 rems to the eye, 250 rads to skin or extremities; a release of radioactive material producing an intake of 5 ALI in 24 hours; or one work week of downtime or property damage over $200,000. Reports are required within 30 days if regulatory dose limits are exceeded for workers, the embryo/fetus of a declared pregnant woman, and members of the public, or other limiting doses stated in the license. In a similar vein, a report is required within 30 days for concentrations or radiation levels that exceed limits stated for restricted areas; are greater than 10 times any limit for an unrestricted area; or that exceed EPA's generally applicable environmental radiation standards. Exemptions can be made by NRC upon application (20.2301); however, the NRC also retains the right to impose additional requirements from time to time to ensure radiation protection and/or safe operations (20.2302).

Radiation Monitoring

Personnel Monitoring is required if adult workers (or employed minors protected by limits that are 10% of those for adults) receive or are liable to receive in one year a dose in excess of 10% of the applicable limits. External radiation dose is typically measured with a film badge or thermoluminescent dosimeter (TLD), typically for a month or a quarter (3 months). These are worn on the trunk, between the waist and the shoulders, at the site of the highest exposure. These devices are very useful in radiation control because their use ensures that all exposures, even unexpected ones, are measured and situations where controls could be improved are identified.

If the hands are exposed to levels significantly higher than the rest of the body because of close work with localized sources, a separate monitoring device should be worn on the wrist or finger. Direct-reading pocket dosimeters or electronic dosimeters are also worn in high-risk areas where it is desirable to have immediate knowledge of integrated exposure.

Area Monitoring may be required if significant radiation levels or airborne contaminants can occur in working areas. Direct radiation level detectors are generally equipped with a recording chart, lights that flash at preset levels, and audible alarms. Continuous air monitors (CAMs) are designed to continuously sample the air by collection of radio-contaminants on a filter, absorbent, or solvent which is measured by a detector. A pump draws the air, a meter determines the flowrate, and lights and audible alarms are usually included along with a continuous recording device.

Airborne Radioactivity Monitoring is used to determine the number of DAC-hours to which a worker is exposed. Control of exposure in airborne radioactivity areas is based on a policy of using engineering controls and confinement of radioactive material to the extent practicable, and air monitoring is used to determine the effectiveness of such controls. If airborne radioactivity is still present after the use of such controls, the licensee is required to increase monitoring and reduce intakes by access control; limiting exposure times; and, as a last resort, use of respiratory protection equipment that has been tested and certified by NIOSH/MSHA, or alternatively by a methodology approved by NRC. Additional requirements for respiratory protection equipment are detailed in Appendix Λ of 10 CFR 20.

Detection of Beta–Gamma Surface-level Contamination is often done with a thin-window G–M tube, typically a 2-inch-diameter "pancake" probe. The 2-inch-diameter pancake probe has a typical background count rate of about 70 c/min, and a counting rate that is approximately twice the background rate is considered a positive indication of contamination, although more sophisticated techniques based on statistical methods can detect lower levels. Measurements of beta–gamma contamination are generally done by interposing a thin shield that just stops beta particles between the contamination source and the sensitive part of the detector; the beta levels can be determined by the difference between readings made with and without the shield. Monitoring for contamination is done by slowly moving the appropriate detector over the suspected surfaces. It is very useful to have an audible signal, such as from earphones or a loudspeaker, since small increases of radiation above the background are indicated by the "click-rate," which is easier and more reliable than watching the meter of the detector.

Alpha Contamination can be monitored with a proportional flow counter equipped with a very thin detector window and set to register only high-energy pulses by alpha particles (see Chapter 8). Scintillation detectors are easier to use, but care must be taken to prevent light leaks in the detector covering which can cause spurious counts.

Contamination Control

Contamination of surfaces and potential removal are always possible when unsealed radioactive materials are used. Of particular concern is contamination of hands and subsequent ingestion of the material, skin contamination, and suspension of the material into the air where it may be inhaled. These issues are addressed by preventing contamination or detecting and removing it before problems occur. Objects or surfaces that have contamination levels that exceed the values in Table 11-3 are generally regarded as contaminated and thus require decontamination.

The Smear or Wipe Test is perhaps the most common method for monitoring loose contamination. A piece of filter paper is wiped over an area of approximately 100 cm^2 and then measured. A thin-window G–M detector is commonly used in the field, but for a large number of smear tests (e.g., periodic surveys of

Table 11-3. Surface Radioactivity Guides in dpm/100 cm^2.

Radioactive material	Removable	Total
All alpha emitters (except natural or depleted uranium and natural thorium); ^{210}Pb; ^{228}Ra[a]	20	100
^{90}Sr; ^{125}I, ^{126}I, ^{129}I, ^{131}I	200	1,000
All other beta/gamma emitters except pure beta emitters with $E_{max} \leq 150$ keV[b]	1,000	5,000
Natural or depleted uranium, natural thorium, and their associated decay products	1,000	5,000

[a]Pb-120 and Ra-228 are not alpha emitters but produce Po-210 and Th-228 that are.
[b]Excluded because detection is not practical.
Source: Nuclear Regulatory Guide 8.23.

work spaces) the filter papers are inserted into liquid scintillation vials and counted in a liquid scintillation counting system.

Leak Tests of Sealed Radioactive Sources must be performed when received and on a regular schedule (typically every 6 months) thereafter. Alpha and beta sources are particularly vulnerable to developing leaks in the covering which must be thin enough to allow penetration of particles. The source is tested by swiping or "smearing" with a filter paper or other medium which is then measured with an appropriate instrument. The presence of removable radioactivity is cause for concern that the source is leaking and followup with the supplier is in order. For sources that produce a high surface dose rate (e.g., greater than 1 rem/min) the swipe should be performed with long-handled tools. For smaller sources, a medical swab moistened with ethanol may be satisfactory. A common limit used by regulatory agencies for removable contamination is 0.005 μCi; however, a leaking source generally yields readily detectable radioactivity on the swipe.

Protective Clothing such as coveralls, laboratory coats, and shoe covers should be worn wherever contamination of skin or personal clothing with radioactive materials is possible. Disposable gloves should be worn whenever hand contamination is likely, and extra care should be exercised to prevent contamination of skin areas where there is a break in the skin. Protective clothing shall not be worn in eating places, and should remain in the area(s) where used unless monitored and determined to be "clean."

Eating, smoking, storing food, and pipetting by mouth is not allowed in areas where work with radioactive materials is being conducted, and especially in rooms containing appreciable loose contamination because of the potential for ingestion of radioactivity. Personnel working in areas containing unsealed sources of radioactivity must "wash-up" before eating, smoking, or leaving work and must use an appropriate detection instrument to monitor hands, clothing, and so on, for possible contamination.

SHIPMENT OF RADIOACTIVE MATERIALS

The basic concept for transportation of radioactive materials is that they move freely in commerce between a shipping licensee and a receiving licensee who, based on package and shipping information provided by the shipper, can receive the material with assurance that the package was provided safe passage during transit while out of the direct control of either licensee (see Figure 11-1). The overall system is designed to assure that packages can be shipped by common carrier with minimal risk of spillage, exposure of personnel, or exposure of unexposed photographic film. Although the NRC has ultimate responsibility for AE Act materials in transit, they have delegated much of this authority to DOT through an agreement by which DOT regulates radioactive shipments, which is a very practical arrangement since DOT regulates shipments of most other hazardous materials. Each licensee who transports licensed material outside of its place of use or who delivers licensed material to a carrier for transport is required (10 CFR 71.5) to comply with applicable DOT regulations.

The NRC has retained various authorities. It continues to regulate Type B and fissile packages, assures transportation safeguards, investigates accidents/incidents, and serves as a technical advisor to DOT. The NRC also provides in Subparts D, E, and F of 10 CFR 71 application procedures for package approval; package approval standards; packaging requirements for special form material; and test requirements for LSA-III packages.

Transportation Regulations that pertain to radioactive materials are detailed primarily in 49 CFR 172 and 173 which place considerable emphasis on packaging, marking and labeling, placarding, accident reporting, and shipping papers. Shipping and packaging of radioactive materials are considered by DOT as class 7 (radioactive) materials, requirements for which are detailed in the following major sections of DOT regulations:

- Requirements for labeling (49 CFR 172.403; 436, 438 and 440), emergency response (49 CFR 172–Subpart G), and training (49 CFR 172–Subpart H)
- Information on the material shipped (49 CFR 172-Subpart C)
- General design and authorized packagings (49 CFR 173.410–419)
- Limited quantities, articles, LSA, and empty packages (49 CFR 173.421–427)
- A_1 and A_2 values (49 CFR 173.431–435) which specify activity limits for Type A and Type B packages
- Limits on radiation levels and operational controls (49 CFR 173.441–448)

The regulations pertaining to the packaging and shipment of radioactive and other hazardous materials are quite complicated; thus, special training is generally required for shipping personnel.

Most shipments of radioactive materials are governed by DOT regulations although some limited items may be mailed through the US Postal Service (see below). Requirements vary for different forms (liquid, solid, gas) and become more stringent as radiation levels and the quantities and concentrations of radionuclides increase. These are keyed to A_1 and A_2 values specified (49 CFR 173.421) for each of 378 radionuclides, values of which are listed alphabetically in Table 11-4 for some of the more common radionuclides.

Table 11-4. Maximum Quantities of Selected Radionuclides that can be Shipped as Type A_1, or Special Form (Solids, Sealed Source, etc.) Shipments, and as Type A_2, or Normal Form (Liquids, Compounds, etc.) Shipments.

Nuclide	Normal Form, A_1		Special Form, A_2		Activity Limit	
	TBq	(Ci)	TBq	(Ci)	TBq/g	(Ci/g)
Ag-110m	0.4	(10.8)	0.4	(10.8)	1.8×10^2	(4.7×10^3)
As-76	0.2	(5.41)	0.2	(5.41)	5.8×10^4	(1.6×10^6)
Au-195	10	(270)	10	(270)	1.4×10^2	(3.7×10^3)
Au-198	3	(81.1)	0.5	(13.5)	9.0×10^3	(2.4×10^5)
Ba-133	3	(81.1)	3	(81.1)	9.4	(2.6×10^2)
Be-7	20	(541)	20	(541)	1.3×10^4	(3.5×10^5)
C-14	40	(1080)	2	(54.1)	1.6×10^{-1}	(4.5)
Ca-45	40	(1080)	0.9	(24.3)	6.6×10^2	(1.8×10^4)
Cd-109	40	(1080)	1	(27)	9.6×10^1	(2.6×10^3)
Ce-141	10	(270)	0.5	(13.5)	1.1×10^3	(2.8×10^4)
Ce-144	0.2	(5.41)	0.2	(5.41)	1.2×10^2	(3.2×10^3)
Cf-252	0.1	(2.7)	1.0×10^{-3}	(2.7×10^{-2})	2×10^1	(5.4×10^2)
Cl-36	20	(541)	0.5	(13.5)	1.2×10^{-3}	(3.3×10^{-2})
Co-57	8	(216)	8	(216)	3.1×10^2	(8.4×10^3)
Co-58	1	(27)	1	(27)	1.2×10^3	(3.2×10^4)
Co-60	0.4	(10.8)	0.4	(10.8)	4.2×10^1	(1.1×10^3)
Cr-51	30	(811)	30	(811)	3.4×10^3	(9.2×10^4)
Cs-134	0.6	(16.2)	0.5	(13.5)	4.8×10^1	(1.3×10^3)
Cs-137	2	(54.1)	0.5	(13.5)	3.2	(8.7×10^1)
Cu-64	5	(135)	0.9	(24.3)	1.4×10^5	(3.9×10^6)
F-18	1	(27)	0.5	(13.5)	3.5×10^6	(9.5×10^7)
Fe-55	40	(1080)	40	(1080)	8.8×10^1	(2.4×10^3)
Fe-59	0.8	(21.6)	0.8	(21.6)	1.8×10^3	(5.0×10^4)
Ga-67	6	(162)	6	(162)	2.2×10^4	(6.0×10^5)
Ga-68	0.3	(8.11)	0.3	(8.11)	1.5×10^6	(4.1×10^7)
Ge-68	0.3	(8.11)	0.3	(8.11)	2.6×10^2	(7.1×10^3)

Table 11-4. (*continued*)

Nuclide	Normal Form, A_1		Special Form, A_2		Activity Limit	
	TBq	(Ci)	TBq	(Ci)	TBq/g	(Ci/g)
Hg-197	10	(270)	10	(270)	9.2×10^3	(2.5×10^5)
Hg-203	4	(108)	0.9	(24.3)	5.1×10^2	(1.4×10^4)
I-123	6	(162)	6	(162)	7.1×10^4	(1.9×10^6)
I-125	20	(541)	2	(54.1)	6.4×10^2	(1.7×10^4)
I-129	Unlimited		Unlimited		6.5×10^{-6}	(1.8×10^{-4})
I-131	3	(81.1)	0.5	(13.5)	4.6×10^3	(1.2×10^5)
In-113m	4	(108)	4	(108)	6.2×10^5	(1.7×10^7)
Ir-192	1	(27)	0.5	(13.5)	3.4×10^2	(9.2×10^3)
K-40	0.6	(16.2)	0.6	(16.2)	2.4×10^{-7}	(6.4×10^{-6})
Kr-85	20	(541)	10	(270)	1.5×10^1	(3.9×10^2)
Mn-54	1	(27)	1	(27)	2.9×10^2	(7.7×10^3)
Mn-56	0.2	(5.41)	0.2	(5.41)	8.0×10^5	(2.2×10^7)
Na-22	0.5	(13.5)	0.5	(13.5)	2.3×10^2	(6.3×10^3)
Na-24	0.2	(5.41)	0.2	(5.41)	3.2×10^5	(8.7×10^6)
Nb-95	1	(27)	1	(27)	1.5×10^3	(3.9×10^4)
Ni-59	40	(1080)	40	(1080)	3.0×10^{-3}	(8.0×10^{-2})
Ni-63	40	(1080)	30	(811)	2.1	(5.7×10^1)
P-32	0.3	(8.11)	0.3	(8.11)	1.1×10^4	(2.9×10^5)
P-33	40	(1080)	0.9	(24.3)	5.8×10^3	(1.6×10^5)
Pb-210	0.6	(16.2)	9×10^{-3}	(0.243)	2.8	(7.6×10^1)
Pm-147	40	(1080)	0.9	(24.3)	3.4×10^1	(9.3×10^2)
Po-210	40	(1080)	2×10^{-2}	(0.541)	1.7×10^2	(4.5×10^3)
Pu-238	2	(54.1)	2×10^{-4}	(5.41×10^{-3})	6.3×10^{-1}	(1.7×10^1)
Pu-239	2	(54.1)	2×10^{-4}	(5.41×10^{-3})	2.3×10^{-3}	(6.2×10^{-2})
Ra-226	0.3	(8.11)	2×10^{-2}	(0.541)	3.7×10^{-2}	(1.0)
Ra-228	0.6	(16.2)	4×10^{-2}	(1.08)	1.0×10^1	(2.7×10^2)
Rb-87	Unlimited		Unlimited		3.2×10^{-9}	(8.6×10^{-8})
Ru-103	2	(54.1)	0.9	(24.3)	1.2×10^3	(3.2×10^4)
Ru-106	0.2	(5.41)	0.2	(5.41)	1.2×10^2	(3.3×10^3)
S-35	40	(1080)	2	(54.1)	1.6×10^3	(4.3×10^4)
Sb-124	0.6	(16.2)	0.5	(13.5)	6.5×10^2	(1.7×10^4)
Sb-125	2	(54.1)	0.9	(24.3)	3.9×10^1	(1.0×10^3)
Sn-113	4	(108)	4	(108)	3.7×10^2	(1.0×10^4)
Sr-85	2	(54.1)	2	(54.1)	8.8×10^2	(2.4×10^4)
Sr-89	0.6	(16.2)	0.5	(13.5)	1.1×10^3	(2.9×10^4)
Sr-90	0.2	(5.41)	0.1	(2.7)	5.1	(1.4×10^2)
Tritium	40	(1080)	40	(1080)	3.6×10^2	(9.7×10^3)
Tc-99m	8	(216)	8	(216)	1.9×10^5	(5.3×10^6)
Tc-99	40	(1080)	0.9	(24.3)	6.3×10^{-4}	(1.7×10^{-2})
Th-230	2	(54.1)	2×10^{-4}	(5.41×10^{-3})	7.6×10^{-4}	(2.1×10^{-2})

(*continues*)

Table 11-4. (*continued*)

Nuclide	Normal Form, A_1		Special Form, A_2		Activity Limit	
	TBq	(Ci)	TBq	(Ci)	TBq/g	(Ci/g)
Th-232	Unlimited		Unlimited		4.0×10^{-9}	(1.1×10^{-7})
Th (natural)	Unlimited		Unlimited		8.1×10^{-9}	(2.2×10^{-7})
Tl-201	10	(270)	10	(270)	7.9×10^3	(2.1×10^5)
Tl-204	4	(108)	0.5	(13.5)	1.7×10^1	(4.6×10^2)
U (enriched)	10	(270)	1×10^{-3}	(2.7×10^{-2})	–	see §173.434
U (depleted)	Unlimited		Unlimited		–	see §173.434
V-49	40	(1080)	40	(1080)	3.0×10^2	(1.1×10^3)
Y-88	0.4	(10.8)	0.4	(10.8)	5.2×10^2	(1.4×10^4)
Y-90	0.2	(5.41)	0.2	(5.41)	2.0×10^4	(5.4×10^5)
Zn-65	2	(54.1)	2	(54.1)	3.0×10^2	(8.2×10^3)
Zr-95	1	(27)	0.9	(24.3)	7.9×10^2	(2.1×10^4)

Adapted from 49 CFR 173.435.

A shipment of radioactive material will come under one of the following major categories listed in order of increasing requirements:

- Exempt: any radionuclide $\leq 0.002\mu\text{Ci/g}(74\,\text{Bq/g})$
- Mailable: radionuclide quantities ≤ 0.0001 of A_1 or A_2 values
- "Limited Quantity": radionuclide quantities ≤ 0.001 of A_1 or A_2 values
- Type A: radionuclide quantities ≥ 0.001 of A_1 or A_2 values but A_1 or A_2 values
- Type B: radionuclide quantities $> A_1$ or A_2 values.

Objects of nonradioactive material that have surface radioactive contamination below $1\,\mu\text{Ci/cm}^2$ (averaged over 1 square meter) for almost all radionuclides can be shipped unpackaged, provided the shipment is for exclusive use and the objects are suitably wrapped or enclosed. Instruments and manufactured articles containing radioactivity are exempt from certain labeling requirements if levels are below prescribed limits (49 CFR 173.421), and articles containing natural uranium or thorium are exempt from various requirements (49 CFR 173.426). Two special categories (49 CFR 173.427) that are defined and regulated specifically are low specific activity (LSA) materials and surface contaminated objects (SCO). LSA materials are primarily high bulk, low activity ones (LSA I), tritium containing materials (LSA II), and various solids (LSA III). Such LSA materials must meet specific packaging requirements or be shipped as exclusive use. SCO materials are objects that are not themselves radioactive but have surface contamination such that the overall activity per gram is low. These materials also have specific packaging and labeling requirements for shipment.

The first three categories of shipment (exempt, mailable, and limited quantity) have the distinct advantage that they are exempt from special packaging and labeling requirements. They can thus be shipped in interstate commerce as ordinary articles as long as certain additional requirements on radiation levels are met and information is provided on shipping papers and inside the package that it contains exempt or limited amounts of radioactive material.

Exempt Materials are those that contain radioactivity in concentrations $\leq 0.002\mu Ci/g$. This *de minimus* level is very convenient in that it provides a regulatory basis for shipping many materials that may potentially contain radioactivity from laboratories and other facilities to another without the need to consider their radioactivity. Even with this lower "cutoff," many users of radioactive material need to address radioactive material shipments either in proper receipt of packages (see above) or shipping radioactive products or wastes.

Packages Mailed through the US Postal Service are required in Postal Service regulations to meet the following requirements:

- Strong, tight packages are to be used that will not leak or release material under typical conditions of transportation.
- If the contents of the package are liquid, enough absorbent material must be included in the package to hold twice the volume of liquid in case of spillage.
- The maximum dose rate on the surface of the package is less than 0.5 mrem/h.
- There is no significant removable surface contamination (i.e., less than 2,200 dpm/100 cm^2 for beta–gamma emitting radionuclides or ≤ 220 dpm/100cm^2 for alpha emitters).
- The outside of the inner container bears the marking "Radioactive Material – No Label Required."

Mailable Quantities of radioactive materials are limited to less than 10^{-4} of the A_1 or A_2 values in Table 11-4; i.e., their radioactivity content is one-tenth the limits for packages designated in DOT regulations as "limited quantity." Such packages generally contain less than 1.0 mCi of the less hazardous beta–gamma emitters in common use, and they are exempt from specific packaging, marking, and labeling regulations. The identity or nature of the radioactive contents must, however, be stated plainly on the outside of the parcel along with the full name and address of both the sender and addressee. The US Postal Service does not allow the mailing of any radioactive materials package that bears a white-I, yellow-II, or yellow-III label in accordance with DOT regulations.

"Limited Quantity" Shipments come under DOT regulations and are required to meet requirements similar to those for mailable packages; i.e., strong, tight packages that will not leak under conditions normally encountered in transportation; radiation levels $\leq 0.5\,\text{mrem/h}$ at any point on the surface; removable contamination on the external surface $\leq 2,200\,\text{dpm}/100\text{cm}^2$ beta–gamma and $220\,\text{dpm}/100\,\text{cm}^2$ alpha averaged over the surface wiped; and the outside of the inner packaging must bear the marking "Radioactive." These are designated as Type A package requirements in DOT regulations, and when used for "limited quantity" shipments the package itself does not require labeling but a notice must be placed inside the package that includes the name of the consignor or consignee and the following statement: "This package conforms to the conditions and limitations specified in 49 CFR 173.421 for radioactive material, limited quantity, n.o.s., UN2910" ("n.o.s." stands for "not otherwise specified"). Maximum quantities which can be shipped as limited quantities depend on the radionuclide shipped; i.e., $0.001A_1$ values for "special form" material or $0.001A_2$ for "normal form." "Special form" materials are solids and encapsulated sources that meet stringent test requirements (49 CFR 173.469), and "normal form" includes all other items, typically liquids and mixtures. Limits for solids are ten times those for liquids, and the limit for most of the beta-gamma radionuclides used in tracer research is 10 mCi for solids and 1 mCi for liquids. There are other exceptions for instruments and articles.

Type A Shipment/"Low-Specific-Activity" Materials

If the amount of radioactivity to be shipped is above the limited quantity cutoff, it is usually classified as "Type A" material and must be shipped as a Type A package. The "A value" refers to the maximum quantity of radioactive material that can be shipped in a "Type A" package. A_1 is the maximum value for "special form" radioactive material where leakage is minimal and the direct radiation level is the only issue for control. Special form material is of a nature that dispersal and contamination is not expected if the contents of the package are released (e.g., durable encapsulation). A_2, "normal form" material (or nonspecial form), is material that is likely to be dispersed and produce contamination and/or the potential for inhalation or ingestion if the contents are released from the package. Neither A_1 nor A_2 can exceed 1000 Ci. Type A packages are limited to the quantities listed as A_1 or A_2 in DOT regulations for each particular radionuclide (see Table 11-4); quantities that exceed the A_1 or A_2 values must be shipped in Type B packaging (see below) which is designed to withstand certain serious accident/damage test conditions (10 CFR 71.51). The NRC, which has retained authority for controlling Type B shipments, requires certification of Type B packaging and maintains a registry of approved Type B containers.

Type A Packages are self-approved by the user, but must be designed and constructed to withstand normal conditions of transport; i.e., they are totally based on performance requirements, not detailed design requirements. Neither the NRC nor DOT approves DOT Specification 7A designs; however, the user must document and retain on file the information supporting the tests and evaluation that demonstrates that the packaging complies with the performance requirements (49 CFR 173.415, 465 and 466). Containers certified to meet Type A requirements are available from commercial suppliers and typically include fiberboard boxes, wooden boxes, and steel drums strong enough to prevent loss or dispersal of the radioactive contents and to maintain the incorporated radiation shielding properties during normal conditions of transport.

Type A Shipments contain radioactive materials above "limited quantities"; i.e., the activity level of the radionuclide is in excess of 0.001 of the A_1 and A_2 values (see Table 11-4). These are controlled in three categories which must be labeled accordingly (49 CFR 172.403):

- Radioactive White-I: if the dose rate at any point on the external surface of the package is less than 0.5 mrem/h and the transportation index is 0 (a measured value ≤ 0.05 mrem/h at 1 m).
- Radioactive Yellow-II: if the dose rate at any point on the exteranal surface is greater than 0.5 mrem/h but less than 50 mrem/h, and the transport index (a measured value 1n mrem/h at 1 m) is ≤ 1.0.
- Radioactive Yellow-III: if the dose rate on the surface is greater than 50 mrem/h but less than 200 mrem/h and the transport index is 10 (transport vehicles for Radioactive Yellow-III require placarding with the sign "Radioactive").

Each of the two yellow labels has an entry for the "transport index," which is the maximum radiation level in millirem per hour (rounded up to the first decimal place) at one meter from the external surface of the package. Packages shipped by passenger-carrying aircraft cannot have a transport index greater than 3.0.

Type B Shipments are required for packages that contain radioactivity in excess of A_1 or A_2 values. Type B packages must be designed and constructed to withstand both normal and hypothetical accident conditions of transport. There are only a few DOT specification Type B packages. Most Type B packages are those which have been issued a certificate-of-compliance by the NRC (10 CFR 71.12 and 49 CFR 173.471). The user is required to have a NRC-approved QA program.

The Department of Energy (DOE) has independent authority to issue Type B package certificates [49 CFR 173.7(d)] for its programs as well as its contractors. Despite this authority, the DOE obtains NRC certification for most of its Type B packages.

In-Transit Requirements

Certain information is required to assure radiological safety while packages are in transit between a licensed shipper and a licensed or authorized receiver. The shipper chooses the proper package to insure integrity of the contents while in transit, communicates essential information via labeling and on the shipping documents, and places the package into the hands of an authorized transporter. In essence, the information thus provided is used along the route to inform others of proper handling, storage, and how to respond to incidents. Vehicles must meet various requirements in accordance with the types of packages (A or B) and provide appropriate information via placarding. In general, placards are not required for LSA Yellow-II categories and below. Vehicles used for Yellow-III shipments require placarding as "radioactive," and in all cases radiation levels at the surface of the vehicle and at a distance of 2 m must meet requirements.

Vehicles used for Transport of radioactive material must meet requirements in accordance with the type of use. If the vehicle is not exclusively for the shipment of radionuclides, the dose rate must be less than 200 mrem/h at the surface of the package and 10 mrem/h at 1 m. If, however, the vehicle is for radionuclides only and the shipment is loaded and unloaded by personnel properly trained in radiation protection, the dose rate can be as high as 1,000 mrem/h at 1 m from the surface of the package, but it cannot exceed 200 mrem/h at any point on the external surface of the vehicle, 10 mrem/h at 2 m from the external surface of the vehicle, and 2 mrem/h in any normally occupied position in the vehicle. Special written instructions must be provided to the driver to assure that exclusive use is assured throughout shipment.

If any category of LSA radioactive material is sent by taxi, the user should ensure that the taxi will not carry passengers and that the package is stored only in the trunk (49 CFR 177.870). Users who are authorized to transport radioactive materials in their own cars should be aware that their insurance policy may contain an exclusion clause with regard to accidents involving radioactive materials.

In the event of a spill, DOT regulations state (49 CFR 173.443) that vehicles may not be returned to service until the radiation dose rate at any accessible surface is less than 0.5 mrem/h and removable contamination levels are less than 2,200 dis/min per 100 cm^2 beta–gamma and 220 dis/min per 100 cm^2 alpha. Incidents that require decontamination (e.g., vehicles, packages, storage areas, streets, roadways, loading areas, etc) must be reported.

Exclusive/Sole Use Shipment

Certain advantages can be gained by shipping Type A radioactive materials as exclusive/sole use. Such shipments are exempted from specific packaging, marking, and labeling requirements; however, the materials must meet the general requirements for Type A packages; i.e., strong, tight packages so that there will be no leakage of radioactive material under conditions normally incident to transportation. The exterior of each package must be stenciled or otherwise marked "Radioactive Type 7A." There must not be any significant removable surface contamination, and external radiation levels must meet limits applicable to radioactive packages.

Shipments sent "exclusive use" or "sole use" originate from a single source and all initial, intermediate, and final loading and unloading are carried out in accordance with directions stated by the shipper or the receiver. Any loading or unloading must be performed by personnel having radiological training and resources appropriate for safe handling of the shipment. Specific instructions for maintenance of exclusive use shipment controls must be issued in writing and included with the information that accompanies the shipment (49 CFR 173.403).

Storage of Packages bearing radioactive yellow-II or radioactive yellow-III labels is limited so that the sum of the transport indexes in any individual group of packages does not exceed 50. Groups of these packages must be stored so as to maintain a spacing of at least 6 m from other groups of packages containing radioactive materials. And, if Type A radioactive materials are part of another shipment, they must be contained in packaging that meets the DOT specifications for Type A packages.

Notification of Incidents is required for events that involve possible body contamination or ingestion of radioactivity by personnel, overexposure to radiation, losses of sources, or significant contamination incidents. Notification of the NRC is required in 10 CFR 20 regulations as a license condition, and immediate notification is required if overexposures occur or property damage exceeds $200,000. Transporters notify local and state agencies and DOT of incidents, and the consigner or consignee or both. Users must report an accident to the radiation protection office at their institution, which in turn will notify the appropriate government agencies.

Shipping Papers and Shipper's Certification

Radioactive material users who act as shippers are responsible for providing shipping papers that describe the radioactive material in a specified format and the shipper's certification. Shipping papers are quite important because they are the information link between the shipper, the transporter, inspectors, emergency personnel during transit, and finally the receiver; shipping papers thus receive considerable scrutiny all along the route. Most returned shipments are due to failures in the paperwork.

For "limited quantity" shipments, the shipping papers must include the name of the consignor or consignee and the statement "This package conforms to the conditions and limitations specified in 49 CFR 173.421 for radioactive material, excepted package-limited quantity, n.o.s., UN2910." Similar wording applies to several other types of excepted articles.

For shipments of radioactive material above "limited quantities" the shipping papers must include the proper shipping name and identification number in sequence; the name of each radionuclide; the physical and chemical form; the terabecquerels (TBq) of activity (conventional units of Ci, mCi, μCi, etc., may only be stated in parenthesis as supplemental information); the category of the label (e.g., radioactive yellow-II); and the transport index. Abbreviations cannot be used unless specifically authorized (abbreviations for nuclides, e.g., Mo-99, are allowed). The following certification must be printed on the shipping paper for Type A materials: "This is to certify that the above named materials are properly classified, described, packaged, marked and labeled, and are in proper condition for transportation according to the applicable regulations of the U.S. Department of Transportation." An authorized shipper, certified by a responsible official as having met DOT training requirements, must sign the statement and the shipping paper before it will be accepted by an authorized and qualified transporter.

Compliance with Regulations

Inspections are conducted by both NRC and DOT to ensure that shipments of radioactive materials comply with regulations. The NRC inspects shipping procedures and licensee records to ensure that packages are shipped and received properly; DOT inspects shipments typically at carrier pickup and transfer terminals. The most common deficiencies in radioactive material shipments noted by DOT inspectors are

- excess radiation levels
- improper packaging and/or handling
- illegible/incorrect label notations
- inadequate provision for liquid contents
- lack of security seal
- carrier acceptance with improper description and/or without shipper's certification
- carriers exceeding 50 TI per vehicle
- failure to placard transport vehicles

SUMMARY

Radiation protection programs are designed and executed to meet radiation protection standards and the regulations that implement them, but are, as a prudent measure, conducted to maintain radiation exposures as low as is reasonably achievable (i.e., ALARA) below regulatory limits. Because of this two-pronged philosophy, most radiation activities are kept at a fraction, and many times a very small fraction, of regulatory limits.

The Atomic Energy Acts of 1946 and 1954 and subsequent amendments gave sweeping powers to the Federal government to own and regulate Atomic Energy (AE) Act materials. X rays and radium and other naturally occurring radioactive material, except uranium and thorium ores containing more than 0.05% uranium or thorium (i.e., source materials), are not regulated by the federal government under the AE Act, but are regulated to varying degrees by the states under broad authorities to protect public health. In general, no private person could possess or use AE Act materials until after the 1954 amendments, but even then only under conditions of a license issued pursuant to the AE Act. In order to have a license to possess and use AE Act material a person must demonstrate in the license application the capability to meet the government's perspective of proper control, especially for workers and the general public.

An extraordinary aspect of radiation protection, in contrast to other potentially hazardous materials, is the dominant role exercised by the NRC through its licensing and inspection authority. Parts of this authority can be delegated to a state that agrees to carry out a program "equivalent in effect" to one conducted by the NRC which retains ultimate responsibility under the AE Act; in non-agreement states the NRC is solely responsible for control of AE Act materials. Thirty-one state programs have been approved, and though not necessarily identical, each is in effect a federal program subject to initial and continuing approval by the NRC.

A major component of a radiation protection program is compliance with regulations contained in 10 CFR 19 and 20, first issued in 1954 after the Atomic Energy Act was amended to allow greater access to radioactive materials for research and industrial applications; these regulations have been revised and/or amended several times since 1954 with the most recent major revision taking effect on January 1, 1994. The purpose of the 1994 revision was to incorporate risk-based standards recommended by key national and international organizations, which represented a substantial change in the underlying philosophy for radiation protection regulations even though the control limits remained essentially the same.

Regulations for transportation of radioactive materials are based on the concept that they move freely in commerce between a shipping licensee and a receiving licensee who, based on package and shipping information provided by the shipper, can receive the material with assurance that the package has provided safe passage during transit while out of the direct control of either licensee. Although the NRC has ultimate responsibility for AE Act materials in transit, they have delegated much of this authority to the US Department of Transportation (DOT) through an agreement by which DOT regulates radioactive shipments, which is a very practical arrangement since DOT regulates shipments of most other hazardous materials. Each licensee who transports licensed material outside of its place of use or who delivers licensed material to a carrier for transport is required (10 CFR 71.5) to comply with applicable DOT regulations.

ADDITIONAL RESOURCES

1   U.S. Code of Federal Regulations—Title 10, Parts 19, 20, 30, 35, and 71
2   U.S. Code of Federal Regulations—Title 49, Parts 172 and 173

PROBLEMS

1 What is the basis for limiting doses to 0.05 Sv/y (5 rem/y) to control stochastic risks?

2 By limiting the total-body effective dose equivalent (TEDE) for workers to 0.05 Sv/y (5 rem/y), what is the best estimate, using modern risk models, of the risk associated with a working lifetime from age 18 to 65?

3 How did the ICRP radiation risk model change from 1977 to 1990 and how did this change affect its basic recommendations issued in 1990 for worker protection? Comment on the rationale and the feasibility/practicality of implementing it.

4 Explain the rationale for limiting nonstochastic risks of radiation exposure and the implementing dose limit.

5 What is the basis for treating the lens of the eye separately in occupational radiation standards and the controlling dose?

6 Explain the effect of a TEDE limit of 0.05 Sv/y (5 rem/y) on assessment of internal and external dose for workers. What practical regulatory scheme was adopted by NRC for this circumstance?

7 Determine the risk benefit of reducing the public dose limit from 0.5 mSv/y (500 mrem/y) to 0.1 mSv/y (100 mrem/y). What risk/cost rationale was used for the reduction?

8 An application for an NRC radioactive materials license (NRC Form 313) requires a statement of training for radiation safety personnel and radioactive material users. Devise the elements and instruction times needed for the training to meet NRC requirements for assuring protection of workers and the public.

9 A researcher who works with Ge-67 has received 3 rem of external dose and has been exposed to airborne Cs-137 compounds equivalent to 200 DAC-h: (a) how much Ge-67 could the person ingest in the same year if nonstochastic values are applied; and (b) if stochastic values are applied.

10 Procedures in a workplace are estimated to produce an average airborne concentration of 30 Bq/m^3 of ^{60}Co: (a) in accordance with 10CFR20 regulations, is monitoring required for workers who spend 6 h a day in the area? and (b) what is the basis for the determination?

11 A university has reached the annual limit for discharges of tritium compounds to the sanitary sewer: (a) how much tritium has been discharged? (b) if I-131 compounds are then used in the work, how much could be discharged to the sewer? and (c) if patients excrete 5 Ci of I-131 during the same year, can research work that involves release of I-131 compounds to the sewer continue?

12 A careful measurement establishes that a surface area has 900 dpm/100 cm^2 of Sr-90, but swipes taken over several 100 cm^2 areas show removable activity of 250 dpm: (a) is the area considered contaminated requiring

remedial action? and (b) if the area is cleaned and new smears contain 150 dpm/100 cm², how would it be considered?

13 If a normal form material contains 0.001 µCi/g, what options are available for transporting it to another user?

14 Mercury metal that contains 3 mCi in several tightly sealed rigid vials is to be shipped in a package with a dose rate of 20 mrem/h at the surface of the package and 0.7 mrem/h at 1 m: (a) what type of packaging is required? (b) how must it be labeled? and (c) what information is required for the shipping-papers?

12

ENVIRONMENTAL RADIOLOGICAL ASSESSMENT

"Persons generally east and northeast of the Hanford reservation who were exposed to releases from the site after 1944 refer to themselves as 'down-winders.'"

– National Cancer Institute, ~ 1995

Various radiological activities have released or have the potential to release radioactive materials to the environment, and an assessment of the radiological significance of such releases is often required. Such assessments consist of several major elements:

- a detailed description of the source term to determine the amounts, form, and rate of release of radioactive materials into a particular environmental pathway;
- determination of the distribution and concentration of released materials in environmental media where they may be inhaled or ingested directly or taken into the body as a result of contamination of food, water, or resuspension of contaminated dust or debris;
- calculation of the environmental levels associated with releases of radio-nuclides from specified sources and the radiation doses to persons exposed; and
- a determination of the health consequences associated with a particular site or activity by use of health risk models.

Radiological Assessments that incorporate these elements (source terms, envirnomental levels and health effects determinations) are generally called environmental radiological assessments; they are of three general types: current, prospective, and retrospective.

Current Radiological Assessments can be based on parameters and conditions that are generally accurate because they apply to existing operations/circumstances, or can be made so by detailed measurements. Such assessments are often calculated in lieu of measured doses/exposures because environmental levels are too low for accurate measurement; for example, doses for atmospheric releases may be based on a long-term average measurement of the amount of a given radionuclide released from a vent or stack and modeling the concentrations in various downwind locations rather than attempting to measure the long-term average air concentration at a point(s) of interest. It is also common practice to make weekly or longer-term measurements to confirm projections and/or to adjust them based on the data.

A Prospective Radiological Assessment requires a forecast of radionuclides to be released and a projection of environmental levels under various routine and accidental conditions. Such assessments are strongly dependent on the estimated source term and require the use of models and the development of site parameters as well as assumptions about populations and their consumption of food/water. Examples of this type of assessment are the siting and licensing of a nuclear facility and EPA's "brownfields" program in which sites with residual contamination may be used for productive purposes if land use controls (i.e., stewardship) are in place. Obviously, the reliability and public trust in such decisions are dependent on good radiological assessment. The National Environmental Policy Act of 1970 (NEPA) requires Federal agencies to file an environmental statement to support major actions or decisions that may affect the environment, and many radiological assessments have been driven by NEPA requirements.

Retrospective Radiological Assessments, or dose reconstructions, are often done for sites that have had long-term radiation-related activities that are questioned for various reasons and/or no longer exist. Some of the large Department of Energy or commercial power reactor sites have been closed, have come under close scrutiny, or are under consideration for decommissioning and/or release for other uses. Similar circumstances exist for various industrial sites that processed naturally occurring radioactive materials (see Chapter 14), each of which may have been committed to other uses. Reconstructing releases, environmental conditions, and exposure conditions of people can be very difficult for such circumstances, especially those that occurred several decades in the past. Such reconstructions require detailed records which may not exist or be available in the most useful form; therefore, major efforts can be expended on detective work to assemble data, and even then it is often necessary to extrapolate sparse data to construct releases and exposure conditions. Even with the most accurate models, dose reconstructions are at best central estimates that can have large uncertainties.

ELEMENTS OF ENVIRONMENTAL MODELS

Perhaps the most demanding aspect of an environmental radiological assessment is modeling the environmental behavior of radionuclides in environmental pathways. Such models account for the distribution and dilution of radioactive materials in the air, surface-water, groundwater, and terrestrial environments, and provide radioactivity concentrations that can then be used to determine human intakes. Such intakes are often governed by a complex inter-linked chain, e.g., downwind air concentrations produced by a source followed by grass deposition, cow/milk/meat accumulation, and exposure of humans due to various patterns of occupancy and food consumption. Environmental pathway models thus require many more assumptions than models for the source (or input) terms and conversion of doses to health risks.

Source Terms   may be calculated or measured either in real time or as periodic (daily, monthly, yearly) averages. Calculated source terms rely on many of the methods that generate radioactivity, as described in Chapter 4, followed by appropriate modifications that may reduce amounts actually released, e.g., holdup or filtration before release to the atmosphere through a vent or stack or perhaps a holdup lagoon before discharge to a receiving stream. Measured source terms are generally easier to obtain for operating facilities and can be quite reliable; for past operations where direct measurements are lacking, source terms are estimated from surrogate measurements or by the use of calculational models. Source terms for future operations must, of course, be based on calculations and assumed conditions.

General Principles for Modeling the Concentration   of a radionuclide in an environmental medium consist of: a source (or input) term in units of activity per unit of time (e.g., Ci/s, g/h, etc), a dispersion (or mixing) term, a velocity term for the rate of input of the diluting medium (windspeed, stream flow rate, etc.), and most important, an expression that describes the area through which the released material flows. These key factors are illustrated in Figure 12-1 in which a substance is introduced into a medium flowing at a linear rate through a projected area where mixing occurs. If the substance is added at a known rate, Q (#/s), the medium flow is u (m/s), and the flow-through area has a width y (m) and height z (m) at a specified location, the concentration (#/m^3) of the discharged material in the respective medium is

$$\text{Conc}(\#/m^3) = \frac{Q(\#/s)}{u(m/s) \cdot y(m) \cdot z(m}) [DF]$$

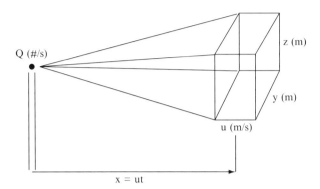

Figure 12-1. A simple box model for dispersion of a contaminant emitted at a rate Q (#/s) into a diluent stream (e.g., air or water) that has a linear flow rate u (m/s) such that uniform mixing occurs as it flows through the area of width y (m), and height z (m) to produce a concentration $\chi = Q/uyz$ (#/m³).

where u is the linear rate of flow of the dilution medium (e.g., air or water) and DF is a dimensionless dispersion factor between 0 and 1.0. The product of u and the area (yz) yields the volume of material that dilutes the pollutant in a unit of time, and if Q is the amount of pollutant released into the flowing diluent in the same unit of time then the equation yields the concentration at the specific point downstream (or downwind for a release into the atmosphere). A decay term is sometimes added to account for diminution of short-lived radionuclides during environmental transport or to account for other changes due to environmental chemistry, e.g., oxidation of the compound in air or perhaps disassociation into separate constituents.

Three Dispersion Factors thus determine the concentration of a radio-contaminant released into a diluting medium: (1) it is directly proportional to the release rate; (2) it is inversely proportional to the flow (velocity) of the mixing medium; and (3) it is inversely proportional to the area through which the contaminant/diluent mixture flows. These three principles are quite evident for discharges to a confined flow medium such as a flowing creek or air flow in a valley and calculations of concentrations are straightforward. They are also evident for dispersion of vented material in the free atmosphere, however, accurate calculation of the atmospheric concentration of material requires dispersion terms to adjust for nonuniform distribution of concentration across the plume both horizontally and vertically. The contaminant concentration in an airborne plume is highest along the centerline but decreases toward the edges such that the concentration profile across the plume (both horizontally and vertically) follows a Gaussian distribution.

WATER TRANSPORT/DISPERSION

Various discharges of radioactive or other materials occur to flowing surface streams and/or locations that result in migration through soils to groundwater.

Surface Water Dispersion   may be represented by mixing of discharged radio-activity in liquids confined by a conduit, channel, culvert, or a naturally flowing stream. A typical example is a known amount of radioactivity in a holding tank that is discharged at a constant rate into flowing water over several hours. In such circumstances the material will form a mixing zone of nonuniform concentration near the discharge point but can be presumed to become uniformly mixed a short distance downstream. The concentration beyond the mixing zone is determined by the input rate, the stream flow rate, and the width and depth of the stream after it leaves the mixing zone as shown in Example 12-1.

Example 12-1: A 12,000 L holding tank containing 0.1 curie of a radionuclide is discharged over a period of 2 hours into a small stream that is 10 meters wide and 3 meters deep that flows past the discharge point at a rate of 0.5 m/s. What is the average concentration (Ci/m^3) of the radiocontaminant over the 2-hour period at a location downstream and beyond the mixing zone?
Solution: The activity concentration in the tank is $0.1 Ci/12,000$ L $= 8.33 \times 10^{-6}$ Ci/L or 8.33 μCi/L. The discharge rate is

$$Q = \frac{12,000 \text{ L}}{2 \text{ h} \times 3600 \text{ s/h}} \times 8.33 \text{ μCi/L} = 13.89 \text{μCi/s}.$$

It can be assumed that the volume of the fluid added by the tank is minimal compared to the volume of stream flow into which the material is diluted and that sufficient turbulence exists in the stream to provide uniform mixing of the radioactivity into the stream volume; therefore, the downstream concentration is

$$\text{Conc.}(\text{μCi/m}^3) = \frac{13.89 \text{ μCi/s}}{0.5 \text{ m/s} \times 10 \text{ m} \times 3 \text{ m}}$$
$$= 0.926 \text{ μCi/m}^3$$

or 9.26×10^{-7} μCi/mL.

Dispersion of Contaminants in Groundwater   is also represented by the three general principles, and can be reasonably modeled using simplified parameters. Such an approach is often sufficient for modeling such releases because they are conservative; however, groundwater flow can be quite complex, and accurate models require sophisticated techniques.

ATMOSPHERIC DISPERSION

Atmospheric dispersion modeling is used to determine concentration profiles when a pollutant is released into the free flowing atmosphere. Such releases are usually assumed to emanate from a point source such as an elevated stack, or a small area that can be reasonably approximated as a "point" emitting material at a steady rate of Q (#/sec). The emitted material enters the streamline flow of the atmosphere that flows past the release point with a velocity u (m/s). If there is uniform turbulence in the atmospheric volume that flows past the point of release, the contaminant emitted in one second will, under these confined circumstances, be diluted in a volume of air equal to a width y (m) and height z (m) multiplied by the translocation rate of air downwind, which is just u (m/s). The volume of diluent that passes the point in one second is thus $u(xy)\text{m}^3$, and the concentration of the contaminant is just the amount released in the same second divided by the volume (in m^3) of flowing air, or

$$\chi(\#/\text{m}^3) = \frac{Q(\#/\text{s})}{u(xy)}$$

where Q is any quantity (mass, activity, etc.) expressed as a release rate (#/s). The concentration downwind is thus directly proportional to the release rate and inversely proportional to the wind speed; i.e., doubling the windspeed halves the concentration if Q remains constant. For a radionuclide release expressed as Ci/s, this general relationship becomes

$$\chi(\text{Ci}/\text{m}^3) = \frac{Q(\text{Ci}/\text{s})}{u(xy)}$$

which is reasonably accurate for plumes released at a steady rate and confined to flow through a specific area such as a plume trapped in a channeled valley of width y and height z.

Gaussian-distributed Plumes represent most releases from "point" sources because the free atmosphere is not confined. The concentration, χ, will still be directly proportional to Q and inversely proportional to the wind speed; however, the concentration varies across an areal segment of the plume at some point, x, downwind and it is necessary to introduce a dispersion term to account for the nonuniform distribution of concentration, or

$$\chi = \frac{Q(\#/\text{s})}{u??}[\text{dispersion term}]$$

Measurements of concentration profiles at distances downwind from elevated point sources have demonstrated that the concentration varies both horizontally and vertically across the plume according to a Gaussian distribution, and because of this the dispersion term has been chosen accordingly. For continuously emitting sources, diffusion in the downwind (or x) direction is completely dominated by the wind speed u; however, both the vertical and horizontal spread of the plume increase with distance downwind (as shown in Figure 12-2) and are represented by the Guassian dispersion coefficients, σ_y and σ_z, which also vary with the downwind distance x.

The general form of the Gaussian probability distribution across the plume (i.e., in the y direction) is

$$P(y) = \frac{1}{\sqrt{2\pi}\sigma_y} \exp\left[-\frac{y^2}{2\sigma_y^2}\right]$$

and vertically (in the z direction)

$$P(z) = \frac{1}{\sqrt{2\pi}\sigma_z} \exp\left[-\frac{z^2}{2\sigma_z^2}\right]$$

If the dispersion term for the y and z dimensions is expressed as the product of these two Guassian probability distributions, the downwind concentration produced by a point source emission is

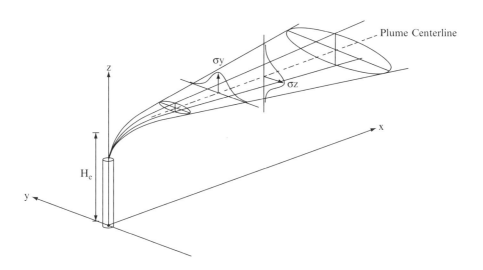

Figure 12-2. Atmospheric dispersion of material released from a point source of height H above the ground that produces Guassian-distributed concentration profiles both horizontally and vertically. Reflection of the plume by the ground surface is accounted for by a "virtual" source located at $-H$.

$$\chi(\#/m^3) = Q \frac{(\#/s)}{2\pi\sigma_y\sigma_z u} \exp\left[-\frac{y^2}{2\sigma_y^2}\right] \exp\left[-\frac{z^2}{2\sigma_z^2}\right].$$

This equation has the same features as the general form equation, but with the exponential terms added to account for plume shape. The product of the two Guassian distribution terms yields a denominator containing $2\pi\sigma_y\sigma_z$ which also has units of area (m^2) since both the horizontal and vertical distribution coefficients, σ_y and σ_z, have units of length (m). Combining this Guassian-distributed area with the wind speed (m/s) yields the volume into which the emitted contaminant is diluted. It is important to note that this equation is for "idealized" free space with no barriers. Most real world situations consist of a release point above the ground and the ground surface represents a barrier that reflects the plume.

Plume Reflection   by the ground is accounted for mathematically by adding a virtual source also emitting contaminant at a rate, Q, at an equal distance, $-H$, below the ground to produce a concentration above the ground surface; this "virtual source" represents the above-ground concentration due to reflection of the "real" plume by the ground. The virtual source emits the contaminant at the same rate, Q, thus producing a general solution for the concentration of the contaminant throughout the free space above the plane $z = 0$ (the reflecting barrier or the ground level); thus

$$\chi(x, y, z, H) = \frac{Q(\#/s)}{2\pi \bar{u}\sigma_y\sigma_z} \exp\left(\frac{-y^2}{2\sigma_y^2}\right)\left[\exp\left(\frac{-(z-H)^2}{2\sigma_z^2}\right) + \exp\left(\frac{-(z+H)^2}{2\sigma_z^2}\right)\right]$$

where
   χ = concentration $(\#/m^3)$ of the contaminant at a distance x downwind and at coordinates y = horizontal distance from the plume centerline and z = height above ground,
   Q = release rate of contaminants $(\#/s)$,
   $\bar{u}$ = mean wind speed, typically measured at 10 m above the ground,
   H = height (+ and −) of release for the "real" and the "virtual" sources,
   σ_y = horizontal dispersion coefficient (m), and
   σ_z = vertical dispersion coefficient (m).
This general equation for the concentration in a ground-reflected plume is the sum of the concentrations, χ, produced by both the "real" and "virtual" sources and is for any value of y and z. Since the ground level concentration (i.e., where $z = 0$) is usually of interest, the solution is further simplified as

$$\chi(x, y, 0, H) = \frac{Q(\#/s)}{\pi\sigma_y\sigma_z u} \exp\left[-\frac{y^2}{2\sigma_y^2}\right] \exp\left[-\frac{H^2}{2\sigma_z^2}\right] \qquad (12\text{-}1)$$

which is twice as large (i.e., the value of 2 is no longer in the denominator) due to reflection of the plume; i.e., it essentially folds over on itself to double the ground level concentration. The concentration at the point $(x, y, 0)$ is a function of the height of release, H.

And, if the main concentration of interest is along the centerline (where $y = 0$), the equation further simplifies to:

$$\chi(x, 0, 0, H) = \frac{Q(\#/s)}{\pi \sigma_y \sigma_z u} \exp\left[-\frac{H^2}{2\sigma_z^2}\right] \tag{12-2}$$

and when the release is at ground level ($H = 0$), the concentration, χ, is very similar to the general flow equation developed earlier, but now with the flow-through area expressed in terms of the Guassian plume dispersion coefficients, σ_y and σ_z:

$$\chi = (x, 0, 0, 0) = \frac{Q(\#/s)}{\pi \sigma_y \sigma_z \bar{u}} \tag{12-3}$$

where the values of σ_y and σ_z that fit a Guassian plume concentration profile are governed by atmospheric properties, principally the vertical temperature profile.

Atmospheric Stability Effects on Dispersion

The values of σ_z and σ_z that fit a Guassian plume concentration profile are governed by atmospheric properties, principally the effect of the vertical temperature profile on the turbulence field which in turn determines both the vertical and horizontal spread of the plume. Classes of atmospheric stability have, therefore, been defined in terms of the temperature profile with height and variations in wind speed/direction, each of which affects the turbulence field. Various combinations of these parameters yield different types of plumes as shown in Figure 12-3 relative to the dry adiabatic lapse rate which is $-9.86°\,C/km$ or about $1°F/100\,m$ or $5.4°F/1000\,ft$.

The vertical temperature profile determines plume shape, as shown in Figure 12-3. If the temperature profile is steeper than the dry adiabatic lapse rate, a displaced parcel of the air/contaminant-mixture in a plume will continue to rise or sink, as shown in Figure 12-3(a). If the atmospheric temperature profile, as illustrated in Figure 12-3(b), is the same as the dry adiabatic lapse rate a fluctuation in the atmosphere that moves the parcel either up or down will change its temperature adiabatically and the parcel will thus be dispersed only as far as the disturbing force moves it. This "neutral" condition (Figure 12-3b) results in plumes that spread in the shape of a cone. And if, as shown in Figure 12-3(c), the vertical temperature profile is less than the dry adiabatic rate, the parcel will tend to be restored to its original position.

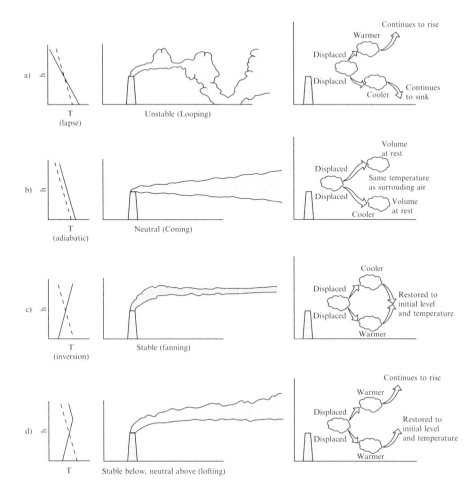

Figure 12-3. Effects of the vertical temperature profile on plume shape: (a) lapse conditions yield looping plumes; (b) neutral conditions produce cone-shaped plumes; (c) inversion conditions result in long thin plumes; and (d) inversion below/neutral above produces a condition known as lofting.

Lapse Conditions occur when the atmospheric temperature profile is steeper than the dry adiabatic lapse rate (Figure 12-3a). If a parcel of the air/contaminant mixture in a plume is quickly displaced upward, its temperature will decrease in absolute terms due to adiabatic cooling but its temperature relative to the surrounding atmosphere will be warmer, causing it to become increasingly buoyant. On the other hand, if the parcel is displaced downward its temperature will fall more rapidly than that of the surrounding atmosphere and its relatively greater density will cause it to settle even further. Strong lapse (superadiabatic temperature) conditions produce looping plumes that spread out over a wide area; thus, they are inherently *unstable* and are highly favorable for the dispersal of pollutants.

Temperature Inversions   yield very thin plumes, as shown in Figure 12-3c, due to highly stable conditions. Fluctuations that displace the parcel upward produce adiabatic cooling such that it is more dense than the surrounding atmosphere, consequently its upward motion stops and it returns to its original level. Likewise, any movement of the parcel downward causes its temperature to increase relative to the atmosphere and its buoyancy causes it to return to its original level. The atmosphere in this case is said to be stable and pollutant dispersion is minimal. This circumstance also occurs for releases into an isothermal profile, but is not as pronounced. Light winds are generally associated with stable conditions and if the wind direction is steady the plume becomes a long, meandering ribbon that extends downwind at the altitude of the stack, but if it fluctuates significantly, the plume spreads out in the horizontal plane like a fan; hence the term *fanning*. A fanning plume is not necessarily an unfavorable condition for the dispersion of effluents since the plume is quite wide and does not touch the ground.

Lofting   is a condition that occurs when a temperature inversion exists below the plume height (Figure 12-3d). It typically occurs just after sunset as radiant cooling of the earth builds up a night-time temperature inversion such that stable air exists below the plume and unstable or neutral conditions continue to exist above it. Lofting is a most favorable condition because effluents can disperse vertically but are kept away from the ground by the stable air below; dispersion thus occurs for great distances throughout large volumes of air.

Atmospheric Stability Classes

Seven categories of atmospheric stability have been defined, each of which can be described qualitatively in terms of the general atmospheric conditions under which they occur:

    Class A: Extremely Unstable Conditions (bright sun, daytime)
    Class B: Moderately Unstable Conditions (sunny, daytime)
    Class C: Slightly Unstable Conditions (light cloudiness, daytime)
    Class D: Neutral Conditions (overcast sky, brisk wind, day or night)
    Class E: Slightly Stable Conditions (early evening, light winds, relatively
             clear sky)
    Class F: Moderately Stable Conditions (late night, light wind, clear sky)
    Class G: Very Stable (predawn, very light wind, clear sky)

These general conditions can also be summarized in terms of wind speed and the amount of sunshine or cloudiness, as shown in Table 12-1, but they are still mostly qualitative.

Table 12-1. Atmospheric Stability Class as a Function of Windspeed and Amount of Solar Insolation.

Windspeed	Daytime Conditions			Nighttime Conditions	
(m/s @ 10 m)	Strong Sun	Moderate Sun	Cloudy	> 4/8 Clouds	Clear Sky
< 2	A	A–B	B	E or F	F or G
2–3	A–B	B	C	E	F
3–5	B	B–C	C	D	E
5–6	C	C–D	D	D	D
> 6	C	D	D	D	D

Table 12-2. Stability Class (or Pasquill Category)[a] vs Vertical Temperature Profiles ($\Delta T/\Delta z$) and the Standard Deviation in the Horizontal Wind Direction, σ_θ.

Stability Class	$\Delta T/\Delta z$ (°C/100 m)	σ_θ (degrees)[b]
A	< −1.9°C	> 22.5°
B	−1.9 ≤ −1.7	17.5 ≤ 22.5
C	−1.7 ≤ −1.5	12.5 ≤ 17.5
D	−1.5 ≤ −0.5	7.5 ≤ 12.5
E	−0.5 ≤ +1.5	3.8 ≤ 7.5
F	+1.5 ≤ +4.0	2.1 ≤ 3.8
G	> +4.0°C	< 2.1°

[a]From Regulatory Guide 1.23 US Nuclear Regulatory Commission, 1980.
[b]Measured at windspeeds ≥ 1.5 m/s

A more quantitative determination of stability class is based on measured values of temperature versus elevation and/or the standard deviation, σ_θ, of the mean wind direction. Stability classes that correspond to measured values of these parameters are listed in Table 12-2.

Calculational Procedure: Uniform Stability Conditions

The first step in performing atmospheric dispersion calculations is to determine the stability class that exists during the period of consideration which in turn establishes the dispersion coefficients, σ_y and σ_z, for a specific downwind distance, x. The stability class can be chosen by one of three methods: (1) a qualitative determination based on the general conditions listed in Table 12-1; (2) measured values of temperature change with elevation as listed in Table 12-2; or (3) the standard deviation, σ_θ, of the mean wind vector, also listed in Table 12-2. Once the stability class is known, dispersion coefficients, σ_y and σ_z, are determined from Figures 12-4 and 12-5 for stability classes A–F. For

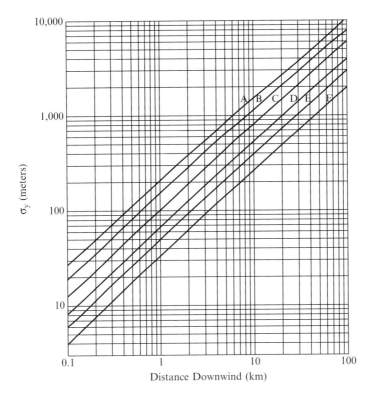

Figure 12-4. Values of the horizontal Guassian dispersion coefficient, σ_y (m), vs distance and atmospheric stability class (A–F).

extremely stable Type G conditions, the dispersion coefficients, σ_y and σ_z, are computed from Type F values since Figures 12-4 and 12-5 do not contain curves for stability class G (which, though not defined in the original work by Hay-Pasquill, was later added by NRC), or

$$\sigma_y(G) = 2/3\sigma_y(F)$$

$$\sigma_z(G) = 3/5\sigma_z(F)$$

The emission rate Q, the stack height, H, and the coordinates x, y, z are usually known and/or can be selected. The general procedure for calculating atmospheric dispersion is illustrated in Example 12-2.

Example 12-2: A 60-m-high stack discharges a radionuclide at a rate of 80 Ci/s into a 6 m/s wind during the afternoon under overcast skies: (a) determine the ground-level concentration at 500 m downwind and 50 m off the plume centerline; and (b) at the same downwind distance but along the centerline.

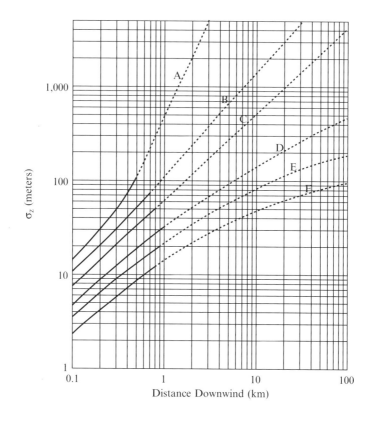

Figure 12-5. Values of the vertical Guassian dispersion coefficient, σ_z (m), vs distance and atmospheric stability class (A–F).

Solution: (a) From Table 12-1, the atmospheric stability is judged to be class D due to overcast conditions and the relatively brisk wind. From Figure 12-4, it is determined that σ_y is 36 m for class D stability at 500 m and σ_z (from Figure 12-5) is 18.5 m; therefore, for $x = 500$ m, $y = 50$ m, and $H = 60$ m,

$$\chi(500, 50, 0, 60) = \frac{80\,\text{Ci/s}}{(3.14)(36\,\text{m})(18.5\,\text{m})(6\,\text{m/s})}\left[\exp-\frac{(60)^2}{2(18.5)^2}\right]\left[\exp-\frac{(50)^2}{2(36)^2}\right]$$

$$= 1.27 \times 10^{-5}\,\text{Ci/m}^3$$

(b) For $x = 500$ m and $y = 0$, the second exponential term is equal to 1.0, and

$$\chi(500, 0, 0, 60) = \frac{80\,\text{Ci/s}}{(3.14)(36\,\text{m})(18.5\,\text{m})(6\,\text{m/s})}\left[\exp-\frac{(60)^2}{2(18.5)^2}\right]$$

$$= 3.34 \times 10^{-5}\,\mathrm{Ci/m^3}$$

yielding, as expected, a higher concentration.

Distance, x_{max}, of Maximum Concentration (χ_{max})

The point at which the maximum ground-level concentration occurs is often of interest especially for design purposes and emergency planning. Solution of the general dispersion equation for the point of maximum downwind concentration indicates that it occurs along the centerline when the vertical dispersion coefficient, σ_z, has the value

$$\sigma_z = 0.707\,\mathrm{H}$$

which is of course dependent on the stability class. This specific value of σ_z is found in Figure 12-5 for the particular stability class and the corresponding value of x (i.e., x_{max}) at which χ_{max} occurs is determined. Once the location, x_{max}, is established, the maximum concentration can then be calculated using the values of σ_y and σ_z that correspond to that distance (it is only necessary to determine σ_y since σ_z has already been calculated based on x_{max} and the stability class). The maximum concentration is

$$\chi_{max} = \frac{Q(\#/s)}{\pi \bar{u}\sigma_{y,\,max}\sigma_{z,\,max}} \times \frac{1}{e}$$

$$= 0.117\,\frac{Q(\#/s)}{\bar{u}\sigma_{y,\,max}\sigma_{z,\,max}}$$

when the values for $\pi = 3.14159$ and $e = 2.7183$ are inserted.

The values of $\sigma_{y,\,max}$ and $\sigma_{z,\,max}$ are determined for the point downwind where the maximum concentration occurs. These relationships are applicable to elevated stack releases only; the maximum concentration for a ground-level release is at the point of release and the downwind concentration gradually decreases with increasing x.

Example 12-3: Estimate: (a) the downwind point of maximum concentration for effluent released at 10 Ci/s at a height of 30 m under Type F stability into a steady wind of 1 m/s, and (b) the maximum concentration at that location.
Solution: (a) For these conditions, $\sigma_z = 0.707H = 21.2\,\mathrm{m}$; therefore, as shown in Figure 12-5, σ_z has the value 21.2 m at about 1900 m (1.2 miles) which is x_{max}.
(b) At $x = 1900\,\mathrm{m}$, $\sigma_y = 60\,\mathrm{m}$ and $\sigma_z = 21.2\,\mathrm{m}$ (already determined); thus the maximum concentration is

$$\chi_{max} = \frac{0.117 \times 10\,\mathrm{Ci/s}}{1\,\mathrm{m/s} \times 21.2\,\mathrm{m} \times 60\,\mathrm{m}} = 9.2 \times 10^{-4}\,\mathrm{Ci/m^3}$$

STACK EFFECTS

The Guassian plume model can be used for stack releases if the release approximates a "point" and the wind field has homogeneous turbulence. These conditions can be assumed for tall slender stacks located such that the turbulence induced by surrounding structures and terrain is minimal: a general rule of thumb is that the top of the stack should be 2.5 times higher than any surrounding buildings or obstructions that may induce turbulence that would produce inhomogeneities at the point of release. Gaussian plume conditions may also exist for shorter stacks if releases occur during moderate to light winds speeds, but they may fail to exist even for tall stacks if wind speeds are very high.

The stack itself exerts effects on plume dispersion, some of which may promote dispersion while others may reduce it. The temperature of the effluent can induce buoyancy (positive or negative) and the momentum of the effluent can affect the height at which the plume levels off. Such effects are considered in terms of effective stack height, stack concentration effects, and stack turbulence effects.

The Effective Stack Height is that elevation above the physical stack height where the plume levels off (see Figure 12-2), often referred to as plume rise. The amount of plume rise, or *dH*, is governed by the efflux velocity which gives momentum to the effluent as it leaves the stack, the horizontal wind speed, and the differential temperature of the effluent versus that of the receiving air stream. For example, a hot plume will be buoyant and if ejected with a high efflux velocity may rise quite a bit above the physical stack height before being leveled off by a combination of the wind speed and the cooling effect of the receiving air. The effective stack height, H_{eff}, is obtained by adding the calculated plume rise, dH, to the physical stack height, or

$$H_{eff} = H + dH$$

Various models have been devised for determining *dH*, one of which was devised by Davidson using Bryant's data.

$$dH = d\left(\frac{v_s}{\bar{u}}\right)^{1.4}\left(1 + \frac{T_s - T_a}{T_s}\right)$$

where
$\quad dH$ = differential plume rise (m)
$\quad d$ = inside diameter of stack (m)
$\quad v_s$ = velocity of effluent in the stack (m/s)
$\quad \bar{u}$ = mean wind velocity (m/s) at the top of the stack
$\quad T_s$ = temperature in the stack (^{o}K)
$\quad T_a$ = temperature of the atmosphere (^{o}K)

The first term in the Davidson-Bryant model accounts for momentum effects due to the velocity of the discharged effluent and the horizontal translation due to the wind field; the second term accounts for the temperature differential between the two fluids.

Example 12-4: An effluent of 100 °C is released from a 100-m-high stack with an inside diameter of 2.44 m and at an efflux velocity of 13.7 m/s into the atmosphere which has a mean windspeed of 3 m/s and a temperature of 18 °C. Calculate the effective stack height, $(H + dH)$.
Solution:

$$dH = 2.44\,\mathrm{m}\left(\frac{13.7\,\mathrm{m/s}}{3\,\mathrm{m/s}}\right)^{1.4}\left(1 + \frac{(100 + 273) - (18 + 273)}{273 + 100}\right)$$

$$= 20.456(1.22) = 25\,\mathrm{m}$$

and,

$$H_{\mathrm{eff}} = H + dH = 100 + 25\,\mathrm{m}$$
$$= 125\,\mathrm{m}$$

Changing the Stack Concentration by increasing the air flow in the stack has little effect on the resulting downwind concentration which is not intuitively obvious. The calculated value of χ is governed primarily by the emission rate Q which would stay the same. The increased air flow may, however, increase the effective stack height, but in general this has minimal influence on χ.

Mechanical Turbulence Induced by a Stack can cause significant tip vortices, and a zone of low pressure may develop on the lee side of the stack, especially during high winds. If the plume is discharged at a relatively low velocity or is not particularly buoyant, it can be drawn into the lee of the stack creating a negative dH. In very high winds, the plume appears to exit the stack several feet below the top. Large stacks often have discoloration several feet below the top due to these combined effects.

Checkpoints

The Gaussian plume model for point source releases is based on very specific conditions; therefore, it is applicable *only* for the following essential conditions:

- for points of release 2.5 times the height of any surrounding structures,

- a steady-state release rate into an atmosphere with homogeneous turbulence,
- unobstructed releases over flat, open terrain where complete reflection of the plume by the ground (i.e., minimal plume depletion) can be assumed,
- diffusion times of 10 min or more, and
- positive values of wind speed (equations fail for zero windspeed).

Despite these restrictions, the models are often used for other conditions because they are the only ones available. It must be recognized, however, that the quality of the results will be influenced by the degree to which conditions deviate from the model, and the results should be used accordingly.

NONUNIFORM TURBULENCE—FUMIGATION, BUILDING EFFECTS

Various conditions can occur in which the turbulence associated with an airborne plume will be nonuniform and it is not appropriate to use a Guassian plume model to determine the plume concentration. Two common situations are fumigation and mechanical turbulence induced by buildings or other obstructions, and a different model is required for each.

Fumigation

A condition of nonuniform atmospheric turbulence often occurs for elevated releases of contaminants due to the breakup of a temperature inversion, at and above the stack. Vertical dispersion of contaminants in the plume is limited by the stable atmosphere associated with the inversion and contaminant concentrations in the thin layer of the confined plume can be quite high. These conditions occur well after sunset during low winds and clear nights which promote radiant cooling of the earth's surface and the air near the ground.

When the sun comes up it heats the ground which in turn produces a zone of turbulent air due to thermal eddies. When the turbulent air zone ascends to the height of the plume, the trapped plume material is rapidly dispersed to ground level, producing relatively high concentrations at ground level, as illustrated in Figure 12-6. Effluents emitted during this period continue to be confined by the inversion overhead, and will be dispersed toward the ground by the turbulence in the newly heated unstable air. Such events are called fumigation, and the ground-level concentration during such conditions is

$$\chi_{fum}(h_e > 10\,m) = \frac{Q(\#/s)}{(2\pi)^{1/2}\bar{u}_h \sigma_y H_e}$$

$$= 0.4\frac{Q(\#/s)}{\bar{u}_h \sigma_y H_e}$$

Figure 12-6. Fumigation breakup of a trapped plume producing a zone of relatively high concentration between the ground and the bottom of the inversion layer, typically at H_e.

where

$\bar{u}_h$ = average wind speed for the layer at H_e (m/s)
σ_y = lateral plume spread (m), typically for class F or G stability
H_e = effective stack height (m)

This fumigation equation cannot be used when H_e becomes small (on the order of 10 m) because calculated χ/Q values will become unrealistically large. For releases < 10 m, it is more realistic to model the dispersion as a ground level release where H_e is replaced by the appropriate value of σ_z.

Example 12-5: If a unit of radioactivity is released from a 100-m stack during Category F stability during which the windspeed is 2 m/s, what is the χ/Q value for: (a) a plume centerline location 3000 m downwind for normal conditions; and (b) during fumigation?

Solution: (a) For non-fumigation (i.e., normal Gaussian plume dispersion), $H_e = 100$ m, $\sigma_y = 100$ m, and $\sigma_z = 280$ m.

$$\chi/Q = \frac{1}{\pi \bar{u} \sigma_y \sigma_z} \exp\left[-\frac{H_e^2}{2\sigma_z^2}\right]$$

$$\chi/Q = \frac{1}{\pi(2\,\text{m/s})(100\,\text{m})(280\,\text{m})} \exp\left[-\frac{(100\,\text{m})^2}{2(280\,\text{m})^2}\right]$$

$$= 5.33 \times 10^{-6}\,\text{s/m}^3$$

(b) For fumigation conditions where $H_e = 100$ m, and $\sigma_y = 100$ m (Figure 12-4),

$$\chi/Q = \frac{1}{(2\pi)^{1/2} \bar{u}_h \sigma_y h_e}$$

$$= \frac{1}{(2\pi)^{1/2}(2\,\text{m/s})(100\,\text{m})(70\,\text{m})}$$

$$= 2.8 \times 10^{-5}\,\text{s/m}^3$$

The ground-level concentration for fumigation effects is, as expected, considerably larger than that calculated for nonfumigation conditions.

An Elevated Receptor may exist at an elevation above the base of the stack but below the effective stack height, and it is necessary to modify the basic dispersion equation for the concentration along the centerline (i.e. y = 0) as follows:

$$\chi = \frac{Q(\#/s)}{\pi \bar{u}_h \sigma_y \sigma_z} \exp\left[\frac{-(H_e - h_t)^2}{2\sigma_z^2}\right]$$

where

$\bar{u}_h$ = wind speed representing conditions at the release height,
H_e = effective stack height above plant grade (m),
h_t = maximum terrain height (m) above plant grade and is the point at which the calculation is made (h_t cannot exceed H_e).
If ($H_e - h_t$) is less than about 10 m, then the condition should be modeled as a ground-level release. An example of the ground-level concentration for an elevated receptor point is shown in Example 12-6.

Example 12-6: Assuming non fumigation conditions, what is the χ/Q value 3000 m downwind for a receptor located 30 m above plant grade if the point of release is a 100-m stack, the wind speed is 2 m/s, and the stability is class F?
Solution: Since the height of the receptor, h_t = 30 m, the effective release point is 100 m − 30 m = 70 m, and σ_y (Figure 12-4) and σ_z (Figure 12-5) are 100 m and 280 m, respectively; thus

$$\chi/Q = \frac{1}{\pi \bar{u}_h \sigma_y \sigma_z} \exp\left[\frac{-(H_e - h_t)^2}{2\sigma_z^2}\right]$$

$$= \frac{1}{\pi (2\,\text{m/s})(100\,\text{m})(280\,\text{m})} \exp\left[\frac{-(70\,\text{m})^2}{2(280\,\text{m})^2}\right]$$

$$= 5.5 \times 10^{-6}\,\text{s/m}^3$$

which is essentially the same result obtained in Example 12-5(a) for level terrain; i.e., a slightly elevated receptor has less effect than other conditions of the dispersion field.

Dispersion from an Elevated Receptor during Fumigation can also be modeled by adjusting the value of H_e in the fumigation equation similar to that done in Example 12-6 to account for the elevated point of interest. This in effect decreases the area through which the plume material flows at the receptor location which in turn increases the ground-level concentration due to the "trapped" plume. For protection purposes, this is a conservative approach because the actual vertical dispersion depth is likely to be greater due to the elevated receptor location, especially if it is on a small hill or due to a gradual change in the terrain.

Building Wake Effects—Mechanical Turbulence

The Guassian diffusion equations were developed for idealized conditions of point-source releases over relatively flat unobstructed terrain where atmospheric turbulence, as characterized by the vertical temperature profile, is fairly homogeneous. These idealized conditions rarely exist, but as a practical matter the point source equations can be used for flat and relatively unobstructed terrain and release points that are 2.5 times higher than surrounding buildings or terrain disturbances (e.g., cliffs or rock outcrops).

Mechanical Turbulence that completely dominates the dispersion pattern can exist due to a combination of moderate to strong winds, a low efflux velocity, and fairly short release points on sizeable structures. A common situation in which near-field mechanical turbulence dominates the dispersion of an effluent is a short stack or vent atop a sizeable building as shown in Figure 12-7. Effluent released under such conditions will be drawn into the building wake produced by mechanical turbulence induced by the building and special considerations are required for determining plume concentrations.

Figure 12-7. Photograph of building-wake entrapment of a plume released in a 6 m/s wind from a stack 3 m above a research reactor building approximately 15 m high with a width projection of about 15 m.

Concentrations of Effluents in Building Wakes are determined by the size of the wake itself. The air in the wake is usually thoroughly mixed because of the turbulence induced by the building and any effluent introduced into it will also be uniformly mixed. A straightforward method for determining the concentration of a plume confined to flow in the lee of a building is to determine the wake volume and calculate the dilution it provides for the emission rate Q, as shown in Example 12-7. The building wake volume is, as shown in Example 12-7, a function of the area the building projects downwind and the wind speed.

Example 12-7: Estimate the street-level concentration at approximately 120 m downwind of the 3 m stack atop the large building in Figure 12-7 ($h = 15$ m; $w = 15$ m) due to a steady-state release of 5 μCi/s of ^{41}Ar when the wind speed is 5 m/s.

Solution: The smoke test in Figure 12-7 was conducted during 5 m/s winds and clearly shows building wake entrapment due to mechanical turbulence induced by the large building; therefore, the plume flows near the ground where its height, as estimated from the figure is about the same as that of the stack. Its width, which is more uncertain from the figure, can be assumed to be the width of the building projected downwind; i.e., no channeling due to windstreaming. The calculated concentration at ~ 120 m is

$$\chi = \frac{Q}{\bar{u} \cdot h \cdot w} = \frac{5\,\mu Ci/s}{5\,m/s \times 18\,m \times 15\,m} 3.7 \times 10^{-3}\,\mu Ci/m^3$$

The ground-level release equation has been modified in various ways to account for mechanical turbulence, one of which has been proposed by Fuquay as

$$\chi = \frac{Q(\#/s)}{u(\pi\sigma_y\sigma_z + cA)}$$

where A, the area of the building on the lee side, is adjusted by the factor c to obtain the area through which the plume flows due to the dominant influence of the mechanical turbulence caused by the building. For downwind distances within about 10 building heights, c ranges between 0.5 and 2.0. Use of small values of c (i.e., $c = 0.5$) yields conservative (i.e., higher) estimates of plume concentrations because of the small flow-through area of the wake. The quantity cA has the greatest effect on the plume concentration at shorter distances downwind because $cA \gg \pi\sigma_y\sigma_z$; however, at larger distances $\pi\sigma_y\sigma_z$ will dominate such that the equation accounts for wake effects both close to the building and at locations further downwind. Values of σ_y and σ_z are determined for distances of x measured from the center of the building.

A Virtual Source Method can also be used to model dispersion of material under the influence of mechanical turbulence. Since the effect of the obstruction is to disperse the released contaminant into the wake causing it to flow along the ground, it can be modeled as though it were a ground-level release. To do so, a ground-level "virtual source" is placed at a location chosen upwind of the point of release such that the plume produced by the "virtual source" would have a height and horizontal spread that just envelops the building as it passes it. The downwind concentration is then modeled as a Guassian-distributed ground-level release at the point of interest downwind as though the building were not present; i.e., the fictitious source would produce a concentration

$$\chi = \frac{Q(\#/s)}{\pi \bar{u} \sigma_{y,x_o} \sigma_{z,x_o}}$$

where σ_{y,x_o} and σ_{z,x_o} are determined for downwind distances measured from the location of the ground-level virtual source. The values of $\sigma_{y,\,x_o}$ and $\sigma_{z,\,x_o}$ that correspond to a plume that just envelops a building of height H_b and width W are

$$\sigma_{z,x_o} = \frac{H_b}{2.14}$$

$$\sigma_{y,x_o} = \frac{W}{4.28}$$

and these calculated values of σ_{y,x_o} and σ_{z,x_o} are used to obtain the upwind virtual-source distances x_{zo} and x_{yo} by entering them separately into the curves in Figures 12-4 and 12-5 for the particular stability class that exists. These two values are added to the downwind distance of interest to establish the value of x at which $\sigma_{y,\,x_o}$ and $\sigma_{z,\,x_o}$ are determined. One can use either of the two distances, or both, for locating the virtual source, but for buildings it is usually based on x_{zo}, as shown in Example 12-8.

Example 12-8: Use the virtual source method to estimate the concentration of radiocesium 200 m downwind of a source that emits 0.5 Ci/s from a 2 m vent atop a building that is 30 m tall and 20 m wide during Type D stability and a wind speed of 6 m/s.
Solution: A ground-level source emitting a plume that would just engulf a 30-m high building when it reaches the building corresponds to a σ_{z,x_o} value of

$$\sigma_{z,x_o} = \frac{30 \, m}{2.14} = 14 \, m$$

Inserting this value into Figure 12-4 for Type D stability corresponds to a downwind distance of 350 m; therefore, the concentration 200 m downwind of the building is modeled as if it were emitted from a ground-level "virtual" source located 350 m + 200 m upwind. At 550 m $\sigma_{y, x_o} = 40$ m and $\sigma_{z, x_o} = 20$ m and the concentration at 200 m downwind of the release point is

$$\chi = \frac{0.5 \, \text{Ci/s}}{\pi \times 6 \, \text{m/s} \times 40 \, \text{m} \times 20 \, \text{m}}$$
$$= 3.32 \times 10^{-5} \, \text{Ci/m}^3$$

Ground-level Area Sources can also be modeled by the "virtual source" method which is very useful for determining downwind concentrations for emissions from contaminated areas such as a waste disposal site, or a treatment pit or lagoon. The virtual source distance for these circumstances is based on the width of the area and it is this dimension that is used to establish x_{yo} and thus σ_{y, x_o} as shown in Example 12-9. The same value (i.e., $x_{yo} = x_{zo}$) is generally used for determining σ_{z, x_o} unless the height of the plume above the area is known.

Example 12-9: Radon gas is exhaled from a roughly circular area 50 m in diameter at a flux of 10 pCi/m² · s. Estimate the concentration of radon in air 400 m downwind for an average wind speed of 4 m/s and an assumed condition of neutral (Type D) atmospheric stability.
Solution: The total emission rate for the area is 19,635 pCi/s which can be modeled as coming from a virtual ground-level point source located upwind such that the plume width during type D stability conditions just envelops the area as it passes over it. This value of σ_{y, x_o} is

$$\sigma_{y, x_o} = \frac{W}{4.28} = \frac{50 \, m}{4.28} = 11.7 \, \text{m}$$

From Figure 12-4 this value of σ_{y, x_o} occurs ~ 100 m downwind of a point source during class D stability; therefore, the concentration at all other distances is modeled as emanating from a virtual point source located 100 m upwind of the center of the area. For a location 400 m downwind, or $x_o = 500$ m, $\sigma_{y, x_o} = 36$ m and $\sigma_{z, x_o} = 18.5$ m from Figures 12-4 and 12-5, respectively, and the radon concentration is

$$\chi = \frac{Q}{\pi u (\sigma_{y, x_o})(\sigma_{z, x_o})}$$
$$= \frac{19,365 \, \text{pCi/s}}{\pi \times 4 \, \text{m/s} \times 18.5 \, \text{m} \times 36 \, \text{m}} = 2.3 \, \text{pCi/m}^3$$

Effect of Mechanical Turbulence on Far-Field Diffusion

The effect of mechanical turbulence on plume concentrations in the far-field of buildings (i.e., at distances greater than 10 building heights) can be determined by use of selection rules applied to dispersion coefficients, χ/Q, calculated by three modified ground-release equations:

$$\chi/Q = \frac{1}{\bar{u}_{10}(\pi\sigma_y\sigma_z + 0.5A)} \tag{12-4}$$

$$\chi/Q = \frac{1}{\bar{u}_{10}(3\pi\sigma_y\sigma_z)} \tag{12-5}$$

$$\chi/Q = \frac{1}{\bar{u}_{10}\pi\Sigma_y\sigma_z} \tag{12-6}$$

where σ_y and σ_z are determined for the downwind distance, x, measured from the center of the building and

$\chi/Q =$ dispersion factor per unit release (s/m³)
$\bar{u}_{10} =$ mean wind speed (m/s) at 10 m above the ground
$A =$ smallest vertical-plane cross-sectional area (m²)
Σ_y lateral plume spread based on meander and building wake effects (m)

The value of Σ_y in equation 12-6 is a calculated quantity that accounts for horizontal (lateral) plume meander, which should be considered during neutral (type D) or stable (type E, F, or G) atmospheric stability conditions when the wind speed at the 10-m level is less than 6 m/s; it is thus a function of atmospheric stability, the wind speed $\bar{u}_{10}$, and the downwind distance. For downwind distances of 800 m or less, $\Sigma_y = M\sigma_y$ where M is a plume meander factor determined from Figure 12-8; for downwind distances greater than 800 m, $\Sigma_y = (M-1)\sigma_{y800\,m} + \sigma_y$ where $\sigma_{y800\,m}$ is the value of σ_y determined from Figure 12-4 at 800 m for the respective stability class. It is not necessary to consider plume meander during unstable atmospheric conditions (type A, B, or C stability) and/or wind speeds of 6 m/s or more because building wake mixing becomes more effective in dispersing effluents than meander effects. This approach, which is described in NRC Regulatory Guide 1.145, eliminates the need to find an upwind virtual source distance; it does, however, rely on **Selection Rules** which are as follows:

- For neutral (type D) or stable (type E, F, G) atmospheric conditions the *higher* of the two values calculated by equations 12-4 and 12-5 is compared to the value from equation 12-6 and the *lower* of these two values is selected as the appropriate dispersion coefficient, or χ/Q value.
- For unstable conditions (type A, B, C) the appropriate χ/Q value is the higher of the two values obtained from equation 12-4 or equation 12-5.

Example 12-10: Determine the appropriate χ/Q value for radioactivity released from a short vent on a building with a cross-sectional area of 2000 m² during Type F atmospheric stability conditions and a wind speed of 2 m/s at: (a) 100 m; and (b) 1000 m downwind.

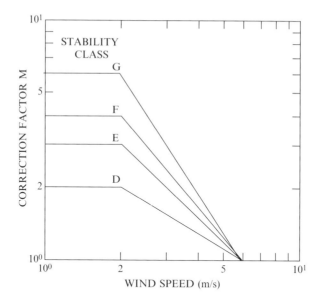

Figure 12-8. The meander factor, M, for determining Σ_y for far-field dispersion downwind of buildings and obstacles.

Solution: This type of dispersion is considered a ground-level release for calculational purposes using equations 12-4, 12-5, and 12-6 in accord with the selection rules. (a) For a downwind distance of 100 m: $\sigma_y = 4$ m and $\sigma_z = 2.4$ m; $M = 4$ (Figure 12-8) and $\Sigma_y = M\sigma_y = 16$ m; and the area-modifying factor is conservatively chosen as c = 0.5. By equation 12-4

$$\chi/Q = \frac{1}{(2\,\text{m/s})[\pi(4\,\text{m})(2.4\,\text{m}) + (2000\,\text{m}^2 \times 0.5)]}$$
$$= 4.9 \times 10^{-4}\,\text{s/m}^3$$

by equation 12-5

$$\chi/Q = \frac{1}{(2\,\text{m/s})[(3)\pi(4\,\text{m})(2.4\,\text{m})]}$$
$$= 5.5 \times 10^{-3}\,s/m^3$$

and by equation 12-6 where $\Sigma_y = 4\sigma_y = 16$ m

$$\chi/Q = \frac{1}{(2\,\text{m/s})[\pi(16\,\text{m})(2.4\,\text{m})]}$$
$$= 4.1 \times 10^{-3}\,s/m^3$$

Application of the selection rules for Type F stability yields a choice of 4.1×10^{-3} s/m^3 for the dispersion factor.

(b) Similar calculations for 1000 m yield χ/Q values of 1.8×10^{-4}s/m^3 (Eq. 12-4), 9.5×10^{-5} s/m^3 (Eq. 12-5) and 8.7×10^{-5} s/m^3 (Eq. 12-6). The selected value based on the selection rules is 8.7×10^{-5} s/m^3.

PUFF RELEASES

Contaminants can also be released in a single burst during emergencies when overpressurization occurs (perhaps an explosion) or as a result of a sudden spillage. Such releases are characterized as puffs of very short duration in which a discrete volume of material travels downwind, expanding in all directions due to atmospheric effects. Such an expansion causes the quantity of the contaminant, Q_p, to be diluted in a continually expanding volume (Figure 12-9). The release, Q_p, is a discrete amount (i.e., it does occur at a continuous rate) and the Guassian distribution of the concentration at any point (x, y, z) measured from the center of the puff is

$$\chi(x, y, z) = \frac{Q_p(\#)}{(2\pi)^{3/2}\sigma_x^*\sigma_y^*\sigma_z^*} \exp - \frac{1}{2}\left[\left[\frac{x - \bar{u}t}{\sigma_x^*}\right]^2 + \left[\frac{y}{\sigma_y^*}\right]^2 + \left[\frac{z - z_0}{\sigma_z^*}\right]^2\right]$$

where $Q_p(\#)$ is the total release of material and values of σ^* are distinct for expansion of the puff as it moves downwind. In contrast to continuous releases, there is no dilution due to wind speed and u does not appear in the denominator of the equation; its only effect is translocation of the puff downwind in which the position of the center of puff downwind is determined by the term ut. Dispersion in the x-direction is accounted for by the dispersion coefficient σ_x^*, and it and the dispersion coefficients σ_y^* and σ_z^* are uniquely determined for a contaminant released in a puff based on data for two downwind distances at 100m and 4,000m:

	Downwind Distance			
	x = 100 m		x = 4,000 m	
Atmospheric Condition	$\sigma_x^* = \sigma_y^*$	σ_z^*	$\sigma_x^* = \sigma_y^*$	σ_z^*
Unstable (A–C)	10 m	15 m	300 m	220 m
Neutral (D)	4 m	3.8 m	120 m	50 m
Very Stable (E–F)	1.3 m	0.75 m	35 m	7 m

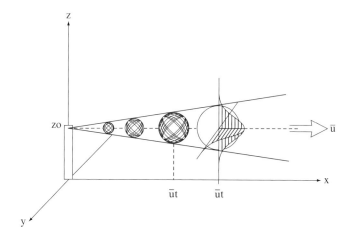

Figure 12-9. Diffusion of a single puff with Guassian concentration distributions, vertically (z), laterally (y), and horizontally (x) from an elevated point where $\bar{u}t$ indicates the distance of travel of the center of the puff in the downwind direction x.

SECTOR-AVERAGED χ/Q VALUES

In practice, most radiological assessments for facilities that release radioactive (or other) materials to the atmosphere are of long duration. Atmospheric conditions, especially wind direction and stability conditions, will obviously vary for extended periods, and it is necessary to account for such changes in determining average concentrations. This is done typically on an annual or quarterly basis by determining a sector-averaged χ/Q value by distributing the released material into several sections of a downwind sector grouped according to wind speed and stability class. The release quantity, Q, is typically expressed as a long-term average, and the χ/Q value is based on accumulated meteorological data.

Many nuclear facilities provide one or more meteorological towers equipped to measure wind speed and direction and the vertical temperature profile from which stability class is determined. These are accumulated in real time and a computer program sorts them into the categories needed for determining long-term sector-averaged $\bar{\chi}/\bar{Q}$ values calculated as

$$\frac{\bar{\chi}}{Q}(\text{s/m}^3) = \left(\frac{2}{\pi}\right)^{1/2} \frac{f_i}{\beta x} \sum_{jk} \frac{F_{jk}}{\sigma_{zj}\bar{u}_k} \exp\left[-\frac{H^2}{2\sigma_{zj}^2}\right]$$

where the quantity βx (m) replaces σ_y in the general dispersion equation. The value of x is the downwind distance to the midpoint of a subsection of the sector of interest (e.g., the subsection extending from 1 to 3 km has a midpoint of 2 km) and when multiplied by the sector width $\beta = 2\pi/n$ radians yields the horizontal width through which the plume flows; it is thus a reasonable surrogate value of σ_y for that distance based on averaged conditions.

The number of downwind sectors for plume averaging is usually chosen to be $n = 16$ which consists of sector arcs of $22.5°$ each extending from the point of release as shown in Figure 12-10. For example, material released when the wind is from due west $\pm 11.25°$ is presumed to diffuse into sector 4 which is ascribed by a pie-shaped wedge extending due east $\pm 11.25°$. Likewise, for winds from due south $\pm 11.25°$ the material is transported into sector 16, a pie-shaped wedge extending due north $\pm 11.25°$. All winds blowing into a given sector are thus recorded for the sector and over time will constitute a fraction f_i of the total amount of material dispersed into a given sector. The fraction of wind blowing into a given subsection of a sector is further subdivided into windspeed groupings (e.g., 0–2 m/s, 2–4 m/s, etc.) which are further subdivided into each of the stability classes recorded for each windspeed grouping yielding another fraction, F_{jk}, If the released material is distributed into $n = 16$ sectors each of $22.5°$ then $\beta = 2\pi/16$ or 0.3927, and each sector-averaged χ/Q is thus (USNRC, Reg. Guide 1.111, 1977)

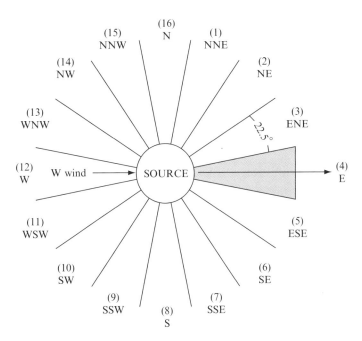

Figure 12-10. Schematic of a 16-sector wind rose for distributing wind direction data into downwind sectors of $22.5°$ each.

$$\frac{\bar{\chi}_i}{Q_i}(s/m^3) = \frac{2.032 f_i}{x} \sum_{jk} \frac{F_{jk}}{\bar{u}_k \sigma_{zj}} \exp\left[-\frac{H^2}{2\sigma_{zj}^2}\right]$$

where

$\bar{\chi}_i/Q_i$ = the sector-averaged dispersion factor for a downwind subsection of sector i;

f_i = the fraction of the time (e.g., month, quarter, year) the wind blows toward sector i;

F_{jk} = the fraction of f_i during which each stability class j exists for wind class k (direction and speed) in sector i;

x = the median downwind distance for a subsection of sector i;

$\bar{u}_k$ = the median wind speed for a chosen group (e.g., $\bar{u}_k = 1\,\text{m/s}$ for winds of 0 to 2 m/s);

σ_{zj} = the vertical diffusion coefficient (m) for each stability class j; and

H = the effective stack height (m)

The interrelationships of these variables are illustrated in very simplified form in Example 12-11 in order to demonstrate the underlying concepts.

Example 12-11: Meteorological data were accumulated for 1 month (720 h) at a facility that releases a radionuclide at an average rate Q (Ci/s) from a 50-m stack. During the 1-month period winds were from due west $\pm 11.25\circ$ for 240 h which blew released material into sector 4. Wind speeds occurred between 0–2 m/s for 60 h, 2–4 m/s for 80 h, and 4–8 m/s for 100 h. Stability classes existed for each wind-speed group as follows:

- For 0–2 m/s winds: Type F for 60 h
- For 2–4 m/s winds: Type B for 40 h; and Type C for 40 h
- For 4–8 m/s winds: Type C for 20 h; and Type D for 80 h.

Determine the monthly sector-averaged χ/Q for a downwind subsection of sector 4 between 1.0 and 3.0 km.

Solution: *Part A:* The median downwind distance for the subsection is 2 km, for which the vertical diffusion coefficients σ_{zj} for each stability class from Figure 12-5 are: 22 m for Type F; 230 m for Type B; 120 m for Type C; and 50 m for Type D. When these σ_{zj} values are combined with the effective stack height of 50 m, the exponential dispersion term, $\exp\left(-\frac{H^2}{2\sigma_{zj}^2}\right)$, for each (F, B, C, and D) is 0.0756, 0.977, 0.917, and 0.607, respectively.

Part B: The median speed $\bar{u}_k$ and the F_{jk} fractions for each wind-speed group and stability class are:

- For Type F stability and wind class 0–2 m/s, $F_{jk} = 60/240 = 0.25$ and the median wind speed $\bar{u}_k = 1\,\text{m/s}$.
- For Type B stability and wind class 2–4 m/s, $F_{jk} = 40/240 = 0.167$ and the median wind speed $\bar{u}_k = 3\,\text{m/s}$.
- For Type C stability and wind class 2–4 m/s, $F_{jk} = 40/240 = 0.167$ and the median wind speed $\bar{u}_k = 3\,\text{m/s}$.

- For Type C stability and wind class 4–8 m/s, $F_{jk} = 20/240 = 0.0833$ and the median wind speed $\bar{u}_k = 6\,\text{m/s}$.
- For Type D stability and wind class 4–8 m/s, $F_{jk} = 80/240 = 0.333$ and the median wind speed $\bar{u}_k = 6\,\text{m/s}$.

Part C: Since the sector-averaged, $\bar{\chi}_i/\bar{Q}_i$, is to be determined for the midpoint of one subsection of sector 4 which is at a median distance $x = 2000\,\text{m}$ and $f_i = 240/720$, the sector-averaged $\bar{\chi}/\bar{Q}$ value for sector 4 is

$$
\frac{\bar{\chi}}{\bar{Q}} = 2.032 \frac{240}{720 \times 2000\,\text{m}} \left[\frac{0.25}{1\,\text{m/s} \times 22\,\text{m}} \times 0.0756 + \frac{0.167}{3\,\text{m/s} \times 230\,\text{m}} \times 0.997 \right.
$$
$$
\left. + \frac{0.167}{3\,\text{m/s} \times 120\,\text{m}} \times 0.917 + \frac{0.0833}{6\,\text{m/s} \times 120\,\text{m}} \times 0.917 + \frac{0.333}{6\,\text{m/s} \times 50\,\text{m}} \times 0.607 \right]
$$
$$
= 7.79 \times 10^{-7}\,\text{s/m}^3
$$

Example 12-11 demonstrates several important observations: first, the fraction f_i merely allocates the average release rate, $\bar{Q}$, into the sector of interest; second, choosing a location of interest within a sector at a particular median downwind distance simplifies the calculation considerably; and third, the complexity of accumulating and allocating large amounts of data into sector averages makes the computer almost indispensable for such tasks. Sector-averaged $\bar{\chi}_i/\bar{Q}_i$ values can often be made more accurate and useful by selecting a particular subsection(s) of interest within a sector (e.g., a pasture or a small community) and calculating specific values for these downwind locations. One or more such locations may in fact be controlling and if it is possible to limit a release(s) to certain periods (e.g., venting only under specified wind conditions) the effects of the release(s) can be optimized to the most favorable sectors and/or meteorological conditions, especially if real-time systems and computer sorting are available.

Long-term sector averaging has become an essential component of environmental radiological assessment. It is common practice in siting and licensing a nuclear facility (e.g., a nuclear power plant) to use meteorological data from local weather stations in projecting the potential impact of estimated releases. These data are then supplemented with several years of onsite meteorological data, prior to and during operations, from towers equipped to measure wind direction and speed and vertical temperature profiles. Many facilities do not have such data for periods of interest (e.g., for a dose reconstruction) and it is necessary to retrospectively construct them, typically from local weather stations at airports or in nearby urban areas. Such reconstructions may be adequate if the data can be shown to be representative of the site conditions, but in any case they are a compromise between no data and that from onsite measurements.

DEPOSITION/DEPLETION—GAUSSIAN PLUMES

As plumes migrate downwind they are subject to various influences, in addition to dispersion and dilution, that tend to reduce the concentration of material in them. Large particles may settle out and plume contaminants may be deposited on vegetation, ground surfaces, or other objects due to physical and chemical processes.

Dry deposition and wet deposition are the two main processes that deplete material from a plume; however, their effects on plume concentrations are overall quite small compared to inputs from the source of release, and they are generally ignored in computations of downwind concentrations. Their more important effect is an accumulation of contaminants on vegetation and the ground where they may become entrained in food pathways and/or become a source of direct exposure from the ground surface. Accumulated areal contamination levels are thus an important input for pathway models used to determine the transfer of radionuclides to food and any resulting dose due to its consumption, or as a source that produces external radiation dose.

Dry Deposition

Dry deposition (fallout) of contaminants onto vegetation and ground surfaces is represented as an aerial contamination, $C_{A, dry}$, that is related to the air concentration by a constant of proportionality, v_d, called the deposition velocity, or

$$C_{A, dry}(\#/m^2) = v_d \frac{\bar{\chi}_i}{\bar{Q}_i} \times Q_{TOT}$$

where Q_{TOT} is the total amount of material released over the period that the average $\bar{\chi}_i/\bar{Q}_i$ factor is determined for a subsection of a downwind sector i. The areal deposition is given in terms of $\bar{\chi}_i/\bar{Q}_i$ because the conditions of interest are relatively long term processes and it is a common practice to determine a weighted average of the dispersion factor for a particular downwind direction, e.g., over a pasture.

The proportionality constant, v_d, is termed the deposition velocity because it has units of m/s; however, it is not a true velocity. Values of v_d have been determined experimentally by measuring the areal contamination (e.g., $\mu Ci/m^2$) during periods of known air concentration over an area; these are provided in Table 12-3 for selected radioelements and for different types of vegetation and soils. Deposition velocities are also a function of wind velocity because the vertical profile of concentration changes with wind velocity and the amount of deposition is greatly influenced by sorption characteristics, roughness of the underlying surface, and the chemical composition of a given effluent. Even the best measured values of deposition velocity have an uncertainty that may vary an order of magnitude or more. Consequently, values of v_d are

Table 12-3. Dry Deposition Velocities, v_d, for Selected Forms of Radioactive Materials on Soils and Vegetation for Several Atmospheric Stability Conditions.

Radioelement	Medium	Atm. Stability	$v_{d(m/s)}$
I_2	Grass/wheat	A, B	1.4×10^{-2}
	Grass	C, D	8×10^{-3}
	Clover	C, D	$1\text{-}2 \times 10^{-2}$
	Dry Grass	E, F	1×10^{-2}
	Grass/Wheat	E, F	3×10^{-3}
	Snow	–	3.4×10^{-3}
	Soil	–	3.3×10^{-3}
CH_3I	Grass	A, B	1×10^{-4}
	Grass	E, F	1×10^{-5}
SO_2	Vegetation	–	$1 \times 10^{-2*}$
CO_2	Grass	–	2×10^{-4}
Ru	Grass	–	1×10^{-5}
	Soil	–	4×10^{-3}
Cs	Grass	–	2×10^{-3}
	Soil	–	4×10^{-4}
Zr-Nb	Soil	–	3×10^{-2}
$1\,\mu$ m aerosols	Grass/soil	–	1×10^{-3}
Fission products	Sagebrush	–	3×10^{-2}
Weapons Fallout	Grass	–	2×10^{-3}
	Sticky ppr	–	$5 \times 10^{-3*}$

* Average value.

limited and must be used with caution for the particular circumstances encountered; e.g., the physical and chemical forms of the radionuclide, the deposition medium (grass, food crops, or ground surfaces), and atmospheric conditions such as stability and wind speed; each of which can markedly influence deposition.

Example 12-12: Estimate the areal contamination due to dry deposition of iodine-131: (a) on the ground downwind of a 100-m stack that releases 10^9 Bq(2.7×10^{-2} Ci of I-131 if the average $\bar{\chi}_i/\bar{Q}_i$ value over the period of release is 1.55×10^{-7} s/m^3; and (b) on grass for the same conditions.

Solution: (a) From Table 12-3, the deposition velocity for elemental iodine on soil is 3.3×10^{-3} m/s; thus the areal contamination is

$$C_A(Bq/m^2) = 3.3 \times 10^{-3}\,m/s \times 1.55 \times 10^{-7}\,s/m^3 \times 10^9 Bq$$
$$= 0.51\,Bq/m^2(27.1\,pCi/m^2)$$

(b) For long-term deposition on grass, it is appropriate to assume an average atmospheric stability condition of Type D (neutral) such that $v_d = 8 \times 10^{-3}$ (Table 12-3), thus

$$C_A(Bq/m^2) = 8 \times 10^{-3} \times 1.55 \times 10^{-7} \, s/m^3 \times 10^9 \, Bq$$
$$= 1.24 \, Bq/m^2 (33.5 \, pCi/m^2)$$

Material deposited onto a surface (e.g., grass or soil) is also subject to removal by weathering and biological processes denoted as a biological half-life T_b. And if the material is radioactive the combined effect of these processes is described by an effective half-life, T_{eff}, values of which are listed in Table 12-4 for selected radionuclides on selected surfaces often considered in radiological assessments. The effective half-life can be used in the usual way to determine an effective removal constant, λ_{eff}, which is calculated as $\lambda_{eff} = \ln 2/T_{eff}$.

For an airborne concentration of some duration the areal contamination, $C_{A, \, dry}(\#/m^2)$ is

Table 12-4. Measured Radiological (T_r), Biological (T_b), and Effective (T_e) Half-lives of Radionuclides on Vegetation and Soil for Spring (SP), Summer (S), Fall (F), and Winter (W) Seasons.

Radionuclide (T_r)	Medium	T_b (days)*	T_e (days)
I-131 (8.02 d)	Pasture (F)	13.9	5.08
	Pasture (W)	34.6	6.51
	Grass (S)	6.8	3.68
	Grass (F)	~ 24	6.01
	Grass (SP, S)	7.5	3.87
	Sagebrush	8.0	4.00
	Hay	16.4	5.39
	Clover (S)	4.6	2.93
	Clover (F)	13.4	5.02
	Dry Soil	18.2	5.57
	Grass washoff	4.1	2.72
Sr-90 (28.78 y)	Rye grass/clover	30.8	30.8
Sr-89 (50.52 d)	Cabbage	24	16.4
	Grass	17	12.7
Cs-137 (30.04 y)	Pasture	31	31.0
	Rye grass/clover	25.9	25.9
Ru-103 (39.27 d)	Grass	8.4	6.92

*Includes multiple environmental/biological processes that diminish deposited radioactivity over time.
Source: Peterson, H.T. NUREG-CR-3332, 1983.

$$C_{A, \text{dry}}(\#/m^2) = v_d \frac{\bar{\chi}}{\lambda_{\text{eff}}}(1 - e^{-\lambda_{\text{eff}} t})$$

which will reach an equilibrium value in 7 to 10 effective half-lives, or

$$C_{A, \text{dry}}(\#/m^2) = v_d \frac{\chi}{\lambda_{\text{eff}}}$$

Example 12-13: If a plume that contains a concentration of $1 \times 10^{-6}\,\mu\text{Ci}/\text{m}^3$ of $^{131}\text{I}(T_{1/2} = 8.02\text{d})$ exists during the fall season over dry pasture grass for 2 h following an incident, estimate: (a) the areal contamination on the pasture for the 2-h period, and (b) the equilibrium value of the areal contamination if the airborne concentration persists over the area.

Solution: (a) From Table 12-3, v_d is $10^{-2}\,\text{m/s}$, and from Table 12-5, the radiological and biophysical effective half-lives are 8.02 d and 13.9 d, respectively. The effective half-life is thus 5.08 d and the effective removal constant is $0.1364\,\text{d}^{-1}$. The areal deposition for the 2-h period is

$$\begin{aligned}
C_{A, \text{dry}} &= 10^{-2}\,\text{m/s} \times \frac{10^{-6}\,\mu\text{Ci}/\text{m}^3}{0.1364\,\text{d}^{-1}} \times 86,400\,\text{s/d}\left(1 - e^{-0.1364 \times \frac{2}{24}}\right) \\
&= 7.16 \times 10^{-5}\,\mu\text{Ci}/\text{m}^2
\end{aligned}$$

(b) The equilibrium value which occurs after about 35 d is

$$\begin{aligned}
C_{A, \text{dry}}(t > 35\,\text{d}) &= 10^{-2}\,\text{m/s} \times \frac{10^{-6}\,\mu\text{Ci}/\text{m}^3}{0.1364\,\text{d}^{-1}} \times 86,400\,\text{s/d} \\
&= 6.33 \times 10^{-3}\,\mu\text{Ci}/\text{m}^2.
\end{aligned}$$

Reaching such an equilibrium condition would be highly unlikely since it is doubtful, because of atmospheric variability, that a constant air concentration would be maintained over the area for 35 d. If, however, the concentration is a weighted average, $\bar{\chi}$, the areal deposition would be of this order.

Air Concentration due to Resuspension

Long-lived (or stable) contaminants that build up over time may also become resuspended to produce elevated concentrations of the material above the immediate area or perhaps translocated to other areas. Resuspension of deposited material that persists near the surface is largely governed by wind speed and surface roughness (e.g., loose soil) and surface creep processes which widen the area of contamination. Of particular interest has been various land areas contaminated with plutonium due to incidents, e.g., leakage of Pu from stored drums at the Rocky Flats Plant in Colorado and partial detonation of nuclear weapons in Spain, the Nevada test site, and Johnston Atoll. A similar circumstance is redistributed uranium tailings that may be re-entrained into the atmosphere. Both are, of course, long-lived insoluble materials that remain near the surface of the ground unless covered in some way.

The air concentration of resuspended long-lived materials is directly related to the areal contamination of the soil and its depth of distribution. Obviously, surface-level contamination is of most significance especially if the soils are loose and dry and the area is open and subject to high winds. The airborne concentration of resuspended materials is thus

$$\chi_{res}(\#/m^3) = k_s C_{A, dry}$$

where k_s is the resuspension factor (m^{-1}) and $C_{A, dry}$ is the areal concentration $(\#/m^2)$. Values of k are scant but a few have been determined, primarily for Pu and U; resuspension factors for relatively short-lived or highly mobile radionuclides in soils would not only be difficult to establish but highly uncertain for air concentration estimations. Several typical values are listed in Table 12-5, but even these should be used with caution.

It is apparent that material will weather due to various environmental effects and that the resuspension factor will thus change with time following deposition of the material. One of the most widely used calculational models for determining a weathered value for k_s is

$$k_s(m^{-1}) = [10^{-9} + 10^{-5} \exp(-0.6769t)]$$

where t is the amount of time in years after deposition. This empirical relationship is based on desert-type soil and should be used with caution. For example, if applied for initial contamination $(t = 0)$, a resuspension factor of $10^{-5}\,m^{-1}$ is

Table 12-5. Resuspension Factors, $k_s(m^{-1})$, for Deposited Radionuclides for Various Soils and Soil Depths.

Radionuclide	Medium	Depth (cm)	$k_s(m^{-1})$
Plutonium	Rocky Flats Soil*	0.02	1×10^{-6}
	Rocky Flats Soil*	20	1×10^{-9}
	Nevada Test Site	3.0	1×10^{-11}
	Moist Soil	1.0	2×10^{-10}
	Mud Flats	0.1	4×10^{-8}
	Fields	5.0	1×10^{-9}
Uranium	Soil-Surrey, UK	1.0	5×10^{-9}
	Soil-NY, USA	1.0	5×10^{-8}
Iodine	Vegetation	–	2×10^{-6}
Fallout	Desert Soil	–	3×10^{-7}

* k_s for Rocky Flats soils varied from 10^{-5} to 10^{-9} for surfaces ranging from vehicular traffic on dusty roads to soil layers contaminated to a depth of 20 cm.

obtained which is a factor of 10 greater than the resuspension observed for semi-arid Rocky Flats soil/vegetation (Table 12-5). Application of this relationship is shown in Example 12-14.

Example 12-14: Calculate: (a) the concentration of plutonium in air above a desert-like plot uniformly contaminated at an initial level of $0.2\,\mu Ci/m^2$; (b) the air concentration at 1 y after the initial deposition; and (c) at 10 y.
Solution: (a) At $t = 0 k_s = 10^{-5} m.1$, and the airborne concentration is

$$C_{A,\,dry} = (10^{-5}\,m^{-1})(0.2\mu Ci/m^2) = 2 \times 10^{-6}\,\mu Ci/m^3$$

(b) At 1 y, $k_s = 5.08 \times 10^{-6}$, and the airborne concentration is 1.02×10^{-6} $\mu Ci/m^3$; and (c) at 10 y, $k_s = 1.25 \times 10^{-8}$, and the resulting airborne concentration is $2.5 \times 10^{-9}\,\mu Ci/m^3$.

For locations with moist soils and vegetation a smaller resuspension factor would normally be chosen and the relationship used in Example 12-14 would not be applicable.

Wet Deposition

Contaminants can also be removed from a plume by precipitation, a process that results in wet deposition of the plume material on vegetation, the ground and other exposed surfaces. Both wet deposition and dry deposition contribute to areal contamination; however, each is considered separately even though both may be occurring during a precipitation period. The amount of wet deposition is similar to that for dry deposition, but in this case the constant of proportionality is a combination of the rainfall rate R and the washout ratio, W_v, values of which are listed in Table 12-6 for selected radionuclides. The amount of washout is a function of the size and distribution of raindrops and the physio–chemical features of the plume. These parameters, themselves functions of the space coordinates (x,y,z), cause the amount of washout to be a space-dependent parameter; however, as a practical matter it is assumed to be constant with respect to space because of scant empirical data.

The areal contamination, $C_{A,\,wet}(\#/m^2)$, due to wet deposition is

$$C_{A,\,wet}(t) = \frac{RW_v e^{-mR}}{\lambda_e + kR} \chi[1 - \exp-(\lambda_e + kR)t]$$

where
χ = the contaminant concentration (Ci/m^3) in the air above the area
R = rainfall rate (mm/h)
W_v = volumetric washout rate (vol of air/vol of rain)
m = the plume depletion constant (h/mm) which is related to rainfall rate
k = vegetation washoff coefficient (mm^{-1})

Table 12-6. Vegetation Washoff Constants, $k(mm^{-1})$, and Volumetric Washout Factors, $Wv(m^3 \, air/m^3 \, rain)$, for Selected Radionuclides.

| Radionuclide | Washoff Constant | | $W_{v(m^3 \, air/m^3 \, rain)}$ | |
	Vegetation	$k(mm^{-1})$	Form	W_v
Sr-89, 90	Grass/cabbage	0.017*	Fallout	5.9×10^5
Zr-95	Cabbage	0.022	Fallout	4.2×10^5
Ru-106	Lettuce	0.063	Fallout	5.6×10^6
	Cabbage	0.026	–	–
I-131	Grass	0.020	Fallout	3.5×10^5
	Cabbage	0.026	I_2	8.3×10^4
			Alkyl (pH 5)	4.2×10^3
Cs-137	Cabbage	0.02	Fallout	4.6×10^5
Ba-140	–	–	Fallout	4.0×10^5
Ce-144	Cabbage	0.025	Fallout	4.6×10^5
Particles ($\sim 1 \, \mu m$)	Grass	0.063	Soluble	1.0×10^6
	Grain	0.069	Insoluble	3.0×10^5

* Median value; reported values range between 0.009 and 0.024.
Source: NUREG/CR-3332, 1983.

λ_e = effective removal constant that accounts for radioactive removal and biological processes other than washoff

t = duration of precipitation (h)

The aerial contamination $C_{A, wet}(t)$ will thus build up to an equilibrium condition as a function of the rainfall rate, R, the removal coefficient, k, and the effective removal constant, λ_e. The areal ground concentration is also a function of dry deposition, which is often ignored for precipitation events.

Example 12-15: Calculate the equilibrium deposition of radioiodine by rainout for an average atmospheric concentration of $10^{-5} \mu Ci/m^3$ when the rainfall rate $R = 2.5 \, mm/h$; the washoff constant, $k = 0.025 \, mm^{-1}$; $m = 0.025 \, h/mm$; the washout coefficient, W_v, is $8.3 \times 10^4 \, m^3 air/m^3$ rain; and the effective removal constant $= 0.1364 \, d^{-1}$.

Solution: Since the equilibrium value is sought, the buildup term $= 1.0$; thus,

$$C_{A, wet}(\infty) = \frac{RW_v e^{-mR}}{(\lambda_e + kR)} \bar{\chi}$$

$RW_v = (2.5 \, mm/h)(24 \, h/day)(10^{-3} \, m/mm)(8.3 \times 10^4 \, m^3 \text{ of air}/m^3 \text{ of rain})$

$$= 4.98 \times 10^3 \, \frac{m^3}{m^2 \cdot d}$$

$$(\lambda_e + kR) = [(0.1364\,d^{-1} + (0.025\,mm^{-1})(2.5\,mm/h)(24\,h/d)] = 1.6364\,d^{-1}$$

and,

$$C_{A,\,wet}(\infty) = \frac{4.98 \times 10^3\,m^3/m^2 \cdot d}{1.6364\,d^{-1}} \times 10^{-5}\,\mu Ci/m^3[\exp(-0.025 \times 2.5)]$$

$$= 0.286\,\mu Ci/m^2$$

It is worth noting that the areal deposition due to precipitation is directly related to the volumetric washout factor, W_v, and its selection is key to such calculations.

Short-term Plume Washout due to precipitation effects on a plume can be determined by use of a washout rate coefficient similar to that for dry deposition. The areal contamination rate using a washout rate is

$$C_{A,\,wet}(\#/m^2 \cdot s) = \frac{\Lambda Q(\#/s)}{(2\pi)^{1/2}\bar{u}\sigma_y}\exp\left[-\frac{y^2}{2\sigma_y^2}\right]$$

where Λ is the washout rate (s^{-1}) for relatively short measurement intervals and single individual precipitation situations, typically on the order of 1 hour. The value of σ_y is, of course, location dependent and should be selected for the specific downwind distance where the precipitation event occurs.

Nominal values of Λ for both aerosols and gases are $10^{-4}\,s^{-1}$ for stable conditions and $10^{-3}\,s^{-1}$ for unstable conditions. If the rainfall rate R is known, a somewhat better value of Λ can be calculated based on data for nuclear power plant releases, which for elemental iodine is

$$\Lambda(s^{-1}) = 8.0 \times 10^{-5}R^{0.6}$$

and for aerosols

$$\Lambda(s^{-1}) = 1.2 \times 10^{-4}R^{0.4}$$

where R is the actual precipitation intensity in mm/h.

Plume Depletion due to Wet Deposition can be significant for persistent precipitation rates and the areal contamination rate will be reduced exponentially as the plume moves downwind. This effect is accounted for by introducing an additional exponential removal term,

$$C_{A,\,wet}(\#/m^2 \cdot s) = \frac{\Lambda \dot{Q}}{(2\pi)^{1/2}\bar{u}\sigma_y} \exp\left[-\frac{y^2}{2\sigma_y^2}\right] \exp\left[-\frac{\Lambda x}{\bar{u}}\right]$$

where the factor $x/\bar{u}$ accounts for the effective time of plume travel and the duration of washout by precipitation.

It is important to recognize that both wet and dry deposition parameters are quite site-specific; thus, the quality of any given estimate of areal contamination is determined by how well the available parameters fit the situation being modeled.

SUMMARY

Environmental radiological assessment has become an important process in recent years for estimating the radiological consequences to persons who live(d) around various sites. The environmental behavior of radionuclides and associated doses and risks to humans or other species can be quite complex, and calculations of environmental levels, doses, and health effects have evolved into specialized sciences/disciplines. And, since it is often difficult to establish exact parameters or to obtain measured values, many assessments are logically based on determining exposure/dose for a hypothetical maximum exposed individual(s) and/or a typical individual rather than attempting to deal with the complexity of environmental conditions for a particular source and persons potentially or actually exposed. Although risk assessments for a hypothetical or typical maximum individual rarely fit any given person potentially or actually exposed, they many times provide a range of exposures/doses that can be used to judge the significance of radionuclides associated with a given activity/facility.

Atmospheric dispersion modeling is often required for determining concentration profiles for pollutants released to the atmosphere. Such releases are usually assumed to emanate from a "point" source such as an elevated stack or a small area, and the concentration, χ, at a downwind distance is directly proportional to Q and inversely proportional to the wind speed; however, the concentration across an areal segment of the plume located at some point, x, downwind fits a Guassian distribution and is modeled accordingly.

The Guassian diffusion equations were developed for idealized conditions of point-source releases over relatively flat undisturbed terrain where atmospheric turbulence, as characterized by the vertical temperature profile, is fairly homogeneous. These idealized conditions rarely exist, but as a practical matter the point source equations can be used for flat and relatively unobstructed terrain and release points that are 2.5 times higher than surrounding buildings or terrain disturbances (e.g., hills or cliffs). The models are also often used for other conditions because they are the only ones available; however, the quality of the results will be influenced by the degree to which conditions deviate from those specified for the model.

Various conditions can occur in which the turbulence associated with an airborne plume will be nonuniform and it is not appropriate to use Guassian plume models to determine air concentration. Two common situations, which require completely different models, are fumigation and mechanical turbulence induced by buildings or other obstructions. Mechanical turbulence can completely dominate the dispersion of an effluent released from a short stack or vent located on a sizable structure during moderate to strong winds, and the emitted effluent will be drawn into the building wake.

The processes of dry and wet deposition are very complex and the coefficients for calculating them are both scant and specific to types of vegetation, soil, wind conditions, atmospheric stability, and precipitation events. Determinations of areal contamination are, because of these highly variable factors, at best estimates that can be expected to have considerable uncertainty. The degree to which these can be supplemented with, or better yet, replaced with measured values offers considerable improvement in the reliability of the results.

ADDITIONAL RESOURCES

1 Randerson, D., *Atmospheric Science and Energy Production*, DOE/TIC-27601, 1984

2 Turner, D.B., *Workbook on Atmospheric Dispersion Estimates*, EPA Publ., AP26, 1970.

3 Hanna, S.R., Briggs, and Hosker, *Handbook on Atmospheric Diffusion*, DOE/TIC-11223, 1982.

4 US Nuclear Regulatory Commission, Regulatory Guides 1.109, 1.111, and 1.145

5 Peterson, H.T., Jr., Terrestrial and Aquatic Food Chain Pathways, in *Radiological Assessment*, eds. Till, J.E., and Meyer, H.R. TID Information Service: NUREG/CR-3332, 1983.

6 Slade, D.H., Meteorology and Atomic Energy—1968, Report TID-24190, Clearinghouse for Federal Scientific and Technical Information, U.S. Department of Commerce, Springfield, VA 22151, July 1968.

PROBLEMS

1 What is the ground level concentration directly downwind of a nuclear facility that emits 80 Ci/s from a 50-m stack over relatively flat terrain during overcast daytime conditions when the wind is 6 m/s: (a) at 500 m along the centerline; and (b) at 1500 m downwind and 50 m off the centerline?

2 For the conditions in Problem 1: (a) find the location of the point of maximum concentration along the plume centerline; and (b) the concentration at that maximum distance.

3 Calculate the effective stack height for a 50-m stack with an inside diameter of 5 m that emits effluent at $200°C$ and at an efflux velocity of 8 m/s into an atmosphere that has an ambient temperature of $20°C$ and a windspeed of 3 m/s.

4 Effluent is emitted at 100 g/s from a 6 ft high vent atop a building that is 80 ft high and 100 ft wide on the lee side. Atmospheric conditions are overcast, late afternoon, and the wind speed is 5 m/s. What is the ground-level air concentration: (a) 50 m downwind; and (b) 300 m downwind?

5 A major reactor accident occurs around noon on a bright sunny June day releasing halogens and noble gases into the wake of a containment building that is 30 m wide and 50 m high. If ^{131}I escapes from the containment building at a rate of 9.0×10^{-4} Ci/s during the first 2 h of the accident and the wind speed is 6 m/s, what is the average air concentration at Joe Panic's house 3000 m directly downwind?

6 Solve the previous problem using the virtual source method and compare the results. Comment on the accuracy of each method and which should be selected.

7 Effluent released from a 50-m stack contains ^{85}Kr released at a rate of 0.1 Ci/s into a 4 m/s wind during stability class F conditions. What is the ground level concentration 5000 m downwind?

8 Calculate the ground level concentration for the conditions in problem 7 if the release occurs during fumigation conditions when the wind speed is 2 m/s at the height of the stack.

9 Estimate the areal contamination of cesium-137 (assumed to form aerosols) on the ground due to dry deposition downwind of a 100-m stack that releases 10^9 Bq(2.7×10^{-2} Ci) of Cs-137 into a sector for which the average value of $\bar{\chi}/\bar{Q} = 1.55 \times 10^{-7}$ s/m³.

10 Using the conditions of Example 12-11, determine the sector-averaged χ/Q value in sector 4 for a subsection with a midpoint located at 1200 m downwind if the effective stack height is 60 m.

11 What would be the χ/Q value at 100 m for the release considered in Example 12-12 for: (a) stability category D instead of category F, and (b) for stability category A conditions?

12 For the conditions in Example 12-7: estimate the ground level concentration of ^{41}Ar at 120 m downwind and during neutral atmospheric stability conditions using: (a) the Fuquay method and a building area correction factor c = 0.8; and (b) the virtual source method.

13 For the conditions in Example 12-7, determine the far-field concentration of ^{41}Ar at 600 m downwind using selection rules: (a) for Type D atmospheric stability conditions; and (b) for Type E stability conditions.

13

RADON—A PUBLIC HEALTH ISSUE

"Hey, Mr. Watras, you left your car lights on."
– Reading Prong, PA (1984)

Radon, which has always been present in the human environment, burst into public view when a nuclear plant worker in Pennsylvania was too contaminated with radon progeny from his home environment to be in a nuclear power plant. Mr. Stan Watras had just entered the Limerick Nuclear Power Station to begin the early shift when the guard informed him that his car lights were still on. To return to the parking lot, he had to pass through portal radiation monitors provided at nuclear plants to ensure that no one leaves the plant with measurable contamination. The monitors alarmed and his clothing was found to contain high levels of radioactivity, but since he had not been in the plant (which had also not yet begun operation) the radiation levels had to be associated with some other source. The Pennsylvania State Health Department eventually established that his home, which was located along a large rock body called the Reading Prong, had very high levels of radon. He moved his family out soon after, the story hit the news media, other localities conducted radon surveys, and national and state radon programs began.

FEATURES OF RADON AND ITS PRODUCTS

Radon is a radioactive transformation product of ^{238}U, ^{235}U, and ^{232}Th, which exists in various concentrations in all soils and minerals. Radon-222 (half-life of 3.82 days) in the ^{238}U series is the most significant isotope; ^{220}Rn (called thoron) produced in the ^{232}Th series, and ^{219}Rn from the ^{235}U series, have very short half-lives (55.6 s and 3.96 s, respectively) and thus are of relatively minor significance. Uranium has a half-life of 4.5 billion years and its intermediate transformation products, ^{230}Th and ^{226}Ra, the immediate parent of ^{222}Rn, have half-lives of 75,380 years and 1600 years, respectively. In natural soils, ^{230}Th and ^{226}Ra are in radioactive equilibrium with uranium, thus a perpetual source of radon exists naturally even though its half-life is only 3.82 days. Sources that contain ^{230}Th and ^{226}Ra separated from uranium (e.g., residues from ore mining and processing) also represent long-term sources of radon, especially ^{230}Th.

Uranium, the precursor of radium and its short-lived product radon, is present in most materials, and many granites, carbonaceous shales, and phosphatic rocks contain elevated concentrations. Uranium concentrations in granites range between 2 to 10 ppm with averages around 3 to 4 ppm; uraniferous black shales often average up to 20 ppm uranium, and phosphate rocks with 100 ppm uranium are quite common. Although no completely reliable prediction method exists, houses built in uranium-rich areas (e.g., the phosphate districts in Florida and along the Reading Prong in Pennsylvania) are likely to have high indoor radon levels. The Reading Prong is a near-surface outcropping of granite that extends from Pennsylvania through New Jersey and into New York.

Whereas uranium and its intermediate products are solids and remain in the soils and rocks where they originate, ^{222}Rn is a radioactive noble gas which migrates through soil to zones of low pressure such as homes. Its 3.82 day half-life is long enough for it to diffuse into and build up in homes unless they are constructed in ways that preclude its entry, or provisions are made to remove the radon. Once radon accumulates in a home it will undergo radioactive transformation; however, the resulting transformation products are no longer gases but are solid particles which, due to an electrostatic charge, become attached to dust particles that are inhaled by occupants, or the particles can be inhaled directly. Because they are electrically charged, these particles readily deposit in the lung, and as they have half-lives on the order of minutes or seconds, their transformation energy is almost certain to be deposited in lung tissue. These unique features of radon and its transformation products represent potential radiation risks to people because they spend a lot of time in enclosed spaces.

The Radon Subseries

As shown in Figure 13-1, ^{222}Rn undergoes transformation by alpha particle emission to produce ^{218}Po (RaA), which in turn emits 6.0 MeV alpha particles with a half-life of 3.11 min to produce ^{214}Pb (RaB). Beta-particle emissions from RaB (^{214}Pb, $T_{1/2} = 26.8$ min) and ^{214}Bi (RaC, $T_{1/2} = 19.9$ min) produce ^{214}Po (RaC′) which quickly ($T_{1/2} = 164\,\mu s$) produces the end product ^{210}Pb (RaD) by emission of 7.69 MeV alpha particles. The alpha transformation of RaC′ is in effect an isomeric alpha transformation of RaC because it occurs so quickly after RaC is formed. The end product ^{210}Pb (RaD) is effectively stable with a half-life of 22.3 years and is treated as such in most calculations of radon

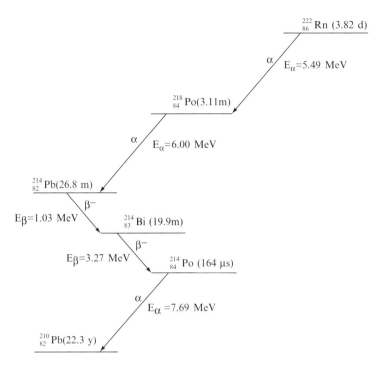

$^{222}_{86}$Rn (3.82 d)

α E_α=5.49 MeV

$^{218}_{84}$Po(3.11m)

α E_α=6.00 MeV

$^{214}_{82}$Pb(26.8 m)

β⁻

E_β=1.03 MeV $^{214}_{83}$Bi (19.9m)

β⁻

E_β=3.27 MeV $^{214}_{84}$Po (164 μs)

α

E_α =7.69 MeV

$^{210}_{82}$Pb(22.3 y)

Figure 13-1. Radioactive series transformation of ^{222}Rn to long-lived ^{210}Pb (minor branching of ^{218}Po to ^{218}At and ^{214}Bi to ^{210}Th not included).

and its progeny. The properties of this important segment of the uranium series are summarized in Table 13-1 (see below) for 100 pCi of ^{222}Rn in radioactive equilibrium with its transformation products.

General expressions (the Bateman equations) for radioactive series transformation can be used to determine the number of atoms of each of the products of radon through ^{214}Po (RaC′) and subsequently the activity of each at any time t after the formation or separation of radon from its progenitors. Since RaC′ is so short-lived, its activity will be exactly the same as RaC, and as a practical matter only the number of atoms of each of the first four members of the radon subseries needs to be calculated. The numbers of atoms of Rn, RaA, RaB, and RaC at any time, t, produced by N_1^o atoms of Rn are:

for RaA, or $N_2(t)$

$$N_2(t) = N_1^o(5.657 \times 10^{-4}e^{-\frac{t}{7936}} - 5.657 \times 10^{-4}e^{-\frac{t}{4.49}}) \quad (13\text{-}1)$$

for RaB, or $N_3(t)$

$$N_3(t) = N_1^o(4.875 \times 10^{-3}e^{-\frac{t}{7936}} + 6.4 \times 10^{-4}e^{-\frac{t}{4.49}} - 5.547 \times 10^{-3}e^{-\frac{t}{38.66}}) \quad (13\text{-}2)$$

for RaC, or $N_4(t)$

$$N_4(t) = N_1^o(3.6176 \times 10^{-3}e^{-\frac{t}{7936}} - 8.81 \times 10^{-5}e^{-\frac{t}{4.49}}$$
$$- 1.59 \times 10^{-2}e^{-\frac{t}{38.66}} + 1.237 \times 10^{-2}e^{-\frac{t}{28.71}}) \qquad (13\text{-}3)$$

And since RaC′ can be thought of as an alpha-isomeric state of RaC ($N_4(t)$), the number of atoms that produce RaC′ emissions, $N_5(t)$, will be the same as $N_4(t)$, or

$$N_5(t) = N_4(t) \qquad (13\text{-}4)$$

The Activity of Radon Progeny for times, t, in minutes after formation of ^{222}Rn can be calculated from the activity of radon, $A_{Rn} = A_1^o$, as
   for RaA, or A_2

$$A_2(t) = A_1^o(1 - e^{-\frac{t}{4.49}})$$

for RaB, or A_3

$$A_3(t) = A_1^o(1 + 0.1313e^{-\frac{t}{4.49}} - 0.1313e^{-\frac{t}{38.66}})$$

and for RaC, or A_4

$$A_4(t) = A_1^o(1 - 0.0243e^{-\frac{t}{4.49}} - 4.394e^{-\frac{t}{38.66}} + 3.4183e^{-\frac{t}{28.71}})$$

And since RaC′ (A_5) is essentially an alpha-isomeric state of RaC, the activity of RaC′ (A_5) will be the same as RaC ($A_4(t)$) or

$$A_5(t) = A_4(t)$$

The activity of radon is assumed to remain constant over the period of calculation; plots of transformation product activity obtained from these equations (Figure 13-2) show that the activity of each product approaches equilibrium with ^{222}Rn in about 3 hours or so.

The Working Level for Radon Progeny

The "working level" (WL) has been defined to deal with the special conditions of exposure to radon and its decay products. It was first defined for exposure of uranium miners as 100 pCi of ^{222}Rn in a liter of standard air in equilibrium with its transformation products RaA, RaB, RaC, and RaC′; however, this condition rarely exists in working environments or in homes. Radon is an inert gas that is quickly exhaled rather than being deposited in the lung; therefore, radiation exposure of the human lung is not caused by radon itself but by

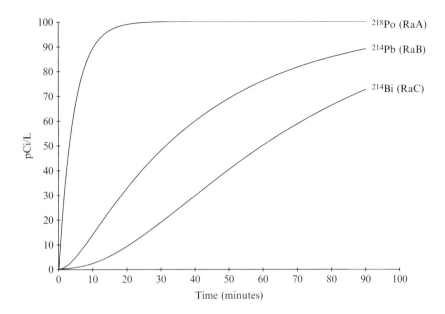

Figure 13-2. Ingrowth of transformation product radioactivity from 100 pCi of pure ^{222}Rn.

deposition of RaA, RaB, RaC, and RaC′ which are charged particles that can be inhaled into and deposited in the lung. Most of the energy deposited in the lung is due to the alpha particles emitted by RaA and RaC′; beta-particle emissions and gamma rays emitted by RaB and RaC make a negligible contribution. Consequently, the working-level is defined as . . . *any combination of the short-lived decay products of radon (RaA, RaB, RaC, and RaC′) in one liter of air that will result in the ultimate emission by them of* 1.3×10^5 *MeV of alpha-ray energy (to lung tissue).*

Calculation of the WL requires first a determination of the number of atoms of each of the radon transformation products and the ultimate alpha energy that each of the product atoms will produce as it undergoes radioactive series transformation to the end product ^{210}Pb. One hundred pCi of ^{222}Rn in one liter of standard air contains 1.77×10^6 atoms (N_1^o) which will remain essentially unchanged over a period of 3 hours or so, the period of interest for most calculations associated with ^{222}Rn. The product atoms in equilibrium with 100 pCi of ^{222}Rn will, as shown in Table 13-1, ultimately emit 1.3×10^5 MeV of alpha energy: each atom of RaA can deliver not only its own characteristic 6.00 MeV energy but also the 7.69 MeV alpha emission from RaC′ because each RaA atom will undergo transition through RaB and RaC and eventually into RaC′ as well (see Figure 13-1), or "ultimately" 13.69 MeV from each atom of RaA. Likewise, every atom of RaB and RaC, even though they are beta emitters, will "ultimately" produce RaC′ which will emit 7.69 MeV of alpha energy.

Table 13-1. Determination of Ultimate Alpha Energy Due to a Concentration of 100 pCi of Radon in Equilibrium with its Four Principal Products, or One Working Level (WL).

Nuclide	MeV/t	$T_{1/2}$	Atoms in 100 pCi	α energy per atom (MeV)	Total energy (MeV/100 pCi)	Fraction of total αE
^{222}Rn	5.5	3.82 d	1.77E06	Excluded	None	None
^{218}Po(RaA)	6	3.11 min	996	13.69	1.36E04	0.11
^{214}Pb (RaB)	0*	26.8 min	8583	7.69	6.60E04	0.51
^{214}Bi (RaC)	0*	19.9 min	6374	7.69	4.90E04	0.38
^{214}Po(RaC$'$)	7.69	164 μs	0.0008	7.69	0	0
				TOTAL	1.3E05	1

* β and γ emissions excluded in WL because of minimal energy deposition in the lung.

Table 13-1 also shows that RaA, in equilibrium with ^{222}Rn, contributes only about 11% of the ultimate or "potential" alpha energy. RaB, however, contributes about 51% because in equilibrium with 100 pCi of ^{222}Rn it constitutes the largest number of atoms (because it has the longest half-life) that will eventually produce 7.69 MeV alpha emissions through RaC$'$ transformations. Analogously, the atoms in 100 pCi of RaC, the other beta – gamma emitter, supply 38% of the ultimate alpha energy through RaC$'$. RaC$'$ produces ^{210}Pb ($T_{1/2} = 22.3$ y) which is unlikely to remain in or undergo transformation in the lung, thus the alpha energy of its subsequent transformation product ^{210}Po (RaF) is not included in the calculation of the WL.

In a practical sense such equilibrium rarely exists, and it is necessary to always determine the number of atoms of the particulate progeny of Rn for determinations of their contribution to personnel exposure, which is very dependent on the time period over which ingrowth occurs. The number of atoms of each radon transformation product at a particular time t can be calculated from the respective Bateman equations (see above) or they can be determined from the curves in Figure 13-3, each of which is a plot of the number of atoms of RaA, RaB and RaC at any given time after formation of a pure source of 100 pCi of radon.

The relative contribution of each of the radon progeny to the WL varies considerably with time as shown in Figure 13-4. During the first few minutes the WL is due almost entirely to the rapid but limited ingrowth of RaA. Between 5 and about 25 minutes the increase in the WL is approximately linear and is due primarily to the ingrowth of RaB. The contribution from RaC is delayed considerably due to the 26.8 minute and 19.9 minute half-lives of RaB and RaC, respectively; thus, the RaB contribution becomes significant only after the radon source is more than 15 minutes old and the RaC contribution only after 40 minutes or so. And, it takes about 40 minutes for the radon progeny to achieve 50% equilibrium with ^{222}Rn, or about half of 1.0 WL.

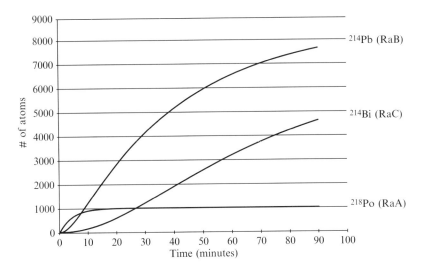

Figure 13-3. Graphical representation of the number of atoms with time of RaA, RaB, and RaC (which is also the number of atoms of RaC′) in a source of 100 pCi of ^{222}Rn, the activity of which does not change appreciably in a few hours after formation.

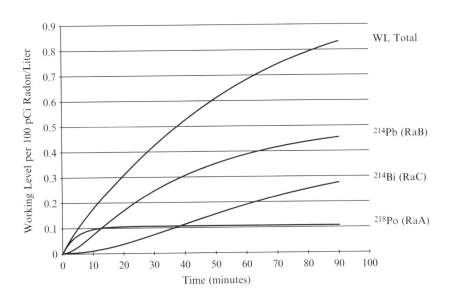

Figure 13-4. Contribution of each of the transformation products of ^{222}Rn to the working level (WL) for various times after the introduction (or separation) of ^{222}Rn into a space.

Clearly, the age of the air has a significant effect on the WL; thus, it is desirable to breathe young air by either filtering it or mixing in fresh air. And, since "dirty air" contains a large percentage of large particles which do not penetrate to the deep reaches of the lung, radon risk may be reduced by breathing "dirty air", but at the expense of other potential effects.

Example 13-1: Determine the WL for air containing 100 pCi/L of radon 10 min after filtration.

Solution: Following the example for defining the WL, the number of atoms present after 10 min of ingrowth are calculated by equations 13-1 through 13-4 or alternatively read from the curves in Figure 13-3; i.e.

Nuclide	Atoms in 100 pCi	α–energy per atom (MeV)	α–energy (MeV/100 pCi)
^{222} Rn	1.77E06	Exhaled	None
^{218}Po (RaA)	876	13.69	1.2E04
^{214}Pb (RaB)	1219	7.69	9.4E03
^{214}Bi (RaC)	114	7.69	8.8E02
^{214}Po(RaC′)	~0	7.69	0
		TOTAL	2.23E04

And the WL is

$$WL = \frac{2.23 \times 10^4 \text{ MeV } \alpha}{1.3 \times 10^5 \text{ MeV } \alpha/WL} = 0.17 \text{ WL}$$

Although the WL unit has shortcomings, it is the most practical single parameter for describing the effect of radon and its transformation products. Since the WL is based on radon, it is, like the Roentgen (for x rays), a unit of exposure, not radiation dose; therefore, exposures of people are controlled by keeping airborne levels of radon and its products below a specified number of WLs.

The **Working-Level Month (WLM)** is used to describe cumulative radon exposure. The average number of exposure hours in a month is 173 (40 × 52/12 = 173); thus, the WLM is calculated by multiplying the exposure in WL by the number of hours of exposure and dividing by the number of working hours in one month, or 173. For example, exposure to one WL for 173 h = 1 WLM; similarly, exposure to 10 WL for 173 h would be 1730 WL · h, or 10 WLM.

Example 13-2: A person is exposed to air containing radon and transformation products which is determined to ultimately emit 10^5 MeV of alpha energy per liter of air. If a person is exposed to this atmosphere for 1000 h, what is the cumulative exposure in WLM?

Solution:

$$WL = \frac{10^5 \text{ MeV}}{1.3 \times 10^5 \text{ MeV/WL}} = 0.77 \text{ WL}$$

and since this exposure occurs for 1,000 h, the number of WLMs is

$$WLM = \frac{0.77WL \times 1000\,h}{173\,h/mo} = 4.45WLM$$

HEALTH EFFECTS OF RADON

Lung cancer has been observed directly in humans exposed to radon, and control policies are based on such observations; i.e., there is no need, as is the case for many other environmental pollutants, to extrapolate risk factors from animal studies with their attendant uncertainties. Over 300 years ago, before radioactivity was known to science, it was common for feldspar miners in Germany and Czechoslovakia to contract and die from a mysterious lung disease called "mountain sickness". It is now known that this mysterious disease was lung cancer caused by the short-lived alpha-emitting transformation products of radon continuously emanating from the uranium in the rocks. From the late 1800s to the early 1900s, about 400 lung cancer deaths occurred among the Schneeberg miners (Holiday 1969; Donaldson 1969). Autopsy reports for Joachimsthal miners from 1929 to 1940 showed that about 50% of the deaths were due to lung cancer, and during a 5-year period from 1939 to 1943, 180 cases of lung cancer were diagnosed in these miners. Radon levels in the Schneeberg mines were determined to be about 2,900 pCi/L or higher, with radon progeny levels of 10 to 180 WL; the radon levels in the Joachimsthal mines were estimated to be similar. A direct epidemiological association has also been established between lung cancer in US and Canadian uranium miners and their exposure to radon, evidence so compelling that vigorous controls have been used to reduce the concentrations of radon and its transformation products in uranium mines.

Radon exposure and lung cancer in the general population have been studied extensively in Sweden (Pershagen and Falk 1984), and these studies show significant associations between cancer and exposure to radon progeny. A study by the National Cancer Institute of lung cancer in women in New Jersey and Missouri also demonstrated a link between lung cancer and indoor radon. Radon concentrations in the 2,000 to 3,000 pCi/L range have been observed along the Reading Prong, levels that are equivalent to 10 to 15 WL from which one could accumulate 200 WLM or more which is comparable to levels that produced lung cancer in miners. Based on this evidence, radon control guides and advisory programs have been developed to inform the public so they may evaluate and control or prevent radon in their home environment.

Radon Risk Coefficients have been derived for the US population using models to relate radon levels and observed effects as listed in Table 13-2. The radon risk coefficients have been stated in terms of the concentration of radon in air or an associated working level rather than lung dose because the actual dose of alpha radiation from inhaled radon and radon progeny is difficult, if not impossible, to determine. Lung morphology, breathing rate, the size distribution of the inhaled particles, and other factors also make it difficult to model the dose, for example

- some of the inhaled particles are deposited in the naso–pharyngeal region and never enter the tracheo–bronchial region where most lung cancers occur;
- particle deposition is nonuniform and the thickness of the bronchial epithelium, which determines how much alpha energy reaches the basal cells, varies with a person's age and from one part of the bronchial tree to the other; and
- the effects of the muco–ciliary layer in relation to the half-life of radon progeny is not well known.

The risk coefficients listed in Table 13-2 are based primarily on lung cancer incidences in miners and generally have not considered other carcinogens in mine air such as diesel fumes, acid fumes, ore dust, and so on. Even though homes do not contain these pollutants, densely populated areas often contain air pollutants that may promote lung cancers.

The Health Effect Mechanism by which radon and its alpha-emitting products induce lung cancer is believed to be due to the deposition of alpha particle energy emitted by radon progeny, which exist either as highly charged unattached particles or attached to dust particles which are readily deposited and retained in the lung. Air containing radon progeny is carried into the lung through the trachea to the bronchial tree and then to the alveoli where alveolar sacs and pulmonary veins meet and gas exchange with the blood takes place. The

Table 13-2. Average Risk of Radon-related Lung Cancer in the General Population, per WLM.

Source/Risk Model	Cases/10^6 Persons
UNSCEAR (1977)/AR*	200–450
BEIR III (1980)/AR	730
ICRP (19810/AR	150–450
NCRP (1984)/AR	130
ICRP (1987)/AR	150
ICRP (1987)/RR**	230
BEIR IV (1988)/RR	350

*AR = absolute risk model; **RR = relative risk model.

bronchial tree branches from the trachea into two main stems to the left and the right sides of the lung where they further divide into many lobar bronchi and further into segmental bronchi and eventually to terminal bronchioles where pulmonary arteries emerge. With such long and winding paths in the lung, radon progeny (either unattached or attached to inhaled airborne particles) will almost certainly be trapped in the lung before the air is exhaled. The 6 and 7.69 MeV alpha particles emitted by RaA and RaC$'$ are massive and highly charged, and this energy is delivered in a huge jolt, which damages and/or kills lung cells.

Health effects associated with alpha emitters are very much a function of the thickness of the epithelium, the size of the inhaled particles, the action of mucus to remove them, and the size and location of the bronchi or bronchioles. The bronchial epithelium, or the moist lining of the bronchi, is 15 to 80 microns thick, exclusive of the mucus layer and the cilia, which are about 7 microns each. It is composed of a thin layer of ciliated cells atop a layer of basal cells. The lining is covered by a thin layer of serous fluid and a thin layer of mucus that clears away any foreign substance deposited on the lining. If the combined thickness of the epithelium including the muco–ciliary layer is less than 40 microns, the basal cell layer can be reached by the 6.0 and 7.69 MeV alpha particles emitted by radon progeny. Killed cells are readily replaced; however, damaged cells can replicate, and cellular defects in them may eventually lead to lung cancer. Autopsy data show a large percentage of lung cancers in the main bronchi of underground uranium miners, and also a large number of lung cancers in the peripheral bronchioles where there are no basal cells.

The Dose Rate to Children from airborne radon progeny is about twice that in adults due to breathing rate and lung morphology. Accordingly, for exposure at age 5 the annual lifetime risk could be twice that due to the same exposure at age 25.

Lifetime Risks Due to Environmental Radon

Because of the strong correlation between high levels of cumulative radon exposure and increased lung cancers in miners, public policies have prudently assumed that the same mechanisms occur at low levels and have extrapolated the data to low levels of radon exposure even though actual incidence at such levels has been difficult to demonstrate.

Most risk models suggest that a lifetime exposure of about 15 WLM (exposures at 4 pCi/L for about 50 y) represents a lifetime lung cancer risk between 1 and 4%. The US Environmental Protection Agency (EPA) has estimated that a lifetime exposure to 4 pCi/L will likely cause 3 fatal lung cancers in a group of 100 typical individuals and at 200 pCi/L this same group could exhibit 60 fatal lung cancers. For normal living conditions, however, exposures in most houses are well below 4 pCi/L, and the lifetime risk for typical occupants under these conditions will be fewer than 0.6 per 100. A chart of radon risk versus exposure is shown in Table 13-3, including a comparison of radon risks to those from chest x rays and cigarette smoking.

Table 13.3. A Chart of Radon Risk VS WL of Exposure and Comparative Risks Associated with Chest X Rays and Cigarette Smoking.

pCi/L	WL	Radon lung cancer deaths per 1,000	Comparable exposure level	Comparable risk
200	1	440-770	1,000 times Avg. outdoor level	>60 times non-smoker risk
				4-pack-a-day smoker
100	0.5	270-630	100 times Avg. indoor level	
			200 times Avg. outdoor level	20,000 chest X rays/yr
40	0.2	120-380		2-pack-a-day smoker
20	0.1	60-210	100 times Avg. outdoor level	1-pack-a-day smoker
10	0.05	30-120	10 times Avg. indoor level	
				5 times non-smoker risk
4	0.02	13-50		200 chest X rays/yr
2	0.01	7-30	10 times Avg. outdoor level	
1	0.005	3-13	Avg. indoor level	Lung cancer risk (non-smokers)
0.2	0.001	1-3	Avg. outdoor level	20 chest X rays/yr

Source: EPA

Since radon is ubiquitous in the environment and if current radon risk coefficients are appropriate, it is reasonable to expect that radon-induced lung cancers also occur in the population. The EPA has estimated that 5230 to 20,894 radon-induced lung cancer deaths occur in the US each year based on: an average indoor radon concentration of 0.004 WL and a correction factor of 1.64 for nonminers; an average life-span of 74 years (888 months) for a population of 226.5 million (1980 census); and a relative lifetime radon lung cancer risk of 1 to 4% per WLM.

Risk to Smokers

Studies of the combined effects of smoking and exposure to radon progeny have produced confounding results: some studies suggest that the combined risks are less than the risks attributed to each separately; some suggest that the combined risks are simply the sum of the two separate risks; and some indicate a synergistic effect between radon and smoking which has been shown to be the major cause for lung cancer. Laboratory studies in rats indicate that the two carcinogens enhance each other such that radiation acts as an "initiator" and tobacco smoke as a "promoter" of the carcinogenic process (Chameaud et al. 1981).

Synergism between radon and cigarette smoke is evident in the data for US and Czechoslovakian miners (Archer et al. 1976). Miners exposed in excess of 1,800 WLM had lung cancer rates of 1.39/1,000 for nonsmokers, 9.4/1,000 for one-pack-a-day smokers and 13.3/1,000 for heavier smokers. On the other hand, Cohen (1982) has applied the risk factor derived from miners, the majority of whom are heavy smokers, to the nonsmoking population, and found that the estimated lung cancer deaths due to indoor radon alone exceeded the actual number that occurred. Despite the potential inconsistencies, the EPA believes that the risks from exposure to radon and cigarette smoke are greater than the sum of the risks from either alone (EPA 1987), and that radon-related lung cancer deaths in smokers (including ex-smokers) are about 5 times greater per unit of exposure than in nonsmokers.

RADON REDUCTION MEASURES

Indoor radon levels are affected not only by the radium content in the nearby soil but the type of house, the permeability of the underlying soil, and the available paths for radon entry. All homes contain some radon, and although the soil surrounding the house has the greatest influence, groundwater and some building materials may also contribute to indoor radon levels. Well water may contain high concentrations of radon which can be released into the house from showers and/or when water is heated or agitated in washers and dishwashers.

Radon from building materials is usually insignificant in the US except for houses that were built with materials containing uranium mill tailings and phosphate slag, e.g., in Grand Junction, Colorado and in certain areas of Florida. In solar houses, the rocks used for heat storage sometimes contain high concentrations of radium and therefore can be a high radon source.

The EPA Screening Level for Radon has been selected, after consideration of average levels and health risk data, as a level above which action to reduce the indoor radon concentration is warranted. It is not a limit, though many use it as such, but a value that suggests that the risk level (a 1 to 4% lifetime risk of lung cancer) is such that home occupants can gain a significant benefit by reducing their exposure to radon, especially because practical means are available to do so. Iowa, North Dakota, Minnesota, and Colorado have the highest percentage of homes that exceed the EPA screening level of 4pCi/L.

Selection of a radon reduction technique is greatly influenced by the radon source itself, the characteristics of a given house, the amount of reduction required to reach recommended levels (currently 4 pCi/L), the amount of money that the homeowner is willing to spend, and the obtrusiveness and appearance of the installed system. The most effective way to reduce the radon concentration in a home is to preclude its migration into the structure by providing barriers to its entry and/or installing systems to route the gas around the home.

Radon ingress into homes can often be retarded by sealing cracks and floor penetrations and/or covering sumps and floor drains; these methods are relatively inexpensive and should be included in any mitigation program. For moderately high radon levels (e.g., 4 to 6 pCi/L), homeowners may be able to correct the problem themselves with these techniques, and although a specific goal may not be achieved in all cases, they are worth trying. For higher levels (≥ 20 pCi/L) reducing the concentration to 4 pCi/L generally requires subslab or block-wall depressurization or, perhaps, use of forced-air ventilation with heat recovery. Other methods involve replacing the indoor air with air that contains much less radon, or filtering radon from the air in the house. Radon removal by charcoal filtration has proven impractical because of problems with saturation and breakthrough even though radon absorbs readily onto charcoal. Key features of the various techniques for radon reduction are summarized in Table 13-4.

Closure Of Radon Entry Routes

Major soil-gas entry routes are illustrated in Figure 13-5; these include exposed soil and rocks in a basement or crawl space, uncapped sumps and drains, and major cracks and penetrations in walls and floors. Exposed soil areas are easy

Table 13-4. Key Features of the Various Techniques for Radon Reduction.

Method	Principle of operation	Reduction
Ventilation (natural or forced air)	Increased movement of fresh outdoor air into the house (or crawl space)	90%
Forced-air ventilation with heat recovery	Movement of fresh outdoor air into the house	50–75%
Avoiding house depressurization	Provides clean makeup air to household appliances that exhaust or consume indoor air	0–10%
Sealing of soil-gas entry routes	Closing openings between the house and the soil	0–90%; very case-specific
Drain-tile ventilation	Fan suction of perforated drain tiles around the house footing	90–99% if complete loop; 40–95% otherwise
Subslab ventilation (Figure 13-6)	Fan suction on pipes inserted through the slab into the soil layer	80–90% if good soil permeability
Block-wall ventilation (Figure 13-7)	Fan suction on the void network inside hollow block foundation	90–99% if walls sealed

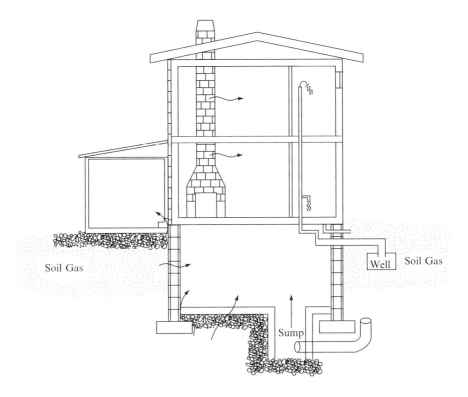

Figure 13-5. Major routes for radon entry are floor-wall joints, basement wall perforations, cracks in slabs in contact with soil, and exposed soil (e.g., sump pumps and crawl spaces), and to a lesser extent building materials and aerated well water that may contain dissolved radon.

to locate and can be sealed off separately; however, hairline cracks in concrete slabs and pores in block walls which may be a primary route of entry are sometimes difficult to identify. Such cracks and pores are tiny and may exist in large numbers, thus requiring collective sealing with a good surface coating. Larger cracks and joints can be sealed with silicone caulk, flowable urethane caulk, gun-grade urethane caulk, nonshrinking grout, and so on. Cracks and joints may need to be enlarged to assure adequate bonding of the sealant.

Mitigation of Radon from Exposed Soil or Rock can be accomplished by various techniques. For example, an earthen floor in a basement can be covered with a concrete slab poured on a 4-inch layer of gravel, which not only provides a foundation for the slab but provides a good permeability zone if an active soil ventilation system is subsequently required. A sheet of 6-mil polyethylene should be used as a vapor barrier on top of the crushed stone and a 2-inch sand layer should be placed between the polyethylene sheet and the concrete slab. A crawl space can be ventilated by passive vents open to the outside and/

or the exposed soil can be covered with polyethelene sheeting to trap radon from the soil, and an active or passive subfilm ventilation system (as shown in Figure 13-9) can be used to route it outside.

Subsoil Depressurization consists of pipes inserted through the foundation slab (Figure 13-6) or through the foundation wall (Figure 13-7); a fan is usually required (versus passive ventilation) to reduce the buildup of radon under the slab. The fan is usually mounted outdoors or in the attic with the exhaust pipe extended above the roof, but may also be located on the

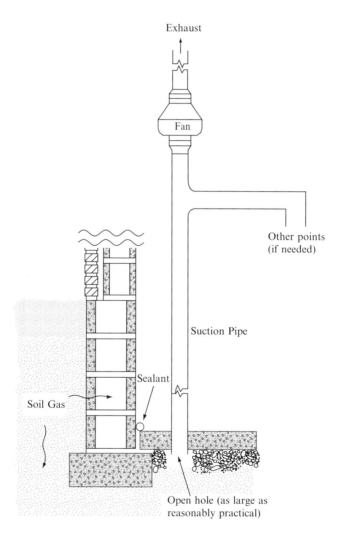

Figure 13-6. Subslab suction through basement floor indicating potential entry routes and places where sealing is indicated. (Adapted from EPA, 1987.)

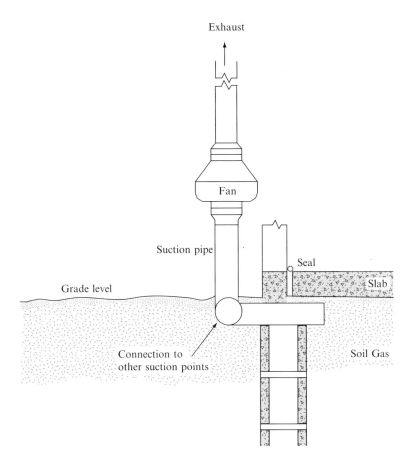

Figure 13-7. Subslab suction via a horizontal pipe through the foundation wall (Adapted from EPA, 1987.)

side of the house. This method has two benefits: radon is vented away before it has a chance to enter the structure, and it reduces the pressure under the slab which in turn reduces the driving force that draws radon indoors. The number of suction points used depends on the permeability of the subslab soil layer; one is often sufficient, but two or three may be needed. Four-inch PVC pipes (schedule 20) are adequate for the application, and joints must be caulked to prevent leaks.

Subslab Pressurization, the inverse of subslab depressurization, is achieved by installing the exact same pipe system except that the fan is used in reverse to blow outdoor air into the soil. This method creates a high-pressure field under the slab that dilutes radon in soil gas and forces it away from the structure. Although not as common as subslab depressurization, good results can be achieved by a pressurization system. Both methods succeed by increasing ventilation of soil gas and the radon it contains.

Drain Tile Suction can also be used as a depressurization system by connecting suction pipes to the existing drain tiles surrounding the house and using a fan to vent soil gas and radon away from the house. Passive ventilation, which relies on the stack effect of differentially heated air to create suction, may provide sufficient ventilation for a depressurization system. To be effective, exhaust pipes should be 6 to 8 inches in diameter and should extend up through the house and above the roofline to assure differential heating.

Sealing of the Void Network in hollow concrete blocks or cinder blocks used in foundations or basements can prevent soil gas entry into the living space. Radon that accumulates in these void spaces can penetrate through pores and cracks in the walls, or if the top course of the wall is left uncapped, from the top of the wall. The top of the concrete block wall can be sealed by pushing crumpled newspaper into the voids to form a base for mortar or expanding foam, which should be applied to a depth of about 2 inches. If leakage is through small voids between the sill plate and the top row of blocks, foam can be injected into holes drilled in the top course of each block or if the gap is very narrow, it can be caulked. And, if entry occurs through the exterior brick veneer of a home, holes can be drilled through the band joist and foam injected into the space.

Block-wall Ventilation may also be used to vent soil gas trapped inside the void network of a concrete block wall by inserting one or two pipes into the void network in each wall. While very effective (radon reduction of up to 99% may be achieved with good closure and sealing of all major wall openings) it is used primarily as a supplement to a subslab depressurization system that fails to achieve the desired reduction, or if poor soil permeability makes subslab depressurization questionable. An effective alternative method of block-wall ventilation is to install a sheet metal baseboard around the entire perimeter of the basement and to vent it outside. This baseboard system must be well sealed and other openings closed to assure that soil gas that might otherwise enter the house is routed to the baseboard vent system.

Block-wall ventilation must be used with great care because it depressurizes the basement (and the house) which may draw other soil gases, or more importantly, combustion gases from a furnace or hot water heater or other devices into the house. If used, venting for carbon monoxide should be provided and a carbon monoxide detector should always be installed.

RADON MINIMIZATION IN NEW CONSTRUCTION

As with radon reduction in existing homes, minimization of potential radon problems in new homes involves: (1) preventing radon entry by using barrier methods, (2) reducing the radon driving force, (3) diverting radon in the surrounding soil away from the structure, and (4) providing features that allow straightforward installation of active removal systems if they are subsequently needed (e.g., a capped subslab penetration that can be connected to an exhaust fan). Fortunately, these techniques are compatible with conventional construction practices. By recognizing the potential for radon entry, extra care can be taken in construction to prevent it, e.g., in the mixing and curing of poured concrete to minimize cracking, choosing optimal materials, sealing penetrations, etc.

Construction of Radon Barriers can be very effective because a major entry route for radon is direct from the soil. Radon barriers are impermeable foundations; poured walls below-grade, and radon-resistant slabs-on-grade, especially if constructed to minimize cracks and voids and elimination of joints and penetrations. Construction of below-grade walls with poured concrete instead of concrete blocks eliminates radon buildup in the void spaces of the blocks which are generally very porous. A radon-resistant poured-concrete wall should have a vapor barrier of 6-mil polyethylene film or equivalent, a drainage system to relieve hydrostatic pressure and ventilate the soil, and a tooled floor/wall joint sealed with nonshrink grout or gun-grade urethane. If, however, masonry block walls are used in below-grade construction, various features can be used to minimize radon infiltration, for example: (a) the bottom row of blocks that contact the foundation or slab should be of solid blocks to prevent radon seepage around the footing; (b) the tops should be sealed and capped by solid blocks or termite caps, or filled with 4 inches of concrete; and (c) the exterior surfaces should be parged and coated with high-quality vapor/water sealants or polyethylene films.

Proper Slab Construction is essential to minimize slab cracking with a particular emphasis on the right amount of mixing water. A plasticizer (water-reducing admixture) or superplasticizer (high-range water-reducing admixture) should be used to achieve the desired pouring consistency; adding more water than specified for the product can promote cracking. Proper curing can be achieved by watering the slab; covering the slab with wet sand, wet sawdust, plastic sheets, or a waterproof paper; and coating the slab with a curing compound such as penetrating epoxy sealer which acts as a curing agent and slab strengthener. Reinforcement of the slab to reduce shrinkage cracking can be achieved by embedding wire mesh or "rebar" in the slab, or adding fiber additives. Although a single rectangular slab is preferred, a reinforced L-shaped slab can be acceptable. The slab should be constructed with the minimum number of pours and as few joints as possible.

Slabs should be placed on a clean coarse aggregate that is highly permeable should a post-construction subslab ventilation system be needed for removal of soil gas. A vapor barrier of 6-mil polyethylene film should be placed between the aggregate and the slab to prevent moisture entry from beneath the slab and to provide an additional radon barrier. It is important to avoid puncturing the film when the slab is poured and to seal plumbing penetrations and other gaps. If the slab intersects the inside edge of the foundation wall, the vertical joints should be sealed; however, it is best to eliminate these joints by a monolithic slab (see below).

Radon Resistant Slab-on-Grade Construction is very important since the slab represents a large area in direct contact with the soil and any radon it contains. The best design is perhaps a monolithic slab foundation that consists of a single slab of concrete poured on grade to form both the floor of the house and the footing for the walls (see Figure 13-8). The house walls rest on the slab and,

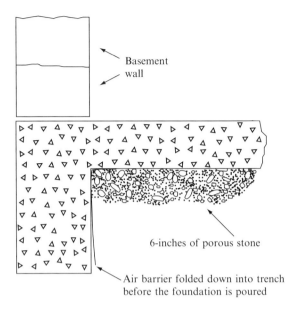

Figure 13-8. Monolithic slab-on-grade in which the building walls rest on the slab instead of the soil (adapted from EPA 1988).

therefore, do not touch the soil, and the usual straight joint around the perimeter of the foundation wall is eliminated. A slab that extends across the top plane of the foundation wall and rests on L-shaped blocks can also provide an acceptable barrier since a tortuous angled joint is created. Regardless of which one is used, the slab should be constructed to minimize pores, cracks, and the number of contact joints through which radon can flow.

Crawl Space construction in new homes should use a vapor barrier of poly-ethylene film or equivalent (as illustrated in Figure 13-9) and anchored in place, typically by a 2-inch layer of sand, to cover any exposed soil. The barrier must be properly glued to the wall, and a perforated pipe should be installed under the barrier and connected to a short "Tee" riser for attachment to a suction system if it is needed later. The riser must be sealed airtight, and electrical, heating, and plumbing penetrations must be properly sealed to preclude radon entry. Screened vents, equally distributed on all sides of the crawl space for good cross-ventilation, should also be installed; these should provide at least 1 sq.ft of vent for every 150 s.ft of crawl space. In cold climates, it is also necessary to in-sulate water pipes and heating pipes, and subfloors should also be well insulated.

Checkpoints

Various radon-resistant techniques can be used in new construction to either prevent radon ingress into a home or to make it more amenable to subsequent

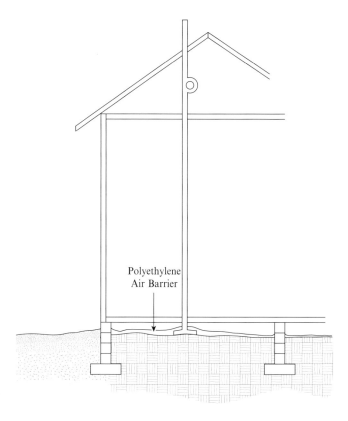

Figure 13-9. Radon removal from exposed soil in a crawl space—soil is covered with polyethylene film sealed to the walls and vented if needed by the stack effect or a fan. (*Source*: EPA, 1988.)

remediation should it become desirable. Among the more important features are

- a vapor barrier of 6-mil polyethylene or equivalent placed under the slab or over any exposed soil and a vent system that is installed and capped; any penetrations for ducts, pipes, or wires should be sealed with high-grade caulks and sealants;
- slab construction that eliminates joints (e.g., by use of a monolithic slab) and minimizes cracking by using recommended amounts of water, a platicizer, and/or reinforcing mesh or rebar;
- floor drains that drain to grade, to a sewer, or to a sump that is sealed at the top and vented; and
- basement/foundation walls of poured concrete with exterior surfaces that are parged and coated with high-quality vapor/water sealants and/or

covered with polyethylene films; or, if masonry walls are used the top and bottom blocks should be solid and the tops should be capped.

RADON MEASUREMENT

Radiation protection recommendations for radon and its transformation products are usually given in terms of the concentration of radon in air (pCi/L) or as a WL based on the potential alpha energy emitted by its transformation products. These two circumstances also largely define the two principal measurement methods: (1) collecting the particles of the transformation products and determining the WL directly, or (2) determining the WL indirectly by absorbing the radon, a noble gas, on an adsorbent and measuring it to determine its concentration in air.

Since both the radon concentration and the WL are related, each can be inferred from the other if certain parameters (e.g., fraction of equilibrium) are known or assumed. Working environments such as mines or uranium mills are usually characterized by a WL measurement which is made by collecting a short-term (5 to 10 min) particulate sample on a filter paper and measuring the alpha-disintegration rate. Home environments are usually sampled over a period of several days (2 to 7) by adsorbing radon onto activated charcoal, which is then measured by counting the gamma-emission rate from RaC (^{214}Bi). The first (work place) requires a particulate air sampler and a field method for measurement; the second (residences) uses passive adsorption and a laboratory technique.

Particulate Sampling for Radon is done with a high-volume (> 2 cfm) air sampler that is operated for 5 to 10 minutes during which the total volume of air (in liters) is measured. It is important to use an efficient membrane or molecular filter to avoid the need to correct for absorption of the alpha particles within the filter medium, which reduces counting efficiency. The alpha disintegration rate (dpm) is measured 40 to 90 minutes later by a laboratory instrument or a portable survey instrument calibrated to a known alpha source. The relative fractions of the alpha-emitting transformation products will change as the sample ages due to the different rates of radioactive transformation of each; therefore, it is necessary to adjust the dpm determination by the Kusnetz factor, as given in Figure 13-10. The Kusnetz factor, named after its developer, converts the dpm rate to one that is weighted according to the alpha energy emission rate thus allowing the WL to be calculated directly from the measured alpha-disintegration rate per liter, as illustrated in Example 13-3.

Example 13-3: A particulate air sample was collected for 5 min at 2 cfm in a tent erected over an excavation site at an old uranium processing site. The filter was measured 50 min later with a portable ZnS alpha scintillation detector and found to contain alpha activity of 10,000 dpm. The air filter is assumed to

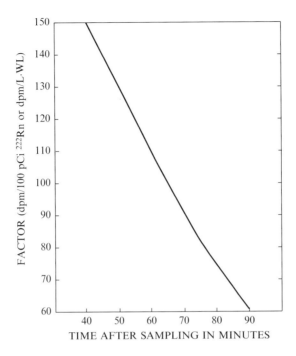

Figure 13-10. Factors of dpm/100 pCi of ^{222}Rn (or dpm/L ·) for times after collection of alpha-emitting particulate progeny of radon.

have 100% collection efficiency and no self-absorption of the emitted alpha particles. (a) Determine the WL in the tent atmosphere; and (b) if a person worked inside the tent 40 h per week for 25 weeks, what would be the accumulated exposure in WLM?

Solution: (a) The WL in the tent is based on the activity per liter of air

$$Vol(L) = 5\,m \times 2ft^3/m \times 28.32\,L/ft^3$$
$$= 283.2\,L$$

From Figure 13-10, the Kusnetz correction factor at 50 min after sample collection is 129 dpm/L · WL, and the working level is

$$WL = \frac{10,000\,dpm}{283.2\,L \times 129\,dpm/L \cdot WL}$$
$$= 0.274\,WL$$

which is just below the recommended exposure guide for uranium miners.

(b) The accumulated exposure in WLM for the worker in the tent is

$$WLM = \frac{0.274WL \times 25\,wk \times 40\,h/w}{173WL \cdot h/WLM}$$

$$= 1.58\,WLM$$

The Radon Absorption Method uses activated charcoal to absorb radon directly from the air. Since this procedure is passive (i.e., no air sampler is used to force air through the charcoal) it requires a standard sampler, usually a small canister that contains about 75 g of charcoal, that is calibrated in a known radon atmosphere. The absorption of radon on the charcoal of the reference canister occurs at a constant rate and is a function of the exposure time; thus, the activity of the absorbed radon is directly proportional to the concentration of radon (pCi/L) in the air surrounding the canister and the exposure period, as given in the following relationship for a 75 g canister (Gray, 1989):

$$Abs(dpm \cdot L/pCi \cdot min) = 35.8511 \times 10^{-3} - (5.1353 \times 10^{-5} \times t)$$

where t is exposure time in hours.

The absorbed radon is determined by counting ^{214}Bi (RaC) gamma rays to determine the ^{214}Bi disintegration rate, and since ^{222}Rn and ^{214}Bi are in equilibrium after 90 min or so, their activities are the same. This activity is then decay-corrected back to the midpoint of the sampling period and reported as pCi of radon per liter of air.

Example 13-4: A 75 g charcoal canister was exposed in the author's home for 189 h (11340 min) with doors and windows closed. A 10-min count taken 40 h later with a detector that has an efficiency of 26.2% for ^{214}Bi gamma rays yielded a net gamma count rate of 119.3 counts per minute (after subtraction of background). What was the radon concentration (pCi/L)?
Solution: The measured activity of ^{214}Bi is

$$\frac{119.3\,cpm}{0.262\,cpm/dpm} = 455.34\,dpm$$

and since ^{222}Rn, the parent of ^{214}Bi, and ^{214}Bi are in equilibrium on the charcoal, the activity rate of ^{222}Rn is also 455.34 dpm. The activity of ^{222}Rn at the midpoint of the sample collection period is a good value of the average ^{222}Rn concentration in the home; therefore, correction of the transformation rate back to the midpoint of the sampling period (189/2 + 40 h, or 5.6 d) yields a ^{222}Rn activity of

$$455.34\,dpm = A_o e^{-\frac{\ln 2}{3.82\,d} \times 5.6\,d}$$

$$A_o = 1259\,dpm$$

The absorbed radon on the canister is

$$\text{Abs} = [35.85 \times 10^{-3} - 5.1353 \times 10^{-5} \times 189\text{h}]$$
$$= 0.0261 \text{ dpm} \cdot \text{L/pCi} \cdot \text{min}.$$

And the average radon concentration over the sampling period is

$$\text{Rn(pCi/L)} = \frac{1259 \text{ dpm}}{11340 \text{ min} \times 0.0261 \text{ dpm} \cdot \text{L/pCi} \cdot \text{min}} = 4.25 \text{ pCi/L}$$

which is just above the EPA screening level requiring the author to think about it (he subsequently covered and vented a crawl space that had exposed soil).

A Working Level (WL) Measurement can be obtained from a measured radon concentration if the degree of radioactive equilibrium between radon and its transformation products is known. For a residence, this equilibrium fraction is about 50% due to the number of air changes that occur; for energy efficient homes it can be as high as 75 to 80%, and for some older homes it could be as low as 20 to 30%. The WL can be defined in terms of the equilibrium fraction as

$$\text{WL} = \frac{\text{Rn(pCi/L)}}{100 \text{ pCi/L} \cdot \text{WL}} \times F$$

where F is the fractional equilibrium between radon and its alpha-emitting transformation products.

Example 13-5: The recommended action level guideline for a residence is 4 pCi of radon per liter of air. (a) If the fractional equilibrium between ^{222}Rn and its particulate, alpha-emitting transformation products is assumed to be 50%, what is the corresponding WL? (b) If 70% occupancy of the residence is assumed, what is the cumulative WLM associated with this exposure level in one year?

Solution: (a) $\text{WL} = \dfrac{4 \text{ pCi}}{100 \text{ pCi/L} \cdot \text{WL}} \times 0.5 = 0.02\text{WL}.$

(b) At 70% occupancy, the annual exposure period is

$$E(\text{h}) = 365 \text{ d/y} \times 24 \text{ h/d} \times 0.7 = 6132 \text{ h}.$$

$$\text{Total exposure} = \frac{6132 \text{ h} \times 0.02\text{WL}}{173\text{WL} \cdot \text{h/WLM}} = 0.71\text{WLM}$$

SUMMARY

Several unique features cause radon, a radioactive noble gas that exists naturally, to be a public health issue: (1) it is a series transformation product of uranium which is extremely long lived ($T_{1/2} = 4.5$ billion y) and widely distributed in nature, therefore it is ubiquitous and perpetual; (2) as a gas with a 3.82 d half-life it persists long enough to migrate from soils into homes, but is short-lived enough to produce even shorter-lived radioactive particulate products that build up in the home; and (3) these highly charged, short-lived alpha-emitting radioactive particles are readily inhaled and retained in the lung where they deposit considerable amounts of alpha particle energy to lung tissue.

In natural soils, ^{230}Th ($T_{1/2} = 75,380$ y), the progenitor of ^{226}Ra ($T_{1/2} = 1,600$ y), and ^{226}Ra, the immediate parent of radon, are in radioactive equilibrium with uranium. Sources that contain ^{230}Th and ^{226}Ra separated from uranium (e.g., residues from ore mining and processing) also represent long term sources of radon, especially ^{230}Th. Radon transformation products or progeny (commonly called daughters) emit alpha particles with energies ranging from 6 to 7.69 MeV, and because alpha particles (as helium nuclei) are massive and highly charged, this energy is delivered in a huge jolt which damages and kills cells on the surfaces of the bronchi and the lung. Killed cells are readily replaced; however, damaged cells can replicate, and the cellular defects in them may eventually lead to lung cancer.

Human response to radon and its progeny is characterized in terms of exposure rather than radiation dose. The unit of exposure to radon is the "working level" (WL) which was first defined for uranium miners as 100 pCi of ^{222}Rn in a liter of standard air in equilibrium with its transformation products RaA, RaB, RaC, and RaC', which produces 1.35×10^5 MeV of alpha energy. Complete equilibrium between radon and its products rarely exists in working environments or in homes, and the WL is used to characterize the amount of energy potentially deposited in the lung from radon/progeny mixtures. The WL is measured by either: (1) collecting the particles of the transformation products and determining the WL directly, or (2) absorbing the radon, a noble gas, on an adsorbent such as activated charcoal and measuring it to determine its concentration in air. Since both the radon concentration and the WL are related, each can be inferred from the other if certain parameters (e.g., fraction of equilibrium) are known or assumed.

Lung cancer has been observed directly in humans exposed to radon progeny (having first appeared in miners 300 years ago in Germany and Czechoslovakia); therefore, control policies for radon are based on human data, in contrast to other environmental pollutants which are largely based on animal studies. Radon control guides for miners are based on cancer incidence, and advisory programs have been developed for non-miners to inform the public so they may evaluate and control or prevent radon in their home environment. Such measures are recommended when radon levels exceed 4 pCi/L with about 50% equilibrium with its transformation products; these measures are based on precluding radon entry into a home or other occupied space, filtration of the air in the home, or dilution with fresh air. Radon-resistant construction uses similar techniques to prevent radon entry by optimal slab construction, sealing and backfilling with appropriate materials, and/or providing systems that can be connected if a remediation system is needed later.

ADDITIONAL RESOURCES

1   US Environmental Protection Agency, *A Citizens Guide to Radon*, Washington, D.C., 20460, Report No 402-K92-001, 1992.
2   Evans, R.D., Engineers Guide to Behavior of Radon and Daughters, *Health Physics*, 17, No. 2 (August 1969).

PROBLEMS

1   The average age of ^{222}Rn in a mine shaft is determined to be 20 min. What would be the WL in the mine shaft for a measured concentration of 100 pCi/L of ^{222}Rn?

2   Calculate the WL for air that contains 100 pCi of radon that is only 3 min old.

3   Choose a radon risk coefficient and use it to determine the expected lung cancer incidence in a group of 1,000 persons exposed to 0.03 WL for 30 y.

4   A charcoal canister was exposed in the basement of a house for 140 h. After 48h a 10 min count on a detector with 25% counting efficiency yielded a net gamma count rate of 96 counts per minute of ^{214}Bi (RaC). Determine the radon concentration (pCi/L) assuming a 50% equilibrium between radon and its progeny in the house.

5   For the previous problem, determine: (a) the WL if the equilibrium fraction between radon and its progeny is 40%; and (b) the cumulative exposure in WLM for a person that lived in the structure for 5 years at an average annual occupancy rate of 80%.

6   Determine the WL for 100 pCi/L of thoron (^{220}Ra) continuously in equilibrium with its short-lived decay products.

14

RADIOACTIVE WASTES

"Radioactive waste: everyone wants it picked up, but no one wants it put down."
— *G. Hoyt Whipple,* ~ *1961*

Radioactive wastes began to emerge as a significant issue during and shortly after World War II. Although legacy wastes associated with the mining and processing of various ores existed well before WW II, they did not receive much consideration until the waste products of fission introduced a host of technical matters intertwined with social and psychological principles of risk, equity, and fairness for a wide range of activities:

- Medical centers and/or universities use radioactive materials in research and public services.
- Various mining and materials processing operations concentrate and distribute naturally occurring radioactive materials.
- Some 20% of electrical energy in the US is produced by nuclear fission,
- National defense uses a dispersed and complex set of industrial processes and facilities to support a nuclear weapons program.

The Atomic Energy Acts of 1946 and 1954 as amended placed source material, special nuclear material, and byproduct materials (known as AE Act materials) under the full control of the federal government. Radioactive wastes that contain AE Act materials are largely grouped into high-level and low-level wastes, naturally occurring and Accelerator-produced Radioactive Materials (or NARM) are outside the AE Act, and if regulated, are done so by the states or other federal statutes. The largest component of

NARM waste is naturally occurring radioactive material (NORM), including a subcategory in which the radionuclide concentrations or potential for human exposure have been increased by human activities above levels encountered in the natural state; i.e., technologically enhanced NORM, or TENORM. Sources of TENORM are abandoned mine lands; radium in scales and sludges from oil and gas and geothermal energy production; sludges from water and sewage treatment plants; ash from burning coal or wood; and phosphogypsum wastes.

Radioactive wastes that are present as site contamination may be subject to the Comprehensive Emergency Response, Compensation, and Liability Act of 1980 (or CERCLA), which is commonly referred to as Superfund. The Act, as amended by the Superfund Amendments and Reauthorization Act (SARA) of 1986, established a national program for responding to releases of hazardous substances into the environment and for remediation of sites. Sites that pose imminent threats to health or the environment are subject to "removal" actions, which are generally limited in duration. Sites that require a longer-term evaluation and response are placed on the Superfund National Priorities List (NPL) and undergo a comprehensive Superfund remediation process. Seventy-five of the sites on the NPL have radioactive contamination that will be remediated by the US Department of Energy (DOE) or the Department of Defense (DOD) in partnership with the respective host state and the US Environmental Protection Agency (EPA), usually under a "Tri-Party Agreement."

A major goal of the National Environmental Policy Act of 1969 (NEPA) is to "fulfill the responsibilities of each generation as trustee of the environment for succeeding generations"; i.e., to ensure that decisions on radioactive waste management and control do not pose an unreasonable risk to human health and the environment now or in the future. Several key issues are: fundamental goals of control; reliance on institutional controls vs engineering and natural barriers; perspectives on risk; and key time periods. These elements are reflected in various government programs for radioactive waste, for example

- regulations for low-level and high-level radioactive wastes limit the dependence on institutional controls to 100 years during which monitoring is required;
- radioactive waste standards are based on analyses of the risks of disposal/ management alternatives;
- long-term risk considerations are reflected in a 1000-year design objective for remediation of uranium mill tailings, and high-level radioactive waste standards (10 CFR 60) require assessment of repository performance for 10,000 years;
- LLRW regulations (10 CFR 61) require a combination of engineered and natural barriers to assure isolation; and

- designs to enhance retrievability of waste and use of markers and other passive measures to inform future generations have been incorporated in sites for high-level and long-lived defense wastes.

LEGACY/PROCESS WASTES

Several major categories of radioactive wastes are associated with the processing of various ores that contained uranium, thorium, and their radioactive progeny, principally radioactive thorium and radium. Legacy/Process wastes include NORM, Manhattan Project wastes, uranium milling wastes, and some defense wastes. High-level and low-level radioactive wastes are a result of fission-related activity during and after World War II; sites that contain these materials are generally not occupied but may contain industrial or commercial activities.

Manhattan Project Wastes exist on some 46 sites due to national defense activities during and shortly after World War II; these and two major thorium process sites added by Congress contain residual materials from processing uranium and thorium ores. The dominant radionuclide at the uranium sites is ^{230}Th (half-life of 75,400 y) because uranium and most of the ^{226}Ra were removed when the original pitchblende ore was processed. The Belgian Minerals Co, the original owner of the pitchblende ore, required the US to separate and hold the residues (known as K-65 residues) for recovery of radium which was quite valuable at the time, but later relinquished it to the US. These radium residues are currently stored in large underground silos at Fernald, OH, and in an earthen disposal site in Niagara Falls, NY. The residues contain ^{230}Th, and future risks will persist and may well increase (unless ^{230}Th is removed) due to the ingrowth of ^{226}Ra and its products (as listed in Table 4-4).

Disposal of Manhattan Project wastes is based on stable near-surface internment, and radiation exposures are low and will remain so if the materials are left alone. A stewardship philosophy in combination with technical systems and land use controls is important for assuring that such materials are in fact left alone. Such a stewardship role began, perhaps unintentionally, with the transfer of large tracts of land to the US Atomic Energy Commission (AEC) for management of these unique reserves to meet national interests. Although operations have yielded to other needs, this basic stewardship responsibility has passed to the DOE to assure that these lands are preserved in the public interest and transferred intact to future generations with intergenerational information about any residual materials on the sites.

Thorium Residues from the processing of thorium ores represent potential radiation exposures due to ^{228}Ra and its gamma-emitting products (see Table 4-5). Over 200,000 cubic yards of soil contaminated with ^{232}Th residues exist beneath active public roadways where potential exposure is limited; active industrial properties contain more than 400,000 cubic yards beneath vegetative or paved surfaces and under buildings with actual human exposures well below

100 mrem/y; and similar inactive industrial and other sites contain more than 600,000 cubic yards of soils contaminated with thorium residues where access controls minimize onsite and offsite exposures. Private property and residential sites generally contain small quantities of material with low concentrations and many have already been cleaned up because of these factors.

Radioactive constituents of thorium residues are distributed in large volumes of dirt and debris; therefore, most management alternatives involve near-surface management of earth-like material in dirt. As long as ^{232}Th remains in the residues it will, because of its 14 billion year half-life, constitute a perpetual source of ^{228}Ra and its products, which are of most concern because they emit high-energy gamma rays. If, however, ^{232}Th is removed, the exposure potential would be greatly reduced allowing onsite or nearby management by land use controls because ^{228}Ra (half-life of 5.7 y) and its gamma-emitting products would diminish to minimal levels in 20 to 30 years and close to zero in 50 to 60 years. Treatment to remove ^{232}Th could be expected to mix any remaining radioactivity into a larger volume (i.e., reduce its concentration) and enhance stability; both endpoints would reduce future radiation exposures and allow more land-use options.

Uranium Mill Tailings are a byproduct of milling uranium ore that began in the 1940s to recover uranium for the buildup of the nuclear weapons program, and these operations expanded considerably as the nuclear power industry grew in the 1970s. The Uranium Mill Tailings Radiation Control Act (UMTRCA) was passed in 1978 primarily to establish a remedial action program for 24 inactive uranium mill sites and some 5300 associated "vicinity properties" that became contaminated with unwise use of tailings as fill materials, etc. Vicinity properties were included in the UMTRCA program if they contained uranium tailings in excess of 5 pCi/g of ^{226}Ra in the first 5 cm of surficial soils or/and 15 pCi/g in the top 15 cm or/and produced gamma exposure in excess of 20 μrem/h. These values were chosen to preclude excess radon levels should structures be present and/or to limit exposures of the public consistent with radiation standards. Similarly, residual levels of ^{230}Th alone in soil were limited to 43 pCi/g to preclude exceeding 15 pCi/g of ^{226}Ra in the 1000-year compliance period.

The UMTRCA remedial action program was carried out by DOE, but in accordance with environmental standards issued by EPA. Typical disposal involved moving the tailings and stabilizing them in a nearby emplacement facility (as shown in Figure 14-1) designed to contain the radioactive constituents up to 1000 years, but for at least 200 years. Completed sites remain under federal/state control to preclude other uses. Similar provisions were also provided in UMTRCA for 27 active uranium mill tailings sites, the operations of which are regulated by the US Nuclear Regulatory Commission (NRC). Each site is required to have closure and stabilization plans by similar methods as a condition for license termination. Both programs are significant because, prior to the passage of UMTRCA, these materials were considered to be naturally occurring and outside the purview of the AE Act.

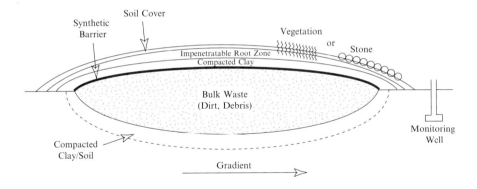

Figure 14-1. Schematic of cut-and-fill method of surface internment of uranium tailings.

Defense Wastes include radioactive materials at nuclear weapons production sites, national research laboratories, and military bases. The DOE facilities, which have been self-regulated, are located in 30 states and territories and include 12 major sites, 10 national laboratories, and 41 other small sites. Enormous quantities of waste materials, some of which are highly radioactive and/or contain hazardous chemicals, exist at the large DOE sites; some of these sites are remote (e.g., the Nevada Test Site and the Hanford reservation). The most important defense wastes are perhaps the transuranic (TRU) wastes from the nuclear weapons complex and the high-level wastes (HLW) produced in the production of plutonium and other nuclear materials. Most of these wastes are currently in storage at several of the major DOE sites; however, disposal of TRU wastes began (in mid 1999) at the Waste Isolation Pilot Plant (WIPP) located in bedded salt more than 2,000 feet below the surface near Carlsbad, NM. TRU wastes are defined as those that contain alpha-emitting transuranic ($Z > 92$) elements with half-lives greater than 20 years and with concentrations in excess of 100 nCi/g.

The National Strategic Materials Stockpile (a multistate reserve of strategic ores for the production of fuel and weapons material) contains thousands of tons of thorium nitrate, and zirconium-bearing ore that contains 0.3 to 0.4% uranium and thorium; some soils and equipment have become contaminated with radium, uranium, and thorium in these materials. Plutonium-contaminated soil also exists due to a chemical explosion and fire involving a nuclear weapon accident that occurred at McGuire AFB, NJ, in 1960. And, several sites are contaminated with low-activity depleted uranium which, because of its high density, is used as an armor supplement and in armor-piercing shells.

Naturally Occurring Accelerator-Produced Radioactive Materials (NARM) are waste forms subject to control by the states and other federal statutes, since they are not AE Act materials. They include naturally occurring radioactive material (NORM) and technologically enhanced NORM (TENORM), which are generally large volumes of soil or debris containing elevated levels of naturally occurring radionuclides. These materials are typically byproducts of the mining and processing of mineral-bearing ores (e.g., uranium, phosphate, aluminum, copper, titanium, etc.) and the most usual constituent is radium which, in addition to potential gamma exposure, can produce elevated radon emissions. The huge piles of phosphogypsum wastes at phosphate plants are an example of such high-bulk technologically enhanced residues. Radium accumulations occur in scales on the inside of pipes (and in piles or barrels of scale removed from pipes) associated with oil and gas pumping operations, and in sludges, resins, or charcoal used to process groundwater for public and other uses. These process materials are usually disposed in local landfills which may, under some circumstances, be considered NORM sites.

Radium wastes were generated from 1912 into the 1940s by the National Radium Institute (NRI) which was located in and around Denver, CO, for access to the Colorado Plateau's deposits of Carnotite ore which contained 1 to 2% uranium. Poor management of the substance itself and the byproducts of its processing and recovery have created long-term radiological problems at some 31 separate locations because no regulations were in place to control radium and its use in the early years of the 20th century. The NRI produced 8.5 grams of radium from approximately 1500 tons of ore until pitchblende ores (containing > 50% uranium) became available from the Belgian Congo in the early 1920s. Other operations produced radium-containing medicines, tonics, cosmetics, pottery glazes, glassware, animal feed, and crop fertilizer. With the exception of the Shattuck Chemical Co., which ceased radium operations in 1941 (but continued to process molybdenum and uranium until 1984), the radium processing and manufacturing plants went out of business in the 1920s and they, along with the radioactive tailings and unprocessed ore, were largely forgotten. The sites, including locations where debris was used as fill and foundation material, were rediscovered in the late 1970s and placed as a single unit (of 31 separate locations) designated as the Denver Radium Sites on the Superfund National Priorities List for remediation.

Cleanup consisted primarily of excavation, backfilling with replacement soil, and disposal of 230,000 cubic yards of contaminated soil. The Shattuck Chemical site was a notable exception, and 50,000 cubic yards of debris were stabilized on site. The deed for the property denotes that the property is a dedicated waste disposal site, and restricts excavation, construction, groundwater use, and agricultural use. Contaminated material underneath city streets and roadways was also left in place; however, the city of Denver was required to develop a health and safety protocol for street and utility workers who might need to dig in the area and to assess and properly dispose of any contaminated material at that time. Remediation of the Denver Radium Sites thus relied on environmental stewardship as a practical and effective measure of managing materials on site rather than their complete removal.

HIGH-LEVEL RADIOACTIVE WASTES

The Nuclear Waste Policy Act of 1982 defines high-level radioactive waste as spent nuclear fuel or the highly radioactive material from the reprocessing of spent nuclear fuel, whether from production reactors or nuclear power plants. The NRC has issued regulations in 10 CFR 60 for licensing DOE to receive and dispose of source, special nuclear, and byproduct material at a geologic repository sited, constructed, and operated in accordance with the regulations. The NRC is in turn required by the Act to implement environmental protection standards (40 CFR 191) issued by the EPA which requires that they be isolated in such a manner that very restrictive release limits not be exceeded for a period of 10,000 years after disposal.

Although it was originally planned that all reactor fuel would be reprocessed to recover unfissioned uranium and byproduct plutonium, this has not occurred for commercial nuclear fuel, primarily for economic reasons. This fuel is currently in storage awaiting an operational federal high-level waste repository, at which point the fuel will be disposed intact (i.e., without processing); its zircalloy cladding is quite stable and assures a high-integrity barrier to release of radionuclides. Consequently, essentially all of the high-level radioactive wastes produced since the 1940s remain in spent fuel storage at commercial nuclear reactor sites or in tanks or other forms at DOE sites. The tank wastes are liquids that contain chemicals and will be vitrified in borosilicate glass for disposal.

Disposal of spent fuel and vitrified high-level radioactive TRU wastes is planned in stable geologic formations to isolate the radioactive constituents from the biosphere for 10,000 years: in a bedded salt formation for TRU wastes and a volcanic tuff formation (Yucca Mountain, NV) for commercial spent fuel and vitrified high-level defense wastes. Since these formations are dry and have remained stable for millions of years, it is presumed that they will remain so for the periods required for the wastes to diminish to risk levels that are not unacceptable should events reintroduce them, presumably at slow rates, into the accessible environment. In essence, deep geologic disposal is a reverse-mining technique (shown schematically in Figure 14-2) in which shafts and tunnels are mined out, but for emplacement of material rather than for extracting it. Shafts and tunnels are backfilled and sealed after emplacement of the waste, and because the formations are hundreds of meters underground, surface radiation levels are nonexistent.

Radioactivity in spent fuel, and to a lesser extent vitrified waste, generates significant amounts of heat which the geologic formation must dissipate without fracturing the surrounding geology to allow migration of radionuclides. Heat generation by TRU wastes is minimal, and the bedded salt formation is expected to reseal any fractures that develop. Volcanic tuff is expected to transfer the heat from spent fuel and vitrified waste with minimal effect; this heat burden can be reduced considerably if stored for a few decades, as is the case for existing wastes. No aquifers exist above the volcanic tuff layer

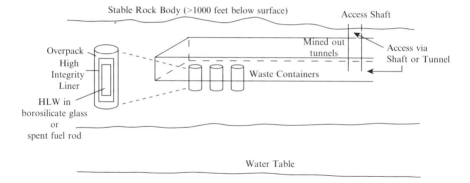

Figure 14-2. Deep geologic disposal of high-level radioactive waste in a mined-out stable rock body which is accessed by a shaft (e.g., bedded salt) or a tunnel (e.g., mountain of volcanic tuff), each of which requires high-integrity sealing after closure.

at Yucca Mountain and the groundwater table is well below the rock body where disposal would occur; however, there is some concern that fracturing in the tuff may allow infiltration of precipitation which is currently just a few inches per year but may change over the 10,000-year performance period.

Mobile long-lived radionuclides represent the largest potential health risks, in particular Tc-99 ($T_{1/2} = 210,000$ y), C-14($T_{1/2} = 5730$ y) and I-129 ($T_{1/2} = 17$ million y) because of their solubility and extremely long lives. Pu-239 and other TRUs are relatively immobile and, despite their toxicity, are of minor consequence to the environment unless disturbed after disposal.

Site investigation and development at Yucca Mountain has experienced considerable delays due to public opposition, legal challenges, and extensive geological studies, and the current schedule for operation is well past 2010, perhaps closer to 2020 unless Congress accelerates it. It appears that some groups are counting on such delays as a means of halting further development of nuclear energy and nuclear weapons. If, however, the Yucca Mountain Site can be shown to provide reasonable assurance that performance standards based on environmental protection criteria (see above) are met, it can be expected to resolve the issue of disposal of high-level defense wastes and spent nuclear fuel from commercial nuclear plants. If not, public trust can be expected to be diminished further and many more decades will probably elapse before final resolution occurs. It is questionable whether the necessary public commitment or resources will exist to deal with accumulated amounts of waste at the point decades from now when such determinations are made. Furthermore, an infrastructure will need to be maintained to safeguard the waste until its proper disposal can be achieved.

LOW-LEVEL RADIOACTIVE WASTES (LLRW)

The Low-Level Radioactive Waste Policy Act (LLRWPA) of 1980 defines LLRW as radioactive material that is not high-level radioactive wastes, spent nuclear fuel, transuranic waste, or byproduct material as defined in Section 11(e)(2) of the AE Act. The major producers of low-level radioactive waste (LLRW) are various research and medical institutions, industrial facilities, processors and handlers of source materials (i.e. uranium), and nuclear power plants.

Low-level waste management was, for various reasons, uniquely confined to federal lands during the period from World War II through the 1950s. A few enterprising companies, with the support of six states, sought privatization of waste disposal as a commercial, and presumably profitable, enterprise and six sites were licensed and became operational between 1961 and 1972. Various events, mostly related to low revenues and public reactions when offsite releases were discovered, caused closure of the NY, KY, and IL sites, thus shifting the entire burden of disposal to the sites in WA, SC, and NV. The NV site closed in the mid 1980s; thus, only the WA site (in the far west), which limits waste disposal to states in the northwest and Rocky Mountain compacts, and the SC site (in the southeast) remain open. Since most low-level radioactive waste is produced in the central regions of the nation, a large portion of it is shipped long distances for disposal. No aspect of this situation echoes the desire to have the best site with optimal geology and arid conditions; access has been the key consideration.

In 1979, the National Governors Association, concerned about a loss of access should existing LLRW sites close—as some were threatening to do— recommended policies that the Congress passed as the "Low-Level Radioactive Waste Policy Act" (PL 96–573) in December 1980. This law is quite short; however, it established sweeping policies with respect to LLRW; i.e.,

- Each state is responsible for the commercial LLRW generated within its borders.
- Regional disposal sites are the most safe and efficient option.
- States may enter into compacts to establish and operate regional disposal sites.
- Regional disposal facilities operated under compacts consented to by Congress may refuse to accept wastes from non-compact states after 1 January 1986.

These major policies were adopted as a means to provide disposal for waste-generating activities in the respective states without the risk of becoming a national disposal site (under the Interstate Commerce Clause of the Constitution, states cannot restrict commercial activities; however, Congressionally approved compacts override this provision).

Major Sources of LLRW

Low-level radioactive wastes are for the most part different in form and content from other wastes. There are, however, several common radionuclides addressed in federal regulations (10 CFR 61) for LLRW primarily because of their half-lives, physical and chemical characteristics, and potential for exposure of persons once disposed in shallow land burial sites.

Institutional/Industrial LLRW generally consist of relatively small quantities of specialized materials. Colleges and universities, medical schools, research facilities and hospitals generate diverse forms of radioactive waste: trash; liquid scintillation vials containing scintillation fluid (shipped with absorbent materials); other liquids (solidified or shipped with absorbent materials); and biological wastes (shipped with absorbent materials and lime). Institutional facilities also generate accelerator targets and sealed sources but these occur in relatively small-volumes.

Industrial activities also produce a wide array of LLRW: tritium; sealed sources and accelerator targets; activated metal and equipment produced by accelerators; activated metal and equipment from research reactors and subcritical assemblies; and activated metal from neutron generators. Tritium production produces wastes that consist of lithium fluoride, trash, plastic, and a small quantity of metal; larger quantities are generated by conversion of tritium gas to tritiated water and by incorporation of tritium into chemical compounds (tritium occurs naturally; however, commercial applications use artificially produced tritium). Various commercial companies use encapsulated radioactive materials as calibration and reference standards (e.g., in gas chromatographs) and these require disposal after uses cease. And, other industries process and fabricate depleted uranium and manufacture chemicals containing uranium, in particular depleted uranium.

LLRW from Nuclear Reactors are produced in various forms, in particular: (1) spent ion-exchange resins used in cleanup of process streams; (2) concentrated liquids from evaporation of collected liquid wastes; (3) sludges from pre-coat filters; (4) trash (compactible and noncompactible) produced by cleanup and control functions; and (5) cartridge filters, which are used much more extensively in PWRs than in BWRs. Many of these wastes are due to efforts to reduce the accumulation of radioactive contaminants within the plant and to preclude or reduce their release to the environment, with some differences between BWRs and PWRs as shown in Figures 4-22, 4-23 and 4-25. These wastes consist of *radioactive corrosion products* (in particular ^{51}Cr, ^{58}Co, ^{60}Co, ^{65}Zn, ^{110m}Ag, ^{55}Fe, ^{59}Fe, ^{59}Ni, and ^{63}Ni) due to activation of metallic components and fission products (^{14}C, ^{90}Sr, ^{137}Cs, ^{95}Zr, ^{144}Ce, ^{103}Ru, ^{99}Tc, and ^{129}I) that, due to imperfections in fuel cladding, leak into the reactor coolant where they are collected on filters and resins. Some small leaks may also occur from valves or process control points, and these are collected in drains and concentrated, typically by evaporation, for disposal as LLRW.

PAST PRACTICES/LESSONS LEARNED

Various radioactive materials, some typical of modern LLRW, have been buried in the US since about 1940 or so. It was generally recognized by people in the 1940s, 50s and 60s that these materials should not be discharged to the environment, and their management, which was begun on federal lands, evolved to six commercial sites that were developed and licensed in the 1960s and early 70s. Four of the original six sites have since been closed, and these provide a record that can be used to gauge the significance of past disposal of LLRW, what it may portend for the future, and lessons that can guide future disposal. Performance at the Sheffield site in Illinois, the West Valley site in New York, and the Maxey Flats site in Kentucky are particularly relevant, especially the Maxey Flats site which has been studied extensively by EPA and the state of Kentucky. These three sites, as well as the existing site in Barnwell, SC, are located in the eastern and midwestern US near generators where moderate rainfall occurs; the arid site near Beatty, NV, has very little precipitation and has not experienced the problems found in IL, NY, and KY.

The Sheffield Site contains 20 acres, is in a region of abandoned coal mines, and is bordered by rolling terrain and a 40-acre hazardous waste-disposal site. Burial of radioactive waste began in August of 1967, and after 10 years of operation some three million cubic feet of waste, including the dismantled Elk River Reactor, had been buried in 21 trenches. Tritium was first detected in a well in the spring of 1978 and thus far is the only radionuclide above background concentrations detected in monitoring wells. The tritium concentrations near the site vary between 200 pCi/L and 2000 nCi/L, which is about two-thirds of the regulatory limit for releases to the general environment (10 CFR 20).

When US Ecology, the most recent owner, wanted to close the site, the state of Illinois enforced an $8 million settlement agreement which required stabilization of the site and a 10-year maintenance period, after which the state assumed long-term custodial care of the site. Clay material was stockpiled for any additional work on the cap, and US Ecology was required to provide equipment for site maintenance for an additional 20 years. US Ecology has also provided a trust fund of $1.9 million to pay for remediation of any significant release(s) of radioactivity.

The West Valley Site was operated in upstate New York by Nuclear Fuel Services (NFS) from 1963 to 1975. The site was located in a region with highly impermeable subsurface soils, mostly clay; however, waste migration by water infiltration through trench covers occurred anyway due to poor management both by NFS and by waste generators. Since no requirements were placed on the density of the waste, many waste drums arrived at the site 30 to 50% empty, which eventually yielded voids equal to 70% of the drum volume due to aerobic and anaerobic decomposition of the degradable contents. Large void spaces were also left when wastes were emplaced, and because of inadequate bulk underneath, the trench caps fractured and collapsed allowing water infiltration.

Since the soils beneath the site were so impermeable, trenches filled up and overflowed carrying the radioactive leachate to surface streams, the so-called "bathtub" effect. And although regulations prohibited the burial of liquid wastes at West Valley, burial of adsorbed liquids on "kitty litter" was permitted; precipitation that fell into the open trenches during burial operations contacted the "kitty litter" and became contaminated.

Water infiltration in disposal trenches was observed soon after operations began, and in 1968 programs to reduce infiltration of surface water and decrease erosion were instituted. These efforts gradually stopped the water level from rising in the older trenches, but ^{3}H, ^{89}Sr, and ^{90}Sr above background levels were measured in surface waters near the burial site, and by 1971–75 the water levels in the older trenches had risen and seeped through trench caps. These water levels were reduced by pumping some 80,000 liters of water into a holding lagoon where it was treated to remove most of the radionuclides and then discharged to a nearby creek. The contaminated rainwater was pumped into local streams and onto the surface of the site which contaminated the surficial soil used to cover the trenches.

The annual average values for tritium and strontium-90 from a water-sampling station are shown in Figure 14-3. These were collected at the converging point for all surface drainage from the burial site. The dose to a public water supply customer has been estimated by EPA to be less than 0.02 mrem/y, considerably below the EPA drinking water standard of 4 mrem/y, and the dose to a fisherman who caught all of his fish in the nearby creek and ate the bones as well as the flesh received at most 0.04 mrem/y. These doses, despite poor management by generators and site operators, have been quite small and have remained so.

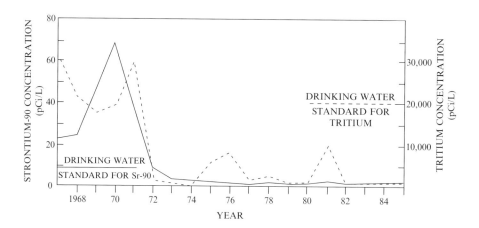

Figure 14-3. Annual average concentrations from 1967 to 85 of ^{90}Sr and tritium in creek water at West Valley compared to National Safe Drinking Water Act standards.

The Maxey Flats radioactive waste site in Kentucky was the largest LLRW site in the US. It operated 13 years (May, 1963 to 1976) and received approximately $121,000 \, m^3$ of waste containing more than 1.92 million Ci of radioactive material, 376 kg of special nuclear material, 154,000 kg of source material, some 80 kg of Pu-239, and additional quantities of other plutonium isotopes. Tritium, the most abundant radionuclide, ranged from less than 1 Ci up to a maximum of 650,000 Ci in trench 37.

Wastes were emplaced in open trenches excavated in impermeable clay soils, allowing rain to infiltrate the trenches and leach radioactivity from the wastes. The radioactive leachate overflowed the trenches and/or moved through the adjacent subsurface soil and rocks. By 1974, small quantities of radioactivity were detected in test wells, and in onsite and offsite soils. Tritium was also measured in surface water just offsite. Onsite radionuclides were ^3H, ^{54}Mn, ^{60}Co, ^{90}Sr, ^{137}Cs, ^{238}Pu, and ^{239}Pu. Offsite soil samples contained ^3H, ^{90}Sr, ^{238}Pu, and ^{239}Pu. Additional capping was done to prevent infiltration of rainfall, the trench caps were recontoured to facilitate run-off, and a cover was installed and planted with vegetation to retard erosion.

Water was pumped from the trenches to large holding tanks, evaporated to reduce the volume, and the residue stored for eventual burial onsite. The evaporator plant effluent was discharged directly to the atmosphere at rates as high as $1.9 \times 10^3 \, mCi/s$ of ^3H and $2.1 \times 10^{-2} \, mCi/s$ for ^{137}Cs. Measured values of ^{60}Co, ^{90}Sr, and ^{137}Cs were also found in every sample, and ^{22}Na, ^{106}Ru, and ^{134}Cs were detected frequently. Annual average offsite doses due to tritium were 2.6 mrem/y and less than 0.1 mrem/y for other radionuclides (^3H, ^{90}Sr, ^{137}Cs, ^{238}Pu, and ^{239}Pu). The other major pathway for offsite migration of radionuclides, as shown in Figure 14-4, was precipitation run-off via the main east wash and the wash on the west side of the site. Stream bed sediments in the main east wash contained ^{54}Mn, ^{60}Co, ^{90}Sr, ^{137}Cs, ^{238}Pu, and ^{239}Pu, and although migration obviously occurred, population exposure was minimal via this pathway. Only tritium was detected above background levels in domestic well water, which at a daily intake of 1 L and an average concentration of 1.7 nCi/L, the highest average concentration measured, would have resulted in a total-body dose equivalent of about 0.1 mrem/y. The highest water concentrations observed for short periods of time were 180 nCi/L of ^3H (or 6% of the maximum permissible concentration) and 0.08 nCi/L of ^{90}Sr (or 27% of the MPC).

Low-level ^3H contamination was measured in cow's milk (0.3–6.5 nCi/L) up to 3.1 km away, and the estimated dose to a child consuming 1 L of milk per day at 6.5 nCi/L of ^3H was approximately 0.4 mrem; other consumers of local milk would receive considerably less due to dilution from other sources in the milkshed. With the exception of low levels of ^3H, radioactivity was undetectable in garden produce; estimated doses from consumption of tomatoes, watermelons, corn, grapes, and cucumbers grown near the site were less than 0.01 mrem/y.

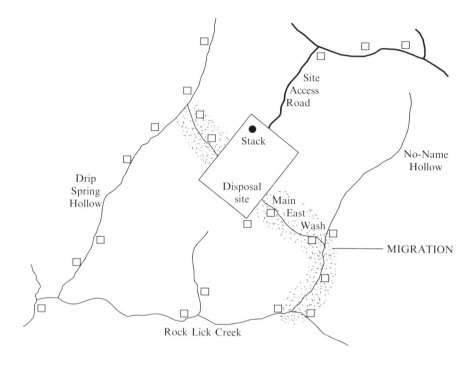

Figure 14-4. Offsite migration of LLRW at the Maxey Flats disposal site.

Lessons Learned from Past Disposal at the Maxey Flats site, as well as the West Valley and Sheffield sites, are: (1) operations should be conducted in areas sheltered from rainfall with strict control on void spaces in containers and backfill; (2) no free liquids can be allowed in the waste because they promote leaching and migration of radionuclides; and (3) a good cap is absolutely essential for keeping wastes where they are disposed. Past disposal methods used sites chosen for good geology but did not have stringent conditions on waste containers, operations, or closure and post-closure care. Inadequate backfill and partially filled containers resulted in settling, cover failure, and water infiltration. Even with such deficiencies, doses to maximally exposed individuals for most routes of potential exposure were found to be fractions of a millirem, even for the Maxey Flats site which had the most wastes and highest releases. What they mean, for now or in the future, can be debated; however, it appears that the presence of radioactivity releases, despite relatively low doses and very poor but correctable management, represents a loss of trust —it was promised and expected that releases would not occur. Perhaps they teach future disposers what to do and what not to do; what they may mean to the general public and regulators will not be clear until open, careful, and fair discussion occurs.

Performance Assessments for LLRW

A myriad of conditions exist for controlling the diversity of low-level radio-active wastes; however, two scenarios appear to bound the possibilities:

- highly radioactive wastes that contain relatively short-lived radionuclides, and
- long-lived persistent radionuclides in relatively low-activity waste forms

In the first, a waste stream could contain highly radioactive components, perhaps ^{60}Co or ^{137}Cs, in a rod, a concentrate, or resins from a nuclear power plant. Dose rates from there wastes could be tens to hundreds of rems per hour at contact due to gamma emission; therefore, controls are based on proper shielding to preclude external dose hazards to handlers (the internal exposure hazard is relatively low from both materials) and the indicated control is to limit the concentrations of each such that shielding can be relied on for protection for the few decades required for ^{60}Co($T_{1/2} = 5.274$ y) and/or the few hundred years required for ^{137}Cs($T_{1/2} = 30.04$ y). A few feet of earth or concrete (or a few inches of lead or steel) will limit exposure rates to minimal levels, and concentration limits can be chosen such that dose levels would meet public dose limits after institutional control periods lapse.

In the other extreme, wastes may contain one or more highly mobile long-lived radionuclides such as ^{14}C($T_{1/2} = 5,715$ y), ^{99}Tc($T_{1/2} = 210,000$ y), or perhaps ^{129}I($T_{1/2} = 17$ million y) which diminish very slowly; thus it takes a lot of them to provide much activity. Each also emits only low-energy beta particles so they present minimal external radiation hazard, but they are readily soluble and quite mobile once they reach ground or surface waters. The main hazard would be internal exposure, but their low activity requires a significant amount to deliver a high dose internally; therefore, their risk is mainly one of chronic and continuous risk to a population of individuals. This situation requires the most deliberation to solve because the materials will persist well beyond institutional controls and the durability of most engineered structures; thus it is important to limit their concentration such that unacceptable risks to future people do not occur. The key questions for these low-hazard, persistent materials are how long to control them and to what level.

Risk Assessments for New LLRW Disposal Systems have been performed that encompass these two examples which should cover most other scenarios. Two typical designs for LLRW disposal are the concrete vault (or bunker), which has been used in France for several years (Figure 14-5), and concrete canisters (Figure 14-6). Both designs are enhanced by segregating wastes according to radioactivity levels (i.e., by class) prior to emplacement and providing surveillance of the disposal site.

Concrete vaults (large sealed rooms) are engineered structures located below natural grade with reinforced concrete floors, walls, and a roof; Class B and C wastes (see below) are placed in the bottom of the vault in successive layers,

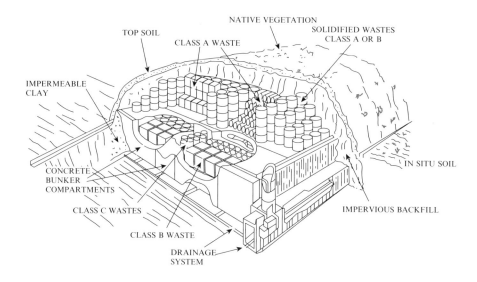

Figure 14-5. Schematic of a below-ground vault unit containing Class A wastes atop Class B and C wastes.

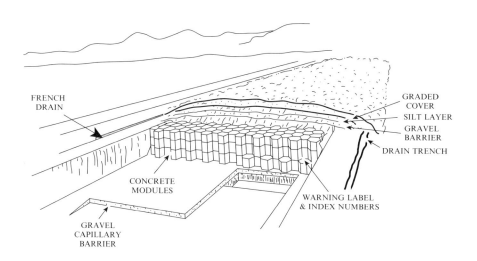

Figure 14-6. Schematic drawing of waste disposal using concrete modules emplaced below ground in constructed disposal cells.

each of which is backfilled with concrete. When the last layer of waste has been placed in a compartment, reinforcing steel is placed on top of the layer and the compartment is completely backfilled with concrete, thereby embedding the waste in one large concrete monolith, creating a buried structure surrounded by fill. Modular concrete canisters, which have good structural stability, are filled with waste and backfilled with grout. These are stacked in shallow trenches slightly below natural grade, as shown in Figure 14-6, and a completed unit resembles a shallow land burial site in that disposal is below grade with an earthen cover. Both designs include a cover that restricts water infiltration into the waste, prevents human, plant, or animal intrusion, and reduces direct radiation exposure at the ground surface.

Estimated cancer deaths in a population of 5000 persons due to LLRW disposal in below-ground vault and modular concrete canisters are

	Expected Fatal Cancers	
Time Period	Below-Ground Vault	Concrete Canister
First 1,000 y	4.5	3.9
First 10,000 y	11.4	9.1

These estimated cancers are based on site parameters chosen for a typical midwestern site for radionuclides that may migrate from a LLRW disposal site for periods of 1,000 and 10,000 years after site closure. A risk coefficient of 500 fatalities/10^6 person-rem (BEIR V, 1992) was used to convert predicted doses to numbers of fatal cancers and serious genetic effects.

Maximum individual dose commitments were also estimated for each disposal method for a 1,000-year period following site closure due to radionuclides in air, food, and water. The peak radiation doses an individual would receive at the site boundary of a reference site are shown in Figure 14-7. The annual dose commitments for both modes of disposal are due to a slow breakout of mobile radionuclides (primarily ^{129}I, ^{14}C, ^{99}Tc, ^{59}Ni, ^{63}Ni, and ^{241}Am) after site closure which declines to low, but prolonged, doses. A key factor in the modeled releases of health effects and dose levels was the assumption that 7.5% of the site cover failed and was not repaired, thus allowing a steady infiltration of precipitation.

REGULATORY CONTROL OF LLRW (10 CFR 61)

The NRC has developed new regulations (10 CFR 61) that incorporate lessons learned from deficiencies in past disposal of LLRW to ensure that current and future disposal are protective of human health and the environment. Near-surface (uppermost 30 m of the earth's surface) land disposal of LLRW is allowed, but with strict controls on the operation and final closure of sites to

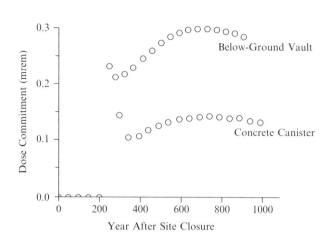

Figure 14-7. Annual dose commitment (mrem) to maximum individuals located at the site boundary for below-ground vault and concrete canister disposal of LLRW.

ensure stability of sites, to lessen the potential for radionuclide migration, and to minimize the need for active long-term maintenance of facilities. The regulations permit LLRW site designs in which waste is disposed in direct contact with the soil (i.e., without a separate liner), but with permeability requirements for soils and stringent requirements on packaging and managing the waste to preclude bathtub effects (current sites, in response to lessons learned, have taken the same steps to preclude the problems experienced at the closed sites).

The regulations encompass two time frames: a short-term operational phase and a much longer post-operational period. They incorporate features that maintain stability of the disposal site, restrict liquids in the waste and the disposal facility, and impose institutional and other controls to reduce the likelihood of inadvertent intrusion. The following performance objectives are to be met:

- an annual dose less than 25 mrem to the whole body, 75 mrem to the thyroid, and 25 mrem to any other organ;
- doses ≤ 500 mrem/y due to inadvertently intruding into and occupying the disposal site and/or inadvertent contact with the waste; and
- compliance with EPA drinking water standards (40 CFR Part 141).

These performance objectives for members of the public and potential inadvertent intruders rely on a system of waste classification followed by disposal requirements based on each LLRW class: LLRW generators are required to classify and prepare LLRW so that LLRW site operators can assure that it is properly emplaced.

The 25 mrem/y limit is derived from 40 CFR Part 190 and is the EPA standard for whole-body and organ doses (except the thyroid which is limited to 75 mrem/y) due to normal releases of material to the general environment from uranium fuel cycle operations. This standard has also been implemented by the NRC in 10 CFR Part 20 for routine releases to the general environment which may expose several individuals. The 500 mrem/y limit for an inadvertent intruder is consistent with national and international recommendations for circumstances which, if justified, allow higher doses due to infrequent exposure of a few individuals in unrestricted areas.

Long-term environmental protection and protection of an inadvertent intruder (but not a deliberate one) are new considerations which recognize the persistence of many radionuclides in LLRW and reflect a policy that such sites be transferred intact to future generations so that potential exposure of the public is minimized, including intrusion into the site without prior knowledge. Potential exposure to an inadvertent intruder from inhalation or ingestion of radionuclides and/or direct gamma exposure would be reduced if the waste is disposed in a form recognizable as something other than dirt, or disposed at a sufficient depth so that contact from normal surface activities is unlikely. Waste classifications (10 CFR 61) are designed to meet these considerations.

Classification of LLRW

Three classes of waste (A, B, and C), as shown in Table 14-1, are defined according to the concentration of specific radionuclides in LLRW. These classes are intended to assure that LLRW is in the proper form to meet stability requirements based on activity, half-life, physical and chemical characteristics, and potential environmental transport. The waste classes determine the waste form, the physical endurance of the waste, and the modes of emplacement in the LLRW site.

Table 14-1. Waste Classification Activity Limits.

Radionuclide	Class A	Class B	Class C
H-3	$40\,Ci/m^3$	$(^2)$	$(^2)$
C-14	$0.8\,Ci/m^3$	–	$8\,Ci/m^3$
Co-60	$700\,Ci/m^3$	$(^2)$	$(^2)$
Ni-63	$3.5\,Ci/m^3$	$70\,Ci/m^3$	$700\,Ci/m^3$
Sr-90	$0.04\,Ci/m^3$	$150\,Ci/m^3$	$7,000\,Ci/m^3$
Tc-99	$0.02\,Ci/m^3$	–	$0.2\,Ci/m^3$
I-129	$0.008\,Ci/m^3$	–	$0.08\,Ci/m^3$
Cs-137	$1\,Ci/m^3$	$44\,Ci/m^3$	$4,600\,Ci/m^3$
Pu-241	$350\,nCi/g$	–	$3,500\,nCi/g$
Cm-242	$2,000\,nCi/g$	–	$20,000\,nCi/g$
Alpha-TRUs[1]	$10\,nCi/g$	–	$100\,nCi/g$

[1]Half-lives greater than 5 years.
[2]There are no limits established for these radionuclides in Class B or C wastes.
Adapted from 10 CFR 61.55

Class A Wastes are wastes that contain either such low levels of long-lived radionuclides that they are not of concern with regard to groundwater contamination, or the content of short-lived radionuclides is such that they will diminish to negligible levels before reaching a supply of ground or surface water. Class A wastes are presumed to be in unstable forms with limited structural stability, and therefore must be disposed of in separate disposal units to preclude compromising the stability of longer-lived wastes due to decomposition of the waste and its packaging (a Class A waste that is stable may, however, be mixed with other classes of waste).

Class B Wastes contain longer-lived radionuclides and/or have activity levels such that they must meet more stringent requirements with regard to waste form and stability than Class A wastes. The activity limits in Class B are chosen to assure that the radionuclides will diminish during a 100-year institutional control period such that the exposure to an inadvertent intruder after a 100-year institutional control period will be less than 500 mrem/y.

Class C Wastes are characterized by those types and quantities of radioisotopes that will not diminish during the 100-year institutional control period and thus could potentially expose an intruder in excess of 500 mrem/y unless disposed at greater depths and in rigorous packaging. Concentrations of radionuclides in Class C wastes are specified at levels which would not pose a danger to the public health and safety nor to an intruder for a period of 500 years after disposal. The waste form and placement must be such that subsequent surface activities by an intruder will not disturb the waste, or if it is disturbed it will be readily recognized. If deeper disposal cannot be achieved, difficult-to-penetrate engineering barriers such as concrete covers must be employed to minimize disturbance and to promote recognition.

Greater Than Class C Wastes contain activity levels above those for Class C, and these wastes are generally precluded for near-surface disposal. No current mechanism exists, however, for their disposal except perhaps storage for eventual management with high-level radioactive waste.

Shipments of LLRW must include the waste classification on the shipping manifests. The manifest must also contain

- the waste volume and its physical description
- the total radioactivity, identity, and amounts of significant radionuclides
- the total quantity of ^{3}H, ^{14}C, ^{99}Tc, and ^{129}I
- a listing of wastes with more than 0.1% of chelating agents and their weight

The federal regulations in 10 CFR 61 can be met by stringent controls by generators and with well-controlled shallow land burial sites that have soils sufficient to contain waste migration, acting in effect as a natural liner. A constructed liner is not required by the regulations; however, most siting/design groups have elected to add sophisticated engineering to stringent site requirements, in effect requiring a constructed liner and drainage system between the waste and the surrounding soil.

POLICY IMPLICATIONS/STATUS OF LLRW

In order to assure continuing access to existing sites, the states began a serious effort in 1980 to implement the LLRWPA and to form interstate compacts. They also began siting processes that attempted to reflect local political interests which for the most part opposed a site in their state. Two forces thus became juxtaposed: providing access for generators, but keeping the site out of one's own state. Eleven different compacts (and several go-it-alone states) were formed (see Figure 14-8), host states were selected, and siting programs were developed, all of which were somewhat easy because nuclear utilities which had a vested interest in assuring access to disposal were tapped to pay the major costs of compact commissions. It soon became evident, however, that no state or compact would develop or license a new site by the 1986 deadline; therefore, Congress, in 1985, amended the act extending the deadline to January 1, 1996 but required sited states to make defined progress or suffer penalties assessed by disposal sites on each unit of LLRW. All states and compacts also missed the extended deadline, and states with existing sites imposed the penalties allowed in the 1985 amendments and reaped a windfall in revenues.

Meanwhile, as shown in Figure 14-9, the volume of LLRW, which had increased steadily up to about 1980, decreased by more than a factor of 10 from 1980 to 1997, from 3.8 to 0.32 million cubic feet. This decrease in volume occurred because the surcharges imposed by the 1985 amendments and site restrictions on generators were so onerous that it became cost-effective to sepa-rate wastes, increase storage time, and provide super-compaction. The activity levels probably decreased very little; i.e., a disposal site would have essentially the same amount of radioactivity, just distributed more compactly, which may be an advantage for environmental protection because of increased stability.

States were quite willing to join compacts to gain its protections; however, as the process proceeded, political interests led to Michigan and North Carolina, both host states, being dismissed by their compacts because of lack of success in siting. Michigan was required to store waste for several years but has since regained access to the Barnwell, S.C. site; however, North Carolina, the original host state for the southeast compact, was (and as of 2002 is still) denied access to the S.C. site which continues to provide disposal to all states including the other southeast compact states. In retrospect, it should have been obvious that these political trends and reduced waste volumes would fail to produce a new site.

The State of South Carolina, which precipitated the low-level radioactive waste policy act of 1980 by threatening closure of its Barnwell site, has, depending on the Governor in office, tightened access, raised fees, and re-stricted access to generators in selected states. The surcharges allowed by the 1985 Act amendments have made the site very profitable to the state and this

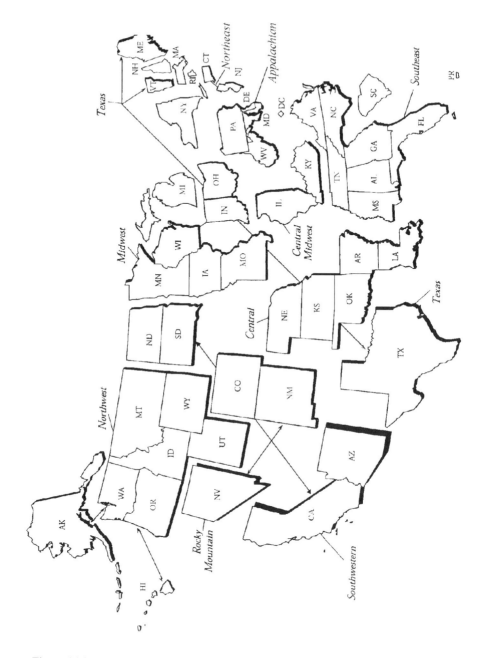

Figure 14-8. Low-level radioactive waste (LLRW) compacts and independent states.

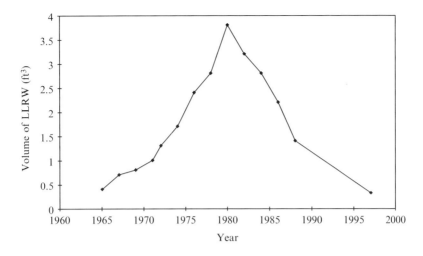

Figure 14-9. Annual volume of LLRW (ft³) disposed in the US from 1965 to 1997.

has probably affected its access policy. South Carolina limited access again in 1999 because of shrinking disposal capacity and is threatening early closure to all but southeast compact members (North Carolina has been dismissed), creating anew the anxiety for a new site(s). Surprisingly, illegal dumping has not occurred, due perhaps to the strict licensing and inspection programs of the NRC and agreement states.

As of 2002, despite expenditures of more than $500 million in site studies and conceptual designs by interstate compact commissions and unaffiliated states (see Figure 14-8), no new site has been designed or constructed. Pennsylvania, California, Illinois, Nebraska, Ohio (preceeded by Michigan), North Carolina, Connecticut/New Jersey (as dual hosts), and Texas designated sites only to be turned away at the final stages, typically by state political actions. Each had proposed disposal designs with supra-isolation features estimated to cost $600 million or more per site; these costs and the reduced waste streams also suggested that they would not be economically viable for a compact region. The reduced waste volume, which is expected to be maintained, further reduces the need for new sites to no more than 2 or 3, certainly not the 13 or so that could be forecast by the number of compacts and individual states that, for various reasons, are expected to go it alone. The costs of the rigorous designs required to satisfy local concerns and the waste volumes, even with high fees, will not support a site in each of the compacts, which further restrains the viability and, unfortunately, the performance of any new site. The first such site, if there is one, would probably need to reach beyond its current compact-protected area to assure a viable revenue stream, a circumstance that circulates back to the original issue—not being a "national site." On the other hand, a single site that would accept LLRW from all states would be enormously profitable because disposal capacity is approaching a crisis. It is intriguing that no state has seized this opportunity to generate enormous funds to solve other state interests by providing stewardship for a few hundred acres of land. The trends in managing LLRW raise the very relevant question of whether LLRW siting and disposal is a technical matter or a socio-political one.

SUMMARY

A myriad of governmental, commercial, and energy-related activities have produced numerous sources of radioactive wastes that range from high-activity, potentially dangerous materials that are for the most part strictly contained, to high-bulk, relatively benign materials that contaminate sites to various degrees. Essentially all of the high-level radioactive wastes produced since the 1940s remain in spent fuel storage at commercial nuclear reactor sites or in tanks or other forms at DOE sites. Continuing the weapons complex (which no longer uses nuclear reactors to produce plutonium) and the nuclear power industry, even with some modest expansion, will not increase the current waste volume, which must be managed with or without additional generation, by more than a factor of about two. Although it is generally agreed that radioactive waste should be isolated from the biosphere for its hazardous lifetime, the means and criteria for doing so remain largely obscure.

Only two of the six LLRW sites licensed in the 1960s and early 1970s remain open; the other four have been closed due to lack of capacity or inability to meet regulatory requirements. Several states have, as provided in the Low-Level Waste Policy Acts of 1980 and 85, assumed host state responsibilities only to have those efforts thwarted when a site location appeared imminent.

Significant aspects of LLRW before and after passage of the LLRW Policy Act of 1980 are (1) even with marginal site management, offsite radiation doses have been minimal; (2) no new sites have been developed since the early 1970s despite the compact protections provided by the Act and new NRC regulations designed to preclude past problems; (3) LLRW siting has proven to be a failure because all proposed sites have been turned away by local and state opposition; and (4) the prospect of a new site(s) is now very remote. Storage, volume reduction, and various onsite procedures have been adopted by generators, driven by costs and disposal restrictions, to reduce the amounts of LLRW.

Despite many assertions about inadequacies of past radioactive waste management, the first and subsequent products of fission were isolated on federal sites and have continued to be largely contained. Although some releases have occurred to the onsite environment and these were not what the public or anyone expected, exposures have been minimal. Considering the circumstances of wartime and the Cold War, it is remarkable that containment of radioactive waste products was established at the outset, a policy that has continued to govern federal oversight and management of these materials. If national defense and/or the nuclear power industry diminish significantly, it is uncertain whether the will or the funds would be available to manage the wastes which are no closer to final disposal than when they were first produced. Society has clearly made much more progress in producing and using radioactive materials than it has in managing the waste byproducts.

Strong opposition exists and can be expected to persist for the next few decades over high-level and low-level radioactive wastes associated with nuclear fission. Parties who oppose current or potential waste sites have emphasized a concern about radiological risk; however, because existing sites have not, even with design deficiencies, produced significant exposures nor are any projected, the concerns are most likely social ones: equity, land values, inadequate information on actual risks, intergenerational concerns, and perhaps just fear of the unknown. This circumstance represents a dilemma: the waste problem must be solved because of its threat to current and future generations, yet its very solution may allow nuclear weapons activities and/or the nuclear power industry to continue. This situation has persisted for several decades and will probably continue until it reaches a crisis and the Congress or the administration dictates a solution, perhaps not an ideal one but one that is practical and reflects the national interest at the time.

Stewardship has been and will continue to be an essential component of radioactive waste management, but it has not been fully recognized and engaged by local and national entities. One possible action could be the setting aside of two or three large tracts of land under federal control in designated regions of the country that assures regional access; provides long-term isolation and federal land stewardship through BLM, the National Forest Service, DOE and/or HHS; and develops a fee structure that pays for optimal technical facilities for recycling and fixation of disposed materials, a trust fund, and shared revenues to states. Such an approach could be taken by the administration or jointly with Congress by simply citing the states' failure to produce any progress in more than two decades.

Acknowledgment

Capt. Floyd L. Galpin† (USPHS-Ret) provided highly prized and appreciated counsel and wisdom on the status of low-level radioactive waste siting and management; his quiet voice and good humor are, and will be, missed.

ADDITIONAL RESOURCES

1   National Academy of Sciences, *Evaluation of Guidelines for Exposures to Technologically Enhanced Naturally-occurring Radioactive Materials*, (Washington, DC: National Academy Press, 1998).

2   US Department of Energy, *Estimating the Cold War Mortgage. The 1995 Baseline Environmental Mangement Report*. Volume II: *Site Summaries*, (Washington, DC: US Department of Energy, DOE/EM-0232, 1995).

3   Wolbarst, A.B., Blom, P.F., Chan, D., Cherry, R.N. Jr., Doehnert, M., Fauver, D., Hull, H.B., MacKinney, J.A., Mauro, J., Richardson, A.C.B., and Zaragoza, L., Sites in the United States Contaminated with Radio-activity, *Health Physics*, 77, No. 3, pp 247–260, September 1999.

4   Vincenti, J.R., Low-Level Radioactive Waste Disposal in the U.S., *Health Physics*, Operational Radiation Safety Supplement, 76, No. 5, pp 91–97, May 1999.

PROBLEMS

1   The cleanup standard for ^{228}Ra($T_{1/2} = 5.7$ y) in the top 15 cm of soil is 15 pCi/g: (a) if a soil cleanup process is applied to remove ^{232}Th which is present at a concentration of 30 pCi/g, how long must the site be controlled such that the replaced soil will be below the ^{228}Ra standard? (b) if 5 pCi/g of ^{232}Th still remains in the soil residue, how will this affect the site-control period?

2   If concentration of ^{228}Ra in the top 5 cm of surface soil is limited to 5 pCi/g, what soil concentration of ^{232}Th could be left such that the standard for ^{228}Ra in the top 5 cm of soil is never exceeded?

3   Some processes applied to uranium ores preferentially removed and/or concentrated ^{230}Th($T_{1/2} = 75,400$ y) which dried out in surface impound-ments leaving a soil residue several inches thick: (a) if the average concen-tration of ^{230}Th in the soil/residue mixture is 40 pCi/g, will the uranium mill standard for fixation of soils containing uranium residues in excess of 15 pCi/g of ^{226}Ra over a period of 1000 years be exceeded? (b) if the standard contained no time period, at what point would the ^{230}Th produce enough ^{226}Ra to exceed the standard?

4   If the carnotite ores processed by the National Radium Institute to obtain 8.5 gram of radium contained 4% uranium, (a) how many tons of ore were processed if the process was 60% efficient; (b) how much radium was left in the residues; and (c) how many tons of residue and how much radium would be left if pitchblende ores (65% U) were processed at 90% efficiency?

5   ^{99}Tc($T_{1/2} = 210,000$ y) is a product of fission with a fission yield of 6.11%. If 70,000 tons of spent uranium fuel is to be stored in a high-level waste repository, (a) what would be the inventory (in Ci) of ^{99}Tc (ignoring radioactive removal) if the fuel burnup is 30,000 MW·d/ton (2.7×10^{21} fissions / MW·d)? and (b) how does this compare to the EPA environmental standard that limits ^{99}Tc releases to the accessible environment to 10,000 Ci per ton of fuel?

6   The estimated dose rate from ingestion of tritium ($T_{1/2} = 12.3$ y) by a maximally exposed person using groundwater from a well near the Maxey Flats, KY LLRW site was 2.6 mrem/y in 1980. If 1000 persons are so

exposed for 100 years from the closed and stabilized site: (a) integrate the person-rems of dose that could occur; and (b) calculate the estimated fatal cancer health risk.

7   If a tritium concentration of 4000 nCi/L is measured in groundwater seepage from a LLRW site, how long would the groundwater source need to be isolated from public use to meet the drinking water standard of 4 mrem/y?

8   Based on an estimated health risk of 4 fatal cancers in 1000 years for a typical Midwest LLRW site and for a chosen risk/benefit criterion, what amount of revenue ($/y) would be required to justify a 50-year period of operation?

9   Determine the degree to which the Class C limit for Ni-63 meets the 10 CFR 61 criteria for Class C waste.

10   If the major contributors to heat generation in spent fuel are Sr-90 and Cs-137 (assume an average half-life of 29 years for each), what effective reduction in heat loading would be gained by a predisposal storage period of: (a) 50 years, (b) 100 years?

11   After doing the technical calculations associated with the previous problems, construct arguments for what role they have (or ought to have) on public policy decisions that affect the management of the various waste types.

APPENDIX A

CONSTANTS OF NATURE AND PARTICLE MASSES

Time
1 y = 3.155760 × 10⁷ s
1 d = 8.6400 × 10⁴ s

Dimensions
1 in = 2.54 cm
1Å = 10⁻⁸ cm
1 m³ = 10⁶ cm³ = 35.315 ft³
1 L = 0.219976 gal
1 ft³ = 28.317 L = 0.028317 m³

Energy
1 eV = 1.602 176 462 (63) × 10⁻¹⁹ J
1 MeV = 1.602 176 462 (63) × 10⁻¹³ J = 1.602 176 462 (63) × 10⁻⁶ ergs
1 J = 10⁷ erg
1 cal = 4.187 J
kT at room temperature = 0.0259 eV

Electrostatics
Electronic charge q = 1.602 176 462 (63) × 10⁻¹⁹ C
1 C cm⁻² = 6.241 509 74 × 10⁻¹⁸ electrons cm⁻²
1 μA cm⁻² = 6.24 × 10¹² electrons cm⁻² s⁻¹
Permeability of free space μₒ = 1.26 × 10⁻⁶ Hm⁻¹
Faraday's constant = 96, 485.309 C/mol

APPENDIX A

CONSTANTS OF NATURE AND PARTICLE MASSES

Time
1 y = 3.155760×10^7 s
1 d = 8.6400×10^4 s

Dimensions
1 in = 2.54 cm
1Å = 10^{-8} cm
$1 \text{ m}^3 = 10^6 \text{ cm}^3 = 35.315 \text{ ft}^3$
1 L = 0.219976 gal
$1 \text{ ft}^3 = 28.317 \text{ L} = 0.028317 \text{ m}^3$

Energy
1 eV = $1.602\ 176\ 462\ (63) \times 10^{-19}$ J
1 MeV = $1.602\ 176\ 462\ (63) \times 10^{-13}$ J = $1.602\ 176\ 462\ (63) \times 10^{-6}$ ergs
1 J = 10^7 erg
1 cal = 4.187 J
kT at room temperature = 0.0259 eV

Electrostatics
Electronic charge q = $1.602\ 176\ 462\ (63) \times 10^{-19}$ C
$1 \text{ C cm}^{-2} = 6.241\ 509\ 74 \times 10^{-18}$ electrons cm^{-2}
$1\ \mu\text{A cm}^{-2} = 6.24 \times 10^{12}$ electrons cm^{-2} s^{-1}
Permeability of free space $\mu_o = 1.26 \times 10^{-6}$ Hm^{-1}
Faraday's constant = 96, 485.309 C/mol

Constants of Nature

Velocity of light in a vacuum $= 2.997\ 924\ 58 \times 10^8$ m s^{-1}

1 N $= 10^5$ dyn $= 1$ kg m s^{-2}

1 mm Hg $= 133.3224$ N m^{-2}

Boltzmann's constant k $= 1.380\ 6503\ (24) \times 10^{-23}$ J K^{-1}

$\qquad\qquad\qquad\quad = 8.617\ 342\ (15) \times 10^{-5}$ eV K^{-1}

Planck's constant h $= 6.626\ 068\ 76\ (52) \times 10^{-34}$ J s

$\qquad\qquad\qquad = 4.135\ 669\ 2 \times 10^{-15}$ eV s

Avogadro's number $= 6.022\ 141\ 99\ (47) \times 10^{23}$ atoms mol^{-1}

Universal Gas Constant $= 8.314\ 510$ J mol^{-1}K^{-1}

Rest Masses

Neutron $= 1.674\ 927\ 16\ (13) \times 10^{-27}$ kg $= 939.565\ 63$ MeV

Proton $= 1.672\ 621\ 58\ (13) \times 10^{-27}$ kg $= 938.272\ 31$ MeV

Electron $= 9.109381\ 88\ (72) \times 10^{-31}$ kg $= 0.510\ 999\ 0$ MeV

Unified Mass Unit $= 1.66054 \times 10^{-27}$ kg $= 931.502$ MeV

1 kg $= 2.205$ lbs.

*From P.J. Mohr and B.N. Taylor (2000) The Fundamental Physical Constants, Physics Today 53 (No. 8, Part 2) BG 6 – BG 13

APPENDIX B

ATOMIC MASSES AND BINDING ENERGIES—SELECTED NUCLIDES

Table B-1. Atomic Masses and Binding Energies for Selected Isotopes of the Elements.

Z	A	Binding Energy (MeV)	Atomic Mass (umu)	Z	A	Binding Energy (MeV)	Atomic Mass (umu)
0	n	1	–	1.008664923		10	64.751	10.012937027
1	H	1	–	1.007825032		11	76.205	11.009305466
		2	2.225	2.014101778		12	79.575	12.014352109
		3	8.482	3.016049268	6	C	10	60.321	10.016853110
2	He	3	7.718	3.016029310		11	73.44	11.011433818
		4	28.296	4.002603250		12	92.162	12.000000000
3	Li	6	31.995	6.015122281		13	97.108	13.003354838
		7	39.245	7.016004049		14	105.285	14.003241988
		8	41.277	8.022486670		15	106.503	15.010599258
4	Be	7	37.6	7.016929246	7	N	13	94.105	13.005738584
		8	56.5	8.005305094		14	104.659	14.003074005
		9	58.165	9.012182135		15	115.492	15.000108898
		10	64.977	10.013533720		16	117.981	16.006101417
		11	65.481	11.021657653		17	123.865	17.008449673
5	B	8	37.738	8.024606713	8	O	14	98.733	14.008595285
		9	56.314	9.013328806		15	111.956	15.003065386

Table B-1. (*continued*)

Z	A	Binding Energy (MeV)	Atomic Mass (umu)	Z	A	Binding Energy (MeV)	Atomic Mass (umu)
	16	127.619	15.994914622		29	245.01	28.976494719
	17	131.763	16.999131501		30	255.62	29.973770218
	18	139.807	17.999160419		31	262.207	30.975363275
	19	143.763	19.003578730		32	271.41	31.974148129
	20	151.371	20.004076150	15 P	29	239.285	28.981801376
9 F	17	128.22	17.002095238		30	250.605	29.978313807
	18	137.369	18.000937667		31	262.917	30.973761512
	19	147.801	18.998403205		32	270.852	31.973907163
	20	154.403	19.999981324		33	280.956	32.971725281
	21	162.504	20.999948921	16 S	30	243.685	29.984902954
10 Ne	18	132.153	18.005697066		31	256.738	30.979554421
	19	143.781	19.001879839		32	271.781	31.972070690
	20	160.645	19.992440176		33	280.422	32.971458497
	21	167.406	20.993846744		34	291.839	33.967866831
	22	177.77	21.991385510		35	298.825	34.969032140
	23	182.971	22.994467337		36	308.714	35.967080880
	24	191.836	23.993615074		37	313.018	36.971125716
11 Na	21	163.076	20.997655099	17 Cl	33	274.057	32.977451798
	22	174.145	21.994436782		34	285.566	33.973761967
	23	186.564	22.989769675		35	298.21	34.968852707
	24	193.523	23.990963332		36	306.789	35.968306945
	25	202.535	24.989954352		37	317.1	36.965902600
	26	208.151	25.992589898		38	323.208	37.968010550
12 Mg	22	168.578	21.999574055		39	331.282	38.968007677
	23	181.725	22.994124850	18 Ar	34	278.721	33.980270118
	24	198.257	23.985041898		35	291.462	34.975256726
	25	205.588	24.985837023		36	306.716	35.967546282
	26	216.681	25.982593040		37	315.505	36.966775912
	27	223.124	26.984340742		38	327.343	37.962732161
	28	231.628	27.983876703		39	333.941	38.964313413
13 Al	25	200.528	24.990428555		40	343.81	39.962383123
	26	211.894	25.986891659		41	349.909	40.964500828
	27	224.952	26.981538441		42	359.335	41.963046386
	28	232.677	27.981910184	19 K	37	308.573	36.973376915
	29	242.113	28.980444848		38	320.647	37.969080107
14 Si	26	206.046	25.992329935		39	333.724	38.963706861
	27	219.357	26.986704764		40	341.523	39.963998672
	28	236.537	27.976926533		41	351.618	40.961825972

Table B-1. (*continued*)

Z	A	A	Binding Energy (MeV)	Atomic Mass (umu)	Z		A	Binding Energy (MeV)	Atomic Mass (umu)
		42	359.152	41.962403059			51	444.306	50.944771767
		43	368.795	42.960715746			52	456.345	51.940511904
20	Ca	38	313.122	37.976318637			53	464.284	52.940653781
		39	326.411	38.970717729			54	474.003	53.938884921
		40	342.052	39.962591155			55	480.25	54.940844164
		41	350.415	40.962278349			56	488.506	55.940645238
		42	361.895	41.958618337	25	Mn	52	450.851	51.945570079
		43	369.828	42.958766833			53	462.905	52.941294702
		44	380.96	43.955481094			54	471.844	53.940363247
		45	388.375	44.956185938			55	482.07	54.938049636
		46	398.769	45.953692759			56	489.341	55.938909366
		47	406.045	46.954546459			57	497.991	56.938287458
		48	415.991	47.952533512	26	Fe	52	447.697	51.948116526
		49	421.138	48.955673302			53	458.38	52.945312282
21	Sc	43	366.825	42.961150980			54	471.759	53.939614836
		44	376.525	43.959403048			55	481.057	54.938298029
		45	387.849	44.955910243			56	492.254	55.934942133
		46	396.61	45.955170250			57	499.9	56.935398707
		47	407.254	46.952408027			58	509.944	57.933280458
		48	415.487	47.952234991			59	516.525	58.934880493
22	Ti	44	375.475	43.959690235			60	525.345	59.934076943
		45	385.005	44.958124349			61	530.927	60.936749461
		46	398.194	45.952629491	27	Co	57	498.282	56.936296235
		47	407.072	46.951763792			58	506.855	57.935757571
		48	418.699	47.947947053			59	517.308	58.933200194
		49	426.841	48.947870789			60	524.8	59.933822196
		50	437.78	49.944792069			61	534.122	60.932479381
		51	444.153	50.946616017	28	Ni	56	483.988	55.942136339
		52	451.961	51.946898175			57	494.235	56.939800489
23	V	48	413.904	47.952254480			58	506.454	57.935347922
		49	425.457	48.948516914			59	515.453	58.934351553
		50	434.79	49.947162792			60	526.842	59.930790633
		51	445.841	50.943963675			61	534.662	60.931060442
		52	453.152	51.944779658			62	545.259	61.928348763
		53	461.631	52.944342517			63	552.097	62.929672948
24	Cr	48	411.462	47.954035861			64	561.755	63.927969574
		49	422.044	48.951341135			65	567.853	64.930088013
		50	435.044	49.946049607	29	Cu	61	531.642	60.933462181

Table B-1. (*continued*)

Z		A	Binding Energy (MeV)	Atomic Mass (umu)	Z		A	Binding Energy (MeV)	Atomic Mass (umu)
		62	540.528	61.932587299	34	Se	72	622.43	71.927112313
		63	551.381	62.929601079			73	630.823	72.926766800
		64	559.297	63.929767865			74	642.89	73.922476561
		65	569.207	64.927793707			75	650.918	74.922523571
		66	576.273	65.928873041			76	662.072	75.919214107
		67	585.391	66.927750294			77	669.491	76.919914610
30	Zn	62	538.119	61.934334132			78	679.989	77.917309522
		63	547.232	62.933215563			79	686.951	78.918499802
		64	559.094	63.929146578			80	696.865	79.916521828
		65	567.073	64.929245079			81	703.566	80.917992931
		66	578.133	65.926036763			82	712.842	81.916700000
		67	585.185	66.927130859			83	718.66	82.919119072
		68	595.383	67.924847566	35	Br	77	667.343	76.921380123
		69	601.866	68.926553538			78	675.633	77.921146130
		70	611.081	69.925324870			79	686.32	78.918337647
		71	616.915	70.927727195			80	694.212	79.918529952
31	Ga	67	583.402	66.928204915			81	704.369	80.916291060
		68	591.68	67.927983497			82	711.962	81.916804666
		69	601.989	68.925580912			83	721.547	82.915180219
		70	609.644	69.926027741	36	Kr	76	654.235	75.925948304
		71	618.948	70.924705010			77	663.499	76.924667880
		72	625.469	71.926369350			78	675.558	77.920386271
		73	634.657	72.925169832			79	683.912	78.920082992
32	Ge	68	590.792	67.928097266			80	695.434	79.916378040
		69	598.98	68.927972002			81	703.306	80.916592419
		70	610.518	69.924250365			82	714.272	81.913484601
		71	617.934	70.924953991			83	721.737	82.914135952
		72	628.685	71.922076184			84	732.257	83.911506627
		73	635.468	72.923459361			85	739.378	84.912526954
		74	645.665	73.921178213			86	749.235	85.910610313
		75	652.17	74.922859494			87	754.75	86.913354251
		76	661.598	75.921402716	37	Rb	83	720.045	82.915111951
		77	667.671	76.923548462			84	728.794	83.914384676
33	As	73	634.345	72.923825288			85	739.283	84.911789341
		74	642.32	73.923929076			86	747.934	85.911167080
		75	652.564	74.921596417			87	757.853	86.909183465
		76	659.892	75.922393933			88	763.936	87.911318556
		77	669.59	76.920647703	38	Sr	82	708.128	81.918401258

Table B-1. (*continued*)

Z		A	Binding Energy (MeV)	Atomic Mass (umu)	Z		A	Binding Energy (MeV)	Atomic Mass (umu)
		83	716.987	82.917555029			100	860.458	99.907477149
		84	728.906	83.913424778			101	865.856	100.910346543
		85	737.436	84.912932689	43	Tc	95	819.152	94.907656454
		86	748.926	85.909262351			96	827.024	95.907870803
		87	757.354	86.908879316			97	836.498	96.906364843
		88	768.467	87.905614339			98	843.776	97.907215692
		89	774.825	88.907452906			99	852.743	98.906254554
		90	782.631	89.907737596			100	859.507	99.907657594
39	Y	87	754.71	86.910877833	44	Ru	94	806.849	93.911359569
		88	764.062	87.909503361			95	815.802	94.910412729
		89	775.538	88.905847902			96	826.496	95.907597681
		90	782.395	89.907151443			97	834.607	96.907554546
		91	790.325	90.907303415			98	844.791	97.905287111
40	Zr	88	762.606	87.910226179			99	852.254	98.905939307
		89	771.923	88.908888916			100	861.928	99.904219664
		90	783.893	89.904703679			101	868.73	100.905582219
		91	791.087	90.905644968			102	877.949	101.904349503
		92	799.722	91.905040106			103	884.182	102.906323677
		93	806.456	92.906475627			104	893.085	103.905430145
		94	814.676	93.906315765			105	898.995	104.907750341
		95	821.139	94.908042739	45	Rh	101	867.406	100.906163526
		96	828.993	95.908275675			102	874.844	101.906842845
		97	834.573	96.910950716			103	884.163	102.905504182
41	Nb	91	789.052	90.906990538			104	891.162	103.906655315
		92	796.934	91.907193214			105	900.13	104.905692444
		93	805.765	92.906377543	46	Pd	100	856.371	99.908504596
		94	812.993	93.907283457			101	864.643	100.908289144
		95	821.482	94.906835178			102	875.213	101.905607716
42	Mo	90	773.728	89.913936161			103	882.837	102.906087204
		91	783.835	90.911750754			104	892.82	103.904034912
		92	796.508	91.906810480			105	899.914	104.905084046
		93	804.578	92.906812213			106	909.477	105.903483087
		94	814.256	93.905087578			107	916.016	106.905128453
		95	821.625	94.905841487			108	925.236	107.903894451
		96	830.779	95.904678904			109	931.39	108.905953535
		97	837.6	96.906021033			110	940.207	109.905152385
		98	846.243	97.905407846			111	945.958	110.907643952
		99	852.168	98.907711598	47	Ag	105	897.787	104.906528234

Table B-1. (*continued*)

Z		A	Binding Energy (MeV)	Atomic Mass (umu)	Z		A	Binding Energy (MeV)	Atomic Mass (umu)
		106	905.729	105.906666431			123	1041.475	122.905721901
		107	915.266	106.905093020			124	1049.962	123.905274630
		108	922.536	107.905953705			125	1055.695	124.907784924
		109	931.723	108.904755514			126	1063.889	125.907653953
		110	938.532	109.906110460	51	Sb	119	1010.061	118.903946460
48	Cd	104	885.841	103.909848091			120	1017.081	119.905074315
		105	894.266	104.909467818			121	1026.323	120.903818044
		106	905.141	105.906458007			122	1033.13	121.905175415
		107	913.067	106.906614232			123	1042.095	122.904215696
		108	923.402	107.904183403			124	1048.563	123.905937525
		109	930.727	108.904985569			125	1057.276	124.905247804
		110	940.642	109.903005578	52	Te	118	999.457	117.905825187
		111	947.618	110.904181628			119	1006.985	118.906408110
		112	957.016	111.902757226			120	1017.281	119.904019891
		113	963.556	112.904400947			121	1024.505	120.904929815
		114	972.599	113.903358121			122	1034.33	121.903047064
		115	978.74	114.905430553			123	1041.259	122.904272951
		116	987.44	115.904755434			124	1050.685	123.902819466
		117	993.217	116.907218242			125	1057.261	124.904424718
49	In	111	945.97	110.905110677			126	1066.375	125.903305543
		112	953.648	111.905533338			127	1072.665	126.905217290
		113	963.091	112.904061223			128	1081.441	127.904461383
		114	970.365	113.904916758			129	1087.524	128.906595593
		115	979.404	114.903878328			130	1095.943	129.906222753
		116	986.188	115.905259995			131	1101.872	130.908521880
50	Sn	110	934.562	109.907852688	53	I	125	1056.293	124.904624150
		111	942.743	110.907735404			126	1063.437	125.905619387
		112	953.529	111.904820810			127	1072.58	126.904468420
		113	961.272	112.905173373			128	1079.406	127.905805254
		114	971.571	113.902781816			129	1088.239	128.904987487
		115	979.117	114.903345973			130	1094.74	129.906674018
		116	988.68	115.901744149	54	Xe	122	1027.641	121.908548396
		117	995.625	116.902953765			123	1035.785	122.908470748
		118	1004.952	117.901606328			124	1046.254	123.905895774
		119	1011.437	118.903308880			125	1053.858	124.906398236
		120	1020.544	119.902196571			126	1063.913	125.904268868
		121	1026.715	120.904236867			127	1071.136	126.905179581
		122	1035.529	121.903440138			128	1080.743	127.903530436

Table B-1. (*continued*)

Z		A	Binding Energy (MeV)	Atomic Mass (umu)	Z		A	Binding Energy (MeV)	Atomic Mass (umu)
		129	1087.651	128.904779458			138	1156.04	137.905985574
		130	1096.907	129.903507903			139	1163.495	138.906646605
		131	1103.512	130.905081920			140	1172.696	139.905434035
		132	1112.447	131.904154457			141	1178.125	140.908271103
		133	1118.887	132.905905660			142	1185.294	141.909239733
		134	1127.435	133.905394504			143	1190.439	142.912381158
		135	1133.817	134.907207499			144	1197.335	143.913642686
		136	1141.877	135.907219526	59	Pr	139	1160.584	138.908932181
		137	1145.903	136.911562939			140	1168.526	139.909071204
55	Cs	131	1102.377	130.905460232			141	1177.923	140.907647726
		132	1109.545	131.906429799			142	1183.766	141.910039865
		133	1118.532	132.905446870			143	1191.118	142.910812233
		134	1125.424	133.906713419	60	Nd	140	1167.521	139.909309824
		135	1134.186	134.905971903			141	1175.318	140.909604800
		136	1141.015	135.907305741			142	1185.146	141.907718643
		137	1149.293	136.907083505			143	1191.27	142.909809626
56	Ba	128	1074.727	127.908308870			144	1199.087	143.910082629
		129	1082.459	128.908673749			145	1204.842	144.912568847
		130	1092.731	129.906310478			146	1212.407	145.913112139
		131	1100.225	130.906930798			147	1217.7	146.916095794
		132	1110.042	131.905056152			148	1225.032	147.916888516
		133	1117.232	132.906002368			149	1230.071	148.920144190
		134	1126.7	133.904503347			150	1237.451	149.920886563
		135	1133.673	134.905682749			151	1242.785	150.923824739
		136	1142.78	135.904570109	61	Pm	143	1189.446	142.910927571
		137	1149.686	136.905821414			144	1195.973	143.912585768
		138	1158.298	137.905241273			145	1203.897	144.912743879
		139	1163.021	138.908835384			146	1210.153	145.914692165
57	La	136	1139.128	135.907651181			147	1217.813	146.915133898
		137	1148.304	136.906465656			148	1223.71	147.917467786
		138	1155.778	137.907106826			149	1230.979	148.918329195
		139	1164.556	138.906348160	62	Sm	142	1176.619	141.915193274
		140	1169.717	139.909472552			143	1185.221	142.914623555
		141	1176.405	140.910957016			144	1195.741	143.911994730
58	Ce	134	1120.922	133.909026379			145	1202.498	144.913405611
		135	1128.882	134.909145555			146	1210.913	145.913036760
		136	1138.819	135.907143574			147	1217.255	146.914893275
		137	1146.299	136.907777634			148	1225.396	147.914817914

Table B-1. (*continued*)

Z		A	Binding Energy (MeV)	Atomic Mass (umu)	Z		A	Binding Energy (MeV)	Atomic Mass (umu)
		149	1231.268	148.917179521			161	1315.912	160.926929595
		150	1239.254	149.917271454			162	1324.109	161.926794731
		151	1244.85	150.919928351			163	1330.38	162.928727532
		152	1253.108	151.919728244			164	1338.038	163.929171165
		153	1258.976	152.922093907			165	1343.754	164.931699828
		154	1266.943	153.922205303	67	Ho	163	1329.595	162.928730286
		155	1272.75	154.924635940			164	1336.269	163.930230577
63	Eu	149	1229.79	148.917925922			165	1344.258	164.930319169
		150	1236.211	149.919698294			166	1350.501	165.932281267
		151	1244.144	150.919846022			167	1357.786	166.933126195
		152	1250.451	151.921740399	68	Er	161	1311.486	160.930001348
		153	1259.001	152.921226219			162	1320.7	161.928774923
		154	1265.443	153.922975386			163	1327.603	162.930029273
		155	1273.595	154.922889429			164	1336.449	163.929196996
64	Gd	150	1236.4	149.918655455			165	1343.1	164.930722800
		151	1242.898	150.920344273			166	1351.574	165.930289970
		152	1251.488	151.919787882			167	1358.01	166.932045448
		153	1257.735	152.921746283			168	1365.781	167.932367781
		154	1266.629	153.920862271			169	1371.784	168.934588082
		155	1273.065	154.922618801			170	1379.043	169.935460334
		156	1281.601	155.922119552			171	1384.725	170.938025885
		157	1287.961	156.923956686	69	Tm	167	1356.479	166.932848844
		158	1295.898	157.924100533			168	1363.32	167.934170375
		159	1301.842	158.926385075			169	1371.353	168.934211117
		160	1309.293	159.927050616			170	1377.946	169.935797877
		161	1314.928	160.929665688			171	1385.433	170.936425817
65	Tb	157	1287.119	156.924021155	70	Yb	166	1346.666	165.933879623
		158	1293.896	157.925410260			167	1353.743	166.934946862
		159	1302.03	158.925343135			168	1362.794	167.933894465
		160	1308.405	159.927164021			169	1369.662	168.935187120
		161	1316.102	160.927566289			170	1378.132	169.934758652
66	Dy	154	1261.749	153.924422046			171	1384.747	170.936322297
		155	1268.584	154.925748950			172	1392.767	171.936377696
		156	1278.025	155.924278273			173	1399.134	172.938206756
		157	1284.995	156.925461256			174	1406.599	173.938858101
		158	1294.05	157.924404637			175	1412.421	174.941272494
		159	1300.882	158.925735660			176	1419.285	175.942568409
		160	1309.458	159.925193718			177	1424.852	176.945257126

Table B-1. (*continued*)

Z		A	Binding Energy (MeV)	Atomic Mass (umu)	Z		A	Binding Energy (MeV)	Atomic Mass (umu)
71	Lu	173	1397.681	172.938926901			185	1476.545	184.954043023
		174	1404.442	173.940333522			186	1484.806	185.953838355
		175	1412.109	174.940767904			187	1491.099	186.955747928
		176	1418.397	175.942682399			188	1499.088	187.955835993
		177	1425.469	176.943754987			189	1505.009	188.958144866
72	Hf	172	1388.333	171.939457980			190	1512.801	189.958445210
		173	13952.94	181.406500000			191	1518.559	190.960927951
		174	1403.933	173.940040159			192	1526.117	191.961479047
		175	1410.642	174.941502991			193	1531.702	192.964148083
		176	1418.807	175.941401828	77	Ir	189	1503.694	188.958716473
		177	1425.185	176.943220013			190	1510.018	189.960592299
		178	1432.811	177.943697732			191	1518.091	190.960591191
		179	1438.91	178.945815073			192	1524.289	191.962602198
		180	1446.298	179.946548760			193	1532.06	192.962923700
		181	1451.994	180.949099124			194	1538.127	193.965075610
73	Ta	179	1438.017	178.945934113	78	Pt	188	1494.208	187.959395697
		180	1444.662	179.947465655			189	1500.941	188.960831900
		181	1452.239	180.947996346			190	1509.853	189.959930073
		182	1458.302	181.950152414			191	1516.29	190.961684653
74	W	178	1429.243	177.945848364			192	1524.966	191.961035158
		179	1436.175	178.947071733			193	1531.221	192.962984504
		180	1444.587	179.946705734			194	1539.592	193.962663581
		181	1451.268	180.948198054			195	1545.697	194.964774449
		182	1459.333	181.948205519			196	1553.619	195.964934884
		183	1465.524	182.950224458			197	1559.465	196.967323401
		184	1472.935	183.950932553			198	1567.022	197.967876009
		185	1478.689	184.953420586			199	1572.578	198.970576213
		186	1485.883	185.954362204	79	Au	195	1544.688	194.965017928
		187	1491.35	186.957158365			196	1551.331	195.966551315
75	Re	183	1464.185	182.950821349			197	1559.402	196.966551609
		184	1470.67	183.952524289			198	1565.914	197.968225244
		185	1478.34	184.952955747			199	1573.498	198.968748016
		186	1484.519	185.954986529	80	Hg	194	1535.495	193.965381832
		187	1491.879	186.955750787			195	1542.395	194.966638981
		188	1497.75	187.958112287			196	1551.234	195.965814846
76	Os	182	1454.06	181.952186222			197	1558.02	196.967195333
		183	14612.71	191.531100000			198	1566.504	197.966751830
		184	1469.919	183.952490808			199	1573.168	198.968262489

Table B-1. (*continued*)

Z		A	Binding Energy (MeV)	Atomic Mass (umu)	Z		A	Binding Energy (MeV)	Atomic Mass (umu)
		200	1581.197	199.968308726	88	Ra	223	1713.828	223.018497140
		201	1587.427	200.970285275			224	1720.311	224.020202004
		202	1595.181	201.970625604			225	1725.213	225.023604463
		203	1601.174	202.972857096			226	1731.61	226.025402555
		204	1608.669	203.973475640	89	Ac	225	1724.788	225.023220576
		205	1614.336	204.976056104			226	1730.187	226.026089848
81	Tl	201	1586.161	200.970803770			227	1736.715	227.027746979
		202	1593.034	201.972090569	90	Th	229	1748.341	229.031755340
		203	1600.883	202.972329088			230	1755.135	230.033126574
		204	1607.539	203.973848646			231	1760.253	231.036297060
		205	1615.085	204.974412270			232	1766.691	232.038050360
		206	1621.589	205.976095321			233	1771.478	233.041576923
82	Pb	202	1592.202	201.972143786	91	Pa	230	1753.043	230.034532562
		203	1599.126	202.973375491			231	1759.86	231.035878898
		204	1607.52	203.973028761			232	1765.414	232.038581720
		205	1614.252	204.974467112	92	U	233	1771.728	233.039628196
		206	1622.34	205.974449002			234	1778.572	234.040945606
		207	1629.078	206.975880605			235	1783.87	235.043923062
		208	1636.446	207.976635850			236	1790.415	236.045561897
		209	1640.382	208.981074801			237	1795.541	237.048723955
		210	1645.567	209.984173129			238	1801.695	238.050782583
83	Bi	207	1625.897	206.978455217			239	1806.501	239.054287777
		208	1632.784	207.979726699	93	Np	236	1788.703	236.046559724
		209	1640.244	208.980383241			237	1795.277	237.048167253
		210	1644.849	209.984104944			238	1800.765	238.050940464
		211	1649.983	210.987258139	94	Pu	238	1801.275	238.049553400
84	Po	207	1622.206	206.981578228			239	1806.921	239.052156519
		208	1630.601	207.981231059			240	1813.455	240.053807460
		209	1637.568	208.982415788			241	1818.697	241.056845291
		210	1645.228	209.982857396			242	1825.006	242.058736847
85	At	209	1633.3	208.986158678	95	Am	241	1817.935	241.056822944
		210	1640.465	209.987131308			242	1823.473	242.059543039
		211	1648.211	210.987480806			243	1829.84	243.061372686
86	Rn	211	1644.536	210.990585410	96	Cm	246	1847.827	246.067217551
		222	1708.184	222.017570472			247	1852.983	247.070346811
87	Fr	212	1646.6	211.996194988			248	1859.196	248.072342247
		223	1713.461	223.019730712	97	Bk	247	1852.246	247.070298533

Table B-1. (*continued*)

Z		A	Binding Energy (MeV)	Atomic Mass (umu)	Z		A	Binding Energy (MeV)	Atomic Mass (umu)
98	Cf	251	1875.103	251.079580056			267	19523.42	268.277400000
99	Es	252	1879.232	252.082972247	108	Hn	263	19183.71	264.287100000
100	Fm	257	1907.511	257.095098635			264	1926.724	264.128408258
101	Md	258	1911.701	258.098425321			265	19333.11	266.300010000
102	No	259	19165.69	260.010240000			266	19413.45	267.300420000
103	Lr	260	19196.21	261.055720000			267	19478.03	268.317740000
104	Db	261	19239.49	262.087520000			268	19555.18	269.321560000
105	Jl	262	19262.06	263.141530000			269	19617.65	270.341140000
106	Rf	258	18940.73	259.131510000	109	Mt	265	19264.13	266.365670000
		259	19007.45	260.146520000			266	19332.05	267.379400000
		260	1909.019	260.114435447			267	19416.62	268.375260000
		261	19154.47	262.161990000			268	19485.32	269.388160000
		262	19232.59	263.164770000			269	19563.33	270.391060000
		263	19296.2	264.183130000			270	19628.98	271.407230000
		264	19371.23	265.189240000			271	19704.98	272.412290000
		265	19431.99	266.210660000	110	Xa	267	19348.9	268.439560000
		266	19504.68	267.219280000			268	19433.59	269.435290000
107	Bh	260	19013.73	261.218030000			269	19499.27	270.451440000
		261	19094.47	262.218000000			270	19584.8	271.446260000
		262	19163.93	263.230090000			271	19651.99	272.460780000
		263	19243.36	264.231460000			272	19730.54	273.463100000
		264	19309.32	265.247300000			273	19783.92	274.492450000
		265	19385.68	266.251980000	111	Xb	272	19655.96	273.534770000
		266	19449.52	267.270090000					

Adapted from Audi and Wapstra, Nuclear Physics, A 595 Vol. 4, p. 409–480, December 25, 1995).

APPENDIX C

RADIOACTIVE TRANSFORMATION DATA

Radioactive transformation diagrams (decay schemes) provided in this Appendix were compiled from resources at the National Nuclear Data Center (NNDC) and are current as of March 2002. The NNDC is operated for the US Department of Energy by Brookhaven National Laboratory. For those circumstances where the most recent authoritative data are desirable, it is prudent to access the NNDC database at *www.nndc.bnl.gov*.

Each decay-scheme drawing shows the radionuclide, its mass number and atomic number in conventional notation, and the half-life in parenthesis to the right [e.g., $^{14}_{6}C$ (5700 y)]; also listed in the drawing is the predominant transformation mode(s) and the associated energy change; the energy of the emitted particle or radiation; the percentage of transformations of the parent, unless noted otherwise, that produce the emission; and the product nuclide and whether it is stable or radioactive (if a half-life is shown). Since radioactive transformation produces an energy change, the various energy levels associated with particle emissions and subsequent emissions of gamma rays and/or conversion electrons (ce) of the product nuclide above the ground state are also shown.

As helpful at it is to have a pictorial decay scheme, it is difficult to display complete information on the emissions from a given radionuclide; therefore, a listing is provided with each decay scheme of each significant particle emission and/or radiation along with the yield (%) of each. Since many of the decay schemes and associated data listings can be quite complex, these have been simplified for radiation protection purposes by eliminating radiations that contribute less than one (and in some noted cases, two) percent of the energy emitted per transformation of the parent unless they are necessary to understand the mode(s) of transformation. The primary particle emissions are β^-, β^+, and $\alpha_i + \alpha_i$ recoils, which designate negatron, positron, and alpha particle emissions, respectively, and their associated average energies in increasing order (the listing only shows average β^- and β^+ energies; their maxima are shown in the decay scheme). Radiations are listed by the same designators used by NNDC: e.g., characteristic x rays are designated as $K_{\alpha 1}$, $K_{\alpha 2}$, $K_{\beta 1}$, K_2, L_α, L_β, etc.; conversion electrons as ce-K γ_i, ce-L γ_i, etc., to designate the specific electron shell that is vacated and the particular gamma ray involved; and auger electrons as Auger-K, Auger-L, etc., to designate the shell vacancy that gives rise to the auger electron. Gamma ray emissions are listed as γ_i by increasing energy and $\gamma\pm$ denotes annihilation photons of 0.511 MeV each.

The mode of production of each radionuclide is shown at the upper left of the table of radiations, and the date (at the upper right) the data were last reviewed by the NNDC. For example, for ^{32}P there are three primary modes of production: an (n,γ) reaction with ^{31}P; irradiation of stable ^{34}S with deuterons to produce ^{32}P by a (d, α) reaction; and an (n, p) reaction with stable ^{32}S. The data were last reviewed by the National Nuclear Data Center in March 2002. The listing shows one beta particle, emitted with a yield, Y_i, of 100% with an average energy per transformation of 0.6949 MeV; the maximum beta energy is provided in the pictorial decay scheme as 1.7104 MeV.

Decay schemes for the following radionuclides are contained in this appendix or in the text as noted:

^{3}H	^{65}Zn	^{203}Hg
^{7}Be	^{68}Ge-Ga	^{204}Tl
^{14}C	^{85}Sr	^{214}Pb (p. 92)
^{18}F	^{90}Sr	^{214}Bi (p. 92)
^{22}Na	^{99}Mo (p. 70)	^{214}Po (p. 92)
^{24}Na	^{99m}Tc (p. 70)	^{218}Po
^{32}P	^{99}Tc (p. 70)	^{222}Rn (p. 92)
^{33}P	^{103}Ru-Rh	^{226}Ra
^{35}S	^{133m}Xe-^{133}Xe	^{228}Ra
^{40}K	^{123}I	^{230}Th
^{51}Cr (p. 76)	^{125}I	^{238}Pu
^{54}Mn	^{129}I	^{239}Pu
^{55}Fe (p. 63)	^{131}I (p. 78)	^{240}Pu (p. 69)
^{58}Co	^{137}Cs (p. 77)	^{241}Am
^{60}Co	^{144}Ce	^{252}Cf (p. 74)
^{59}Ni	^{195}Au	
^{65}Ni	^{198}Au	

Acknowledgments

This Appendix is due to the patient and dedicated efforts of Chul Lee and Patricia Ellis, and J.I. Tuli and the staff at NNDC.

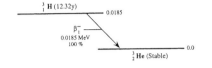

^{6}Li(n,α); Natural [^{14}N(n,t); ^{16}O(n,t)]   July, 2000

Radiation	Y$_i$ (%)	E$_i$ (MeV)
β-1	100.0	0.0057*

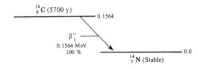

^{14}N(n,p); Natural [^{14}N(n,p)]   October,

Radiation	Y$_i$ (%)	E$_i$ (MeV)
β-1	100.0	0.0495*

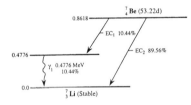

^{6}Li(d,n); ^{10}B(p,α); ^{12}C(^{3}He,2α); Natural   May, 2001

Radiation	Y$_i$ (%)	E$_i$ (MeV)
γ1	10.44	0.4776
ce-K, γ1	8.04E-06	0.4776
ce-L1, γ1	1.18E-07	0.4776[a]
Kα1 X-ray	0.0163	0.0001

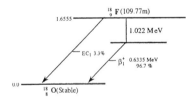

^{18}O(p,n); ^{16}O(t,n); ^{16}O(^{3}He,p); ^{19}F(n,2n); ^{19}F(d,t); Ne(d,α)   November, 1996

Radiation	Y$_i$ (%)	E$_i$ (MeV)
β+1	96.73	0.2498*
$\gamma\pm$	193.46	0.5110
K X-ray	0.018	0.0005*
Auger-K	3.07	0.0005*

* Average Energy
a Maximum Energy for Subshell

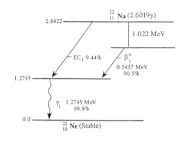

^{19}F(α,n); ^{24}Mg(d,α)   April, 2000

Radiation	Y_i (%)	E_i (MeV)
β^+_1	90.5	0.2155*
$\gamma\pm$	181	0.5110
γ_1	99.9	1.2745
K X-ray	0.125	0.0008*
Auger-K	9.18	0.0008*

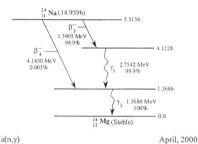

^{23}Na(n,γ)   April, 2000

Radiation	Y_i (%)	E_i (MeV)
β^-3	99.9	0.5541*
β^-4	0.0030	1.8650*
γ_2	100.0	1.3686
γ_3	99.9	2.7542

* Average Energy

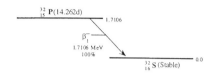

^{31}P(n,γ); ^{34}S(d,α); ^{32}S(n,p)   March, 2002

Radiation	Y_i (%)	E_i (MeV)
β^-1	100.0	0.6949*

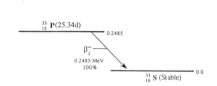

^{33}S(n,p); ^{37}Cl(γ,α)   March, 2002

Radiation	Y_i (%)	E_i (MeV)
β^-1	100.0	0.0764*

^{34}S(n,γ); ^{37}Cl(d,α)   March, 2002

Radiation	Y_i (%)	E_i (MeV)
β^-1	100.0	0.0486*

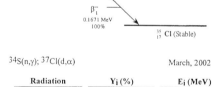

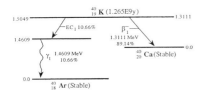

Natural   July, 1998

Radiation	Y_i (%)	E_i (MeV)
γ_1	10.66	1.4609
β^-_1	89.14	0.5606*
K X-ray	0.938	0.0030*
Auger-K	7.22	0.0027*

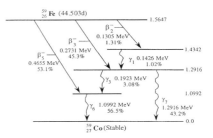

^{56}Fe(d,α); ^{51}V(α,n); ^{53}Cr(d,n); ^{54}Cr(p,n)   July, 2001

Radiation	Y_i (%)	E_i (MeV)
γ_1	100.0	0.8348
ce-K, γ_1	0.022	0.8289
Kα_1 X-ray	15.0	0.0054
Kα_2 X-ray	7.66	0.0054
K X-ray	3.06	0.0060*
L X-ray	0.37	0.0006*
Auger-K	63.2	0.0048*

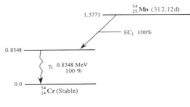

^{58}Fe(n,γ)   November, 1993

Radiation	Y_i (%)	E_i (MeV)
β^-_2	1.31	0.0356*
β^-_3	45.3	0.0808*
β^-_5	53.1	0.1491*
γ_1	1.02	0.1426
γ_3	3.08	0.1923
γ_6	56.5	1.0992
γ_7	43.2	1.2920

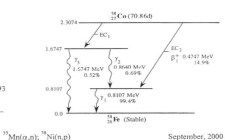

^{55}Mn(α,n); ^{58}Ni(n,p)   September, 2000

Radiation	Y_i (%)	E_i (MeV)
β^+_1	14.9	0.2011*
$\gamma\pm$	29.8	0.5110
γ_1	99.4	0.8107
γ_2	0.69	0.8640
γ_3	0.52	1.6750
Kα_1, X-ray	15.7	0.0064
Kα_2, X-ray	8.0	0.0064
K, X-ray	3.24	0.0071*
Auger-K	48.8	0.0056*

* Average Energy

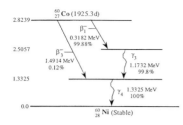

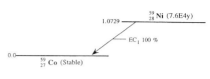

^{58}Ni(n,γ); ^{59}Co(d,2n)   November, 1993

Radiation	Y_i (%)	E_i (MeV)
Kα$_1$ X-ray	20.0	0.0069
Kα$_2$ X-ray	10.3	0.0069
Kβ, X-ray	4.16	0.0077*
L X-ray	0.975	0.0008*
Auger-K	54.3	0.0061*

^{59}Co(n,γ)   September, 2000

Radiation	Y_i (%)	E_i (MeV)
β$^-$1	99.88	0.0958*
β$^-$3	0.12	0.6259*
γ3	99.8	1.1732
γ4	100.0	1.3325

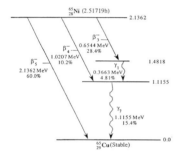

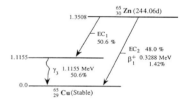

^{64}Zn(n,γ)   September, 2000

Radiation	Y_i (%)	E_i (MeV)
β$^+$1	1.42	0.1430*
γ±	2.84	0.5110
γ3	50.6	1.1160
Kα$_1$ X-ray	22.9	0.0080
Kα$_2$ X-ray	11.8	0.0080
Kβ X-ray	4.83	0.0089*
L X-ray	1.24	0.0009*
Auger-K	47.5	0.0070*

^{64}Ni(n,γ)   August, 1993

Radiation	Y_i (%)	E_i (MeV)
β$^-$3	28.4	0.2205*
β$^-$4	10.2	0.3717*
β$^-$5	60.0	0.8754*
γ1	4.81	0.3663
γ7	15.4	1.1155
γ8	23.6	1.4818

* Average Energy

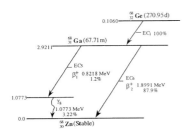

68Ge: 66Zn(α,2n) 68Ga: Daughter of 68Ge;
65Cu(α,n); 68Zn(p,n); 57Zn(d,n)   August, 2000

Radiation	Y_i (%)	E_i (MeV)
68Ge		
Kα1 X-ray	25.8	0.0093
Kα2 X-ray	13.3	0.0092
Kβ X-ray	5.68	0.0103*
L X-ray	1.52	0.0011*
Auger-K	41.7	0.0080*
68Ga		
β+1	1.2	0.3526*
β+2	87.9	0.8360*
γ±	178	0.5110
γ8	3.22	1.0770

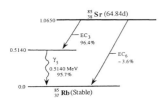

84Sr (n,γ); 85Rb (p,n); 85Rb (d,2n)   April, 1991

Radiation	Y_i (%)	E_i (MeV)
γ5	95.70	0.5140
ce-K, γ5	0.603	0.4988
ce-L, γ5	0.068	0.5119a
Kα1 X-ray	33.1	0.0134
Kα2 X-ray	17.2	0.0133
Kβ X-ray	8.98	0.0150*
L X-ray	2.59	0.0017*
Auger-K	28.6	0.0114*

* Average Energy
a Maximum Energy for Subshell

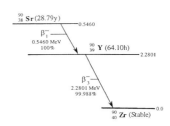

FP; 89Sr(n,γ)90Sr; 89Sr (n,γ)90Y   January, 1998

Radiation	Y_i (%)	E_i (MeV)
90Sr		
β−1	100.0	0.1958*
90Y		
β−3	99.988	0.9337*

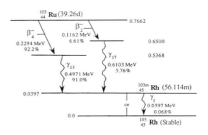

102Ru(n,γ); FP   August, 2001

Radiation	Y_i (%)	E_i (MeV)
β−2	6.61	0.0307*
β−4	92.2	0.0641*
γ1b	0.068	0.0398
ce-K, γ1	9.89	0.0165
ce-L, γ1	73.3	0.0364a
ce-M, γ1	14.5	0.0391a
γ14	91.0	0.4971
γ18	5.76	0.6103
Kα1 X-ray	4.88	0.0202
Kα2 X-ray	2.58	0.0201
Auger-K	2.12	0.0170*

b Gamma from 103mRh

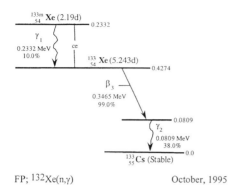

FP; ^{132}Xe(n,γ)   October, 1995

Radiation	Y_i (%)	E_i (MeV)
β⁻3	99.0	0.1005*
γ2	38.0	0.0809
ce-K, γ2	55.1	0.0450
ce-L, γ2	8.21	0.0753ᵃ
ce-M, γ2	1.69	0.0798ᵃ
Kα1 X-ray	26.1	0.0310
Kα2 X-ray	14.1	0.0306
Kβ X-ray	9.47	0.0350*
L X-ray	6.06	0.0043*
Auger-K	5.86	0.0255*

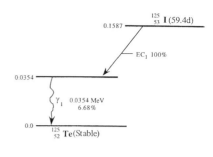

^{123}Sb(α,2n); ^{125}Xe product; ^{125}Te(d,n)   July, 1995

Radiation	Y_i (%)	E_i (MeV)
γ1	6.68	0.0354
ce-K, γ1	80.2	0.0037
ce-L, γ1	10.8	0.0306ᵃ
ce-M, γ1	2.15	0.0345ᵃ
Kα1 X-ray	74.4	0.0275
Kα2 X-ray	40.0	0.0272
Kβ X-ray	25.9	0.0310*
L X-ray	14.9	0.0038*
Auger-K	20.0	0.0227*

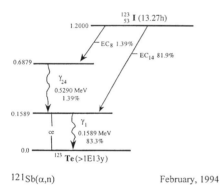

^{121}Sb(α,n)   February, 1994

Radiation	Y_i (%)	E_i (MeV)
γ1	83.3	0.1589
ce-K, γ1	13.6	0.1272
ce-L, γ1	1.77	0.1540ᵃ
γ24	1.39	0.5290
Kα1 X-ray	45.9	0.0275
Kα2 X-ray	24.2	0.0272
Kβ X-ray	16.0	0.0310*
L X-ray	8.98	0.0038*
Auger-K	12.3	0.0227*

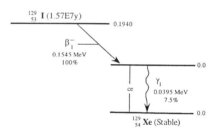

FP; ^{128}Te(n,γ); ^{129}Te(β⁻)   May, 1990

Radiation	Y_i (%)	E_i (MeV)
β⁻1	100.0	0.0409*
γ1	7.5	0.0395
ce-K, γ1	78.8	0.0050
ce-L, γ1	10.7	0.0341
ce-M, γ1	2.16	0.0384
Kα1 X-ray	36.9	0.0298
Kα2 X-ray	19.9	0.0295
Kβ X-ray	13.2	0.0336
L X-ray	7.98	0.0041
Auger-K	8.79	0.0246
Auger-L	73.9	0.0034

* Average Energy
ᵃ Maximum Energy for Subshell

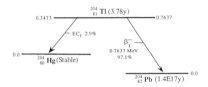

203Tl(n,γ)   November, 1994

Radiation	Y_i (%)	E_i (MeV)
Kα$_1$ X-ray	0.817	0.0708
Kα$_2$ X-ray	0.476	0.0689
Kβ X-ray	0.355	0.0803*
L X-ray	0.802	0.0100*
β−1	97.1	0.2440*

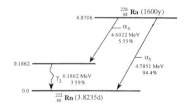

238U series   January, 1996

Radiation	Y_i (%)	E_i (MeV)
α4	5.55	4.6022
α recoil	5.55	0.0822
α5	94.4	4.7851
recoil	94.4	0.0855
γ2	3.59	0.1862

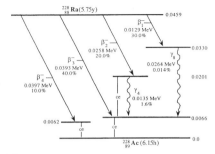

232Th series   April 1997

Radiation	Y_i (%)	E_i (MeV)
β−1	30.0	0.0032*
β−2	20.0	0.0065*
β−3	40.0	0.0099*
β−4	10.0	0.0100*
ce-M, γ1	7.5	0.0013ᵃ
ce-M, γ2	37.5	0.0017ᵃ
γ4	1.60	0.0135
ce-M, γ4	7.31	0.0085ᵃ
ce-L, γ5	2.21	0.0066ᵃ
L X-ray	1.03	0.0127*
Auger-K	1.18	0.0093*

238U series   April 1996

Radiation	Y_i (%)	E_i (MeV)
α8	23.4	4.6213
α recoil	23.4	0.0811
α9	76.3	4.6878
α recoil	76.3	0.0822
γ1	0.377	0.0676
ce-L, γ1	17.0	0.0484ᵃ
ce-M, γ1	4.6	0.0629ᵃ
ce-N+, γ1	1.66	0.0665ᵃ
L X-ray	7.73	0.0123*

* Average Energy
ᵃ Maximum Energy for Subshell

ANSWERS TO SELECTED PROBLEMS

(Additional/revised answers are available on the website:
www-personal.umich.edu/~jemartin)

CHAPTER 1

1. (a) 6p, 8n; (c) 54p, 79n
3. 5.9737×10^{23} atoms
4. 3.34707×10^{-24} g
5. (b) 3.9 fermi
6. 6.507×10^{21} atoms
7. (c) 700 keV or 1.215×10^{-13} J
9. $v = 0.866c$
10. (a) 37.6 MeV or 5.37 MeV/nucleon; (c) 1801.72 MeV or 7.7 MeV/nucleon

CHAPTER 2

1. 2.254 MeV
2. 1.98 eV; 1.98 eV; 4.784×10^{14} /s
3. (a) 1.226×10^{-9} m; (b) 1.4×10^{-12} m (relativistic); (c) 1.798×10^{-12} m; (d) 1.59×10^{-38} m

CHAPTER 3

1. 2403 Ci
3. 1.81 h
4. 209 h
5. 8.88×10^{-4} g; (b) 276 d
6. (a) 1.23×10^{6} Ci; (b) 7.3×10^{-11} g
9. 11.5 μCi of K-40
10. 304 mCi
11. (a) 532 mCi; (b) 14.93 d
12. (b) 0.637 of original
15. 5.28 min
17. (a) 1.94 mCi; (b) 9.97 mCi; (c) 9.98 mCi at 766 h
18. (b) 146.9 mCi at 5.65 d

CHAPTER 4

1. 3.58×10^{9} t/s
2. 1.74×10^{-9} g of Mn
5. 1.765×10^{8} Ci
6. 1.4×10^{6} Ci
8. (a) 1.09×10^{8} Ci; (b) 1.05×10^{8} Ci
9. (a) 1.54×10^{4} Ci
10. 1.86×10^{6} Ci (assuming thermal neutron fission)

CHAPTER 5

1. 6.5 MeV
3. (a) $\phi = 2.62 \ \beta/cm^2 \cdot s$; (b) $\phi_{\beta, E} = 183 \ MeV/cm^2 \cdot s$; (c) $D_{\beta, Sh} = 97 \ mrad/hr$
6. (a) 1.77 rads/hr; (b) 0.996 rads/hr

CHAPTER 6

2. Changes from 1.38×10^{-3} to 4.14×10^{-3} for 30 yr
4. 3.78×10^{-2} fatal cancers
6. (a) 7.2×10^{-5} fatal cancers per typical adult

CHAPTER 7

1. 0.7176
3. 5.4 cm

4. 2.1 mR/hr
5. 11.6 cm (includes buildup)
7. 0.04 mrem/hr
10. 32.3 rem/hr

CHAPTER 8

1. ce-K $= 12$ keV; ce-L $= 35$ eV
3. 0.796 MeV
8. 100 ± 10 c or 50 ± 5 c/m at 68.3% confidence; 100 ± 19.6 c or 50 ± 9.8 c/m at 95% confidence
10. (a) 1955 to 2045 c; (b) 1912 to 2088 c
12. 25 ± 1.94 c/m

CHAPTER 9

1. (a) 7.29 d
3. (a) 30.53 rad/h; (b) 7.7×10^3 rads
5. (a) 466 mrads to lung and 48.3 mrads total body due to photons
7. (a) 4.95×10^{12} t; (b) 0.5 rads
9. 6220 DAC $\cdot$ hrs (non-stochastic)
11. (a) ALI $= 1.4 \times 10^8$ Bq, DAC $= 5.82 \times 10^4$ Bq/m^3; (b) ALI $= 7.66 \times 10^6$ Bq, DAC $= 3.2 \times 10^3$ Bq/m^3
13. 1.2×10^{-4} Sv
15. 34.3 Sv
17. 1.8×10^6 Bq

CHAPTER 10

1. 4.02×10^{-2}/yr (1925) and 2.87×10^{-3}/yr (2002)
4. 174 mrem/yr for 1 yr; 15 mrem/yr for 11.6 yr; 5.8 mrem/yr for 30 yr

CHAPTER 11

2. 0.135 (13.5%) lifetime cancer risk
6. Monitoring not required unless one or the other exceeds 10% of the respective limit
9. (a) 0.3ALI or 9×10^3 μCi non-stochastic; (b) 0.3ALI or 11.5×10^3 μCi stochastic
11. (a) 5 Ci of ^{3}H; (b) 1 Ci; (c) Yes, no limit for patient excreta

CHAPTER 12

1. 2.24×10^{-4} Ci/m^3; (b) 4.38×10^{-4} Ci/m^3
3. 71 m
4. (a) 2.7×10^{-2} g/m^3 (or 2.0×10^{-2} g/m^3 by virtual source); (b) 1.57×10^{-2} g/m^3 by selection rule
7. 5.29×10^{-7} Ci/m^3
9. 0.62 Bq/m^2

CHAPTER 13

1. 0.315WL
4. 3.9 pCi/L

CHAPTER 14

1. (a) 5.7 yr; (b) 7.54 yr
3. Ra-226 = 14.06 pCi/g (or < 15 pCi/g)
5. (a) 14 Ci/ton; (b) Tc inventory 0.0014 of EPA standard
10. (a) 30% of initial loading; (b) 9% of initial loading

INDEX

Absolute Risk Model, 190
Absorbed Fraction AF (T←S), 281,
  283–285, 287
Absorption Edges, 208
Activation Products, 121–129
Acute Radiation Effects, 183
Agreement States, 353
Alpha Particles
  Emissions of, 65
  Interactions, 153
  Range in Matter, 154
  Scattering Experiments, 42
  Shielding, 199
  Sources of, 199
Ankylosing Spondylitis, Studies of, 187
Annihilation Radiation, 61, 247–249
Annual Limit on Intake (ALI), 294–295,
  300–304, 361–368
Applicable Relevant and Appropriate
  Requirements (ARARs), 346, 347
As Low as Reasonably Achievable
  (ALARA), 338
Atmosphere Stability 399–404
Atmospheric Dispersion, 396–407
  Atmospheric Stability Classes, 401
  Building Wake Effects – Mechanical
    Turbulence, 411
  Distance, x_{max}, of Maximum
    Concentration (χ_{max}), 405

Fumigation, 408
Puff Release, 417
Sector-Averaged χ/Q Values, 418
Stack Effects on Atmospheric
  Dispersion, 406
Atomic Bomb Survivors, Study of, 186
Atomic Energy Act, 347, 352, 467
Atomic Energy Act Materials, 354, 463
Atomic Mass Unit (u), 11, 26
Atomic Masses, 493–504
Atoms, Structure of, 2, 39
Attenuation
  of Beta Particles, 200
  Photon Coefficients, μ, 165
Avogadro's Number, 11, 40, 41

Background Radiation, 106
Backscatter, 160
Becquerel, 32, 79
Beryllium-7, 110
Beta Particles
  Absorption Coefficients, 157
  Attenuation of, 200
  Backscatter Factors, 160
  Bremsstrahlung Production, 155, 203
  Interactions, 155
  Radiation Dose, 157–159
  Range in Matter, 156–157, 202–203
  Shielding, 199, 200–203, 205

Beta Particles (*cont.*)
  Sources of, 200–203
Binding Energies, 493–504
Binding Energy of Nuclei, 22
Bioassay, 314, 323
Bioeffects Studies, 186–189, 334
Biokinetics of Radionuclides, 295
Bohr Model of the Atom, 43–46
Boiling Water Reactor (BWR), 139
Buildup Factor, 216–222

Carbon-14, 57–60, 109
Carcinogenic Effects of Radiation, 185
Cathode Rays, 30
CERCLA, 344, 464
Cesium-137, 59, 72, 77, 153, 170–171, 231, 291, 319, 368, 475, 481
Characteristic X Rays, 104–106
Chart of Nuclides, 8–10, 56, 58, 75, 132–134
Chi-Square Test of a Detector System, 271
Class A, B, C Radioactive Wastes, 482
Cobalt-60, 59, 84
Committed Effective Dose Equivalent, 361–367
Compton Effect, 39, 163
Constants of Nature, 491
Cosmic Radiation, 107
Cosmogenic Radionuclides, 107
Counting Statistics, 257–264
Cross Section, 121
Crystalline Detectors and Spectrometers, 237
Curie, Unit of Radioactivity, 79
Curie, Marie S., 32
Cyclotron, 126–128

Davisson-Germer Experiment, 47
de Broglie, Louis
  Postulate, 47
  Waves, 49
deBroglie Waves, 47
Decay Schemes, 75–78, 505–514
Denver Radium Sites, 346–347
Deposition/Depletion – Gaussian Plumes, 422
  Dry Deposition, 422
  Resuspension, 425
  Wet Deposition, 427

Derived Air Concentration (DAC), 294–295, 300–306, 361–368
Disc and Planar Sources, Dose from, 225
Dose and Dose Rate Effectiveness Factor (DDREF), 192
Dose Limits – Occupational, 358
Dose Limits – Public, 359
Dose Reciprocity Theorem, 282

Effective Dose Equivalent, 293
Effective Half Life, 277, 291
Einstein, 17–21, 29, 35
Electromagnetic Force, 3
Electromagnetic Radiation, 33
Electromagnetic Spectrum, 36
Electron Capture (EC), 56, 63–65
Electron Volt (eV) , 20
Electron, 3
Endoergic Nuclear Reactions, 122
End-Window G-M Counters, 255
Energy Absorption Coefficient (μ_{en}/ρ), 166
Energy
  in Atoms, 12
  Mass Equivalence, 21
  Relativistic, 16
Environmental Models, Elements of, 393
Environmental Radiological Assessment, 391–393
Exoergic Nuclear Reactions, 122

Faraday, 40–41
Film Badge, 242
Fission
  Discovery of, 135
  Products of, 135, 142–146
Fission Products, 135, 142–146
Fumigation, 408, 410

Gamma Rays
  Emission, 69, 70
  Gamma-ray Spectra: $h\nu \leq 1.022$ MeV, 244
  Gamma-ray Spectra: $h\nu \geq 1.022$ MeV, 247
  Gamma-ray Spectroscopy of Positron Emitters, 248
  Interactions, 161–165
Gas-Filled Detectors, 223–226

Geiger-Muller (GM) Detector, 236, 239
Genetic Effects of Radiation, 184
GI Tract Model, 305
Gray, 150

Half Life
  Curve of $A_o/A(t)$ vs. Half Life, 84
  Determination of, 85–87
  Effective, 277, 291
Half Value Layer (Thickness), 214
Health Effects of Radiation, 189–194
  Leukemia, 189
High-Level Radioactive Wastes, 469

Ingested Radionuclides, Doses of, 297
Ingestion Dose Factors, 305–311
Ingestion Model, 305
Inhalation Dose Factors, 300–306
Inhaled Radionuclides, Doses of, 297
Intake of Radioactive Material, 295–314,
  369
Intake Retention Fractions, 314–320
Intake Retention Functions, 290
Interactions
  Alpha Particles, 124, 153
  Beta Particles, 155
  Neutrons, 128
  Photons, 161–165
  Protons and Deuterons, 125
Internal Conversion, 71–73
Internal Radiation Dose Factors, 280–287
  Absorbed Fractions, 281–287
  Energy Emission per Transformation,
    280
  Initial Activity, 280
  Tissue Masses, 284
International Commission on
    Radiological Protection (ICRP), 352
Iodine-125, 83, 85
Iodine-131, 78, 145, 311, 314–320, 368
  Kinetic Energy, 18, 67, 101
Internal Radiation Dose, 275–325
  Accumulated Dose, 275
  Bioassay Estimates, 314
  Deposition and Clearance Data, 288
  Dose Rate, 275
  Medical Uses, 279
Internal Transition, 70
Inverse Square Law, 152

Ion Chamber, 236–239
Ionization, 155
Isomeric States, 70
Isotope, 6

Legacy/Process Wastes, 465
Leukemia, Incidence of, 189
Line Sources, Dose from, 223
Linear Accelerators, 126
Linear Attenuation Coefficient, (μ),
  165–245
Liquid Scintillation Analysis, 250
Lower Limit of Detection (LLD), 265–268
Low-Level Radioactive Waste Policy Act,
  471
Low-Level Radioactive Wastes (LLRW),
  471–486
  Classification of LLRW, 482
  Control for (10 CFR 61), 479
  Modes of Disposal, 477
  Risks of LLRW Disposal, 481
Lung Model, 297–299

Manhattan Project Wastes, 465
Mass-Energy Equivalence, 21
Maxcy Flats, KY Site, 475
Maximum Activity Radioactive Series
    Transformation, 95
Mercury-203, 73, 147
Metastable States, 70
Micro-Rem Meter, 241
Minimum Detectable Concentration,
  268–269
Minimum Detection Levels, 270
Multi-Compartment Clearance/Retention
    of Radionuclides, 286, 289

National Council on Radiation Protection
    (NCRP), 352
National Governors Association, 471
National Radium Institute, 347
Natural Radioactivity, 106–120
Naturally Occurring or Accelerator-
    Produced Radioactive Materials
    (NARM), 343, 464, 467
Naturally Occurring Radioactive Material
    (NORM), 116–121, 464
Neutrino, 58
Neutron Activation, 128–134

Neutron, 4, 50, 114
Neutron-Rich Radionuclides, 7
Non-Stochastic Radiation Effects, 182
Normal (Gaussian) Distribution, 258–259
Nuclear Force, 2
Nuclear Reactions
  Deuteron-Alpha (d, α), 128
  Neutron-Alpha (n, α), 130
  Neutron-Gamma (n, γ), 129
  Proton-Alpha (p, α), 127
  Proton-Gamma (p, γ), 128
  Proton-Neutron (p, n), 127
Nuclear Power Plants, 130–140, 339, 462
Nuclear Reactor Systems, 136–140, 462
Nuclear Shell Model, 51
Nuclides, Chart of, 8–10, 56, 58, 75,
  132–134

Pair Production, 164
Particle Masses, 491
Personnel Dosimeters, 242
Phosphate Ores, 119
Phosphorus-32, 204, 276
Photoelectric Effect, 162
Photon Sources in "Good Geometry", 209
Photon Sources in "Poor Geometry",
  215
Photons
  Attenuation of, 207–214
  Bremsstrahlen, 165, 171
  Compton-Effect, 163
  Detection of, 244–250
  Dosimetry, 167–171
  Gamma Constant, Γ, 171
  Gamma-Ray Emission, 69, 70
  Interactions, 162–164
  Pair Production Interactions, 164
  Photoelectric Effect, 162
  Properties of, 36–39
  X-Ray Production, 101–104
Planck Radiation Law, 34
Pocket Dosimeters, 243
Poisson Distribution, 257–259
Population Dose Limits, 338
Portable Field Instruments, 239
Positrons
  Annihilation, 61, 247–249
  Emission of, 60
  Gamma Spectra of, 248

Positron Emission, 56
Pressurized Water Reactor (PWR), 139
Primordial Radionuclides, 115
Propagation of Error, 260–263
Proportional Counter, 236, 253–255
Proton, 3
Proton-Rich Radionuclides, 60

Quantum Hypothesis, 35
Quantum Mechanics, 47
Quench, 252
Q-Value, 23, 59, 67, 122–124

Radiation Absorbed Dose (Rad/Gray),
  150
Radiation Biology, 175–186
  Chemical Effects of Radiation, 177
  Radioprotectors, 178
  Radiosensitivity, 177
Radiation Dose Equivalent (Rem/Sievert),
  150
Radiation Dose, 149–173
Radiation Monitoring Regulations,
  374–376
Radiation Protection Standards, 329–348
  Historical Trends, 330
Radiation Regulations, 10 CFR 19 and
  20, 354–360, 362–367
Radiation Risk Calculations, 194
Radiation Risk Factors, 192, 193
Radiation Risk Models
  Absolute Risk, 190
  Linear Quadratic, 190
  Non-Threshold, 190
  Relative Risk, 190
Radiation Shielding, 199–231
Radiation Units, 150–152
Radiative Capture, 129
Radioactive Equilibrium, 93
Radioactive Series, 89, 110
Radioactive Transformation, 56
  Calculations of, 80–85
  Discovery of, 32
  Series Transformation, 89, 110
  Transformation Modes, 56, 505–514
  Wastes, 463–486
Radiological Assessment, 391–433
Radiological Health, 1
Radium-Dial Painters, 188

Radon and its Products, 435–438
  Guides, 342
  Measurement of, 456–458
  Radon Subseries, 92, 111–114, 436
  Risk Factors, 443–446
  Working Level for Radon Progeny,
    438–442, 459
Radon Reduction Measures, 448–455
Radon Subseries, 92
Radon, Health Effects of, 443–446
Reactor Systems, 136–140
Regulation of Radiation and Radioactive
    Material
  Good Practice Requirements, 370
  Radioactive Waste Disposal, 372
  Records and Reports, 373
  Radiation Monitoring, 374
Relative Risk Model, 190
Relativistic Energy, 20
Relativity, Special Theory of, 17
Resource Conservation and Recovery Act
    (RCRA), 344, 346
Respiratory Tract Model, 297–299
Risk Assessment
  LLRW Disposal, 477
  Radiation Exposure, 191–195
Risk Factors, Radon, 446
Radiation Protection Standards
  Design Objectives for Nuclear Power
    Plants (Appendix I, 10 CFR 50), 339
  Radionuclides as Hazardous Air
    Pollutants – 40 CFR60, 341
  Radionuclides in Drinking Water, 341
  Radon Guides, 342
  Uranium Fuel Cycle Standards – 40
    CFR 190, 340
  Uranium Mill Tailings, 343
Risk-Based Standards, 332, 335–336
Roentgen, W, 103, 151, 152
Rutherford, Ernest
  Alpha Scattering, 42
  Atom, 42–43
  Radioactivity, 33, 55

Schrödinger, E, 50
Secular Equilibrium, 94
Semiconducting Detectors, 237
Shattuck Chemical Site, CO, 347
Sheffield, IL Site, 473

Shipment of Radioactive Materials,
    377–386
  Type A Shipments, 382
Sievert, 150
Site Cleanup Standards/Criteria, 344
Sodium-22, 110
Somatic Effects of Radiation, 182
Specific Activity, 87, 88, 90
Specific Effective Energy (SEE), 279
Spontaneous Fission, 73
Statistics of Radiation Measurements,
    257–263
Stochastic Effects, 182
Strontium-90, 94, 172, 474
Submersion Dose, 311–313
Surface Barrier Detectors, 256

Technologically Enhanced Naturally
    Occurring Radioactive Material
    (TENORM), 464, 467
Tenth-Value Layer (Thickness), 214
Teratogenic Effects of Radiation, 184
The Roentgen (R), 151
Thermoluminescent Dosimeters (TLDs),
    242
Thorium Ores, 120
Thorium Residues, 465
Thoron, 435
Thorotrast Patients, 188
Tissue Masses, 281
Tissue Weighing Factors W_T, 294
Total Effective Dose Equivalent (TEDE),
    336, 360
Transient Equilibrium, 94, 96
Transmutation of Nuclei
  by Alpha Particles, 124
  by Neutrons, 128
  by Protons and Deuterons, 125
Transuranic (TRU) Radionuclides, 56, 68
Tritium
  Bioassay of, 323
  Biokinetics, 320
  In LLRW, 481
  In the Environment, 109, 474
  Natural Sources, 109

UMTRCA, 343, 346
Underground Miners, 188
Uranium Mill Tailings, 466

Uranium Ores, 117
U.S. Department of Energy, 354, 464
U.S. Department of Transportation, 353, 377
 Shipping Regulations, 377–386
U.S. Environmental Protection Agency (EPA), 347, 353
U.S. Nuclear Regulatory Commission (NRC), 352–357
U.S. Postal Service, 381

Virtual Source, Atmospheric Dispersion, 412

Volume Sources, Dose from, 226

Water Transport/Dispersion, 395
Wave Mechanics, 47
West Valley, NY Site, 473
Working Level, Radon Progeny, 438–442
Working Level Month, Radon Progeny, 442

X rays, 31, 101–103

Yucca Mountain, 469